Modeling Coastal Hypoxia

Dubravko Justic · Kenneth A. Rose
Robert D. Hetland · Katja Fennel
Editors

Modeling Coastal Hypoxia

Numerical Simulations of Patterns, Controls and Effects of Dissolved Oxygen Dynamics

Springer

Editors
Dubravko Justic
Department of Oceanography and Coastal Sciences
Louisiana State University
Baton Rouge, LA
USA

Kenneth A. Rose
Department of Oceanography and Coastal Sciences
Louisiana State University
Baton Rouge, LA
USA

Robert D. Hetland
Department of Oceanography
Texas A&M University
College Station, TX
USA

Katja Fennel
Department of Oceanography
Dalhousie University
Halifax, NS
Canada

ISBN 978-3-319-54569-1 ISBN 978-3-319-54571-4 (eBook)
DOI 10.1007/978-3-319-54571-4

Library of Congress Control Number: 2017933669

Printed on acid-free paper

This Springer imprint is published by Springer Nature
The registered company is Springer International Publishing AG
The registered company address is: Gewerbestrasse 11, 6330 Cham, Switzerland

Preface

The number of coastal hypoxic zones has been increasing at an exponential rate since the 1960s, and there are currently more than 600 documented hypoxic zones in the estuarine and coastal waters worldwide. These include the well-studied large systems affected by riverine inputs, as well as increasingly important deeper-water ocean shelf oxygen minimum zones and small, shallow coastal systems. Hypoxia causes the loss of habitat and the spatial displacement of plankton, large invertebrates, and fish, and is often associated with mass mortalities of marine organisms that live on or near the bottom. In addition to being a widespread environmental problem, coastal hypoxia influenced by riverine inputs is also of great socioeconomic and political interest because of its association with agricultural fertilizer activities in the watershed. There will be increasing demands for predicting the ecological responses to hypoxia in order to quantify the ecological benefits and costs of management actions and to express the simulated effects of coastal management and climate change in terms of direct relevance to managers and the public. Numerical models can provide the needed information for understanding hypoxia and ensuring effective management, and this book provides a snapshot of representative modeling analyses of coastal hypoxia and its effects.

This book is a collection of case studies presented at hypoxia modeling sessions that editors have organized over the past four years at meetings of the Association for the Sciences of Limnology and Oceanography and the Coastal and Estuarine Research Federation. Chapter authors include senior scientists who have studied hypoxia for many years, but also, importantly, many junior scientists who bring specialized knowledge on selected hypoxia modeling topics. This book consists of 15 chapters that are broadly organized around three main topics: (1) Modeling of the physical controls on hypoxia, (2) Modeling of biogeochemical controls and feedbacks, and (3) Modeling of the ecological effects of hypoxia. The final chapter is a synthesis chapter that draws generalities from the earlier chapters, highlights strengths and weaknesses of the current state-of-the-art modeling, and offers recommendations on future directions. We hope that the "physics to fish" approach of this book will make it a useful reference for oceanographers, environmental scientists, resource managers, and graduate students.

All chapters in this book underwent standard peer review typical of scientific journals, with each receiving multiple, anonymous reviews and with subsequent revisions monitored by one of the editors. We thank the many people that provided reviews on the chapters. We are grateful to Janet Slobodien, executive editor, Ecology and Evolutionary Biology, Springer, for the invitation to write a book on coastal hypoxia modeling and Center for Sponsored Coastal Ocean Research, National Oceanic and Atmospheric Administration, for supporting many of the hypoxia modeling studies discussed in this book.

Baton Rouge, USA — Dubravko Justic
Baton Rouge, USA — Kenneth A. Rose
College Station, USA — Robert D. Hetland
Halifax, Canada — Katja Fennel

Contents

Editors and Contributors

About the Editors

Dr. Dubravko Justic is Texaco Distinguished Professor in the Department of Oceanography and Coastal Sciences at Louisiana State University (LSU). Previously, he was Eric L. Abraham Distinguished Professor in Louisiana Environmental Studies and director of LSU's Coastal Ecology Institute. His research interests include ecosystem modeling, coastal eutrophication, hypoxia, and potential impacts of climate change on coastal ecosystems. He has studied extensively low-oxygen zones in the northern Adriatic Sea and northern Gulf of Mexico and has employed various types of numerical simulation models to describe controls of environmental factors on hypoxia and predict the consequences of management actions. He is presently working on characterizing connectivity among wetland, estuarine, and shelf ecosystems in the northern Gulf of Mexico and evaluating tradeoffs associated with different Mississippi River management alternatives.

Dr. Kenneth A. Rose is professor in the Department of Oceanography and Coastal Sciences, and associate dean for Research in the College of the Coast and Environment, at Louisiana State University. He will soon be joining the faculty at Horn Point Laboratory, part of the University of Maryland Center for Environmental Science, as the Franz Merrick Professor in Sustainable Ecosystem Restoration. Dr. Rose's research centers on using mathematical and computer simulation modeling to predict and better understand fish population and food web dynamics in estuaries, lakes, reservoirs, and oceans. He was recently awarded the Award of Excellence (for lifetime achievement) from the American Fisheries Society. He has been a member of multiple steering and advisory committees providing scientific guidance and oversight, including several National Academy of Sciences' committees. Dr. Rose has published over 150 papers on topics related to ecological and fisheries modeling and analysis, and has served on multiple editorial boards. He is presently working on the scaling of hypoxia effects from individual fish to the population level.

Dr. Robert D. Hetland is a professor in the Department of Oceanography, Texas A&M University. His research focuses on understanding and predicting circulation in buoyancy-driven flows in estuaries and the coastal ocean. Dr. Hetland's primary research tool is the Regional Ocean Modeling System (ROMS) which he uses to perform numerical simulations of coastal and estuarine circulation, with applications that include oil spill trajectory prediction, harmful algal bloom formation, and coastal hypoxia. Dr. Hetland's simulations of circulation over the Texas–Louisiana shelf have provided new insights into the mechanisms that cause seasonal hypoxia in that region. Dr. Hetland is an editor for the Journal of Geophysical Research.

Dr. Katja Fennel is professor in the Department of Oceanography at Dalhousie University. As head of the Marine Environmental Modeling Group (http://memg.ocean.dal.ca), she leads the development of marine ecosystem and biogeochemical models at Dalhousie. For over two decades, Dr. Fennel has developed and applied numerical models of marine ecosystems and biogeochemistry with particular focus on continental shelf systems and the cycling of nitrogen, carbon, and oxygen. In addition to implementing biogeochemical models, Dr. Fennel has developed and applied methods for the assimilation of observations into these models in order to improve their predictive capabilities. She serves as co-editor in chief of the high-impact journal Biogeosciences, and has served on the editorial boards of three other scientific journals and on several international science advisory bodies including the IMBER/LOICZ Continental Margins Task Team and the CLIVAR Working Group on Ocean Model Development. Currently, she is science team member of GODAE OceanView, co-chairs the GODAE Marine Ecosystem Analysis and Prediction Task Team, and serves on the science advisory boards of the Copernicus Marine Environment Monitoring Service in Europe, the Ocean Frontier Institute at Dalhousie, and the international Biogeochemical Argo steering committee.

Contributors

Aaron T. Adamack Department of Oceanography and Coastal Sciences, Louisiana State University, Energy, Coast, and Environment Building, LA, USA; *Present address*: Institute for Applied Ecology, University of Canberra, Bruce, ACT, Australia

Mohammad Nabi Allahdadi Department of Marine, Earth and Atmospheric Sciences, North Carolina State University, Raleigh, USA

Harold P. Batchelder College of Earth, Ocean, and Atmospheric Sciences, Oregon State University, Corvallis, USA; North Pacific Marine Science Organization, Sidney, BC, Canada

Damian C. Brady School of Marine Sciences, University of Maine, Walpole, ME, USA

Hector R. Bravo Department of Civil and Environmental Engineering, University of Wisconsin, Milwaukee, USA

Denise L. Breitburg Academy of Natural Sciences Estuarine Research Center, Leonard, MD, USA; *Present address*: Smithsonian Environmental Research Center, Edgewater, MD, USA

Christopher W. Brown Center for Satellite Applications and Research, National Oceanic and Atmospheric Administration, College Park, MD, USA

Mark J. Brush Virginia Institute of Marine Science, College of William and Mary, Gloucester Point, VA, USA

Carl F. Cerco U.S. Army Engineer Research and Development Center, Vicksburg, MS, USA

Villy Christensen Institute for the Oceans and Fisheries, University of British Columbia, Vancouver, B.C, Canada

James H. Cowan Department of Oceanography, Louisiana State University, Baton Rouge, LA, USA

Sean Creekmore Department of Oceanography and Coastal Sciences, Louisiana State University, Baton Rouge, LA, USA

Dominic M. Di Toro Department of Civil and Environmental Engineering, University of Delaware, Newark, DE, USA

Katja Fennel Department of Oceanography, Dalhousie University, Nova Scotia, Canada

Sajad A. Hamidi Department of Physics, Indiana University of Pennsylvania, Indiana, PA, USA

Robert D. Hetland Department of Oceanography, Texas A&M University, College Station, TX, USA

Haosheng Huang Department of Oceanography & Coastal Sciences, College of the Coast & Environment, Louisiana State University, Baton Rouge, LA, USA

Raleigh R. Hood Horn Point Laboratory, University of Maryland Center for Environmental Science, Cambridge, MD, USA

Dubravko Justic Department of Oceanography and Coastal Sciences, Louisiana State University, Baton Rouge, LA, USA

W. Michael Kemp Horn Point Laboratory, University of Maryland Center for Environmental Science, Cambridge, MD, USA

Dong S. Ko Naval Research Laboratory, Oceanography Division, Stennis Space Center, Bay St. Louis, MS, USA

Andrey O. Koch Department of Marine Science, University of Southern Mississippi, Hattiesburg, USA

Sarah E. Kolesar Chesapeake Biological Laboratory, University of Maryland Center for Environmental Science, Solomons, MD, USA; *Present address*: Oregon Sea Grant, Oregon State University, Corvallis, OR, USA

Elizabeth LaBone Department of Oceanography & Coastal Sciences, College of the Coast & Environment, Louisiana State University, Baton Rouge, LA, USA

Arnaud Laurent Department of Oceanography, Dalhousie University, Nova Scotia, Canada

Younjoo J. Lee Department of Oceanography, Naval Postgraduate School, 1 University Circle, Monterey, CA, USA

John C. Lehrter Gulf Ecology Division, US EPA, Office of Research and Development, National Health and Environmental Effects Laboratory, Gulf Breeze, FL, USA; *Present address*: University of South Alabama and Dauphin Island Sea Lab, Dauphin Island, AL, USA

Chunyan Li Department of Oceanography and Coastal Sciences, Coastal Studies Institute, Louisiana State University, Baton Rouge, USA

Ming Li Horn Point Laboratory, University of Maryland Center for Environmental Science, Cambridge, MD, USA

Yun Li College of Marine Science, University of South Florida, Saint Petersburg, FL, USA

Lisa L. Lowe Leidos, Research Triangle Park, Durham, NC, USA

Kim de Mutsert Department of Environmental Science and Policy, George Mason University, Fairfax, VA, USA

Scott W. Nixon (Deceased) Graduate School of Oceanography, University of Rhode Island, Narragansett, RI, USA

Bradley Penta Naval Research Laboratory, Oceanography Division, Stennis Space Center, Bay St. Louis, MS, USA

Kenneth A. Rose Department of Oceanography and Coastal Sciences, Louisiana State University, Energy, Coast, and Environment Building, Baton Rouge, LA, USA

Shaye Sable Dynamic Solutions, LLC, Baton Rouge, LA, USA

Yvette H. Spitz College of Earth, Ocean, and Atmospheric Sciences, Oregon State University, Corvallis, USA

Jeroen Steenbeek Ecopath International Initiative, Barcelona, Spain

Jeremy M. Testa Chesapeake Biological Laboratory, University of Maryland Center for Environmental Science, Solomons, MD, USA

J. Val Klump School of Freshwater Sciences, University of Wisconsin, Milwaukee, USA

Lixia Wang Department of Oceanography & Coastal Sciences, College of the Coast & Environment, Louisiana State University, Baton Rouge, LA, USA

James T. Waples School of Freshwater Sciences, University of Wisconsin, Milwaukee, USA

Jerry D. Wiggert Division of Marine Science, School of Ocean Science and Technology, University of Southern Mississippi, Stennis Space Center, MS, USA

Xiaoqian Zhang Department of Oceanography, Texas A&M University, College Station, TX, USA

Chapter 1
Numerical Experiment of Stratification Induced by Diurnal Solar Heating Over the Louisiana Shelf

Mohammad Nabi Allahdadi and Chunyan Li

Abstract The effect of diurnal solar heating on the stratification of waters over the Louisiana shelf was examined using the 3-dimensional Finite Volume Community Ocean Model (FVCOM). The simulation was for June 2009 to examine the effects of solar heating on summertime stratification. The input components of solar radiation to the FVCOM model were calculated using available relationships for shortwave, longwave, latent heat, and sensible heat radiation and using Metocean field data obtained from WAVCIS stations. Simulation results showed a continuous increase in water temperature and stratification during June 2009 with daily fluctuations of sea surface temperature as large as 0.9 °C. The corresponding stratification strengthening was quantified by an increase in the gradient Richardson number and buoyancy frequency. Development of shelf-wide stratification coincided with a significant decline in bottom water oxygen concentration. Our results demonstrate how, under certain conditions, solar heating can significantly contribute to vertical stratification and may also create conditions conducive to the formation and persistence of hypoxia.

Keywords Numerical modeling · Solar radiation · Stratification · Hypoxia · FVCOM · Louisiana shelf · Gulf of Mexico

---

M.N. Allahdadi (✉)
Department of Marine, Earth and Atmospheric Sciences,
North Carolina State University, Raleigh, USA
e-mail: nabiallahdadi@gmail.com

C. Li
Department of Oceanography and Coastal Sciences, Coastal Studies Institute,
Louisiana State University, Baton Rouge, USA
e-mail: cli@lsu.edu

D. Justic et al. (eds.), *Modeling Coastal Hypoxia*,
DOI 10.1007/978-3-319-54571-4_1

1.1 Background

Hypoxia is a condition of low dissolved oxygen concentration less than 2 mg/l (Rabalais et al. 2002). Over the Texas–Louisiana shelf, hypoxia is a seasonal phenomenon that starts in late spring or early summer and persists in bottom waters until the early fall (Wiseman et al. 1997). The seasonal occurrence of hypoxia is the result of influence from several biological and physical factors. The springtime floods of the Mississippi and the Atchafalaya rivers greatly increase nutrient loadings to the shelf area (Wiseman et al. 1997). Biological processes supported by this increased nutrient load contribute to oxygen depletion in bottom waters, thereby producing hypoxia 1–2 months after the peak in river discharge (Justic et al. 1996). Freshwater associated with river flooding also increases the buoyancy of shelf waters and thereby strengthens the stratification, which allows for the hypoxia formation (Wiseman et al. 1997). However, a significant portion of stratification can be induced by solar heating during summertime and can act as primary contributing factor to seasonal hypoxia development in the northern Gulf of Mexico (Hagy and Murrell 2007). The role of water column mixing in re-oxygenating bottom waters and mitigating hypoxic events, as well as the positive feedback between stratification and hypoxia and its role in mediating pollutant distribution, have been addressed by several studies for the Texas–Louisiana shelf (Wiseman et al. 1997; Wang and Justic 2009; Allahdadi et al. 2013; Chaichitehrani 2012; Tehrani et al. 2013). Allahdadi et al. (2013) compared the relative Richardson numbers for stations located to the east and west of the Mississippi Delta and concluded that for the western stations, the Richardson numbers were an order of magnitude larger than those for the eastern stations.

Buoyancy is dependent on temperature and salinity gradients in the vertical dimension. Solar radiation increases in the beginning of the summer. This causes the water column to stratify with higher sea surface temperature. The stratification varies periodically due to diurnal variations of solar heat (Chen et al. 2003).

Increasing solar radiation has both positive and negative impacts on reducing bottom water oxygen. Although the stratification caused by excess solar heating inhibits bottom water re-oxygenation, it can also enhance photosynthesis and affect the balance between photosynthesis and respiration rate. In very shallow estuarine waters, where stratification is not dominant, hypoxic events generally correspond to low solar radiation. An example of such phenomenon is the hypoxic events in the Upper Newport Bay Estuary, California (water depth less than 5 m) that tended to occur more frequently on cloudy days (Nezlin et al. 2009). However, for the deeper shelf waters of Louisiana west of the Mississippi Delta, we expect that stratification is dominant and there is negative feedback between solar radiation and bottom water oxygen concentration.

Many studies have addressed different physical and biogeochemical causes of hypoxia over the Louisiana shelf (for example Justic et al. 2003, Rabalais and Louisiana Universities Marine Consortum 1991, Rabalais et al. 2004, Turner and

Rabalais 1994, Turner et al. 2008). However, the effect of summer solar heating on stratification and hypoxia formation has not been studied.

The present study examines the effects of solar radiation on stratification for the Louisiana shelf west of the Mississippi Delta using FVCOM simulations. The effect of solar heating on water column temperature is included as a heat source term in the model heat balance equation. The approach has been used by others to study the effect of atmospheric heat budget on the water column stratification and mixing (e.g., Elsberry et al. 1976; Price 1981; Bender et al. 1993; Zedler et al. 2002). We followed the approach of the Chen et al. (2003) here, who applied the ECOM-Si model with different components of radiation to study stratification and circulation over the Georges Bank in the Gulf of Maine. Determining parameters for the vertical structure of downward heat flux in the water column was a challenge in their study. They used the distinction between coastal and ocean waters to specify the attenuation of different wavelengths in the water column. They compared their simulated sea surface temperature (SST) with satellite-derived values, and their results showed a diurnal fluctuation of SST with approximately 1 °C difference between day and night temperatures superimposed on a general increasing trend. In the present study, we used the same approach with the FVCOM model.

1.2 Numerical Model

In order to study the effect of solar heating on shelf stratification, solar radiation components were introduced to an existing implementation of FVCOM (Fig. 1.1, Allahdadi 2015). FVCOM is a prognostic, unstructured grid, finite volume, free surface, three-dimensional (3-D) primitive equations ocean model (Chen et al. 2006). The main equations solved by the model are momentum, continuity, salt transport, heat transport, and density equation. The equations representing heat and salt transport and density contribution are as follows:

$$\frac{\partial T}{\partial t} + u\frac{\partial T}{\partial x} + v\frac{\partial T}{\partial y} + w\frac{\partial T}{\partial z} = \frac{\partial}{\partial z}\left(K_h\frac{\partial T}{\partial z}\right) + F_T \tag{1.1}$$

$$\frac{\partial S}{\partial t} + u\frac{\partial S}{\partial x} + v\frac{\partial S}{\partial y} + w\frac{\partial S}{\partial z} = \frac{\partial}{\partial z}\left(K_h\frac{\partial S}{\partial z}\right) + F_s \tag{1.2}$$

$$\rho = \rho(T, S) \tag{1.3}$$

where T is water temperature, S is salinity, and ρ is water density. u, v, and w are current velocity components in the horizontal (x and y), and vertical (z) directions. K_h is the coefficient of vertical diffusivity for salt and heat. F_T and F_s represent horizontal thermal and salt diffusion terms, respectively. Solar radiation is applied to the model equation through the surface boundary condition for temperature:

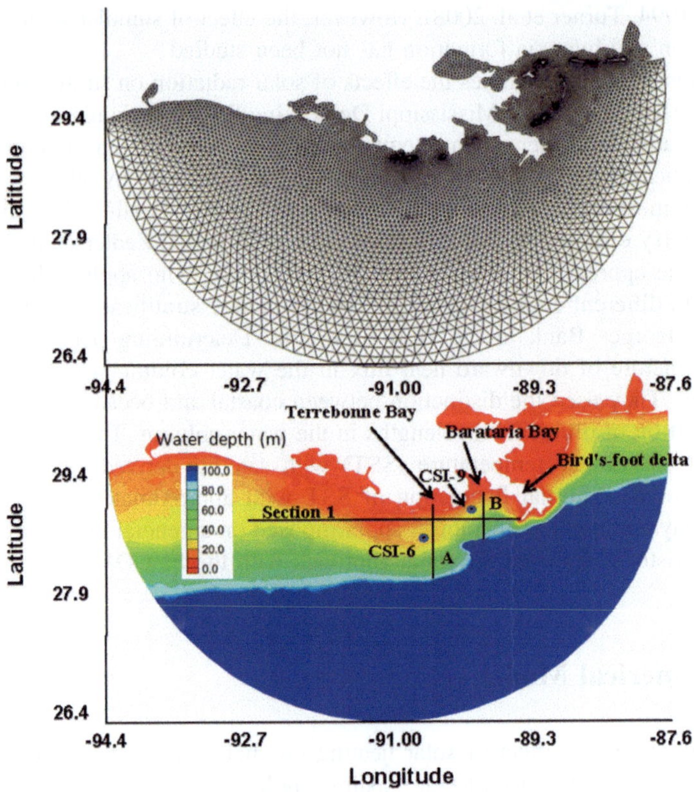

Fig. 1.1 Computational mesh and model bathymetry for the FVCOM model

$$\frac{\partial T}{\partial z} = \frac{1}{\rho c_p K_h}[Q_n(x,y,t) - SW(x,y,\xi,t)] \tag{1.4}$$

In the above equation, $Q_n(x,y,t)$ is the surface net heat flux that includes four components: shortwave flux, longwave flux, sensible heat flux, and latent heat flux. $SW(x,y,t)$ is the shortwave flux at sea surface, and C_p is the specific heat of seawater.

Governing equations of flow, and salt and heat transport, are discretized using finite volumes and solved with the Runge–Kutta method. Triangular grid elements are used for computation. This offers more flexibility, including islands and complicated coastline geometry that are common west of the Mississippi Delta. Furthermore, mesh flexibility allows finer spatial resolution where needed. For the present simulation, a region including the coastline from Mobile-AL to the west of the Sabine Bank, TX was used as the modeling area (Fig. 1.1). The model has a circular open boundary extended to the outer-shelf and deep water to the depths beyond 2000 m. Mesh resolution was the highest for the area associated with the inner-shelf west of the Mississippi Delta (about 500–1000 m, Fig. 1.1), which is the area of focus in this study. This region includes the shallow shelf off Terrebonne Bay, outside of the Atchafalaya Bay, and outside of the Barataria Bay, whose bottom water becomes hypoxic during the summertime (Rabalais et al. 2002).

1.3 Model Specification

1.3.1 Modeling Period and Data

The simulation covered June 1–June 30, 2009, with several days of spin up in May because the summer-related thermal stratification usually starts in June (Allahdadi et al. 2013). Field measurements of the required Metocean data (current speed, wind speed, sea surface temperature, air temperature, air pressure) were available for June 2009 from the WAVCIS (www.wavcis.lsu.edu) station CSI-6, located off Terrebonne at 20 m water depth. The data were used for model setup and validation. Meteorological measurements, including air pressure, air temperature, wind speed, and relative humidity, as well as oceanographic data and sea surface temperature (SST), were used in calculating different components of input heat flux to the model (see Sect. 1.3.2.1). SST data are also used to compare with the simulated SSTs.

SST observed at CSI-6 increased from May to the maximum value in August and decreased thereafter (Fig. 1.2). SST values for a longer time period from May to December are shown in the inset of Fig. 1.2. The time series of June through July SST shows a generally increasing trend with an average initial SST value of about 26 °C during the first 5 days. SST increased to a maximum of 31 °C by June 20 with daily fluctuations of ~0.9 °C. SST variations during this time period (from June 10 to 20) are all oscillatory, presumably due to diurnal variation of heat flux, and these continued till mid-July, when diurnal fluctuations of SST almost

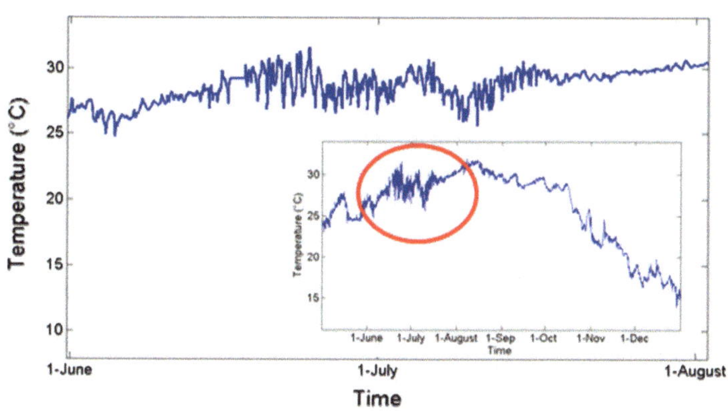

Fig. 1.2 Time series of measured SST at WAVCIS station CSI-6

disappeared and SST showed a gradual (almost smooth) increase. The nighttime mitigation of SST, and thereby reduced stratification that could contribute to re-oxygenation in the water column, was absent after mid-July and therefore more severe hypoxic events are expected.

1.3.2 Model Inputs

1.3.2.1 Heat Flux

The effect of solar insolation on water temperature and density was incorporated into the FVCOM model through two components of the surface boundary condition: net surface heat flux and short wave radiation flux. Although these parameters can be obtained from atmospheric model databases, low temporal resolution of the available model outputs (time steps of 6 h or larger) limits their use for a detailed study of diurnal shelf heating and stratification. Hence, the two heat flux components were calculated using hourly Metocean data.

Net surface heat flux is the algebraic sum of four different components:

$$Q_N = Q_s + Q_{LW} + Q_L + Q_{sn} \tag{1.5}$$

where the quantities in the right-hand side of the equation are shortwave, longwave, latent heat, and sensible heat fluxes, respectively. For each of them, the equations are presented in Appendix A.

The hourly Metocean data used for the calculation of heat flux components provided an adequate temporal resolution of shortwave radiation and net heat flux. Following the approach presented in Appendix A and using Eq. 1.5, we calculated shortwave radiation and net heat flux for the simulation period using the Metocean data from CSI-6 (Fig. 1.3). The peak values show the daytime maximum insolation

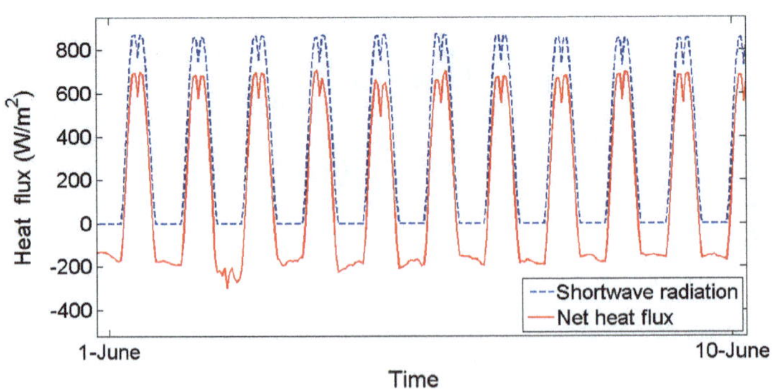

Fig. 1.3 Variation in calculated shortwave radiation and net surface heat flux for June 1–June 10, 2009

occurring two times between 1 and 4 PM. The two distinct peaks of heat flux correspond to two maxima of the function for cosine of the zenith angle used in the calculation (Appendix A).

Shortwave radiation is the only component of net heat flux that penetrates through the water column. Attenuation of shortwave flux versus water depth is presented by the following equation (Chen et al. 2003):

$$SW(z,t) = SW(0,t)\left[Re^{\frac{z}{a}} + (1-R)e^{\frac{z}{b}}\right] \quad (1.6)$$

where $SW(0,t)$ is the shortwave radiation at the water surface, t is time, and $SW(z,t)$ is the shortwave radiation at water depth z. Parameters a and b are the attenuation lengths for longer and shorter (waveband blue-green) wavelengths, and R is the portion of shortwave flux associated with the longer wavelengths. Appropriate values for a, b, and R should be considered based on the clarity of water over the modeling area. Paulson and Simpson (1977) suggested values of $R = 0.78$, $a = 1.4$, and $b = 7.9$ for coastal waters. In our study, a sensitivity analysis was implemented on each parameter to obtain the optimal agreement with SST measurements. The final applied values were consistent with Chen et al. (2003).

1.3.2.2 Wind Data

We considered the effect of wind, and the associated mixing, in order to evaluate the model performance and to properly interpret the effects of solar heating on stratification. Wind data for June 2009 were obtained from CSI-6 and reduced to the standard level of 10 m above sea surface. The average wind speed during this month was less than 6 m/s and rarely reached 10 m/s (Fig. 1.4). This implies a weak wind effect on mixing over the shelf.

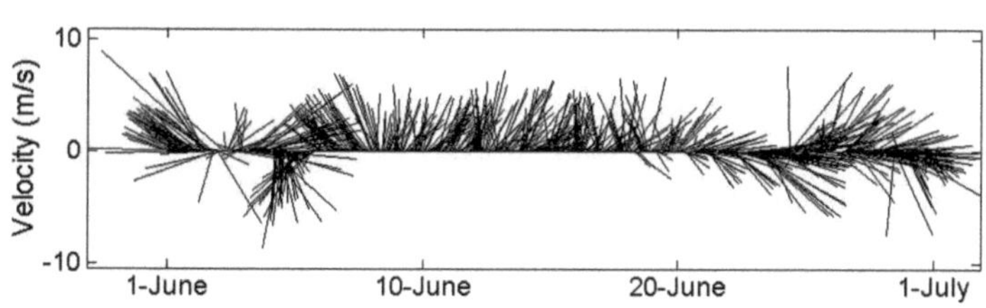

Fig. 1.4 Time series of wind speed measurements at station CSI-6 for June 2009

1.3.2.3 Initial Temperature Profile

We used the climatological profile for May over the Louisiana shelf from NOAA as the initial temperature profile for FVCOM. Site-specific measurements were not sufficient to obtain temperature profiles for the initial conditions.

1.3.3 Boundary Conditions

A dissipative internal boundary condition was used to damp the waves reflected from the boundary to the model domain. No temperature or salinity values were applied to the open boundary, and also no tide was assumed at the boundary.

1.4 Simulation Results

1.4.1 Model Evaluation

Simulated currents under wind forcing were compared with available measurements at WAVCIS stations CSI-6 and CSI-9 (Fig. 1.5, see Fig. 1.1 for locations). Small values of simulated currents for this time period are consistent with measurements and previous studies (Allahdadi et al. 2013; Chaichitehrani et al. 2014) Current measurements during simulation period were available for only several days. Results for SST were compared to field measurements at CSI-6 (Fig. 1.6a). An optimal agreement was achieved for the case that used the short/longwave attenuation lengths suggested by Chen et al. (2003). FVCOM reproduced the increasing SST trend from 26.5 to 30 °C for June 3–June 15.

Simulated daily fluctuations of temperature are more or less in phase with measurements showing the mid-day temperature peak and nighttime minimum (Fig. 1.6a). The simulated time series of SST was de-trended to show the fluctuations associated with diurnal variations of solar insolation. The resultant time series of daily temperature fluctuations (Fig. 1.6b) show a maximum day–night temperature difference of 0.9 °C with an average of 0.5 °C for the first 20 days of June 2009.

1.4.2 Sea Surface Temperature

Simulated spatial variation of SST was complicated due to complex shelf bathymetry and the dynamics of circulation on the shelf (Fig. 1.7). SST maps of the study area for different times on July 15 show the effects of including the minimum

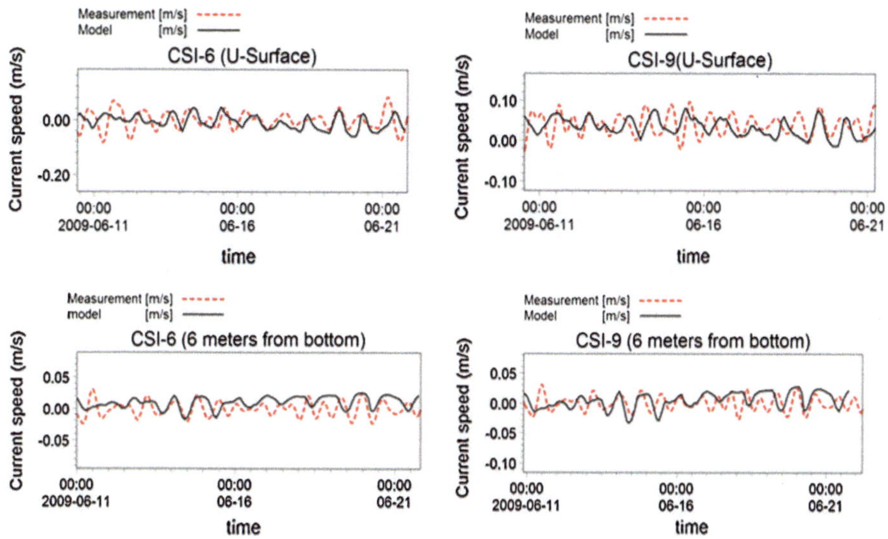

Fig. 1.5 Comparison of measured and simulated east–west current component (U-component) for surface and bottom at stations CSI-6 and CSI-9

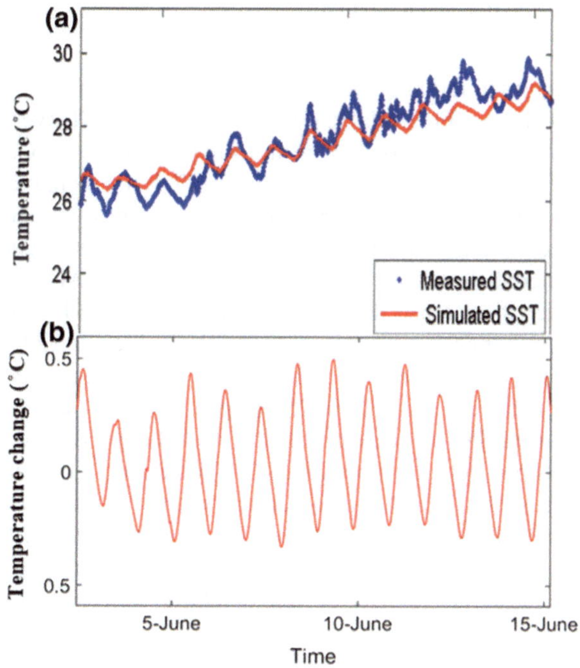

Fig. 1.6 **a** Comparison of simulated and measured SST at station CSI-6; **b** day–night fluctuations in SST

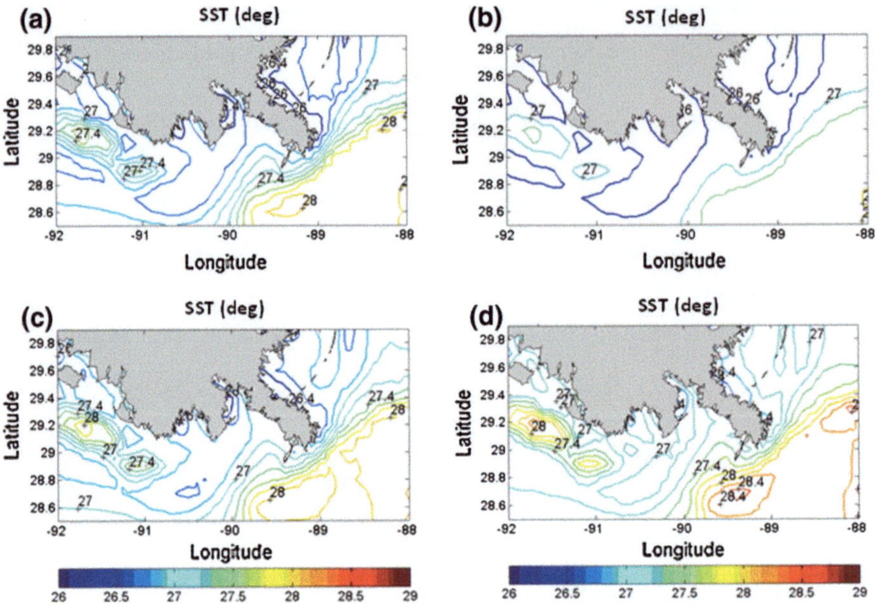

Fig. 1.7 Shelf-wide variation in simulated SST for June 15, 2009; **a** 12:00 AM, **b** 3:00 AM, **c** 12:00 PM, and **d** 3:00 PM

nighttime radiation and peak daytime radiation. Shelf-wide SST map at 12:00 AM local time (Fig. 1.7a) shows that simulated SST over the shelf area off the Barataria and Terrebonne Bays was uniform at about 26.5 °C. Over the shallow shelf off the Atchafalaya Bay, SST was higher (27.4 °C), which can be partly due to the smaller depths and advection of warm water from outer-shelf to this area. SST over the deep waters off the Mississippi Delta was higher (28 °C) compared to the inner-shelf waters. Three hours later at 3:00 AM (Fig. 1.7b), which is several hours after the intense daytime solar insolation, SST distribution off the Barataria and Terrebonne bays was similar to the values at 12:00 AM, but SST off the Atchafalaya Bay and the Mississippi Delta decreased to about 27 and 27.1 °C, respectively. At 12:00 PM (Fig. 1.7c), the daytime increase of solar insolation caused SST to increase off the Atchafalaya Bay to about 28 °C and off the Mississippi Delta to about 28.2 °C. Temperature distribution over the shelf just west of the Mississippi Delta was similar to other two time steps (12:00 AM and 3:00 AM), but with temperature increased to about 26.8 °C. Corresponding to the peak of insolation at 3:00 PM (Fig. 1.7d), SST over the shelf west of the Mississippi Delta increased to 27–27.2 °C and similarly increased in other areas over the shelf.

1.4.3 Vertical Distribution of Temperature

The simulated increase in SST as a result of solar insolation affected the development of temperature stratification. Furthermore, the temperature difference during daytime and nighttime produced diurnal differences in water column stratification. The general behavior of induced shelf stratification based on simulation results is presented along an east–west cross section (Sect. 1.1 in Fig. 1.1). The section extends from the southwest Pass to Sabine Bank that is about 400 km west of the Mississippi Delta and represents the vertical variation in simulated temperature for the inner-shelf region. Figure 1.8 shows the depth profiles of temperature across this section for two different times (12:00 AM and 3:00 PM) on June 15 (almost 15 days after the SST started to increase). Solar heating induced stratification along the initially mixed shelf, except for a shallow region of 30–50 km width (water depth less than 10 ms) located west of the Terrebonne Bay over which the water column was well mixed (Fig. 1.8a). However, the upper part of the water column (upper 7 m) for all points along the section remained well mixed. Water temperature increased from the initial value of 25 °C to highest value of about 27 °C just west of the Mississippi Delta off Barataria Bay, as well as over the shallow and well-mixed region west of the Terrebonne Bay. As compared to 12:00 AM, the 3:00 PM profiles occurred at the peak daytime SST values of 28 °C (Fig. 1.8b). This peak followed the maxima in solar insolation and was associated with a much stronger stratification compared to nighttime (12:00 AM) values. At other time points during the simulation period, the daytime and nighttime stratification patterns were similar to daytime and nighttime patters as shown in Fig. 1.8. Temperature distribution at depths greater than 10 m remained unchanged from day to night, consistent with the attenuating effect of water column depth on short wavelength radiation. Throughout the simulation period, isotherms were tilted upward in the shelf area between the Terrebonne and Barataria bays in response to upwelling resulting from the southwesterly winds.

The diurnal evolution of water column stratification is illustrated with vertical profiles on transect A (Fig. 1.9) on the shelf in front of the Terrebonne Bay (see Fig. 1.1 for location). Water temperatures are shown at four different times on June 15 for the shelf waters up to 50 m. At night (Fig. 1.9a), the mixed layer depth for the shoreward region of the transect (depths smaller than 20 m) was 7–10 m, while the SST was about 26.3 °C. For the deeper region, the mixed layer depth was shallower (less than 5 m) and SST was about 27 °C. Three hours later at 3:00 AM (Fig. 1.9b), the overall pattern of temperature distributions in both shallower and deeper water was similar. However, the mixed layer depth in the shallower region decreased to about 5 m or less, and the associated SST increased to 26.6 °C. At 12:00 PM (Fig. 1.9c), when there was a substantial increase in solar radiation compared to that of morning time over the major part of the shallow shelf, isotherms shifted upward making mixed layer depths less than 5 m. SST over the shallow area increased to 26.9 °C, while the deeper locations had SST values of about 27 °C. The strongest stratification occurred at 3:00 PM (Fig. 1.9d) and

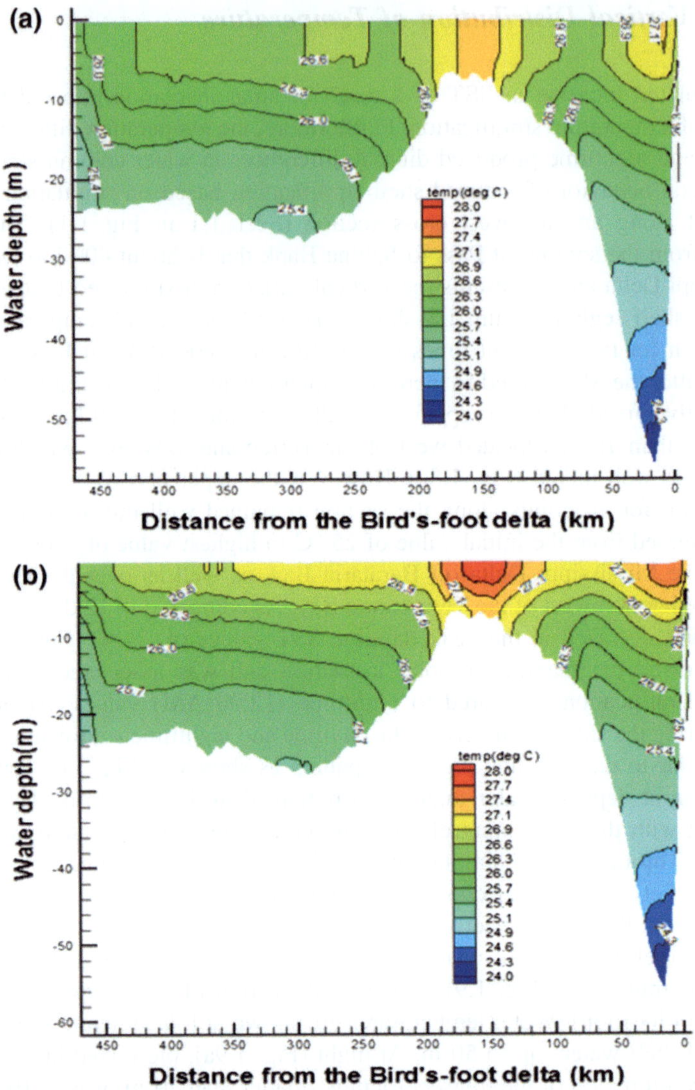

Fig. 1.8 Simulated **a** nighttime and **b** daytime temperature distributions along east–west transect (Sect. 1.1 in Fig. 1.1) for June 15, 2009

extended across the entire transect length with SST increased to 27 °C. Isotherms were closer to each other compared to other times, demonstrating larger vertical temperature gradient and stronger stratification at 3:00 PM compared to the other times.

Development of stratification during night and daytimes is also illustrated using temperature profiles along transect B west of the Mississippi Delta in front of the

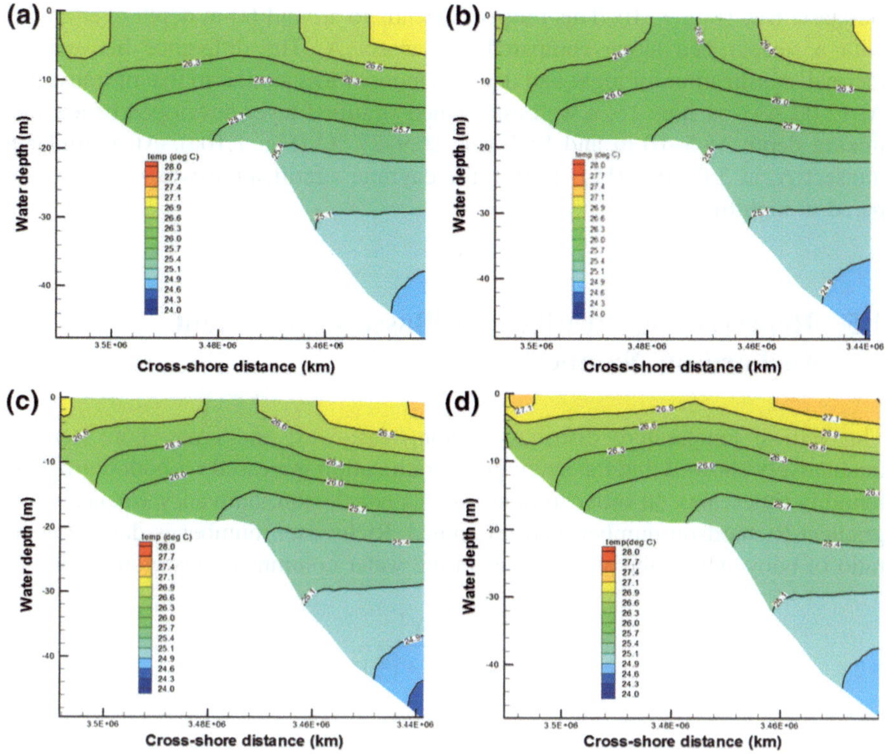

Fig. 1.9 Simulated changes in stratification along transect A for June 15, 2009; **a** 12:00 AM, **b** 3:00 AM, **c** 12:00 PM, and **d** 3:00 PM

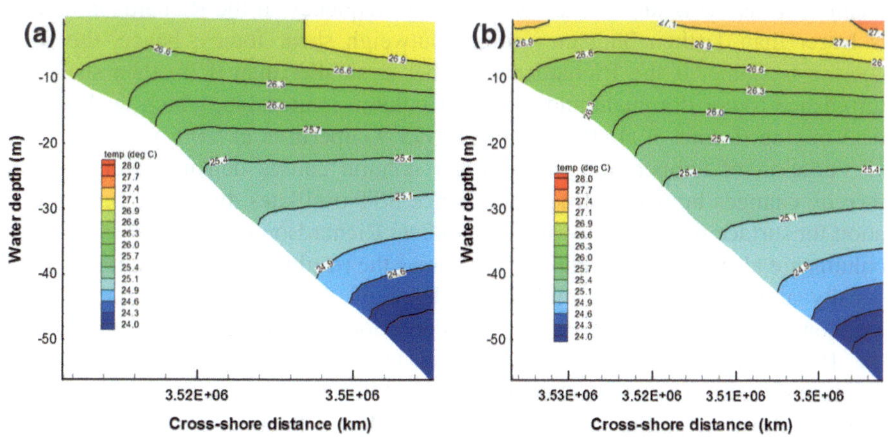

Fig. 1.10 Simulated temperature values across transect B on July 15, 2009; **a** 12:00 AM and **b** 3:00 PM

Barataria Bay (Fig. 1.10). This transect extends to around 60-m depth over an area with a steeper bed slope compared to transect A. The difference between the thermal stratification for night and day was more or less similar to that of transect A. At 12:00 AM on June 15, a well-established stratified layer developed beneath the mixing depth of 5–10 m and SST was 26.9–27 °C (Fig. 1.10a). Also similar to transect A, at 3:00 PM (Fig. 1.10b), the daytime stratification was stronger compared to nighttime.

1.5 Representing Stratification Based on Gradient Richardson Number

The vertical temperature distribution along east–west and cross-shore transects qualitatively showed that the strength of stratification increased on both diurnal and monthly timescales. In order to quantify this conclusion, buoyancy frequency and gradient Richardson number were examined. Richardson number is defined as the ratio of buoyancy to the shear forces in the water column (Lyons et al. 1964):

$$R_i = \frac{N^2}{\left(\frac{\partial u}{\partial z}\right)^2 + \left(\frac{\partial v}{\partial z}\right)^2} \tag{1.7}$$

$$N^2 = -\frac{g}{\rho_0}\frac{\partial \rho}{\partial z} \tag{1.8}$$

in which u and v are the horizontal velocity components varying in the vertical (z) direction, ρ is water density, and g is gravitational acceleration (m/s^2). The quantity N^2 is the Brunt-Väisälä or buoyancy frequency. If the Richardson number is larger than 1, then buoyancy forces outweigh shear forces; hence, the water column is stable. If the Richardson number is smaller than 0.25, then shear and turbulence forces dominate and the water column is unstable, which leads to vertical mixing (Lyons et al. 1964; Turner 1973; Galperin et al. 2007).

Variation of temperature in the water column causes density to change and thereby changes buoyancy frequency. Since SST oscillates diurnally, similar variation for surface water density and the gradient Richardson number across the water column are also expected. Figure 1.11a shows the results for time variation of water density at the surface and mid-depth (depth of 10 m from the surface) at the location of CSI-6. The surface and mid-depth locations were selected to calculate the Richardson Number. The decreasing trend of density is consistent with increasing trend of water temperature during the simulation period. Surface water density follows a similar diurnal pattern as SST. No fluctuations were present for water density at the 10-m depth (mid-water). During the nighttime between June 5 and June 10, surface water density increased to about the mid-depth density due to the nighttime minimum heating. However, after June 10, the difference in densities

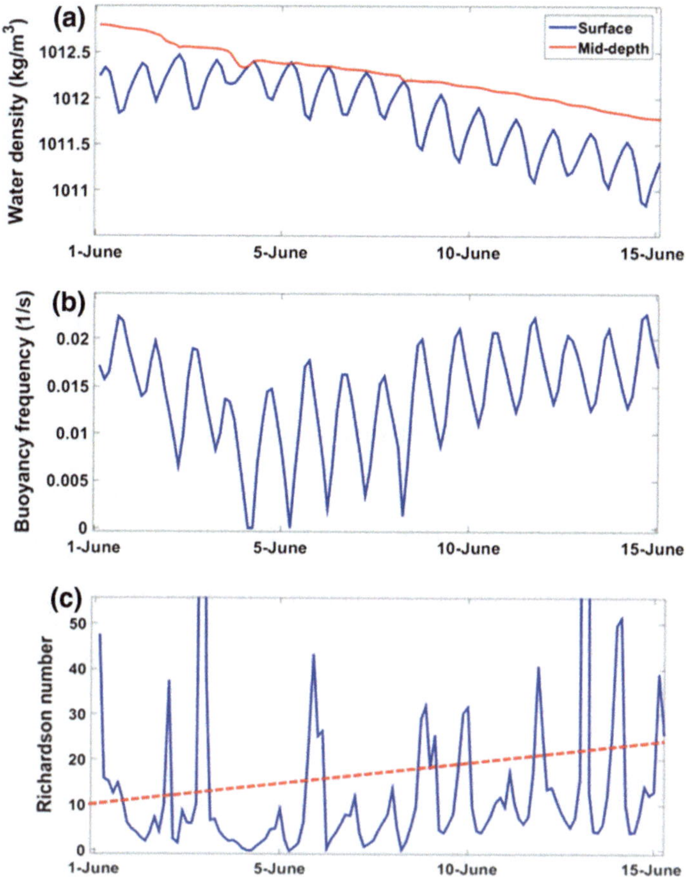

Fig. 1.11 Temporal changes in **a** simulated water density, **b** calculated buoyancy frequency, and **c** calculated Richardson number during June 1–15, 2009, of the simulation. *Dashed line* in Fig. c. denotes the temporal trend in Richardson numbers

increased as a result of faster heating of the surface water compared to the mid-water. Note that the mid-depth density experienced a drop caused by the mixing event of 4 June.

Vertical gradient of water density at the same location was quantified to obtain buoyancy frequency (Fig. 1.11b). The variation over time in buoyancy frequency was similar to SST and surface density on both monthly and diurnal scales. Before June 5, buoyancy frequency experienced a general declining trend caused by a northern wind-induced mixing. After this date, buoyancy effect increased as shelf waters were exposed to higher solar heating.

The strength of stratification was quantified based on the gradient Richardson number at a fixed point, which is the mid-depth water at CSI-6 location (water depth of 10 m). Vertical gradients of current components at the mid-depth were calculated based on the results for wind-induced currents. Since wind speed at the time was generally weak (less than 7 m/s), its impact on mixing was small. The resultant Richardson number was therefore large, showing the dominant effect of buoyancy. The time variation of Richardson number in the mid-depth waters at CSI-6 showed an increasing trend (Fig. 1.11c). The small values of Richardson number prior to June 5 show the effect of wind mixing events. Variations of buoyancy frequency and Richardson number elsewhere over the shelf were similar.

1.6 Diurnal Heating/Stratification and Measured Bottom Oxygen Concentration

Our simulation results showed that stratification becomes stronger as a result of summertime solar insolation. Hence, it is expected that bottom water oxygen concentration would decrease during June 2009 as a stratified water column blocked water re-oxygenation. Time series of measured oxygen concentrations at the bottom of CSI-6 (Fig. 1.12a) during June 2009 verified this. Oxygen concentration was about 4 mg/l on the first day of June 2009, followed by a decline, presumably due to increasing solar radiation and the consequent stratification as well as enhanced biological processes. On June 4, the northerly winds increased

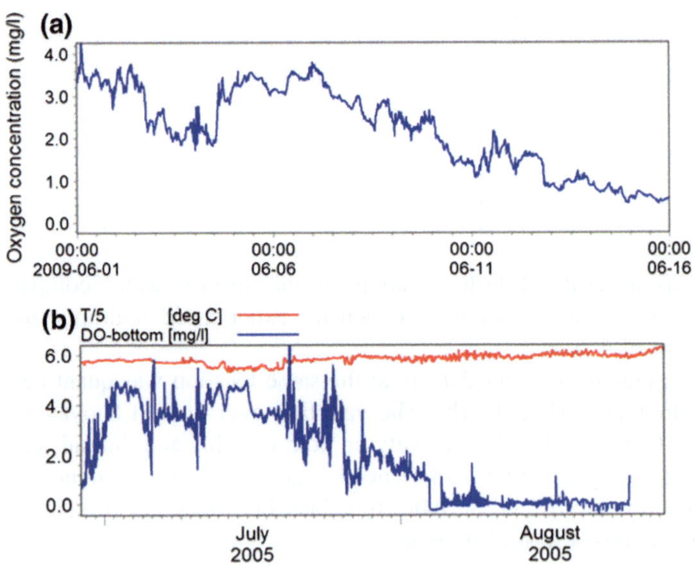

Fig. 1.12 **a** Measured bottom water oxygen concentration at station CSI-6 for June 1–June 15, 2009. **b** Measured SST and bottom water oxygen concentration for July and August 2005

oxygen concentration from 2 mg/l to about 4 mg/l. After this, with the reinforcement of stratification (see Fig. 1.11), oxygen concentration started a longer term decline, depleting the bottom oxygen to less than 1 mg/l. Declining measured bottom oxygen concentrations were consistent with the increase in the simulated SST and the associated buoyancy frequency (implying that the wind-induced mixing was not significant). The consistency was also examined for measurements of dissolved oxygen at CSI-6 during the summer of 2005 (Fig. 1.12b). Time series of bottom oxygen concentration during July 2005 had an average of 4 mg/l for the first 20 days when the average SST was 29 °C. During the next two weeks, SST increased to greater than 30 °C, coincided with declining oxygen concentration. Oxygen concentration then decreased from 4 mg/l to about 1.6 mg/l between July 20 and August 3 when SST increased to 31 °C. This hypoxic bottom water persisted on the shelf until the middle of the last week of August 2005, when mixing produced by Hurricane Katrina broke down the stratification and re-oxygenated the bottom water.

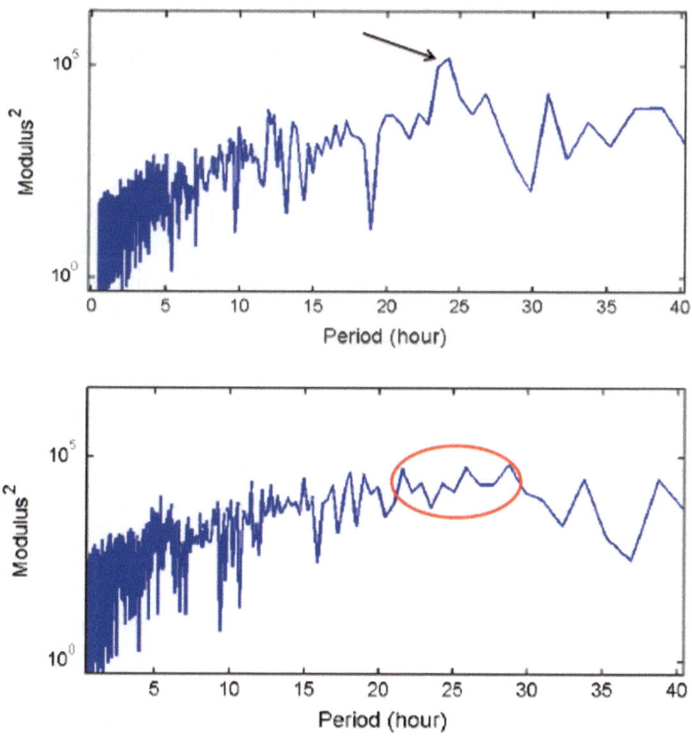

Fig. 1.13 *Upper panel,* energy spectrum of measured SST at station CSI-6 during summer 2005. The *arrow* shows the daily dominant spectral energy for this parameter. *Lower panel,* the corresponding energy spectrum for measured bottom oxygen concentration at station CSI-6. The dominant range of spectral energy is denoted by an *ellipse*

The coincident variation of SST (which is mainly produced by solar heating) and bottom water oxygen concentration are further elaborated by examining the energy spectrum of each parameter during the measurement period of July and August 2005 (Fig. 1.13). The peak of energy for SST was daily (24 h), and bottom oxygen concentration peaked within a narrow band from 22 to 28 h. This accounts for the strong daily variation in bottom oxygen concentration. Based on these time series comparisons, diurnal heating of water column and the resulted stratification can be a main contributor. However, it should be noted that strengthening stratification is not the only effect of the heating induced by solar radiation. Enhanced biological process as a result of higher water temperature can significantly contribute to the oxygen depletion of bottom water. The present study did not address the oxygen consumption by biological processes, and our main focus was demonstrating the role of diurnal solar heating during summertime as a barrier for re-oxygenation from the water surface.

1.7 Summary and Conclusion

Development of daily stratification, caused by summertime increases in solar radiation over Louisiana shelf, was studied using the 3-D FVCOM circulation model. SST over the shelf west of the Mississippi Delta increased from 25 °C on June 1, 2009, to about 28.5 °C on June 20, 2009. In addition to the diurnal variation of solar radiation, SST increased steadily during the month. Model simulation showed that the steady increase in the temperature of the surface layer caused the enhancement of stratification. As a result, the stratification was widespread on the shelf west of the Mississippi Delta and was stronger during the daytime and weaker during the nighttime. The stratification developed more over the shelf area off the Terrebonne Bay to the Mississippi Delta. This is the area that historically has had the most severe hypoxic events.

Analysis showed that the buoyancy frequency followed the same increasing trend of SST in the absence of significant mixing. The Richardson number exhibited an increase after the initial mixing over the shelf. Stratification was consistent with the measured bottom water dissolved oxygen during the simulation period. Oxygen concentration decreased with the increasing SST and stratification strength during period with little mixing.

Acknowledgements The authors are grateful to Nancy Rabalais for sharing the dissolved oxygen data from WAVCIS-CSI-6 and Changsheng Chen for providing the FVCOM.

Appendix A: Formulation of Different Surface Heat Components

(All parameters are described in Table 1.1)

Shortwave Radiation: Relationships presented by Guttman and Matthews (1979), Ivanoff (1977), and Cotton (1979) were used to calculate shortwave radiation flux:

$$Q_o = \frac{S\cos^2 Z}{(\cos Z + 2.7)e \times 10^{-5} + 1.085 \cos Z + 0.10} \tag{1.9}$$

The cosine of the zenith angle is computed using the formula:

$$\cos Z = \sin\phi \, \sin\delta + \cos\phi \, \cos\delta \, \cos HA \tag{1.10}$$

Table 1.1 Parameters used for the formulation of surface heat components

Variable	Value	Description
(a, b)	(9.5, 7.66)	Vapor pressure constants over ice
(a, b)	(7.5, 35.86)	Vapor pressure constants over water
C		Cloud cover fraction
C_E	1.75×10^{-3}	Transfer coefficient for latent heat
C_H	1.75×10^{-3}	Transfer coefficient for sensible heat
c_p	1004 J kg^{-1} K^{-1}	Specific heat of dry air
Δ		Declination
E		Vapor pressure in pascals
e_s		Saturation vapor pressure
∈	0.622	Ratio of molecular weight of water to dry air
HA		Hour angle
L	2.5×10^6 J kg^{-1}	Latent heat of vaporization
L	2.834×10^6 J kg^{-1}	Latent heat of sublimation
Φ		Latitude
Q_o		Incoming radiation for cloudless skies
q_s		Surface specific humidity
q_{10m}		10 m specific humidity
ρ_a		Air density
S	1353 W m^{-2}	Solar constant
Σ	5.67×10^{-8} W m^{-2} K^{-4}	Stefan–Boltzmann constant
T_a		Air temperature
T_d		Dew point temperature
T_{sfc}		Surface temperature of the water/ice/snow
V_{wg}		Geostrophic wind speed
Z		Solar zenith angle

The declination is $\delta = 23.44^{\circ} \times \cos\left[(172 - day\,of\,year) \times 2\pi/365\right]$, and the hour angle is $HA = (12\,h - solar\,time) \times \pi/12$. The correction for cloudiness is given by

$$SW\downarrow = Q_o(1 - 0.6c^3) \tag{1.11}$$

The cloud correction is optional since some sources of radiation contain it already.

Longwave Radiation: The clear sky formula for incoming longwave radiation is given by Wyrtki (1965):

$$F\downarrow = \sigma T_a^4\left\{1 - 0.261\exp\left[-7.77 \times 10^{-4}(273 - T_a)^2\right]\right\} \tag{1.12}$$

while the cloud correction is given by:

$$LW\downarrow = (1 + 0.275c)F\downarrow \tag{1.13}$$

Sensible Heat: The sensible heat is given by the standard aerodynamic formula (Imberger and Patterson 1981):

$$H\downarrow = \rho_a c_p C_H V_{wg}\left(T_a - T_{sfc}\right) \tag{1.14}$$

Latent Heat: The latent heat depends on the vapor pressure, and the saturation vapor pressure given by Imberger and Patterson (1981):

$$e = 611 \times 10^{a(T_d - 273.16)/(T_d - b)} \tag{1.15}$$

$$e_s = 611 \times 10^{a(T_{sfc} - 273.16)/(T_{sfc} - b)} \tag{1.16}$$

The vapor pressures are used to compute specific humidity according to:

$$q_{10m} = \frac{\in e}{p - (1 - \in)e} \tag{1.17}$$

$$q_s = \frac{\in e_s}{p - (1 - \in)e_s} \tag{1.18}$$

The latent heat is also given by a standard aerodynamic formula:

$$LE\downarrow = \rho_a L C_E V_{wg}(q_{10m} - q_s) \tag{1.19}$$

References

Allahdadi M (2015) Numerical experiments of hurricane impact on vertical mixing and re-stratification of the Louisiana shelf. Baton Rouge, Louisiana, Louisiana State University, PhD dissertation, 178p

Allahdadi MN, Jose F, Patin C (2013) Seasonal hydrodynamics along the Louisiana coast: implications for hypoxia spreading. J Coastal Res 29:1092–1100

Bender MA, Ginis I, Kurihara Y (1993) Numerical simulations of tropical cyclone-ocean interaction with a high-resolution coupled model. J Geophys Res-Atmos 98:23245–23263

Chaichitehrani N (2012) Investigation of colored dissolved organic matter and dissolved organic carbon using combination of ocean color data and numerical model in the Northern Gulf of Mexico. Baton Rouge, Louisiana, Louisiana State University, Master's thesis, 112p

Chaichitehrani N, D'Sa EJ, Ko DS, Walker ND, Osburn CL, Chen RF (2014) Colored dissolved organic matter dynamics in the Northern Gulf of Mexico from ocean color and numerical model results. J Coastal Res 30(4):800–814

Chen C, Cowles G, Beardsley RC (2006) An unstructured grid, finite volume coastal ocean model: FVCOM user manual, 2nd edn. SMAST/UMASSD Technical Report-06-0602, p 315

Chen CS, Beardsley RC, Franks PJS, Van Keuren J (2003) Influence of diurnal heating on stratification and residual circulation of Georges Bank. J Geophys Res-Oceans 108

Cotton GF (1979) ARL models of global solar radiation. In: Quinlan FT (ed) SOLMET volume 2: hourly solar radiation—surface meteorological observations. National Oceanic and Atmospheric Administration, Asheville, NC, pp 165–184

Elsberry RL, Fraim TS, Trapell RN (1976) Mixed layer model of oceanic thermal response to hurricanes. J Geophys Res-Oceans Atmos 81:1153–1162

Galperin B, Sukoriansky S, Anderson PS (2007) On the critical Richardson number in stably stratified turbulence. Atmos Sci Lett 8:65–69

Guttman NB, Matthews JD (1979) Computation of extraterrestrial solar radiation, solar elevation angle and true solar time of sunrise and sunset. In: Quinlan FT (ed) SOLMET volume 2: hourly solar radiation—surface meteorological observations. National Climatic Center, US Department of Commerce, Asheville, NC, pp 48–54

Hagy JD, Murrell MC (2007) Susceptibility of a Northern Gulf of Mexico estuary to hypoxia: an analysis using box models. Estuar Coast Shelf Sci 74:239–253

Imberger J, Patterson JC (1981) A dynamic reservoir simulation model—DYRESM: 5. In: Fischer HB (ed) Transport models for inland and coastal waters. Academic Press, pp 310–361

Ivanoff A (1977) Oceanic absorption of solar energy. In: Kraus EB (ed) Modelling and prediction of the upper layers of the ocean. Pergamon, New York, p 326

Justic D, Rabalais NN, Turber RE (2003) Simulated responses of the Gulf of Mexico hypoxia to variations in climate and anthropogenic nutrient loading. J Mar Syst 42:115–126

Justic D, Rabalais NN, Tutner RE (1996) Effects of climate change on hypoxia in coastal waters: a doubled CO2 scenario for the Northern Gulf of Mexico. Limnol Oceanogr 41:992–1003

Lyons R, Panofsky HA, Wollaston S (1964) The critical Richardson number and its implication for forecast problems. J Appl Meteorol 3:136–142

Nezlin NP, Kamer K, Hyde J, Stein ED (2009) Dissolved oxygen dynamics in a eutrophic estuary, Upper Newport Bay, California. Estuar Coast Shelf Sci 82:139–151

Paulson CA, Simpson JJ (1977) Irradiance measurements in the upper ocean. J Phys Oceanogr 7:952–956

Price JF (1981) Upper ocean response to a hurricane. J Phys Oceanogr 11:153–175

Rabalais NN, Atilla N, Normandeau C, Turner RE (2004) Ecosystem history of the Mississippi river-influenced continental shelf revealed through preserved phytoplankton pigments. Mar Pollut Bull 49:537–547

Rabalais NN, Turner RE, Scavia D (2002) Beyond science into policy: Gulf of Mexico hypoxia and the Mississippi River. Bioscience 52:129–142

Rabalais NN, Louisiana Universities Marine Consortum (1991) Fate and effects of nearshore discharges of OCS produced waters New Orleans (1201 Elmwood Park Boulevard, New Orleans 70123-2394), U. S. Dept. of the Interior, Minerals Management Service, Gulf of Mexico OCS Regional Office

Tehrani NC, D'Sa EJ, Osburn CL, Bianchi TS, Schaeffer B (2013) A chromophoric dissolved organic matter and dissolved organic carbon from sea-viewing wide field-of-view sensor (SeaWiFS), Moderate resolution imaging spectroradiometer (MODIS) and MERIS sensors: case study for the Northern Gulf of Mexico. Remote Sens 5:1439–1464

Turner RE, Rabalais NN, Justic D (2008) Gulf of Mexico hypoxia: alternate states and a legacy. Environ Sci Technol 42:2323–2327

Turner RE, Rabalais NN (1994) Coastal eutrophication near the Mississippi river delta. Nature 368:619–621

Turner JS (1973) Buoyancy effects in fluids. Cambridge University Press, London, Cambridge

Wang LX, Justic D (2009) A modeling study of the physical processes affecting the development of seasonal hypoxia over the inner Louisiana-Texas shelf: circulation and stratification. Cont Shelf Res 29:1464–1476

Wiseman WJ, Rabalais NN, Turner RE, Dinnel SP, Macnaughton A (1997) Seasonal and interannual variability within the Louisiana coastal current: stratification and hypoxia. J Mar Syst 12:237–248

Wyrtki K (1965) The average annual heat balance of the North Pacific ocean and its relation to ocean circulation. J Geophys Res 70(18):4547–4559

Zedler SE, Dickey TD, Doney SC, Price JF, YU X, Mellor GL (2002) Analyses and simulations of the upper ocean's response to Hurricane Felix at the Bermuda Testbed Mooring site: 13–23 August 1995. J Geophys Res-Oceans 107

Chapter 2
Physical Drivers of the Circulation and Thermal Regime Impacting Seasonal Hypoxia in Green Bay, Lake Michigan

Hector R. Bravo, Sajad A. Hamidi, J. Val Klump and James T. Waples

Abstract The physical processes that drive the circulation and the thermal regime in the bay largely control the duration and persistence of hypoxic conditions in Green Bay. A review of previous studies, existing field data, our own measurements, hydrodynamic modeling, and spectral analyses were used to investigate the effects on the circulation and the thermal regime of the bay by the momentum flux generated by wind, the heat flux across the water surface, the Earth's rotation, thermal stratification and the topography of the basin. Stratification and circulation are intimately coupled during the summer. Field data show that continuous stratification developed at regions deeper than 15–20 m between late June and September and that surface heat flux is the main driver of stratification. Summertime conditions are initiated by a transition in the dominant wind field shifting from the NE to the SW in late June and remain in a relatively stable state until bay vertical mixing in early September. It is during this stable period that conditions conducive to hypoxia are present. Wind parallel to the axis of the bay induces greater water exchange than wind blowing across the bay. During the stratified season flows in the bottom layers bring cold water from Lake Michigan to Green Bay and surface flows carry warmer water from the bay to Lake Michigan. Knowledge of the general patterns of the circulation and the thermal structure and their variability will be essential in producing longer term projections of future water quality in response to system scale changes.

H.R. Bravo (✉)
Department of Civil and Environmental Engineering,
University of Wisconsin, Milwaukee, USA
e-mail: hrbravo@uwm.edu

S.A. Hamidi
Department of Physics, Indiana University of Pennsylvania, Indiana, PA, USA
e-mail: sajad.hamidi@iup.edu

J. Val Klump · J.T. Waples
School of Freshwater Sciences, University of Wisconsin, Milwaukee, USA
e-mail: vklump@uwm.edu

J.T. Waples
e-mail: jwaples@uwm.edu

D. Justic et al. (eds.), *Modeling Coastal Hypoxia*,
DOI 10.1007/978-3-319-54571-4_2

Keywords Physical drivers • Circulation • Thermal regime • Hydrodynamic modeling • Hypoxia • Green Bay • Lake Michigan

2.1 Introduction

Green Bay is a large, elongated gulf (190 km × 22 km) in the northern part of the Lake Michigan basin. Physically and biogeochemically the bay can be considered to be divided into two portions, a northern and a southern bay, defined by a dividing line at the Chambers Island cross section. The northern or "upper" bay, open to Lake Michigan, is relatively deep (max. depth >50 m) with water quality similar to that of the open lake and is mesotrophic in character, while the southern or "lower" bay is both shallower (mean depth ~10 m) and more confined, resulting in a steep trophic gradient from south to north. The bay is fed by a number of rivers draining a watershed that represents approximately one-half of the total drainage basins for the entire lake (Bertrand et al. 1976). The riverine filling time is approximately 6 years, but because of the large exchange of water with Lake Michigan, the actually flushing time for the bay is less than a year (Mortimer 1979; Miller and Saylor 1993). The largest of these rivers, and the major tributary to the bay with ~50% of the inflow, is the Fox River, entering the bay at the extreme southern end (Fig. 2.1). Once considered one of the most heavily industrialized rivers in the USA as a result of a major regional paper industry and an extensive, largely dairy, agricultural industry within its watershed, the Fox River has a history of excessive loading of nutrients and suspended sediments to lower Green Bay (Klump et al. 1997). As a result, the lower bay has experienced hyper-eutrophic conditions for more than 70 years. The morphology of the lower bay makes it a highly efficient nutrient and sediment trap, with organic-rich sediments that accumulate at rates up to a cm per year and organic carbon contents in excess of 10% by weight (Klump et al. 2009). These sediments quickly become anaerobic and high rates of benthic respiration drive recurring summertime bottom water hypoxia during periods of thermal stratification, generally from late June to early September. The physical processes of circulation and the thermal regime of the bay drive the mixing of tributary nutrients and sediment loads with Lake Michigan waters, affecting the duration and persistence of these hypoxic conditions. The physical drivers of circulation and the thermal regime in the bay are the focus of this chapter.

A review of previous studies, existing field data, our own measurements, hydrodynamic modeling, and spectral analyses were used to investigate the effects on the circulation and the thermal regime of the bay by the momentum flux generated by wind, the heat flux across the water surface, the Earth's rotation, thermal stratification, and the topography of the basin. Stratification and circulation are intimately coupled during the summer.

A brief summary of previous studies on the effects of momentum flux generated by the wind on circulation and thermal regime provides a foundation for our research. Miller and Saylor (1985) analyzed currents and temperatures measured in 1977 in the passages between Green Bay and Lake Michigan (herein referred to as the mouth) and within the bay proper. They found that the direction of circulation

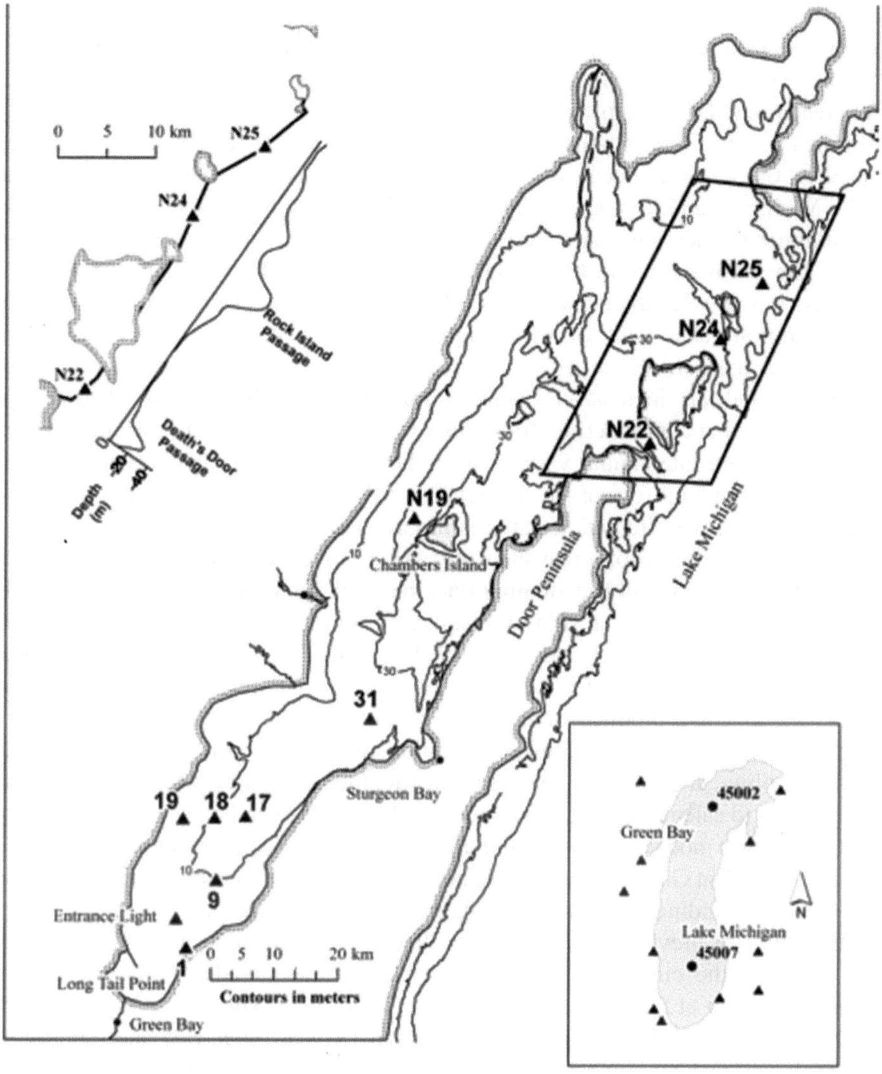

Fig. 2.1 Bathymetric map of Green Bay showing locations of 1989 NOAA measurements, 2011 current and temperature measurements. The *top left inset* shows map and cross section of the passages between Lake Michigan and Green Bay. The *bottom right inset* shows the locations of ASOS meteorological stations and NBDC buoys

reverses with a SW-NE (along-bay) wind direction. Gottlieb et al. (1990) measured currents and temperatures in Green Bay during the 1988–1989 winters and the summer and fall of 1989. Their monthly averaged summer currents showed that the direction of circulation in the bay varied with the wind direction. Gottlieb et al. (1990) found that the power spectrum of their measured currents showed the surface-mode oscillations computed by Rao et al. (1976), and they attributed the 8-day mode to wind forcing with that frequency. Mortimer (2004) reasoned that: "[t]hese (winds) occurred at intervals ranging from 6 to 12 (average 8) days, apparently regular enough to generate and maintain the internal seiche during the stratified season from mid-July to September." Waples and Klump (2002) analyzed the effect of the wind direction on the water mass exchange between Green Bay and Lake Michigan, and hypoxia in the southern bay. They found a significant shift in the summer surface wind direction over the Laurentian Great Lakes from 1980 to 1999. They showed that in Green Bay the new wind field most likely resulted in a decrease in the water mass exchange with Lake Michigan, leading to an increase in vertical mixing as a result of a less highly stratified water column, a concomitant decrease in the persistence of bottom water hypoxia, warmer bottom water temperatures, and an increase in benthic microbial metabolism, as observed in enhanced methane production from bottom sediments. Recently, Hamidi et al. (2015) analyzed interannual variability in the summer wind direction and the circulation patterns and discussed the relation between multi-year monthly average wind fields and the circulation patterns in Green Bay.

The effects of heat flux across the water surface on the circulation and the thermal regime has been the subject of important research. Beletsky and Schwab (2001) applied a three-dimensional primitive equation numerical model to Lake Michigan for the periods 1982–1983 and 1994–1995 to study the seasonal and interannual variability of the lake-wide circulation and the thermal structure in the lake. The model was able to reproduce all of the basic features of the thermal structure in Lake Michigan: spring thermal bar, full stratification, deepening of the thermocline during the fall cooling, and finally, overturn in the late fall. They used a bulk aerodynamic formulation to calculate heat and momentum flux fields over the water surface for the lake circulation model. Lately Hamidi et al. (2015) analyzed the relation between the thermal regime in Green Bay, heat flux across the water surface and heat transport by circulation, including the water exchange with Lake Michigan.

Examples of previous studies on the effects of stratification and topography of the basin on the circulation and the thermal regime include Miller and Saylor (1985), Gottlieb et al. (1990), and Saylor et al. (1995). Miller and Saylor (1985) analyzed currents and temperatures measured in 1977 in Green Bay and found flow in two layers and the opposite direction through the mouth of the bay during the stratified period. They reasoned that cold hypolimnetic water maintains stratification and promotes flushing. They also observed oscillations in current records and analyzed coherence between currents and water level fluctuations. Gottlieb et al. (1990) measured currents and temperatures in Green Bay during the 1988–1989 winters and the summer and fall of 1989. Their July and August 1989 current and temperature measurements revealed a strong, persistent, well-defined 8-day-long oscillation associated with seiching of the thermocline. The power spectrum of their measured currents also showed the surface-mode oscillations computed by Rao et al. (1976). Gottlieb et al. (1990) added that the observed period agreed with that of a free standing internal wave (i.e., an internal seiche). Saylor et al. (1995) used summer 1989 measurements (Gottlieb et al. 1990) to investigate near-resonant wind forcing of internal seiches in Green Bay. Saylor et al. (1995) found persistent oscillations of the thermocline at the period of the bay's lowest-mode, closed basin internal seiche, an 8-day long period.

This chapter is organized as follows. Section 2.2 presents the research methods used and Sect. 2.3 presents the research results. Hamidi et al.'s (2015) results on the effects of surface heat flux and momentum flux generated by the wind on the circulation and the thermal regime are summarized in Sects. 2.3.1 and 2.3.2 of this chapter. In Sect. 2.3.3, we present the results of a numerical experiment on the effect of the wind direction on the water exchange between Lake Michigan and Green Bay. Section 2.3.4 presents a calculation of the water exchange through a midbay section at Chambers Island and the mixing time that shows the effects of stratification and the bay and lake topography. In Sect. 2.3.5, we investigate the effects of wind, stratification, Earth's rotation, and the bay and lake topography on two-layer flows using frequency domain analysis of along-the-bay wind, and top and bottom currents. In Sect. 2.3.6, we analyzed field data on rotating currents to investigate the effects of stratification, Earth's rotation, and the bay and lake topography on the direction of currents. Section 2.4 presents the conclusions of this research study. The main questions explored in this chapter on physical drivers of hypoxia are the relation between the surface heat flux and stratification, the relation between multi-year monthly averages of wind fields and circulation pattern, the relation between the wind direction and the water exchange between Green Bay and Lake Michigan, the residence time in lower Green Bay, and the effect of the Earth's rotation on the currents in the top and bottom layers during the stratified season.

2.2 Methods

2.2.1 New Field Measurements

Our data collection program during the summers of 2011, 2013, and 2014 focused on southern Green Bay. Currents were measured at three stations using Nortek Aquadopp acoustic Doppler profilers (2 MHz) (stations 1, 18, and 19, see Fig. 2.1 and Table 2.1). Continuous measurements of the water temperature at 1–3 m depth intervals were collected at stations 9, 17, 31 and Entrance Light—EL using Onset Hobo temperature data loggers (±0.21 C). The Aquadopp ADCPs were deployed between June 17 and October 5, 2011, and the settings included a sampling frequency of 2 Hz, a cell size 0.5 m, an averaging interval 180 s, and a horizontal velocity precision 0.5 cm/s. Our 2014 field measurements included time series profiles of horizontal velocity and temperature measured at a GLOS Buoy (NOAA # 45014) located at Station 17 (see: glos.us). The 2011ADCP measurements were done in cooperation with NOAA GLERL as part of an effort to improve nearshore wave climate and beach forecast models within the bay. The 2011 current measurements were obtained at sites with a depth up to 10 m because that is the effective range of the ADCPs employed.

Table 2.1 Locations and characteristics of new and historical field data collection sites

Station	°Lat.	°Lon.	Depth (m)	Data	Year
1	44.60	87.87	4	Current	2011
9	44.70	87.82	9	Temp.	2011
17	44.79	87.76	13	Current/Temp.	2014
18	44.78	87.82	10	Current	2011
19	44.87	87.88	4	Current	2011
31	44.93	87.51	25	Temp.	2011
Entrance Light (EL)	44.65	87.90	6.8	Temp/DO	2011
N19	45.21	87.41	33	Current/Temp.	1989
N22	45.29	86.97	30	Current	1989
N24	45.43	86.80	45	Current/Temp.	1989
N25	45.40	86.75	37	Current	1989

2.2.2 *Historical Observations*

NOAA GLERL (Hawley, personal communication, 2012) provided summer 1989 current and temperature data (Gottlieb et al. 1990) for 21 moorings, including stations N22, N24, and N25 at the boundary between Green Bay and Lake Michigan at the tip of the Door Peninsula, and station N19 west of Chambers Island (Fig. 2.1 and Table 2.1). The model validation used 1989 historical measurements that were made at sites that are critical to the understanding of the water exchange between the lake and the bay at the northern tip of the Door Peninsula (Stations N22, N24, and N25 at depths larger than 30 m) and between southern and northern Green Bay at the Chambers Island east and west channels (Station N19). Great Lakes surface water temperature data obtained from NOAA Coast Watch (http://www.coastwatch.glerl.noaa.gov/ftp/glsea/) were used to validate modeled temperatures.

2.2.3 *Meteorological Forcing*

Meteorological data sets were developed to run the Lake Michigan model for years in which new and/or historical observations existed, thus allowing validation against measured currents and temperature data. For 2011, we used a data set prepared by NOAA GLERL, consisting of wind velocities, air temperature, dew point, and cloud cover. For years with historical observations, we used the method described by Beletsky and Schwab (2001) to interpolate to the model grid from the meteorological data available at stations around the lake (Fig. 2.1). To calculate the overwater meteorological fields from land observation data, we first interpolated the data to an hourly basis, then carried out height adjustment and overland/overwater adjustments, and finally interpolated the adjusted data for all stations spatially over the 2 km grid. Data from 11 NOAA Automatic Surface Observing System (ASOS) stations on land were used. After interpolation over water, results were crosschecked with data at the locations of NBDC buoys 45007 (south) and 45002 (north) in the lake. The accuracy of this interpolation method was verified by comparing interpolated meteorological data with measured data. The comparison (not shown) demonstrated the good accuracy of the method (Hamidi et al. 2015).

2.2.4 Modeling

We employed two hydrodynamic models in this study, namely a Lake Michigan model that is a version of the Great Lakes Coastal Forecasting System (GLCFS) developed and operated by NOAA GLERL, and a high resolution nested model of Green Bay developed for this study. The GLCFS model is based on a Princeton Ocean Model (POM) version adapted to the Great Lakes (Schwab and Bedford 1994). Hydrodynamic models numerically solve the governing equations to predict currents and temperature that result from the combined effects of meteorological forcing functions, the Earth's rotation, and bathymetry. The Princeton Ocean Model (POM) (Blumberg and Mellor 1987) is a time-dependent 3D, non-linear, finite difference model that solves the conservation of heat, mass, and momentum equations, considering the combined effects of the physical drivers listed above. The surface heat flux, hf, is calculated as

$$hf = swr + shf + lhf + lwr \tag{2.1}$$

where swr is shortwave radiation from the sun, shf is sensible heat transfer, lhf is latent heat transfer and lwr is long wave radiation. The equations governing the dynamics of coastal circulation contain fast moving external gravity waves and slow moving internal gravity waves. It is desirable in terms of computer economy to separate the vertically integrated equations (external mode) from the vertical structure equations (internal mode). This technique, known as mode splitting (Mellor 2002), permits the calculation of the free surface elevation with little sacrifice in computational time by solving the velocity transport separately from the three-dimensional calculation of the velocity and the thermodynamic properties. The nested model used time steps of 1 s and 20 s for the external and internal time steps, respectively. In this study, simulations started in March–April with Lake Michigan well mixed from top to bottom at temperatures near the temperature of maximum density for freshwater, about 4 °C. For applications to the Great Lakes, the salinity is set to a constant value of 0.2 parts per thousand. The turbulence closure scheme characterizes the turbulence by equations for the turbulence kinetic energy and a turbulence macroscale, according to the Mellor and Yamada 2.5 model (Mellor and Yamada 1982). Full details are given in Beletsky and Schwab (2001).

The Lake Michigan model uses a grid size of 2 km and 131 × 251 cells in the EW and the NS directions, respectively, and 20 vertical depth intervals (sigma layers). The model is driven by the meteorological conditions of wind, air temperature, dew point, and cloud cover developed as described above. Meteorological forcing is distributed over the lake and varies in time every hour. The nested model for Green Bay uses a grid size of 300 m and 132 × 644 cells in the directions along the bay and across the bay, respectively, and 20 vertical depth intervals or sigma layers (Fig. 2.2). The nested model obtains its meteorological forcing, initial and boundary conditions between Green Bay and Lake Michigan from the Lake

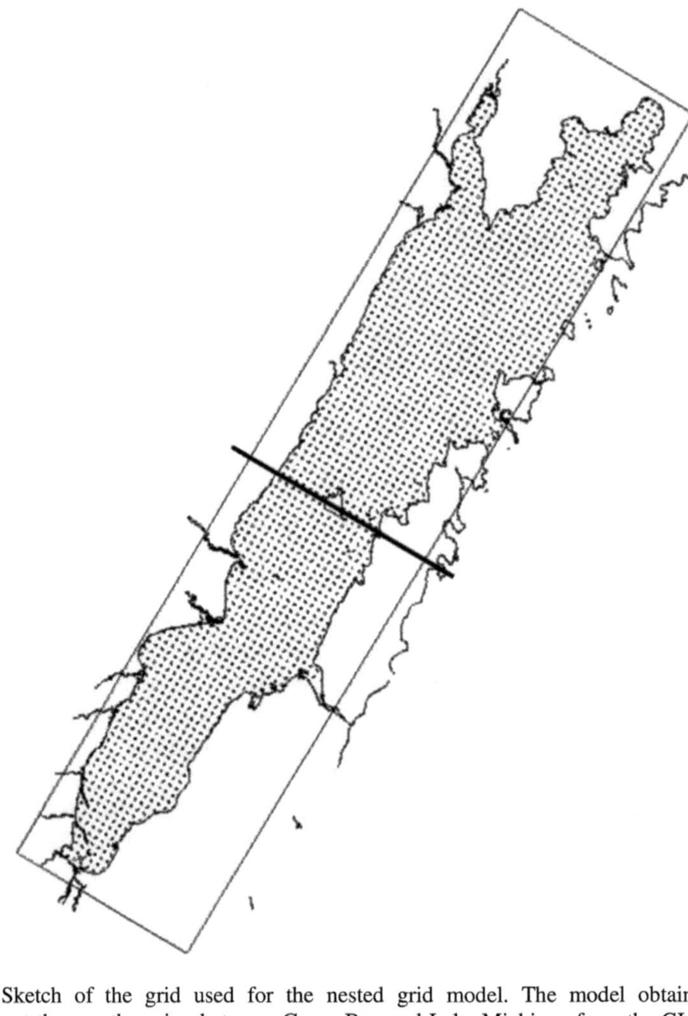

Fig. 2.2 Sketch of the grid used for the nested grid model. The model obtains boundary conditions at the mouth region between Green Bay and Lake Michigan from the GLCFS model, and tributary inflows from USGS Web sites. The *dark line* across Chambers Island delineates the grid used for Lower Green Bay model simulations

Michigan model, and tributary inflows from USGS Web sites (http://www.waterdata.usgs.gov/wi/nwis/sw). Monthly averaged winds and currents were calculated for up to five years within the 2004–2008 period from GLCFS results provided by NOAA GLERL.

2.2.5 Model Validation

As described by Hamidi et al. (2015), the hydrodynamic model results were validated against our own 2011 and the historical 1989 measurements, i.e., for two years with available field data and quite different meteorological forcing. As explained in Sect. 2.3.2, we found significant interannual variability in meteorological forcing and circulation patterns. Once the model was validated under different meteorological forcing, we used it in Sect. 2.3.2 to analyze the relation between multi-year monthly average wind forcing and circulation patterns. The goodness of fit between measurements of currents or temperature and model predictions was quantified in terms of the estimated root mean square error (RMSE), the normalized root mean square error (NRMSE), and the correlation coefficient. The NRMSE is equivalent to the Fourier norm, F_n, used by Beletsky and Schwab (2001) and can be thought of as the relative percentage of variance in the observations unexplained by model calculations. The correlation coefficient between the measured and the model-predicted time series is their covariance divided by the product of their individual standard deviations. Blumberg et al. (1999) assessed the skill of their model of the New York Harbor region in terms of the RMSE and the correlation coefficient. We investigated the sensitivity of the model with respect to model parameters and decided to use default values. The sensitivity analysis included varying the horizontal and vertical grid resolution, changing the model parameters such as the dimensionless coefficient (C) that relates velocity gradients to horizontal kinematic viscosity in Smagorinsky's horizontal diffusivity and the turbulent Prandtl number (the relation between diffusivity and viscosity), varying the description of the shortwave radiation penetration, and varying the turbulent vertical mixing. Table 2.2 shows a comparison between the observed and the measured currents at station 1. Verification of the temperature profiles was presented in detail in Hamidi et al. (2015).

Table 2.2 Statistics of the comparison between observed and modeled currents at station 1

Depth avg. currents	Min	Max	Average
RMSE (cm/s)	1.10	2.19	2.10
NRMSE	0.52	0.95	0.82
Correlation coefficient	0.47	0.85	0.62

2.2.6 Spectral Analysis

The time series data collected in Green Bay during 1988–1989 (Gottlieb et al. 1990) and 2011 showed oscillatory patterns in temperatures and currents. Frequency analysis, including the estimation of power spectra, coherency and phase, was applied to uncover the main frequencies in the oscillations of currents and temperature isotherms. Recall that the spectrum of a time series or signal is a function of a frequency variable, which has dimensions of power or energy per frequency unit (such as cycles per day). Intuitively, the spectrum decomposes the content of the time series into different frequencies present in that process and helps identify periodicities. The SSA toolkit was used to estimate the power spectra.

Coherence analysis, or cross-spectral analysis, was used to identify variations that have similar spectral properties (high power in the same spectral frequency bands), i.e., if the variability of two distinct, detrended time series is interrelated in the spectral domain (Von Storch and Zwiers 1999). Squared coherency, the frequency domain analogue of correlation, was estimated in this study following Jenkins and Watts (1968) and Bloomfield (1976). Values of coherency estimates were considered significant at the 95% level of confidence when they were larger than the critical value T derived from the upper 5% point of the *F*-distribution on (2, *d*-2) degrees of freedom, where *d* is the degrees of freedom associated with the univariate spectrum estimates.

2.2.7 Effects of Earth's Rotation

Inertial currents are consequences of the Coriolis effect, caused by the rotation of the Earth and the inertia of the mass experiencing the effect. The effects of the Coriolis force generally become noticeable only for motions occurring over large distances and long periods of time, such as a large-scale movement of water in the ocean. This force causes moving objects on the surface of the Earth to be deflected in a clockwise sense (with respect to the direction of travel) in the Northern Hemisphere. Inertial currents occur in all large stratified basins and oceans, and the theoretical inertial period is 24/(2 sin ϕ), i.e., 17.3 h, for a latitude $\phi = 44°$. Inertial oscillations observed in Lake Michigan do not ordinarily reach the theoretical inertial limit. The observed periods are slightly less (but never more) than the theoretical limiting period. Thus one may speak of near-inertial oscillations (Mortimer 2004). The Rossby radius of deformation, or simply the Rossby radius $a = c/f$ (c is the wave speed and f is the Coriolis parameter), is the length scale at which rotational effects become as important as buoyancy or gravity wave effects in the evolution of the flow about some disturbance. For Green Bay, representative values of c for surface and internal long waves are $c_s = 17$ m/s and $c_i = 0.27$ m/s. Therefore, for $f = 10^{-4}$ rad/s, the Rossby radius for surface and internal waves are $a_s = 168$ km and $a_i = 2.6$ km, respectively. Green Bay is "fairly small" for surface

Kelvin waves but "quite large" for internal waves. In Sect. 2.3.6, currents measured west of Chambers Island by Gottlieb et al. (1990) were analyzed using spectral analysis to investigate possible evidence of the Coriolis effect.

2.3 Results and Discussion

2.3.1 Relation Between the Surface Heat Flux and Stratification

Gottlieb et al.'s (1990) temperature measurements and our own 2011 measurements had a resolution sufficient to characterize stratification. The net heat flux across the water surface of Green Bay was calculated for those two years in order to explore its relation with stratification. Figure 2.3a, c show the calculated net heat flux, and its components of shortwave radiation, sensible heat transfer, latent heat transfer, and long wave radiation, for 1989 and 2011, respectively. The meteorological forcing files have an hourly time step, and the calculated heat flux varies hourly. Heat flux terms are displayed in Fig. 2.3 with a daily time step for readability reasons. The figure shows that the net heat flux is consistently positive between mid-June and September. Shortwave radiation from the sun was the largest single component of the surface heat flux. The calculated heat fluxes compared very well with heat flux measurements made in 2011 (Grunert 2013), and with Beletsky and Schwab's (2001) calculated annual cycles of the net heat flux for Lake Michigan during 1982–1983 and 1994–1995, which ranged from −400 W m^{-2} in winter to 200 W m^{-2} in summer.

Figure 2.3b, d show measured temperature profiles at Stations N19 and 31 (see locations in Fig. 2.1 and Table 2.1) during the summers of 1989 and 2011, respectively. Continuous stratification developed at regions deeper than 15–20 m between late June and September. A positive surface heat flux heats the surface waters, and Fig. 2.3 shows that stratification in the measured temperature profiles follows the surface heat flux cycle, indicating that the surface heat flux is the main driver of stratification in Green Bay.

2.3.2 Relation Between Wind Fields and Circulation Pattern

Previous descriptions of the general circulation in Green Bay were based on the field measurements made within one-year period (Miller and Saylor 1985; Gottlieb et al. 1990). Examining the more general description of circulation patterns contained in the 1988–1989 data supplied by NOAA GLERL, our own 2011 current measurements, and model simulations for 1989 and 2011, revealed considerable differences between the 1989 monthly averaged wind field and circulation patterns,

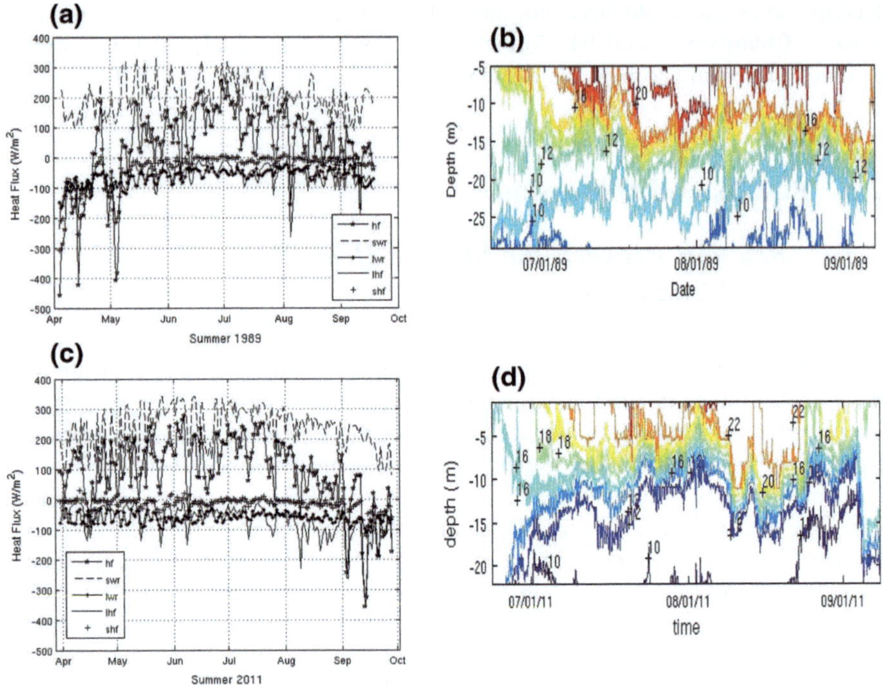

Fig. 2.3 **a** and **c** calculated net heat flux across the water surface of Green Bay, and its components, for 1989 and 2011, respectively. **b** and **d** measured temperature profiles (C) at Stations N19 and 31 during the summers of 1989 and 2011, respectively

and those in 2011. The monthly averaged wind was northerly in August 1989 and westerly in August 2011. The circulation pattern in August 1989 was very different than that in August 2011, particularly across the region between Green Bay and Lake Michigan. Results based on the single-year measurements can give incomplete descriptions of longer term conditions.

The relation between summer wind fields and circulation was therefore studied using multi-year averages. Mean summer circulation patterns were determined by calculating multi-year averages of wind shear fields and depth-averaged currents in the bay, for each month from May to September, starting with two-year averages and successively increasing the number of years included. The circulation patterns remain unchanged beyond a four-year average. Five-year averages of wind shear fields and currents are illustrated in Figs. 2.4 and 2.5, respectively, for the 2004–2008 period. The circulation in the bay depends on the wind field over the whole lake. Figures 2.4 and 2.5 show the wind and current fields in Green Bay and the immediately adjacent area of Lake Michigan.

Even though there is significant variability from one year to another, maps of climatological circulation in Green Bay are useful because they describe the general circulation patterns in the bay and can show the patterns of seasonal progression in

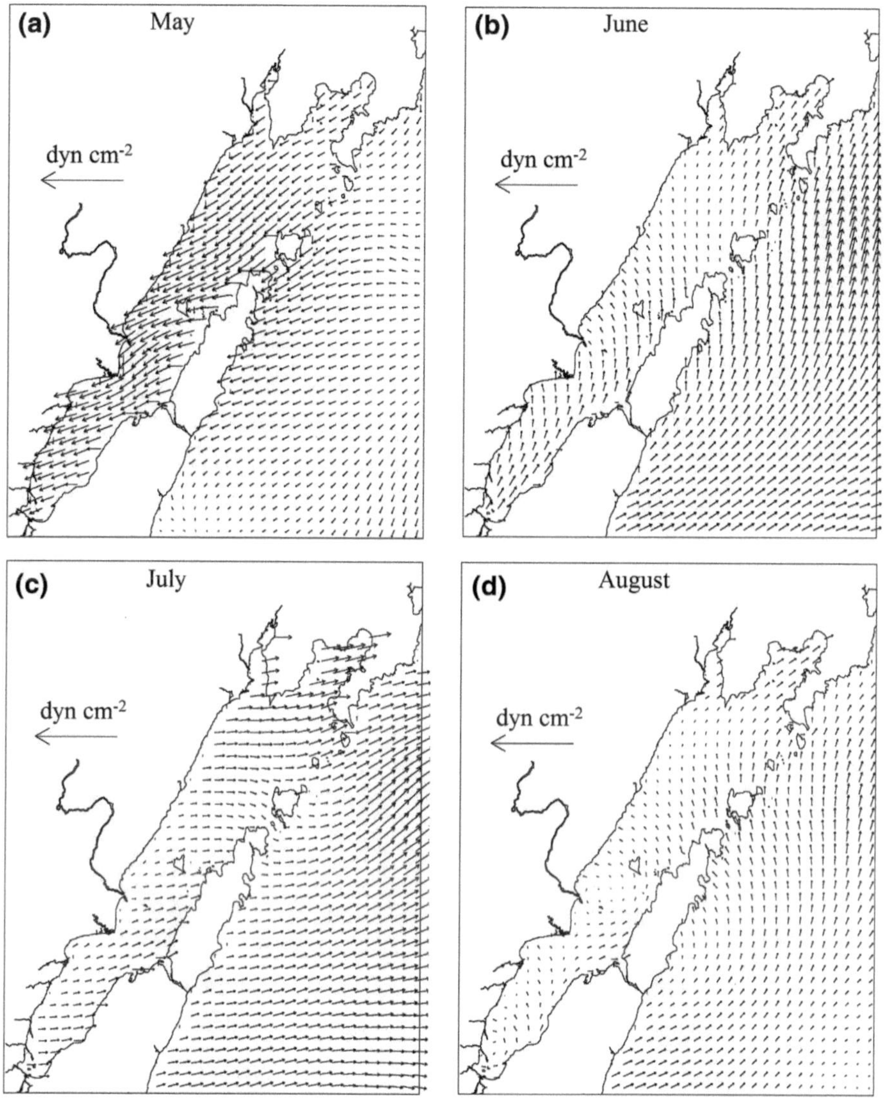

Fig. 2.4 Monthly averaged wind stress for May through August for the 2004–2008 five-year period

the wind field and the resulting mean flows. Beletsky and Schwab (2008) used 10-year averages of model results combined with measurements that validated the model, to map basin scale, climatological circulation patterns in Lake Michigan. They pointed out that maps of climatological circulation are extremely useful for a variety of issues ranging from water quality predictions to sediment transport and ecosystem modeling.

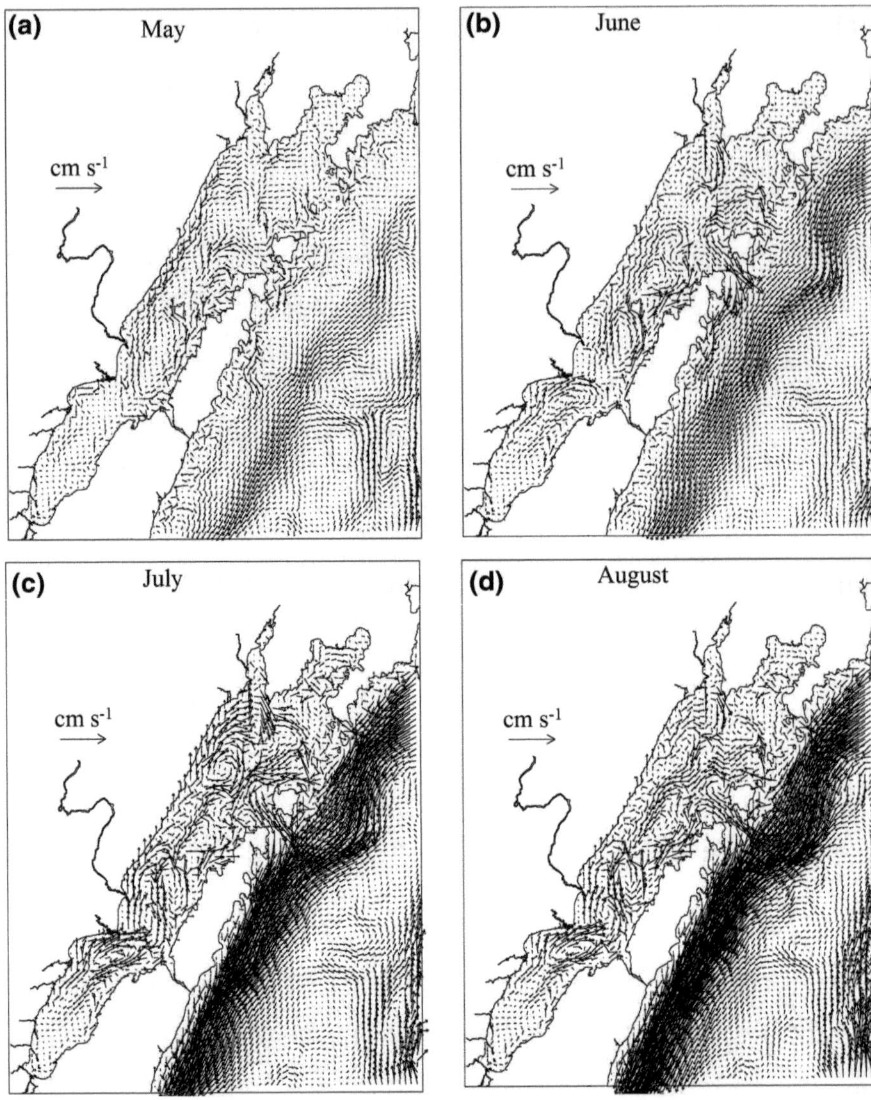

Fig. 2.5 Monthly averaged circulation for May through August for the 2004–2008 five-year period

In May, the monthly averaged wind is predominantly NE. In June, it rotates to the SW and keeps the same monthly average direction for the rest of the summer. The switch in wind direction in May and the consistent wind direction during the summer occurs over the whole lake. Analysis showed a consistent relationship between the monthly average wind shear direction and the monthly average currents and circulation. At the beginning of summer in May, the average circulation pattern responds to a changing wind climate, which transitions from a predominantly NE to a SW wind direction in June. During July, August, and September while the wind blows from the SW the circulation patterns remain similar (Figs. 2.4 and 2.5). Figure 2.4 also shows that the average magnitude of wind shear in June is smaller than that later in the summer as the variability in the direction of the winds cancels one another and reduces the mean. Water circulation patterns change in June as a result of changes in the wind direction, and as more persistent wind conditions become established, a more stable circulation pattern develops during the rest of the summer. Hypoxia is typically observed from late June to early September. Results are shown for the months of May to August to demonstrate the May to June switch in the dominant winds and the subsequent period of consistent average wind associated with stratification. Figure 2.5 shows depth-averaged currents to illustrate general circulation patterns. Figure 2.3b, d show the relevance of bottom currents during the inception of stratification, when the transport of colder water by into-the-bay bottom currents induces a decrease in bottom temperatures.

Circulation varies from year to year depending on wind forcing, yet it was possible to find persistent, monthly average patterns in wind forcing and consequent circulation patterns. Schwab and Beletsky (2003) clearly explained the relation between transport vorticity and the curl of the wind stress in Lake Michigan. When the dominant wind field shifts from NE to SW in late June there is a change in the wind curl, which in turn results in a change in transport vorticity and circulation patterns, as shown in Fig. 2.5. Furthermore, conditions conductive to hypoxia are present during the July–September period of consistent circulation.

This description complements the climatological maps of summer and winter circulation in Lake Michigan developed by Beletsky and Schwab (2008), who found consistent overall cyclonic circulation in Lake Michigan. Maps of climatological circulation are useful for understanding water quality and ecosystem issues and patterns of sediment transport.

2.3.3 Relation Between Wind Direction and Water Exchange Between Green Bay and Lake Michigan

The water exchange between Green Bay and Lake Michigan varies continuously and depends on several of the physical drivers considered in this chapter. The effect of wind direction alone was explored by estimating the water exchange between Lake Michigan and Green Bay under idealized conditions consisting of summer 2011 atmospheric forcing, except for steady uniform wind with a velocity of 5 m s^{-1} parallel or perpendicular to the main Green Bay axis. This experiment tested the effect of the wind direction alone and it did not reflect real time variability in wind speed and direction. The purpose of the experiment was to compare the water exchange induced by along-bay and cross-bay winds. An estimation of the water exchange with a historical unsteady wind is described in Sect. 2.3.4.

This idealized condition is intermediate between the barotropic model with a steady uniform wind used by Beletsky (2001) to study the circulation in Lake Ladoga and the general case of a baroclinic model with a spatially variable wind. In this idealized condition, the wind curl is zero because the wind field is steady and spatially uniform (Schwab and Beletsky 2003). The model was run in each case for about one month, when it was verified that the circulation and the thermal regime approached steady conditions. A steady uniform wind parallel to the axis of the bay (NNE and SSW) induces roughly a 25% greater water exchange (11,670 and 10,080 $m^3\ s^{-1}$, respectively) than wind blowing across the bay (ESE and WNW, 9,900 and 7,200 $m^3\ s^{-1}$, respectively). In addition, wind blowing from the lake to the bay (NNE) induces a greater water exchange than wind blowing from the bay to the lake (SSW).

A water exchange through the mouth region of 11,550 $m^3\ s^{-1}$ was estimated for wind from the WSW (30° south from W), which approximates the monthly average wind direction during August 1994. The ESE wind direction described above approximates the monthly averaged wind direction in August 1995. Across-bay wind from the ESE induced a substantially smaller water exchange (9,900 $m^3\ s^{-1}$) than wind from the WSW. Waples and Klump (2002) found that SW wind (parallel to the major axis of the bay) in August of 1994 produced decreases in bottom temperature and oxygen concentration, while SE (cross-axial) winds in August of 1995 caused increases in bottom temperature and oxygen concentration. They explained the 1995 effect by saying that the water mass exchange with Lake Michigan slowed under more easterly winds. The comparison performed in this study is not based on modeling the circulation and thermal regime during August 1994 and 1995 using the actual meteorological forcing, yet confirms Waples and Klump's (2002) finding about the important effect of the wind direction on the water exchange between Green Bay and Lake Michigan. One of the long-term impacts hypothesized as potentially important for this system is a regime change in the propagation of storm tracks through the Laurentian Great Lakes basin in response to large-scale climate change patterns that have pushed summertime storm tracks further to the south.

Miller and Saylor (1985) estimated at 3,300 $m^3\ s^{-1}$ the average water exchange during June, July, and August of 1977, and at 0.6 yr the "emptying time" for Green Bay. The monthly average wind direction was from the SWW in June and from the SW in July and August of 1977. Our estimate for idealized steady uniform wind shown above is almost three times larger than Miller and Saylor (1985) estimate. The water exchange rates presented above are realistic estimates because they were obtained using a model that was positively tested against 1989 measurements (Hamidi et al. 2013), assuming an idealized uniform wind with normal speed. The water exchange flow rates presented here are therefore comparable with Miller and Saylor's (1985) estimates based on 1977 measurements.

2.3.4 Estimation of Water Transport Between Lower and Upper Green Bay

The water transport between lower and upper Green Bay was estimated based on model simulation of the circulation and the thermal regime during the summer of 1989, using actual meteorological forcing. This water transport was calculated by integrating the modeled currents through the cross section at Chambers Island (Fig. 2.2). During the stratified season, a complex two-layer transport oscillated out of and into lower Green Bay. Figure 2.6 shows that summer 1989 net transport varied continuously over time. The three-month (June–August) average in and out flows were practically equal with a value of 1,500 m^3 s^{-1}. This means that exchange between the upper and lower bay is ~15% of the exchange of the upper bay with the open lake, i.e., the flushing or exchange attenuates significantly as water penetrates into the southern bay. The model estimation is almost double the water transport values of 790 m^3 s^{-1} and 830 m^3 s^{-1} out of and into lower Green Bay, respectively, estimated by Miller and Saylor (1993) for the 93 days with baroclinic currents, June 22–September 22, 1989. Miller and Saylor's (1993) estimation was based on actual current measurements along four vertical profiles. The model estimation presented here is based on simulation of the circulation and the thermal regime in the whole bay, using actual summer 1989 meteorological forcing. Dividing the volume of lower Green Bay (23.7 km^3) by the estimated average flow rate of 1,500 m^3 s^{-1} yields a "mixing time" for the lower bay across the Chambers Island section of ~0.5 year. Concomitantly, a mixing time for the upper bay would be on the order of 50 days.

The values of average transport estimated from the data and model are within a factor of two. The comparison of estimations seems reasonable given that the former value resulted from measurements made along four vertical profiles, while the latter values were estimated independently by a model driven by reconstructed meteorological conditions, as explained in Sect. 2.2.3. Spectral analysis was used to investigate additional evidence of the physical drivers investigated in this chapter. Figure 2.6b shows the estimated spectrum for the net transport out of or into lower Green Bay during summer 1989 across the Chamber Island cross section. The spectrum shows peaks for the first surface mode of Green Bay GB_1 (0.097 h^{-1}), the first surface mode of Lake Michigan LM_1 (0.108 h^{-1}), and the first internal mode of Green Bay, GB_{i1} (0.005 h^{-1} or 0.12 d^{-1}). This means that the water exchange between upper and lower Green Bay shows relevant timescales that demonstrate the effects of stratification (GBi1), and the bay and lake topography (GB_1 and LM_1). This result confirms the important contribution to the transport of the 8-day period oscillations reported, based on a qualitative analysis of the currents' data, by Miller and Saylor (1993).

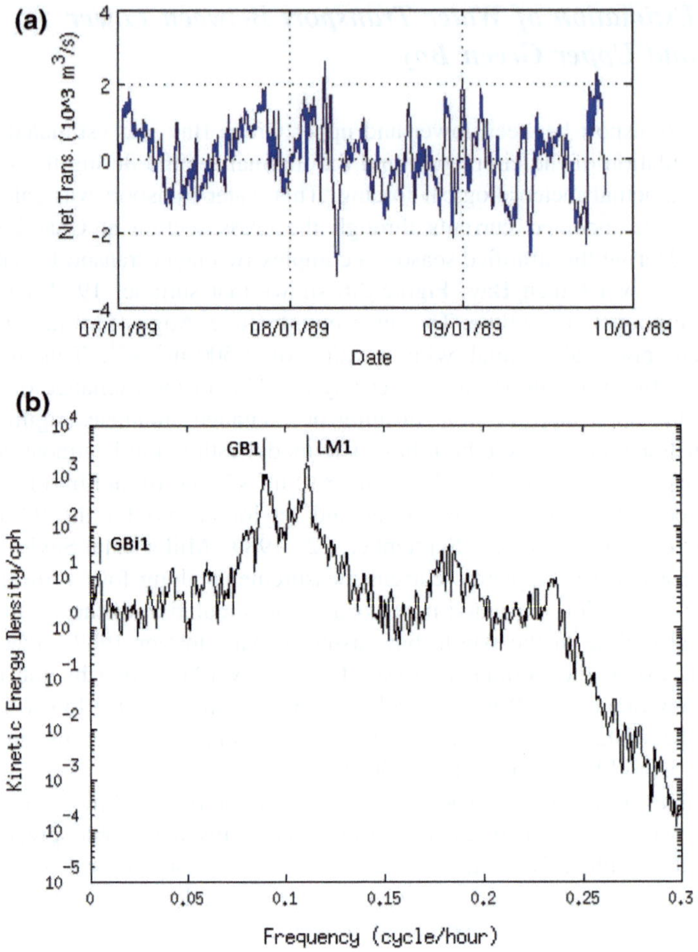

Fig. 2.6 **a** Model-estimated net transport out of (+) or into (−) lower Green Bay during summer 1989 across the Chambers Island midbay cross section. **b** Spectrum of the net flow out or into lower Green Bay during summer 1989

2.3.5 *Effects of Wind, Stratification, Earth's Rotation, and the Bay and Lake Topography on Two-Layer Flows*

We used frequency domain analysis of field data to further investigate the effects of wind, stratification, the Earth's rotation, and the topography of the bay and the lake topography on the observed two-layer flows. Specifically, we investigated common timescales of top and bottom currents and along-bay wind, and whether the relevant timescales indicate evidence of the effects of basin topography, the Earth's rotation, and stratification.

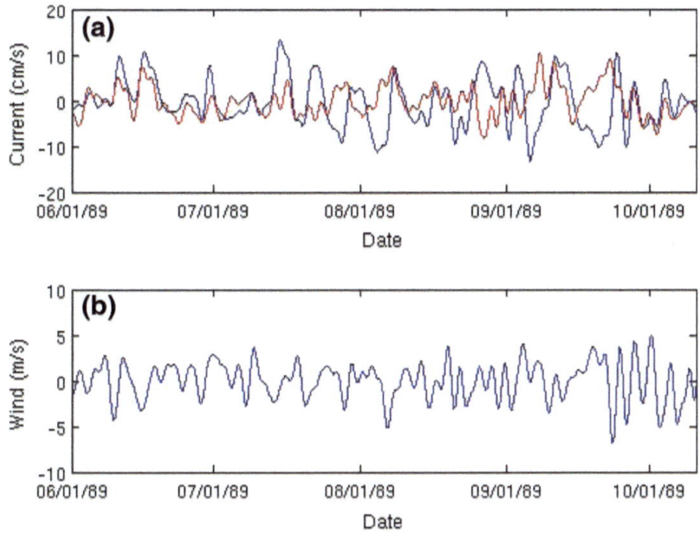

Fig. 2.7 **a** along-bay currents measured at station N19 (Chambers Island west) in the top (*blue*) and bottom (*red*) layers (Gottlieb et al. 1990), and **b** along-bay wind speed during summer 1989. *Positive values* denote currents and wind flowing out of the bay

Typical two-layer flows occur during the stratified season. Flows in the bottom layers bring cold water from Lake Michigan to Green Bay and surface flows carry warmer water from the bay to Lake Michigan. Figure 2.7a shows along-bay currents measured at Station N19 (Chambers Island west) in the top and bottom layers (Gottlieb et al. 1990), and Fig. 2.7b shows along-bay wind speed during the summer of 1989. The figure shows that during the stratified season (between late June and September) along-bay surficial and bottom currents vary out of phase.

The time series were analyzed in the frequency domain to uncover relevant timescales. Figure 2.8a–c show the power spectral density estimates for summer 1989 along-bay currents measured at station N19 (Chambers Island west) in the top (a) and bottom (b) layers, and along-bay wind (c). The frequencies of the significant peaks in Fig. 2.8a (top currents) and b (bottom currents) include the inertial frequency f (0.058 h^{-1}), the M_2 tide (0.081 h^{-1}), the first surface mode of Lake Michigan, LM_1 (0.108 h^{-1}), the first surface mode of Green Bay GB_1 (0.097 h^{-1}), and the first internal mode of Green Bay, GB_{i1} (0.005 h^{-1} or 0.12 d^{-1}). The M2 tidal constituent, the "principal lunar semi-diurnal," was observed in field data and is not considered by the POM model.

Figure 2.8d presents the squared coherency and phase between top and bottom currents. The figure shows that surface and bottom currents are significantly coherent (above the 95% confidence level shown in the figure) at several frequencies, including the first internal mode of Green Bay, GB_{i1}, the inertial frequency *f* caused by the Earth's rotation, the M_2 tide, the first surface mode of Green Bay GB_1, and the first surface mode of Lake Michigan LM_1. In other words, the

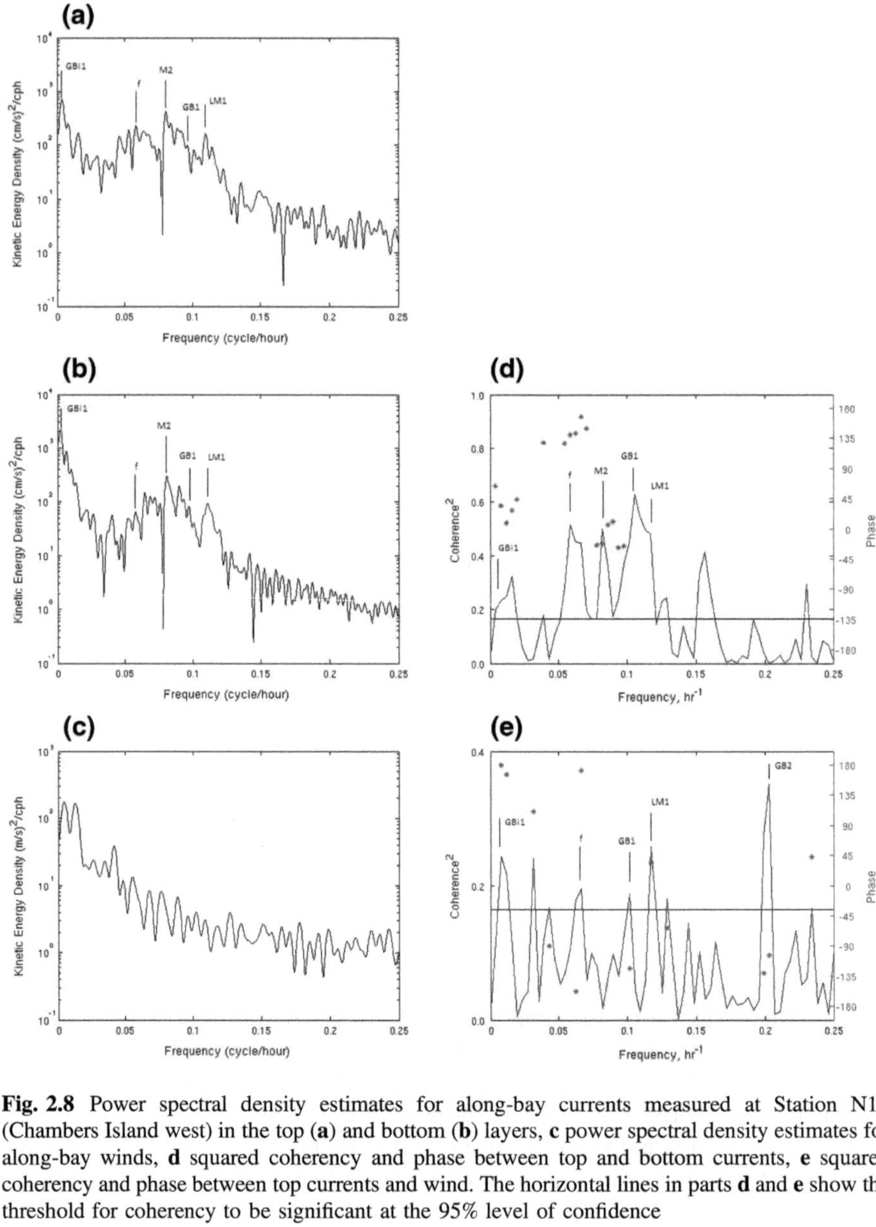

Fig. 2.8 Power spectral density estimates for along-bay currents measured at Station N19 (Chambers Island west) in the top (**a**) and bottom (**b**) layers, **c** power spectral density estimates for along-bay winds, **d** squared coherency and phase between top and bottom currents, **e** squared coherency and phase between top currents and wind. The horizontal lines in parts **d** and **e** show the threshold for coherency to be significant at the 95% level of confidence

squared coherency in Fig. 2.8d confirms the information conveyed by the spectra in Fig. 2.8a and b, showing that top and bottom currents have common relevant timescales that demonstrate the effects of stratification (GBi1), the Earth's rotation (f), and the bay and lake topography (GB_1 and LM_1). Figure 2.8d also shows that at the inertial frequency top and bottom currents are out of phase by about 135°.

The squared coherency and phase between wind and top currents (Fig. 2.8e) shows that wind and surface currents are significantly coherent at several frequencies, including the first internal mode of Green Bay, GB_{i1}, the inertial frequency f caused by the Earth's rotation, the first surface mode of Lake Michigan LM_1, and the first two surface modes of Green Bay GB_1 and GB_2 (0.207 h^{-1}). In other words, wind and top currents show common relevant timescales that demonstrate the effects of stratification (GB_{i1}), the Earth's rotation (f), and the bay and lake topography and the Earth's rotation (GB_1, GB_2, and LM_1).

2.3.6 *Effects of Stratification, Earth's Rotation, and the Bay and Lake Topography on the Direction of Currents*

Currents measured in the top and bottom layers, west of Chambers Island during summer of 1989, were analyzed to investigate evidence of the effect of the Earth's rotation. Top and bottom currents rotated clockwise (as shown by Fig. 2.9a for the

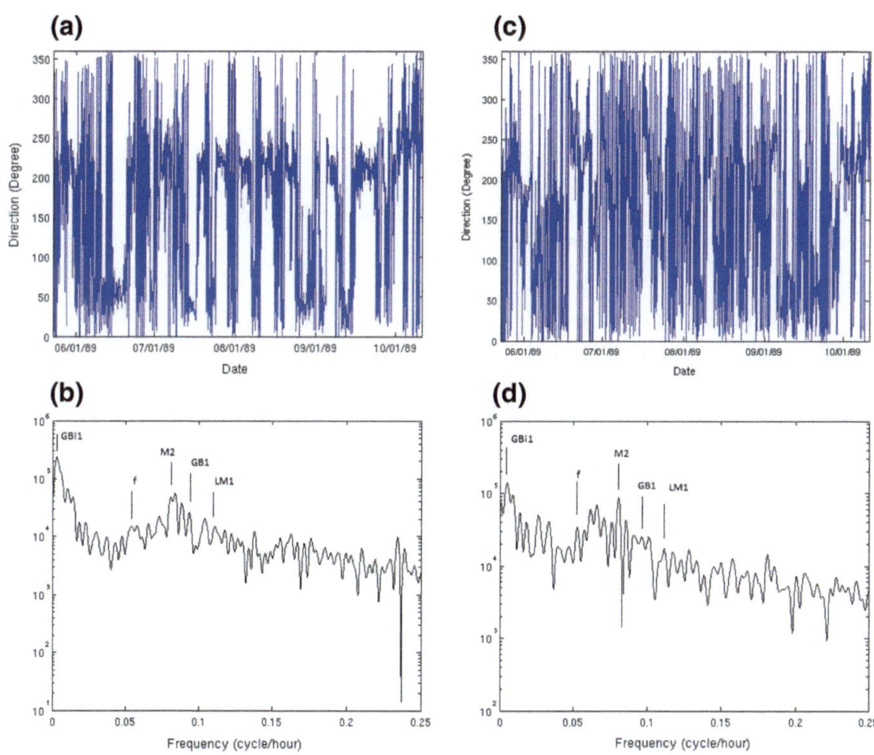

Fig. 2.9 Direction of **a** bottom and **c** top currents measured at Chambers Island west during summer 1989; spectra of **b** bottom and **d** top currents direction

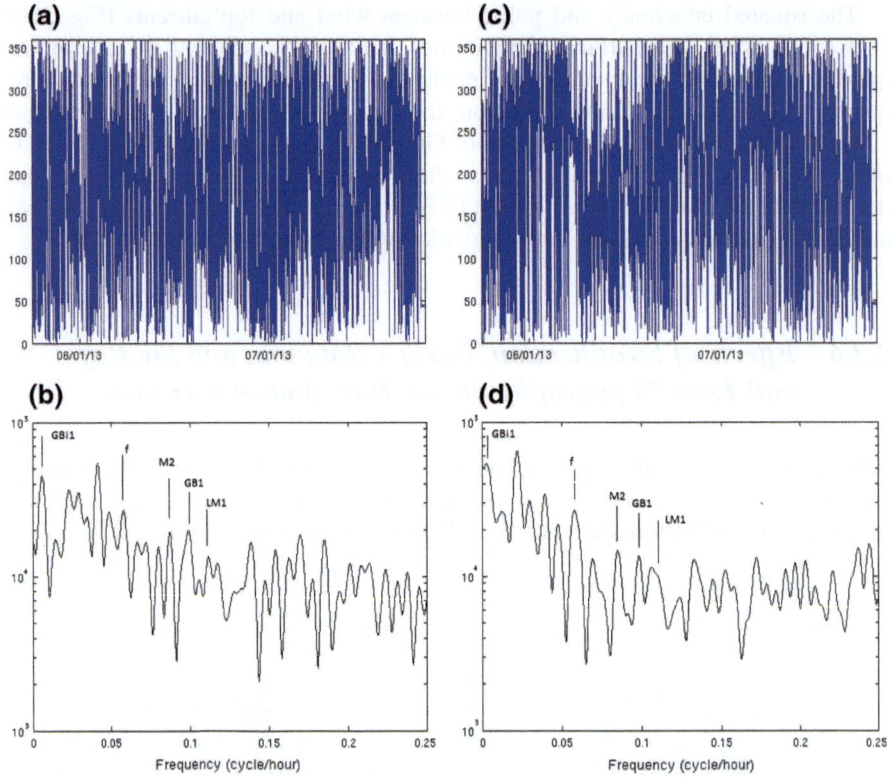

Fig. 2.10 Direction of **a** bottom and **c** top currents measured at Station 17 during summer 2013; spectra of **b** bottom and **d** top currents direction

near bottom currents) with distinct periodicity during the stratified season (between late June and September).

Figure 2.9b, d show the power spectral density estimates for the direction of summer 1989 bottom and top currents, respectively, measured at station N19 (Chambers Island west). The frequencies of the significant peaks in Fig. 2.9b, d include the first internal mode of Green Bay, GB_{i1} ($0.005\ h^{-1}$ or $0.12\ d^{-1}$), the inertial frequency f ($0.058\ h^{-1}$), the M_2 tide ($0.081\ h^{-1}$), the first surface mode of Lake Michigan, LM_1 ($0.108\ h^{-1}$), and the first surface mode of Green Bay GB_1 ($0.097\ h^{-1}$). In other words, the rotation of top and bottom currents shows time-scales that demonstrate the effects of stratification (GBi1), the Earth's rotation (f), and the bay and lake topography and the Earth's rotation (GB_1 and LM_1). The first internal mode of Green Bay, GBi1, is the frequency of persistent oscillations of the thermocline at the period of the bay's lowest-mode, a closed basin internal seiche, an 8-day long period found by Saylor et al. (1995). The observed inertial period f was very close to the theoretical limiting period. Mortimer (2004) used the term near-inertial oscillations to describe this phenomenon. The passage west of

Chambers Island is about four times the Rossby radius for internal waves (estimated at 2.6 km); i.e., it is wide enough to observe the effect of the Earth's rotation.

A similar analysis of our 2013 current measurements at Station 17 showed results that are very similar to those just described. These results, shown in Fig. 2.10, demonstrate that the effects of the Earth's rotation are also observable in the lower Green Bay circulation.

2.4 Conclusions

The thermal regime in Green Bay is determined by the combination of the heat flux across the water surface and heat transport by circulation, including the water exchange with Lake Michigan. The analysis of the calculated heat flux across the water surface and measured, stratified temperature profiles showed a clear cause-effect relationship in regions deeper than 15–20 m between late June and September.

Green Bay exhibits significant interannual variability in the summer wind direction and the circulation patterns. Hence, descriptions of circulation based on single-year field data or modeling may represent incomplete pictures or may miss the range of variability present. The existence of persistent, monthly average patterns in wind forcing and consequent circulation patterns is an important finding of this study. The monthly average wind direction in May is from the NE and rotates to the SW in June, producing changes in the circulation pattern. Summertime conditions are initiated by this transition in the dominant wind field shifting from the NE to the SW in late June, and conditions remain relatively stable until bay vertical mixing in early September. It is during this stable period that the stratified conditions conducive to hypoxia are present. Wind direction has a significant effect on the water exchange between Lake Michigan and Green Bay. A computational experiment for an idealized steady uniform wind parallel or perpendicular to the bay showed that the former condition induces a greater water exchange than the latter condition, with a difference as much as 60%.

The water exchange between upper and lower Green Bay yields a measure of the mixing between those water bodies. The hydrodynamic model was used to simulate, fairly closely, the water exchange measured by Miller and Saylor (1993) and to estimate a "mixing time" for the lower bay across the Chambers Island section of ~0.5 year. Spectral analysis of the net water exchange between upper and lower Green Bay showed relevant timescales that demonstrate the effects of stratification, the bay and the lake topography and the Earth's rotation.

Frequency domain analysis of wind and measured currents' data revealed the effects of wind, stratification, the Earth's rotation, and the topography of the bay and the lake topography on the observed two-layer flows in Green Bay. This two-layered flow, particularly the southerly propagation of cooler, denser waters, establishes and reestablishes the conditions most susceptible to the onset and persistence of hypoxia in the lower bay. This cooler hypolimnetic water mass becomes progressively depleted of oxygen as it moves south, at times reaching complete anoxia. Top and bottom currents rotated clockwise with distinct periodicity during the stratified season. Frequency domain analysis of top and bottom currents clearly showed the effects of stratification, Earth's rotation, and the bay and lake topography on the direction of currents.

The general patterns of the circulation and thermal structure and their variability as related to the changing wind regimes and the thermal climate will be essential in producing longer term projections of future water quality in response to system scale changes, e.g., changes in seasonality, duration of summertime conditions, climate forcing mechanisms, nutrient loading, habitat restoration and alteration, and long-term changes in predicted lake levels.

References

Beletsky D, Schwab DJ (2001) Modeling circulation and thermal structure in Lake Michigan: annual cycle and interannual variability. J Geophys Res 106(C9):19745–19771

Beletsky D, Schwab DJ (2008) Climatological circulation in Lake Michigan. Geophys Res Lett 35: L21604. doi:10.1029/2008GL035773):5

Beletsky D (2001) Modeling wind-driven circulation in Lake Ladoga. Boreal Environ Res 6:307–316

Bertrand G, Lang J, Ross J (1976) The Green Bay watershed: past, present, future. Univ Wisconsin Sea Grant Technical Report WIS-SG-76-229

Bloomfield P (1976) Fourier analysis of time series: an introduction. Wiley, New York

Blumberg AF, Mellor GL (1987) A description of a three-dimensional coastal ocean circulation model. Coast Estuar Sci 4:1–16

Blumberg AF, Ali Khan L, St. John JP (1999) Three-dimensional hydrodynamic model of New York harbor region. J Hydraul Eng 125(8):799–816

Gottlieb ES, Saylor JH, Miller GS (1990) Currents and water temperatures observed in Green Bay, Lake Michigan. Part I: Winter 1988–1989, Part II: Summer 1989, (1990), NOAA GLERL TM-073

Grunert B (2013) Evaluating the summer thermal structure of Southern Green Bay, Lake Michigan, MS Thesis, University of Wisconsin-Milwaukee

Hamidi SA, Bravo HR, Klump JV (2013) Evidence of multiple physical drivers on the circulation and thermal regime in the Green Bay of Lake Michigan. In World environmental and water resources congress 2013: showcasing the future, pp 1719–1726

Hamidi SA, Bravo HR, Klump JV, Waples JT (2015) The role of circulation and heat fluxes in the formation of stratification leading to hypoxia in Green Bay, Lake Michigan. J Great Lakes Res 41(2015):1024–1036

Jenkins GM, Watts DG (1968) Spectral analysis and its applications. Holden-Day, San Francisco

Klump JV, Sager P, Edgington DN, Robertson D (1997) Sedimentary phosphorus cycling and a phosphorus mass balance for the Green Bay ecosystem. Can J Fish Aq Sci 54:10–26

Klump JV, Fitzgerald SA, Waples JT (2009) Benthic biogeochemical cycling, nutrient stoichiometry, and carbon and nitrogen mass balances in a eutrophic freshwater bay. Limnol Oceanogr 54:792–812

Mellor GL (2002) Users guide for a three-dimensional, primitive equation, numerical ocean model. Princeton University

Mellor GL, Yamada T (1982) Development of a turbulence closure model for geophysical fluid problems. Rev Geophys 20(4):851–875

Miller GS, Saylor JH (1985) Currents and temperatures in Green Bay, Lake Michigan. J Great Lakes Res 11(2):97–109

Miller GS, Saylor JH (1993) Low-frequency water volume transport through the midsection of Green Bay, Lake Michigan, calculated from current and temperature observations. J Great Lakes Res 19(2):361–367

Mortimer CH (1979) Water movement, mixing and transport in Green Bay, Lake Michigan. Univ Wisconsin Sea Grant Institute No. WI-SG-78-234

Mortimer CH (2004) Lake Michigan in motion—responses of an inland sea to weather, Earth-spin, and human activities. The University of Wisconsin Press

Rao DB, Mortimer CH, Schwab DJ (1976) Surface normal modes of Lake Michigan: calculations compared with spectra of observed water level fluctuations. J Phys Oceanogr 6(4):575–588

Saylor JH, Miller GS, Gottlieb ES (1995) Near-resonant wind forcing of internal seiches in Green Bay, Lake Michigan, NOAA GLERL Contribution Number 790

Schwab DJ, Bedford KW (1994) Initial implementation of the Great Lakes forecasting system: a real-time system for predicting lake circulation and thermal structure. Water Pollut Res J 29:203–220

Schwab DJ, Beletsky D (2003) Relative effects of wind stress curl, topography, and stratification on large scale circulation in Lake Michigan. J Geophys Res 108(C2):26–1 to 26-6

Von Storch H, Zwiers FW (1999) Statistical analysis in climate research. Cambridge University Press, Cambridge, 484pp, ISBN 0521 450713

Waples JT, Klump JV (2002) Biophysical effects of a decadal shift in summer wind direction over the Laurentian Great Lakes. Geophys Res Lett 29:1201. doi:10.1029/2001GL014564

Chapter 3
Interannual Variation in Stratification over the Texas–Louisiana Continental Shelf and Effects on Seasonal Hypoxia

Robert D. Hetland and Xiaoqian Zhang

Abstract A numerical dye is used to track freshwater released in May and June from the Mississippi and Atchafalaya rivers using a hydrodynamic model. These months are chosen because discharge and nutrient load in May and June is significantly correlated with an area of the Texas–Louisiana continental shelf affected by seasonal bottom low dissolved oxygen. Results show that the two different river sources influence different parts of the region affected by hypoxia, so that both rivers appear to contribute to forming the hypoxic region. Analysis shows that both nutrient loading and stratification caused by freshwater fluxes from the rivers are consistent with the distribution of dyed freshwater in late July.

Keywords Freshwater discharge · Stratification · Hypoxia · Modeling · Mississippi River · Texas–Louisiana shelf · Gulf of Mexico

3.1 Introduction

The Mississippi–Atchafalaya river system drains 41% of the continental USA, supplying the northern Gulf of Mexico annually with 530 km^3 of freshwater, 210 million tons of sediments, and 1.5 million tons of nitrogen (Milliman and Meade 1983; Goolsby et al. 2001). This large flux of carbon and nitrogen, combined with the stratifying effects of the freshwater, create a large region of near-bottom hypoxia south of the Louisiana coast. This layer is typically a few meters thick, with the lowest oxygen concentrations most commonly observed in the benthic nepheloid layer. The affected area is generally confined between the 10 and 50 m isobaths and may extend into Texas waters during years with a very large hypoxic area. In years with a small hypoxic area, low oxygen conditions are typically found in the vicinity of the two large river mouths, west of the Mississippi Delta and south of Atchafalaya Bay.

Many previous studies have found significant statistical relationships between either the freshwater discharge from the Mississippi and Atchafalaya rivers (Wise-

R.D. Hetland (✉) · X. Zhang
Department of Oceanography, Texas A&M University, College Station, TX, USA
e-mail: hetland@tamu.edu

D. Justic et al. (eds.), *Modeling Coastal Hypoxia*,
DOI 10.1007/978-3-319-54571-4_3

man et al. 1997; Bianchi et al. 2010) or the nitrogen load carried by these rivers (Scavia et al. 2003; Turner et al. 2005; Greene et al. 2009; Forrest et al. 2011). The highest correlations are found between the May–June average nitrogen load and the hypoxic area in late July. However, questions remain about the processes that drive these correlations.

Nutrient loads and freshwater discharge are significantly correlated ($r^2 = 0.71$, $p = 2.2 \times 10^{-7}$); that is, the *concentration* of riverine nitrogen is roughly constant between years and is uncorrelated to the larger relative variations in freshwater discharge and load. Because of this, the causal relationships between both nutrient load and freshwater flux that create interannual variations in hypoxic area are confounded. It is not clear if observed interannual changes in hypoxic area are caused by the stratifying effects of the freshwater, or the eutrofying effects of the increased nitrogen load.

The goal of this paper is to trace the river water released onto the shelf during May and June in a number of different years, to examine the relationships between the fate of this water on the shelf and hypoxic area. May and June are chosen because of the significant statistical relationship between freshwater flux and nutrient load in these months with the subsequent extent of hypoxia in July. The numerical simulations are accomplished by adding a numerical dye to each large river source during each month. This results in four separate dyes, one for both the Mississippi and Atchafalaya rivers during both May and June. Distributions of these dyes are then compared to the extent of hypoxia in late summer.

3.2 Model Setup

We use the Regional Ocean Modeling System (ROMS, Shchepetkin and McWilliams 2005; Haidvogel et al. 2008) configured for the Texas–Louisiana shelf for this study. This model has been described in previous studies of circulation and freshwater budgets by Zhang et al. (2012a, b). Briefly, the model extends roughly from Laguna Madre in Mexico to Mobile Bay in Alabama. The model has 30 vertical layers, and $\sim$1 km horizontal resolution over the Louisiana shelf. The model domain is shown in Fig. 3.1. The model is forced with inputs from the six major rivers in Louisiana and Texas, with the Mississippi and Atchafalaya rivers contributing the most to the riverine freshwater inputs. The model is nudged to results from the GOM-HYCOM operational model to include the effects of deep ocean currents on shelf circulation. The North American Regional Reanalysis (NARR) model is used for surface momentum, heat, and freshwater fluxes; heat fluxes are calculated through a bulk formulation with a Q-correction of 50 W m^2.

The primary addition to the present set of simulations is that the freshwater from the Atchafalaya and Mississippi rivers is dyed for each river in both May and June of each simulation year. Freshwater entering the domain from the rivers is tagged with a concentration of 1 m^{-3}, so that the dye concentration represents the fraction of dyed freshwater at a particular point in the domain. The manner in which the dyes are

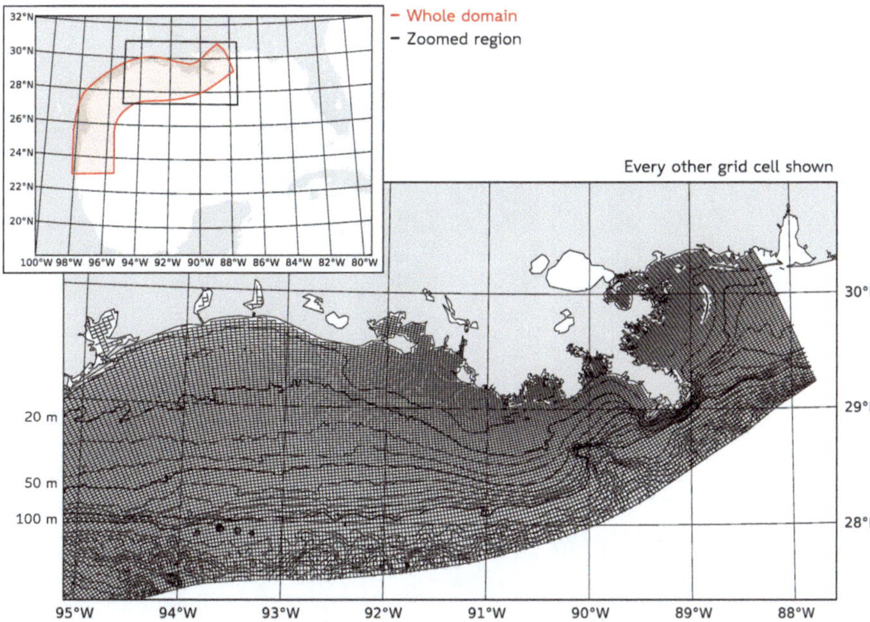

Fig. 3.1 The model domain and grid are shown in the two maps. The model domain covers the entire Texas and Louisiana shelves from the coast past the shelf break. The grid resolution in the region just west of the Mississippi River delta is less than 1 km

added to the freshwater inputs is similar to the method used in Zhang et al. (2012b), the primary difference being that only water released in either May or June is dyed.

3.3 Results

The year 2008 is presented here as an example of the distribution of the dye from the Mississippi and Atchafalaya rivers in the months of May and June. The year 2008 is chosen as an example because it had a very high discharge, relatively typical summertime winds, and was the second largest hypoxic area recorded during the late July annual survey (see http://gulfhypoxia.net). Figure 3.2 shows the surface concentration of dye from each river, released during each of May and June over the summer. As the dye is introduced at a concentration of one, the dye may be considered as a proxy for dilution of freshwater over the shelf. Thus, the dye represents the fraction of river water in a given model cell and is thus unitless. The dye is plotted on a logarithmic scale, so that each gradation indicates an order of magnitude dilution. For this year, where the discharge was above average, by the end of summer essentially the entire Louisiana shelf is covered with fresh river water that has been diluted less

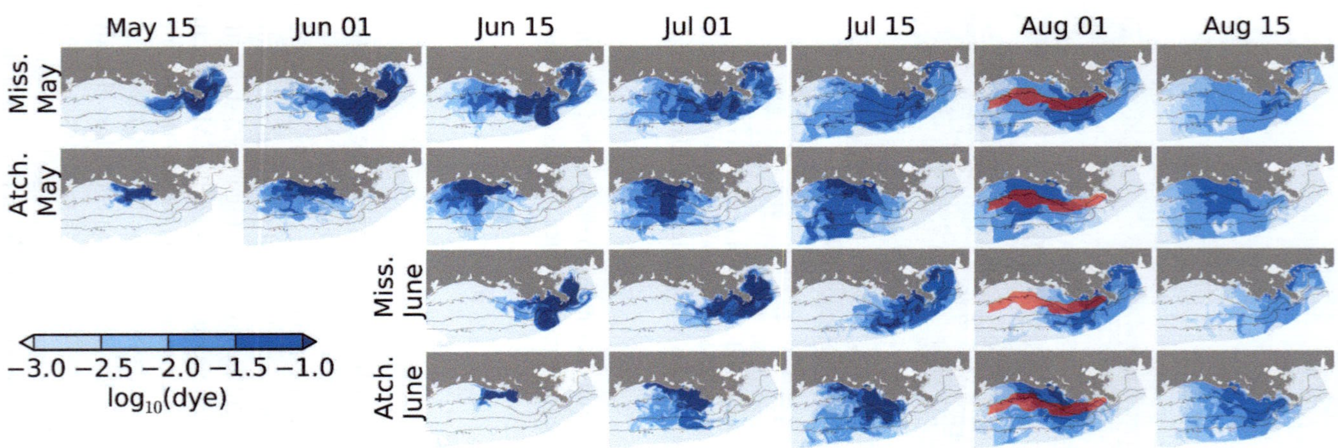

Fig. 3.2 Surface dye concentrations are shown every 2 weeks during summer, 2008. The dye is separated into the river source, either Mississippi or Atchafalaya, and the month of the dye release, either May or June. Concentrations are plotted on a logarithmic scale. The observed area of hypoxia is shown in the four August 1 frames

than one thousand times, with significant regions near the source that have dilution factors of less than ten.

As the undiluted dyed freshwater has a concentration of one, the dye may be considered as a proxy for volume of freshwater per unit volume ocean water. Thus, the integral

$$h_f = \int_{-H}^{\eta} d_i \, dz \tag{3.1}$$

represents the freshwater thickness, h_f, associated with a particular dye, d_i. The freshwater thickness is the thickness of the dyed freshwater, if the water column "unmixed" into purely dyed freshwater with a dye concentration of one and completely undyed water. The remaining undyed water may contain some freshwater, but this freshwater was introduced at times when dye was not included in the discharge of freshwater to the ocean. Distributions of vertically integrated dye show similar patterns (Fig. 3.3); the highest concentrations of integrated dye show that some regions of the shelf have over three meters of riverine freshwater mixed though the water column.

Figure 3.4 shows vertical profiles of dye centered about the 20 m isobath in locations where surface concentrations are high, diluted by less than a factor of 100. Profiles indicate that the dye is typically concentrated at the surface and has the strongest concentrations in the upper half of the water column. The dye released in June has a particularly strong surface signature that persists throughout the month. However, all the dye profiles indicate that by the end of July, all of the dye has been significantly diluted, with concentrations in the upper water column about double those in the lower water column. Thus, while surface dye concentrations are stronger in the upper half of the water column through the entire summer, there is a relatively significant fraction of the dye that penetrates into the lower layer.

The differences in the character of each source can be found by examining the relationship between the dye and other oceanic tracers. The relationship between dye and salinity, shown in Fig. 3.5, shows that the highest dye concentrations are found at intermediate salinity ranges, between fresh riverine water ($S = 0.0$) and ambient Gulf water ($S \simeq 36.0$). The relationship between dye and salinity is controlled primarily by the dye source; the two river sources appear distinct, regardless of the month in which the dye was released. There are some common patterns in each dye release. Initially, a mixing line is formed between the freshwater in which the dye is introduced into the domain, and the ambient Gulf water that initially contains no dye. Points below this line are filled in as the dye mixes with freshwater that was introduced earlier, and that contains no dye. Since the freshwater released from the Mississippi Delta mixes quickly, dye concentrations are not found at salinities much fresher than about 10 g kg^{-1}. As this dye interacts with the Atchafalaya plume, dye is found at even lower salinities. The Atchafalaya discharge, on the other hand, is released at the edge of a broad, shallow shelf. As such, there is a pool of freshwater that separates the Atchafalaya plume water at the beginning of each dye release from the ambient Gulf water. Because this dye is present at very freshwaters, even early in the Atchafalaya

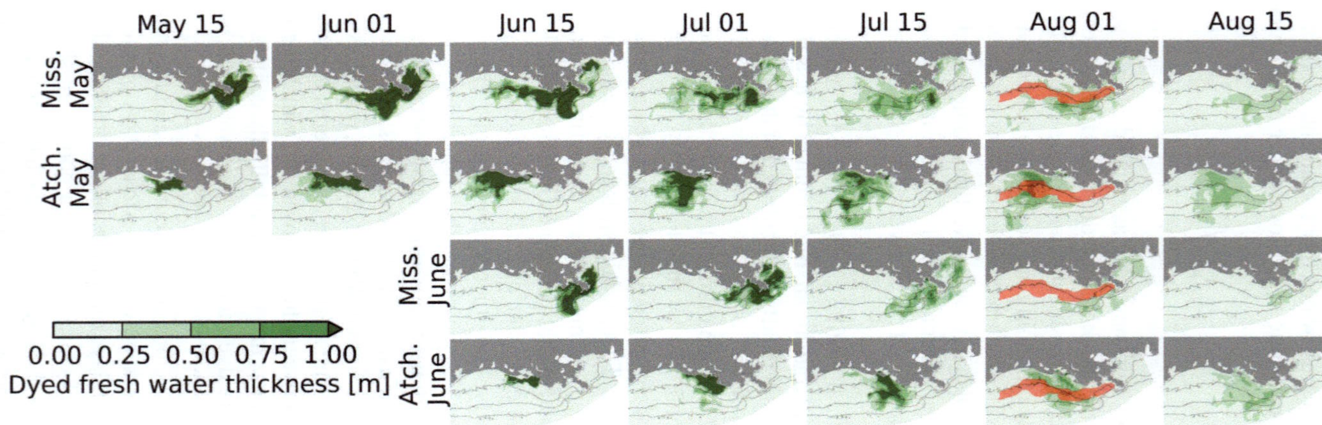

Fig. 3.3 Integrated dye, equivalent to the freshwater thickness associated with each source and month, is shown every 2 weeks during summer, 2008. The observed area of hypoxia is shown in the four August 1 frames

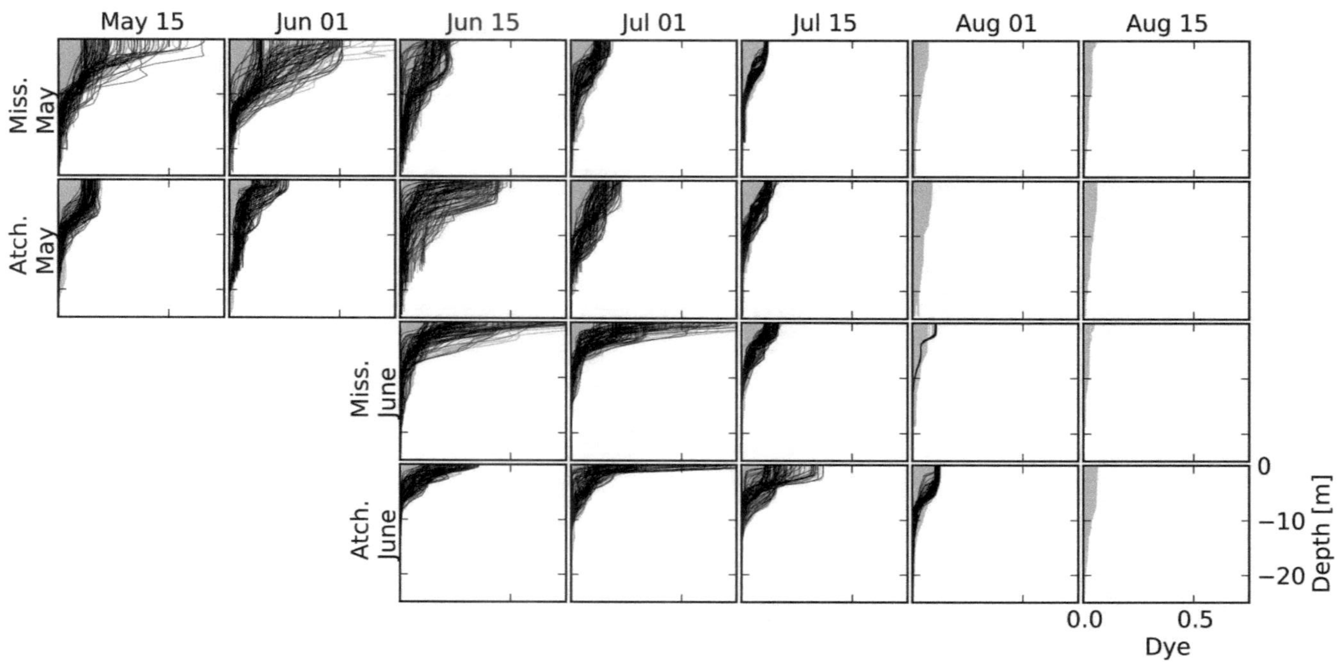

Fig. 3.4 Random profiles of dye concentration are shown in locations between 15 and 25 m deep, where the dye is diluted by a factor of 10 (*black lines*) and 100 (*gray lines*). Profiles are shown for dye from each source and month every 2 weeks during summer, 2008

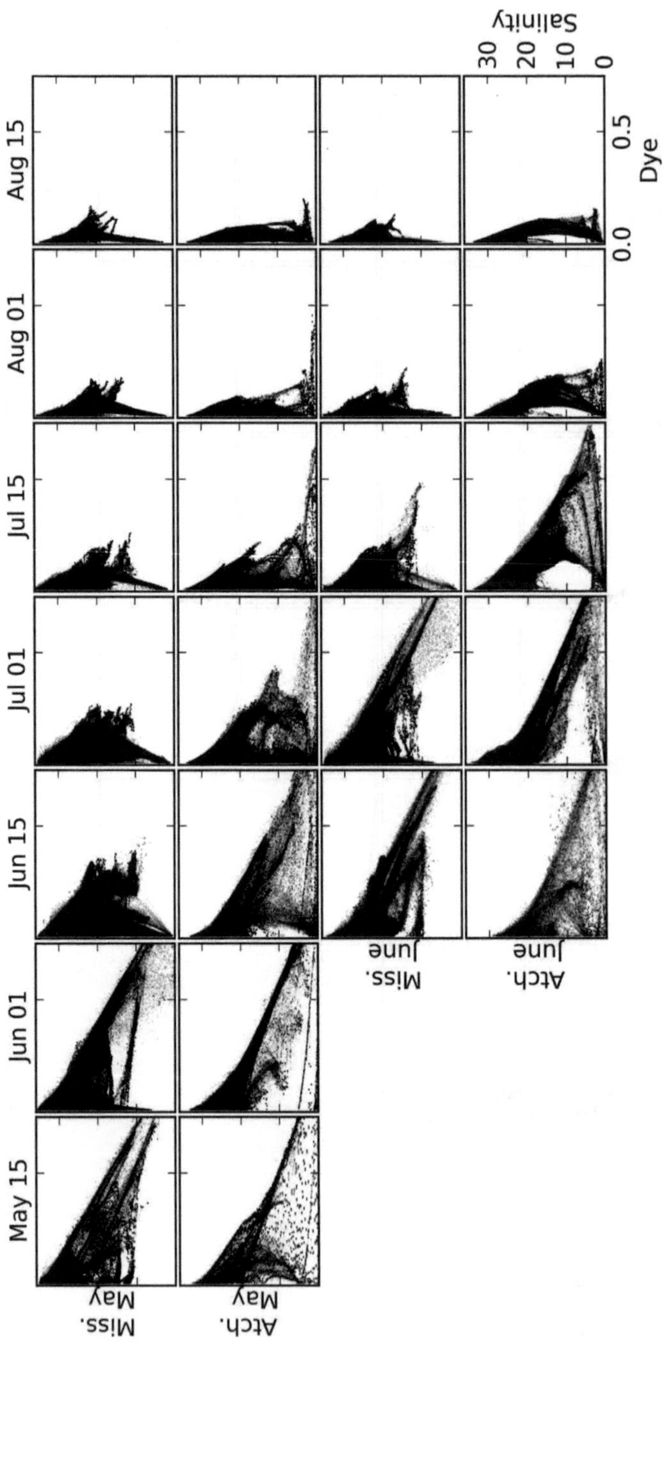

Fig. 3.5 The relationship between dye and salinity is shown for all the model points for each source and month every 2 weeks during summer, 2008

dye releases, the mixing line at dye concentrations higher than about 0.2 indicates mixing with waters that have a salinity fresher than 30 g kg^{-1}.

3.4 Discussion

In both the surface dye concentrations (Fig. 3.2) and the dyed freshwater thickness (Fig. 3.3), the Mississippi and Atchafalaya plumes cover distinct regions on the shelf. The Atchafalaya plume is generally westward and inshore of the Mississippi plume water. However, it is clear from the overlays of hypoxic area that the region of the shelf that is affected by hypoxia is associated with neither plume in particular. In Table 3.1, the cubic kilometers of dyed freshwater is integrated in the regions of the shelf that are affected by hypoxia in a given year. Generally, it is clear from this table, as well as from Figs. 3.2 and 3.3, that no one, particularly freshwater source—Mississippi or Atchafalaya—or month of dye release—May or June—is the primary contributor to stratification or nutrients in the regions associated with low dissolved bottom oxygen.

Figure 3.6 shows the dye thickness from years 2003 to 2011, with the observed hypoxic area overlaid. On average, there is about 1 m of freshwater over the hypoxic zone in each year, ranging from about 0.7 m in years with smaller areas (2003 and 2009), and about 1.2 in years with large areas (2007, 2008, and 2010). A stoichiometric analysis suggests that 1 m of freshwater could supply enough organic material to fuel hypoxia in these regions. Assuming that the nitrogen to oxygen ratio is 1:130,

Table 3.1 Integrated freshwater (km^3) in the regions associated with bottom hypoxia on July 28 of each year. The final column shows the observed hypoxic area for each year. The correlations between each integrated dye and associated p-values are found in the bottom rows. The correlation between the sum of all the integrated dyes is $r = 0.90$, $p = 0.0011$

Year	May Miss (km^3)	May Atch (km^3)	June Miss (km^3)	June Atch (km^3)	Hypoxic area (km^3)
2003	1.34	0.91	1.90	1.76	8,560
2004	1.94	3.09	2.31	4.75	15,040
2005	1.90	1.40	2.74	2.62	11,840
2006	2.59	3.90	4.17	3.08	17,280
2007	10.24	4.37	3.69	4.64	20,500
2008	8.18	8.20	3.95	6.30	20,720
2009	1.23	2.16	0.38	2.33	8,000
2010	5.43	6.88	5.95	7.71	20,000
2011	7.05	10.39	2.85	7.99	17,520
	$r = 0.83$	$r = 0.73$	$r = 0.83$	$r = 0.77$	
	$p = 0.0061$	$p = 0.024$	$p = 0.0059$	$p = 0.015$	

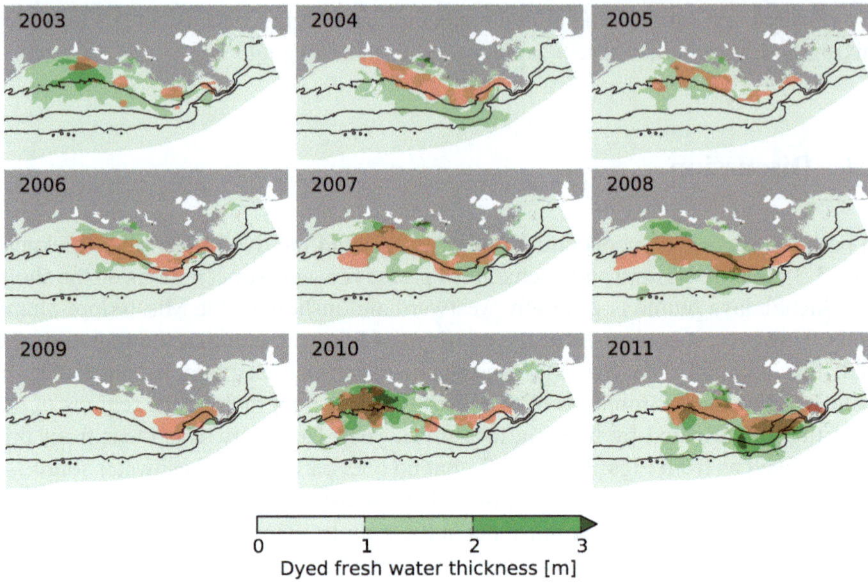

Fig. 3.6 Dye thickness, summing together all four dyes associated with the two sources and months are shown in late July for nine different years, corresponding to the period when the annual hypoxia survey occurs. The observed area of hypoxia is shown as a *red shaded region*

that the nitrogen concentration of river water is 120 μM, that the apparent oxygen utilization in hypoxic waters is 250 μM, and that the efficiency of converting nitrogen into oxygen utilization is 0.2, 1 m of freshwater would convert roughly to 6 m of hypoxic water. This is of course assuming (1) that there are no other mechanisms that reduce efficiency such as lags in oxygen utilization, (2) that the organic matter is delivered roughly evenly over the hypoxic area, and (3) that there is no ventilation of the bottom waters. Also, it is not clear what the timescales and processes of organic matter creation and conversion are using this simple conceptual model. For this, one would need to use a full model of biological processes, such as the NPZD model described by Fennel et al. (2011). Even so, this model suggests that it is plausible that a significant fraction of organic matter required to form hypoxia may be delivered by the two river systems during May and June.

However, each dye is not spread evenly over the hypoxic area, rather different rivers contribute differently to different regions. For example, the Mississippi is concentrated more to the east, the Atchafalaya more to the west. This may have important consequences for the formation of hypoxia in different region of the shelf, because the character of the water introduced to the shelf is very different between the Mississippi and Atchafalaya. For example, The Atchafalaya River Basin may be a small source for inorganic nitrogen, but a sink for organic nitrogen (for a total 14% reduction in total nitrogen) (Xu 2006; Scaroni et al. 2010). Other properties, such as sediments, phosphorous, and organic carbon, may be similarly altered as river water passes through the swampy region that defines the Atchafalaya River Basin. Thus, the Mississippi and Atchafalaya form water masses with very different properties, and freshwater distributions over the shelf (Hetland and DiMarco 2008).

The dye experiments also suggest that the freshwater delivered to the shelf during May and June from both river sources may create stratification in the regions affected by hypoxia. The vertical structure of the dye in August suggests that the dye is stratified (see Fig. 3.4). The dye is associated with freshwater, and freshwater is the primary determinant of density over the Texas–Louisiana shelf. Also, the horizontal distribution of freshwater is roughly co-located with the westward termination of the hypoxic zone. This implies that it is indeed freshwater in the months of May and June that are primarily associated with determining the areal extent of hypoxia.

3.5 Conclusions

The water released from the Mississippi and Atchafalaya rivers during May and June appears to roughly correlate with the regions of the Texas–Louisiana shelf affected by seasonal bottom hypoxia. Different regions of the area affected by hypoxia are influenced by different rivers and different release times; the sum total of water released during May and June from the two sources extends roughly across the entire hypoxic area, and the along-shore extent of this water appears to be roughly correlated with the along-shore extent of hypoxia.

However, this analysis is not able to differentiate between the organic material flux to the benthos due to nitrogen inputs from the river, and the stratifying effects of the fresh river water. Both interpretations are consistent with the model results.

Acknowledgements This work was funded by the NOAA Coastal Ocean Program (NA09N0S 4780208) and the Texas General Land Office (664 10-096-000-3927).

References

Bianchi TS, DiMarco SF, Cowan JH Jr, Hetland RD, Chapman P, Day JW, Allison MA (2010) The science of hypoxia in the Northern Gulf of Mexico: a review. Sci Total Enviorn 408(7):1471–1484. doi:10.1016/j.scitotenv.2009.11.047

Fennel K, Hetland R, Feng Y, DiMarco SF (2011) A coupled physical-biological model of the Northern Gulf of Mexico shelf: model description, validation and analysis of phytoplankton variability. Biogeosciences 8:1881–1899. doi:10.5194/bg-8-1881-2011

Forrest DR, Hetland RD, DiMarco SF (2011) Multivariable statistical regression models of the areal extent of hypoxia over the Texas–Louisiana continental shelf. Environ Res Lett 6:045002

Goolsby DA, Battaglin WA, Aulenbach BT, Hooper RP (2001) Nitrogen input into the Gulf of Mexico. Environ Qual 30(2):329–336

Greene RM, Lehrter JC, III JDH (2009) Multiple regression models for hindcasting and forecasting midsummer hypoxia in the Gulf of Mexico. Ecol Appl 19(5):1161–1175

Haidvogel DB, Arango H, Budgell WP, Cornuelle BD, Curchitser E, Di Lorenzo E, Fennel K, Geyer WR, Herman AJ, Lanerolle L, Levin J, MacWilliams JC, Miller AJ, Moore AM, Powell TM, Shchepetkin AF, Sherwood CR, Signell RP, Warner JC, Wilkin J (2008) Ocean forcasting in terrain-following coordinates: formulation and skill assessment of the regional ocean modeling system. J Comput Phys 227:3595–3624

Hetland RD, DiMarco SF (2008) How does the character of oxygen demand control the structure of hypoxia on the Texas–Louisiana continental shelf? J Marine Syst doi:10.1016/j.marsys.2007.03.002

Milliman JD, Meade RH (1983) World-wide delivery of river sediment to the ocean. J Geol 91(1):1–21

Scaroni AE, Lindau CW, Nyman JA (2010) Spatial variability of sediment denitrification across the Atchafalaya River Basin, Louisiana, USA. Wetlands 30(5):949–955

Scavia D, Rabalais NN, Eugene Turner R, Justić D, Wiseman WJ Jr (2003) Predicting the response of Gulf of Mexico hypoxia to variations in Mississippi River nitrogen load. Limnol Oceanogr 48(3):951–956

Shchepetkin AF, McWilliams JC (2005) The regional oceanic modeling system (ROMS): a split-explicit, free-surface, topographically-following-coordinate oceanic model. Ocean Model 9:347–404

Turner RE, Rabalais NN, Swenson EM, Kasprzak M, Romaire T (2005) Summer hypoxia in the northern Gulf of Mexico and its prediction from 1978 to 1995. Marine Environ Res 59:65–77

Wiseman WJ Jr, Rabalais NN, Turner RE, Dinnel SP, MacNaughton A (1997) Seasonal and inter-annual variability within the Louisiana coastal current: stratification and hypoxia. J Marine Syst 12:237–248

Xu JY (2006) Total nitrogen inflow and outflow from a large river swamp basin to the gulf of mexico. Hydrol Sci J 51(3):531–542

Zhang X, Marta-Almeida M, Hetland RD (2012a) A high-resolution pre-operational fore cast model of circulation on the Texas–Louisiana continental shelf and slope. J Oper Oceanogr 5(1):19–34

Zhang X, Hetland RD, Marta-Almeida M, DiMarco SF (2012b) A numerical investigation of the Mississippi and Atchafalaya freshwater transport, filling and flushing times on the Texas-Louisiana shelf. J Geophys Res 117:C11009, 21 pp. doi:10.1029/2012JC008108

Chapter 4
A Reduced Complexity, Hybrid Empirical-Mechanistic Model of Eutrophication and Hypoxia in Shallow Marine Ecosystems

Mark J. Brush and Scott W. Nixon

Abstract Numerical simulation models have a long history as research tools for the study of coastal marine ecosystems, and are increasingly being used to inform management, particularly related to nutrient-fueled eutrophication. Demand for modeling assessments is rapidly increasing, and managers need generally applicable tools that can be rapidly applied with limited resources. Additionally, a variety of calls have been made for the development of reduced complexity models for use in parallel with more complex models. We propose a simplified, empirically constrained modeling approach that simulates the first-order processes involved in estuarine eutrophication, contains a small number of aggregated state variables and a reduced set of parameters, and combines traditional mechanistic formulations with robust, data-driven, empirical functions shown to apply across multiple systems. The model was applied to Greenwich Bay, RI (USA), a subestuary of Narragansett Bay, and reproduced the annual cycles of phytoplankton biomass, dissolved inorganic nutrients, and dissolved oxygen, events including phytoplankton blooms and development of hypoxia, and the rate of annual primary production. While the model was relatively robust to changes in parameter values and initial conditions, sensitivity analysis revealed the need for better constraint of the phytoplankton carbon-to-chlorophyll ratio, temperature dependence of phytoplankton production, and parameters associated with our formulations for water column respiration and the flux of phytoplankton carbon to the sediments. This reduced complexity, hybrid empirical-mechanistic approach provides a rapidly deployable modeling tool applicable to a wide variety of shallow estuarine systems.

Dr. Nixon is deceased.

M.J. Brush (✉)
Virginia Institute of Marine Science, College of William and Mary,
Gloucester Point, VA 23062, USA
e-mail: brush@vims.edu

S.W. Nixon
Graduate School of Oceanography, University of Rhode Island,
Narragansett, RI 02882, USA

D. Justic et al. (eds.), *Modeling Coastal Hypoxia*,
DOI 10.1007/978-3-319-54571-4_4

Keywords Ecosystem model • Estuarine eutrophication • Phytoplankton • Nutrients • Hypoxia • Greenwich Bay

4.1 Introduction

Dynamic simulation models have a long history as heuristic and synthetic research tools in the study of coastal marine ecosystems (Brush and Harris 2010, 2016; Ganju et al. 2016). In recent decades, these models have been increasingly applied to guide management of coastal systems, particularly with respect to the effects of nutrient loading on eutrophication (EPA 1999; NRC 2000; Giblin and Vallino 2003; Harris et al. 2003). In the USA, large investments over many years have been made in development of high resolution, biogeochemically complex ecosystem models of major estuarine and coastal systems including Chesapeake Bay (Cerco and Noel 2004a), Long Island Sound (HydroQual 1991), and Massachusetts Bay/Boston Harbor (Jiang and Zhou 2008; Chen et al. 2010).

While many modeling efforts are focused on these large estuaries, it is often the shallow, fringing systems around the perimeter of these estuaries or those lying outside of larger systems that are subject to the highest rates of nutrient loading and the most extreme symptoms of eutrophication, including phytoplankton blooms, proliferation of nuisance macroalgae, and development of hypoxia/anoxia (Valiela et al. 1992; Boynton et al. 1996; Nixon et al. 2001; Bricker et al. 2007; McGlathery et al. 2007). These smaller systems are seldom the focus of large, intensive modeling programs, yet they are often at the scale at which local management decisions are made. With the increasing demand for modeling to inform decisions across a wide range of estuarine and coastal systems (e.g., total maximum daily loads for all impaired waterbodies in the USA; EPA 1999), managers need readily applied, generally applicable modeling tools that can be applied quickly to a variety of systems with limited resources.

In addition to the need for general, widely applicable models, the last two decades have seen a growing body of work examining the role of complexity and spatial resolution in models (Baird et al. 2003; Denman et al. 2003; Fulton et al. 2003, 2004; Friedrichs et al. 2006; Raick et al. 2006; Ménesguen et al. 2007). Multiple calls have been made for development of simpler, "reduced complexity" models for application to management (Rigler and Peters 1995; NRC, 2000; Pace 2001; Duarte et al. 2003; Ganju et al. 2016), and a wide variety of alternative modeling approaches have recently been developed (e.g., Scavia et al. 2006, 2013; Swaney et al. 2008; Obenour et al. 2014). When used in combination with both empirical and complex mechanistic models, these alternative models have provided good examples of the use of multiple modeling approaches (or "ensembles") to inform coastal management (Stow et al. 2003; Scavia et al. 2004; Van Nes and Scheffer 2005).

Given the ever-increasing demand for application of models to an increasing number of systems, the need for generally applicable tools that can be rapidly applied to inform management, and recent calls for development of reduced complexity models, our objective was to develop a general, reduced complexity model for shallow marine ecosystems. We include only those state variables and rate processes of primary importance to the process of estuarine eutrophication as reflected in the dynamics of phytoplankton, macroalgae, inorganic nutrients, dissolved oxygen, and sediment organic accumulation. Selected state variables and associated rates were aggregated into bulk compartments which reflect processes at the system level and result in a reduced number of parameters. In the case of four particularly important rate processes, we developed aggregated formulations based on robust, empirical relationships that have been shown to apply across multiple temperate estuaries. Our goal was to develop a model generally applicable to shallow temperate estuaries and capable of rapid implementation in new systems.

4.2 Methods

4.2.1 Study System

Greenwich Bay, RI (USA) is a subestuary of Narragansett Bay and typical of shallow temperate estuaries subject to elevated nutrient loading (Fig. 4.1). The bay is shallow (average depth = 2.6 m at mean sea level) and grades from a deeper, open embayment directly connected to Narragansett Bay to restricted, shallow coves (1–2 m) which contain approximately 8% of the total water volume but receive over

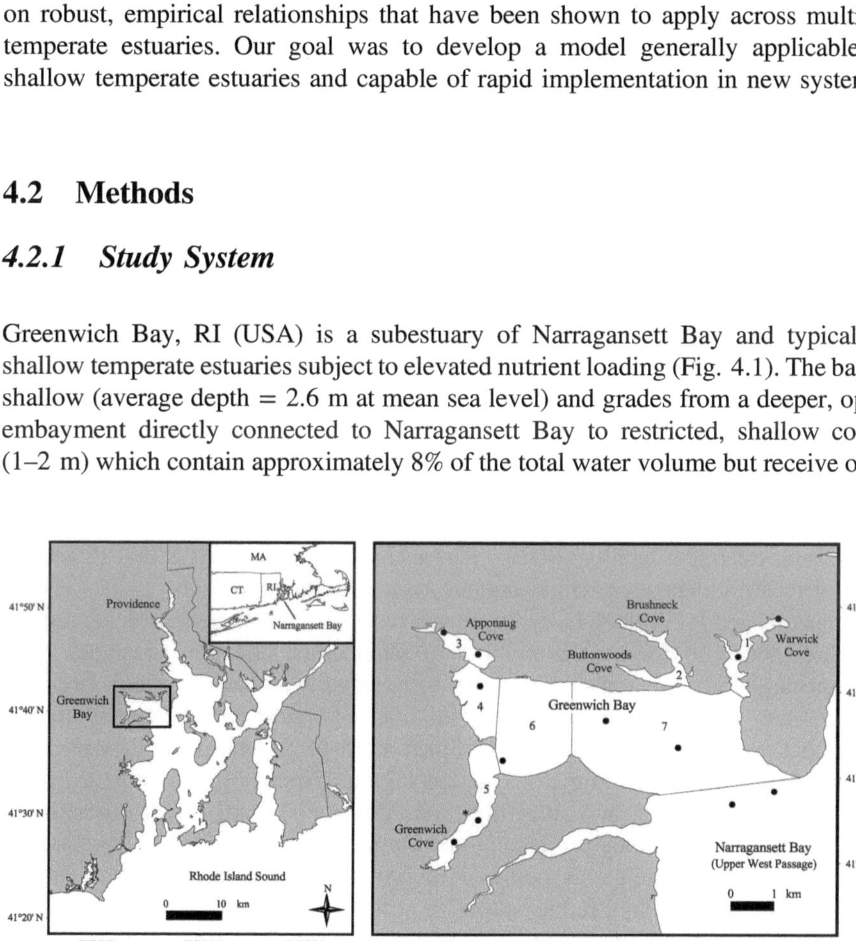

Fig. 4.1 Map of Greenwich Bay and the spatial elements (*1–7*) used in the model. Closed circles show the locations of the Granger et al. (2000) monitoring stations; no samples were collected in Brushneck Cove because a shoal at the mouth prevented routine boat access. The *asterisk* (*) shows the location of the East Greenwich WWTF

90% of the watershed nitrogen load (Granger et al. 2000; Brush 2002). Mean daily total nitrogen loading to the system is approximately 3.1 mmol m^{-2} d^{-1}, which increases to 8.7–21 mmol m^{-2} d^{-1} in the inner coves. The system exhibits several symptoms of nutrient enrichment, including concentrations of dissolved inorganic nitrogen and phosphorus up to 140 mmol m^{-3} and 17 mmol m^{-3}, respectively, frequent phytoplankton blooms with chlorophyll-*a* (chl-*a*) concentrations up to 60 mg m^{-3}, hypoxic events in which concentrations of dissolved oxygen in bottom waters fall below 2 mg l^{-1} and at times approach anoxia, and accumulations of the macroalgae *Ulva lactuca* and *Gracilaria tikvahiae* to densities up to 400 g dry weight m^{-2} in the coves (Granger et al. 2000). Granger et al. (2000) conducted an extensive monitoring program at a series of stations throughout the system during 1995–1997 (Fig. 4.1); the model developed here was run for 1 year between May 1996 and May 1997, the period during which Granger et al. (2000) conducted monthly monitoring cruises.

4.2.2 Ecosystem Model Kinetics

The ecosystem model includes only those state variables, rate processes, and parameters of primary importance to eutrophication in shallow coastal marine ecosystems (Fig. 4.2). State variables include the pools of organic carbon (C), nitrogen (N), and phosphorus (P) in phytoplankton (*PHY*), the macroalgae *Ulva lactuca* (*ULVA*) and *Gracilaria tikvahiae* (*GRAC*), decaying macroalgal biomass (*DEC*), water column pools of dissolved inorganic nitrogen and phosphorus (*DIN*, *DIP*), dissolved oxygen (O_2), biological oxygen demand from the East Greenwich wastewater treatment facility (BOD_w, Fig. 4.1) and its associated N and P, and the pool of recently deposited, labile organic carbon in the sediments (SED_C) and its associated N and P (Table 4.1). Several state variables and processes are aggregated into bulk terms to reduce the number of parameters, and key processes (e.g., phytoplankton primary production) are formulated with cross-system empirical formulations which further reduce the number of model parameters relative to traditional mechanistic formulations. Since these empirical formulations were derived using data from numerous temperate estuaries, they confer a degree of generality to the model, and since they were developed using direct observations (e.g., ^{14}C production), they produce model output that can be directly compared to routine observations (e.g., as opposed to growth rates predicted by traditional mechanistic formulations). In the following sections, we detail the main features of the model, particularly our formulation of key rate processes; the macroalgal formulations are presented elsewhere (Brush and Nixon 2010).

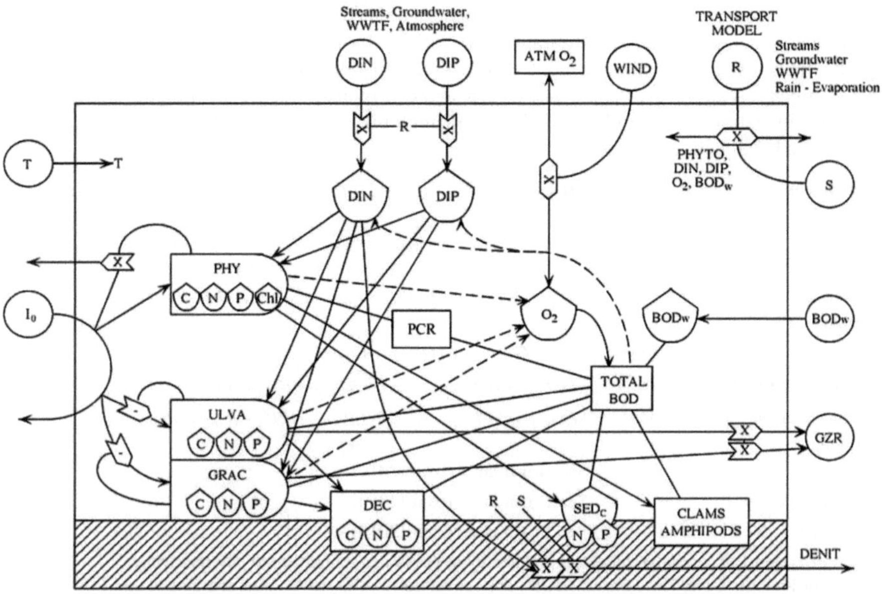

Fig. 4.2 Schematic of the estuarine model. State variables, major flows (*with arrows*), and major connections (*without arrows*) are depicted. Flows which consume material (e.g., nutrient uptake, O_2 consumption, loss of biomass) are shown with *solid lines*. Flows which produce material (e.g., remineralization, photosynthetic O_2 production) are shown with *broken lines*. To reduce diagram complexity, all respiratory demands are shown as being integrated into an estimate of total BOD, which draws from the O_2 pool and remineralizes into N and P. The effect of temperature (T) on most state variables and flows has likewise been excluded from the diagram. *ATM* atmospheric, *DENIT* denitrification, *GZR* macroalgal grazers, *PCR* plankton community respiration, *S* salinity. All other terms are defined in the text. Symbols are from Odum (1994)

4.2.2.1 Phytoplankton Biomass and Production

A single aggregated phytoplankton pool was modeled in carbon units, with the associated pools of N, P, and chl-*a* computed stoichiometrically. Brush et al. (2002) highlighted the diversity of temperature-dependent formulations for phytoplankton growth rate in existing models and a tendency for these functions to underestimate the fundamental rate of primary production; they suggested use of the empirical "light biomass" (BZ_pI_o) approach which has been shown to apply across a wide range of estuarine systems. This approach computes daytime phytoplankton net production (NPP_d, g C m^{-2} d^{-1}) as a function of chl-*a* biomass (B, mg chl-*a* m^{-3}), photic depth (Z_p, m), and incident irradiance (photosynthetically active radiation, PAR) (I_o, E m^{-2} d^{-1}) (e.g., Cole and Cloern 1987; Keller 1988; Kelly and Doering 1997):

Table 4.1 Model governing equations. Terms are defined in the text. Inputs and losses include net primary production (*NPP*), respiration (*R*), grazing by filter feeders (G_{FF}), exchanges among spatial elements (*exch*), external inputs (*ext*), remineralization (*rem*), and air-sea diffusion (*diff*). For simplicity, N_{rem}, N_{NPP}, P_{rem}, P_{NPP}, O_{2NPP}, and O_{2R} represent aggregated inputs and losses due to remineralization, NPP, and respiration from multiple sources; e.g., O_{2R} includes both water column and sediment respiration

Phytoplankton carbon[a]	g C m^{-2}	$PHY_{C(t)} = PHY_{C(t-dt)} + \left((1 - fNPP_{SED}) \cdot NPP_d^* - R_{WC} - G_{FF} \pm exch\right) \cdot dt$
Dissolved inorganic N	g N m^{-2}	$DIN_{(t)} = DIN_{(t-dt)} + (N_{ext} + N_{rem} - N_{NPP} - N_{den} \pm exch) \cdot dt$
Dissolved inorganic P	g P m^{-2}	$DIP_{(t)} = DIP_{(t-dt)} + (P_{ext} + P_{rem} - P_{NPP} \pm exch) \cdot dt$
Dissolved oxygen	g O_2 m^{-2}	$O_{2(t)} = O_{2(t-dt)} + \left(O_{2NPP} - O_{2R} \pm O_{2diff} \pm exch\right) \cdot dt$
Wastewater BOD[b]	g C m^{-2}	$BOD_{w(t)} = BOD_{w(t-dt)} + (BOD_{wext} - R_{BODw} \pm exch) \cdot dt$
Labile sediment carbon[a]	g C m^{-2}	$SED_{C(t)} = SED_{C(t-dt)} + \left(fNPP_{SED} \cdot NPP_d^* - R_{SED}\right) \cdot dt$
Ulva C, N, P	g m^{-2}	See Brush and Nixon (2010)
Gracilaria C, N, P	g m^{-2}	See Brush and Nixon (2010)
Decaying C, N, P	g m^{-2}	See Brush and Nixon (2010)

[a]Mass of N and P in the phytoplankton and sediment pools are computed stoichiometrically
[b]Inputs of biological oxygen demand (BOD) are converted from O_2 to C units; mass of N and P is computed stoichiometrically

$$NPP_d = \left[b_{BZI} + m_{BZI}\left(BZ_pI_0\right)\right] \cdot 10^{-3} \tag{4.1}$$

where b_{BZI} and m_{BZI} are the fitted y-intercept and slope, respectively. The factor of 10^{-3} converts from units of mg C m^{-2} d^{-1} (which are the units of published BZ_pI_o models) to g C m^{-2} d^{-1}. Photic depth is defined as the depth of the 1% light level and is computed as:

$$Z_p = \frac{4.61}{k_D} \tag{4.2}$$

where k_D (m^{-1}) is the vertical attenuation coefficient for light which was modeled as a function of predicted chl-*a* (*B*) according to the data of Granger et al. (2000):

$$k_D = k_0 + k_{Chl} \cdot B \tag{4.3}$$

where k_O and k_{Chl} were empirically determined to be 0.82 m^{-1} and 0.018 m^2 mg^{-1} chl-*a* for the coves and 0.72 m^{-1} and 0.019 m^2 mg^{-1} chl-*a* for the bay, respectively. The primary strengths of the empirical BZ_pI_o formulation are its basis in direct measurements of productivity and consistent ability to explain the majority of the variability in daily production across a wide range of systems with remarkably consistent slopes (see Brush et al. 2002 for a review).

Keller's (1986) ^{14}C productivity data from the Marine Ecosystems Research Laboratory (MERL) nutrient gradient experiment in which 12 tanks enriched from zero to 32 times the ambient DIN and DIP loading to Narragansett Bay were used to fit an m_{BZI} of 0.76 with b_{BZI} forced to zero. The y-intercept is an artifact of model fitting and predicts positive production in the absence of chlorophyll and/or light. The intercept is frequently not significantly different from zero (Keller 1988; Cole 1989), and when it is the lower 95%, confidence limit is often just above zero (Cole and Cloern 1987). Our fitted slope of 0.76 was within the 95% confidence limits of the slope without forcing the intercept to zero (0.71).

Z_p is a theoretical construct which predicts the depth of 1% I_o as a function of k_D, regardless of the actual depth of the system. The BZ_pI_o models have been developed in relatively deep estuarine systems where the computed Z_p is generally less than the actual depth of the system (Z). However, in shallow systems where light reaches the bottom, the theoretical value of Z_p will often be greater than Z. Brush and Brawley (2009) presented a depth-correction factor for the BZ_pI_o relationship to correct predicted production when $Z < Z_p$:

$$pNPP_d = 5.95 + (2.31pZ_p) - \left(9.95E^{-3}pZ_p^2\right) - \left(9.35E^{-5}pZ_p^3\right) + \left(5.13E^{-7}pZ_p^4\right) - (0.12I_0) - \left(7.34E^{-3}pZ_pI_0\right) + \left(9.46E^{-5}pZ_p^2I_0\right) \tag{4.4}$$

where $pNPP_d$ is the percent of predicted NPP_d (Eq. 4.1) that occurs over the water column and:

$$pZ_p = MIN\left[100, \left(\frac{100 \cdot Z}{Z_p}\right)\right] \tag{4.5}$$

where Z is water column depth (m). $pNPP_d$ is converted to a fraction and multiplied by the predicted production from Eq. 4.1.

The BZ_pI_o models have been developed in relatively nutrient replete, light-limited estuaries; application in a broader range of systems requires constraining the predicted amount of daily production by the available pool of nutrients. The total demand for both *DIN* and *DIP* from uptake by phytoplankton, *Ulva*, and *Gracilaria*, as well as losses to denitrification and mixing, is computed each time step (N_{demand}, P_{demand}) and compared to the available pool plus inputs during that time step (N_{avail}, P_{avail}). If the availability of both nutrients exceeds the total demand, neither is limiting and NPP_d is not reduced. If availability of either nutrient is less than the total demand, the nutrient with the lower ratio of availability to demand is taken as the limiting nutrient. The ratio of phytoplankton demand to total demand is then computed for both N and P ($NUTLIM_N$ and $NUTLIM_P$) and used to scale potential NPP_d to actual production (NPP_d^*):

$$\begin{aligned} &\textit{If N is limiting:} && NPP_d^* = NPP_d \cdot \tfrac{pNPP_d}{100} \cdot NUTLIM_N \\ &\textit{If P is limiting:} && NPP_d^* = NPP_d \cdot \tfrac{pNPP_d}{100} \cdot NUTLIM_P \\ &\textit{If neither are limiting:} && NPP_d^* = NPP_d \cdot \tfrac{pNPP_d}{100} \end{aligned} \tag{4.6}$$

The same approach is used to reduce all other nutrient-consuming processes (*Ulva* and *Gracilaria* uptake, denitrification, and exchange) in the case of limiting nutrients.

Since phytoplankton biomass is directly included in the equation for daily production in absolute terms (i.e., g C m^{-2} d^{-1}, Eq. 4.1), it is appropriate to incorporate computed daily production as a zero-order input to the phytoplankton governing equation (Table 4.1). While models generally rely on multiplication of instantaneous rates (i.e., d^{-1}) by modeled biomass (i.e., first-order kinetics), conversion of our BZ_pI_o predicted production to an instantaneous rate resulted in an underestimation of daily production according to the empirical model. This numerical artifact necessitates incorporating BZ_pI_o predicted rates as a zero-order term.

As Greenwich Bay is a nutrient-rich system, we selected the median carbon: chlorophyll (*C:Chl*) ratio of 42 g g^{-1} from nutrient sufficient cultures in Cloern et al. (1995) compilation of culture data (see Brush et al. 2002). Phytoplankton production was converted to oxygen evolution (O_{2NPP}, g O_2 m^{-2} d^{-1}) using a photosynthetic quotient (PQ) of 1. Production was converted to demand for N and P (N_{NPP}, P_{NPP}, g m^{-2} d^{-1}) using molar ratios from the original Narragansett Bay Model (Kremer and Nixon, 1978) of 7:1 (C:N) and 85:1 (C:P).

4.2.2.2 Pelagic Respiration

Phytoplankton biomass in the model is lost to pelagic heterotrophic processes, flux to the sediments, filter feeders, and physical exchange. Since the BZ_pI_o model computes daytime net production, pelagic heterotrophic processes include respiration by phytoplankton at night, grazing and subsequent respiration by zooplankton and other consumers, and incorporation and subsequent respiration of phytoplankton exudates by the microbial loop. These are difficult to formulate completely as they are influenced by a variety of factors and carried out by several trophic groups for which model formulations and parameter values are uncertain. Regardless of these numerous sinks for phytoplankton biomass within the water column, however, biomass that is not consumed by filter feeders, deposited on the bottom, or lost due to exchange will eventually be respired by the pelagic community. The standing stock of phytoplankton carbon is thus continually returning to a quasi-steady state, with system respiration often tightly coupled to net daytime production (Nixon et al. 1986; Smith and Kemp 1995; Caffrey 2003).

A survey of literature equations for predicting plankton community respiration (PCR) indicates that rates are generally functions of temperature and/or chlorophyll (Table 4.2; Hopkinson and Smith 2005). All pelagic, heterotrophic loss terms were

Table 4.2 Predictive empirical equations from the literature for computing plankton community respiration (PCR) as a function of temperature (T, °C), chlorophyll-*a* (Chl, mg m^{-3}), and daytime net production (P_d, see footnotes for units)

Source	Location	PCR units	Equation	r^2
Nixon and Oviatt (1973)	Bissel Cove, RI	g O_2 m^{-2} h^{-1}	$PCR = 10^{(0.0575T-2.74)}$	0.42
Turner (1978)[a]	Georgia creek	mg-at O_2 m^{-3} d^{-1}	$ln(PCR) = \beta_0 + \beta_r\ ln(T)$	0.23-0.68
Nowicki (1983)	Potter Pond, RI	mg O_2 m^{-2} h^{-1}	$PCR = 0.39\ e^{0.21\ T}$	0.54
Holligan et al. (1984)[b]	English Channel	mg O_2 m^{-3} d^{-1}	$PCR = 56.2 + 25.8\ Chl$	0.91
Jensen et al. (1990)	Roskilde Fjord	mg O_2 m^{-3} h^{-1}	$PCR = 21 + 0.87\ Chl$	0.63
Iriarte et al. (1991)[c]	North Sea	μmol $O_2 1^{-1}$ h^{-1}	$PCR = 0.038\ Chl + 0.088$	0.76
Howarth et al. (1992)[d]	Hudson River	g C m^{-2} d^{-1}	$PCR = 0.77\ e^{0.0693\ T}$	0.62
Kemp et al. (1992)[e]	Chesapeake Bay	g O_2 m^{-3} d^{-1}	$PCR = \beta_0 + \beta_1\ T$	0.23, 0.69
			$PCR = \beta_2 + \beta_3\ P_d$	0.49, 0.76
Sampou and Kemp (1994)[f]	Chesapeake Bay	μg $O_2 l^{-1}$ h^{-1}	$PCR = 3.14\ e^{0095\ T}$	0.62
Smith and Kemp (1995)[g]	Chesapeake Bay	mg $O_2 l^{-1}$ h^{-1}	$PCR = \beta_0\ e^{\beta 1\ T}$	0.28-0.87
Fourqurean et al. (1997)	Tomales Bay, CA	μmol $O_2 1^{-1}$ h^{-1}	$PCR = -0.178 + 0.033\ T + 0.059\ Chl$	0.75
Caffrey et al. (1998)[h]	San Francisco Bay	mmol O_2 m^{-3} d^{-1}	$PCR = 4.0\ ln(Chl) + 2.6$	0.50
Moncoiffe et al. (2000)[i]	Ria de Vigo, Spain	uM O_2 d^{-1}	$PCR = -21.21 + 1.75\ T + 0.77\ Chl$	0.44
Hopkinson and Smith (2005)	Literature synthesis	mmol C m^{-3} d^{-1}	$\log(PCR) = 0.89 + 0.08\ T$	0.28
			$\log(PCR) = 1.19 + 0.63\ \log(Chl)$	0.38
Goebel and Kremer (2007)	Long Island Sound	mmol O_2 m^{-3} h^{-1}	$PCR = 0.39 + 0.09\ Chl$	0.36
Brush (unpublished)[j]	MERL mesocosms	g O_2 m^{-2} d^{-1}	$PCR = (4.2 \times 10^{-11})\ Chl\ e^{0.88\ T} + (0.72\ Pd)$	0.78

[a]Regressions reported for five different sites; parameter ranges: β_0 (−6.81 to −0.88), β_1 (1.16–3.11)
[b]Regression fit to data in authors' Table 4.2
[c]Regression fit to data in authors' Table 4.1. Multiple linear and nonlinear (see MERL equation) regressions produced identical r^2 values
[d]Regression estimated from authors' Fig. 4.5
[e]Regressions fit for upper and lower water column layers, with higher r^2 in the upper layer. Pd has units g O_2 m^{-2} d^{-1}
Coefficients for upper and lower layers, respectively: β_0 (−0.16, 0.16), β_1 (0.04, 0.02), β_2 (−0.15, −0.03), β_3 (0.48, 0.41)
[f]Regression is from authors' in situ data; similar regressions are given for separate May and August experiments
[g]Regressions fit for upper and lower layers in three regions of the bay; parameter ranges: β_0 (0.002–0.006), β_1 (0.032–0.104)
[h]Regression fit to data in authors' Table 4.1
[i]Regression for entire water column is given. Similar regressions were reported for the surface, depth of 1% *Io,* and 40 m
[j]Data from the 1981–1983 eutrophication gradient experiment at the Marine Ecosystems Research Laboratory (MERL) (Frithsen et al. 1985a, b, c). Pd has units g O_2 m^{-3} d^{-1}

aggregated into a single term recognizing that (1) phytoplankton biomass is the ultimate substrate for pelagic heterotrophy, (2) PCR is generally exponentially related to temperature, and (3) heterotrophic responses typically lag behind phytoplankton biomass dynamics. To achieve a lag time in heterotrophic response, our model computes a 10-day moving average of predicted phytoplankton biomass (PHY_{10}). The model then consumes a temperature (T)-dependent fraction of this moving average biomass each time step to reflect water column respiration (R_{WC}, g C m^{-2} d^{-1}):

$$R_{WC} = PHY_{10} \cdot R_{WC_0} \cdot e^{R_{WC_k} \cdot T} \tag{4.7}$$

The exponent (R_{WC_k}, 0.095 °C^{-1}) was obtained from Sampou and Kemp's (1994) relationship between temperature and PCR in Chesapeake Bay and reflects a Q_{10} of 2.6. The 0 °C value (R_{WC_0}) was set by calibration at 0.055 d^{-1} to achieve the best visual fit to the observations. This respiratory rate is used as the loss term to the phytoplankton state variable and converted into an oxygen demand ($O_{2R_{WC}}$, g O_2 m^{-2} d^{-1}) using a respiratory quotient (RQ) of 1. Nitrogen and phosphorus are remineralized stoichiometrically ($N_{R_{WC}}$, $P_{R_{WC}}$, g m^{-2} d^{-1}) using the C:N and C:P ratios for phytoplankton.

4.2.2.3 Carbon Deposition and Sediment Fluxes

Shallow marine ecosystems are typified by strong benthic–pelagic coupling, with the key rates being the flux of pelagic carbon to the sediments and subsequent remineralization. The traditional formulation for this flux is to apply a sinking rate as a loss term to the phytoplankton state variable(s); however, these rates are highly variable and change with the dominant phytoplankton group (Riebesell 1989; Reckhow 1994). Alternatively, Nixon (1986) reported a strong linear relationship between annual benthic carbon remineralization and the total annual input of carbon from primary production and allochthonous inputs in 16 coastal marine systems. The slope of the relationship was 0.238, which suggests that roughly 25% of annual phytoplankton production (ignoring external carbon inputs, which are comparatively low for Narragansett Bay) is deposited to and respired in the sediments. This formulation thus gives an empirically based, cross-system relationship independent of the particular species present. Though the slope surely varies throughout a given year, as a first approximation our model takes 25% ($fNPP_{SED}$) of daily phytoplankton production and adds it directly to the sediment carbon pool (SED_C), with the remaining NPP entering the phytoplankton state variable (NPP^*_{SED} and NPP^*_{PHY}, respectively):

$$NPP^*_{SED} = fNPP_{SED} \cdot NPP^*_d \tag{4.8}$$

$$NPP^*_{PHY} = (1 - fNPP_{SED}) \cdot NPP^*_d \tag{4.9}$$

The state variable SED_C therefore reflects the pool of labile organic carbon which accumulates seasonally from settling of phytoplankton and heterotrophic processes (e.g., fecal pellets) rather than recalcitrant carbon which may build up over longer time scales.

Respiration of the sediment carbon pool (R_{SED}, g C m^{-2} d^{-1}) was driven as an exponential function of temperature, using an identical formulation to that for water column respiration:

$$R_{SED} = SED_C \cdot R_{SED_0} \cdot e^{R_{SED_k} \cdot T} \tag{4.10}$$

where R_{SED_0} was determined by calibration to be 0.10 d^{-1} for the main bay and 0.15 d^{-1} for the coves which contain sediments with a higher organic content, and the exponent R_{SED_k} was set at 0.14 °C^{-1} based on measurements from Narragansett Bay (Nixon et al. 1976; Kremer and Nixon 1978). Calibration values were determined as those that kept the carbon pool in steady state on an annual basis. The calculated rate of respiration is converted to an oxygen demand ($O_{2R_{SED}}$, g O_2 m^{-2} d^{-1}; RQ = 1) and used to remineralize nitrogen and phosphorus ($N_{R_{SED}}$, $P_{R_{SED}}$, g m^{-2} d^{-1}) using the C:N and C:P ratios for phytoplankton.

The other key flux associated with the sediments is the loss of nitrogen to denitrification. Traditional mechanistic approaches for computing this loss range from relatively simple functions of water column or porewater nitrate concentrations and water temperature to detailed sediment flux models; as above, these include a number of parameters which are often not well constrained by the available data, or are highly variable among systems. Alternatively, Nixon et al. (1996) presented an empirical regression using data from several systems showing a linear relationship between flushing time in months (FT_{mo}) and the percent of nitrogen inputs from land and the atmosphere which are lost to denitrification (pN_{den}):

$$pN_{den} = 20.8 \cdot log(FT_{mo}) + 22.4 \tag{4.11}$$

As with our formulations for production, respiration, and carbon flux to the sediments, this equation represents a robust, data-driven relationship which applies across a wide range of estuarine systems and contains only two parameters. Denitrification (N_{den}, g N m^{-2} d^{-1}) loss is computed from Eq. (4.11) using flushing times from the transport model (see below) and the daily input of nitrogen from the watershed (N_{wshd}) and atmosphere (N_{atm}):

$$N_{den} = (N_{wshd} + N_{atm}) \cdot pN_{den} \tag{4.12}$$

4.2.2.4 Remaining Formulations

The rest of the model follows a traditional mechanistic approach; details are available from the authors and Brush and Nixon (2010). Briefly, macroalgal growth is mechanistically driven by temperature, light, and nutrients, and losses include respiration, grazing, and decay. Macroalgae take up nutrients when in excess (i.e., luxury uptake) and are able to use both internal and external nutrients to drive primary production. Decay of dead biomass is a function of tissue C:N ratio and temperature. The sediments of Greenwich Bay contain high densities of hard clams (*Mercenaria mercenaria*) (Ganz et al. 1994) and amphipods (*Ampelisca* sp.) (Stickney and Stringer 1957); mean sizes (5 classes for clams) and densities were forced and used to compute grazing rates on the phytoplankton using the formulations of Loosanoff (1939), Doering and Oviatt (1986), and Beatty (1991), and increased sediment fluxes of oxygen and nutrients using the formulations of Hibbert (1977) and a literature synthesis by Brush (2002). All photosynthetic production (phytoplankton, *Ulva*, *Gracilaria*) is used to stoichiometrically produce oxygen and consume nutrients; all respiratory processes are used to stoichiometrically consume oxygen and release nutrients. Air-sea diffusion of oxygen is computed as a function of mean daily wind speed (Marino and Howarth 1993).

4.2.3 Forcing Functions

Model forcing functions were derived from a combination of cosine fits to the data of Granger et al. (2000) and daily to weekly data derived directly from local sources (Fig. 4.3). Watershed and atmospheric inputs of DIN and DIP were computed using watershed yields from Nixon et al. (1995) and atmospheric data from Fraher (1991) and Nowicki and Oviatt (1990). Boundary concentrations of chlorophyll-*a*, DIN, DIP, and O_2 in the surface and bottom waters of Narragansett Bay were forced from the Granger et al. (2000) data for two stations just outside the mouth of Greenwich Bay (Fig. 4.1).

4.2.4 Spatial Elements and Transport Model

Greenwich Bay was divided into a series of spatial elements based on morphology, water quality, and the distribution of survey stations (Fig. 4.1). While boxed schemes are relatively coarse, the purpose of the current work was to evaluate the ability of our formulations to reproduce system dynamics; the boxed version makes possible multiple fast runs during model testing. A second goal was to develop a model that could be rapidly applied to new systems; again, the boxed version accomplishes that goal. Recent work has confirmed the utility of boxed approaches (Ménesguen et al. 2007; Testa and Kemp 2008; Kremer et al. 2010). Boxes were divided into surface and bottom layers based on mean pycnocline depth using salinity profiles (Erikson 1998; Granger et al. 2000). Area, volume, mean depth, and the fraction of bottom area in the surface and bottom layers were determined from interpolated bathymetric data (NGDC 1996).

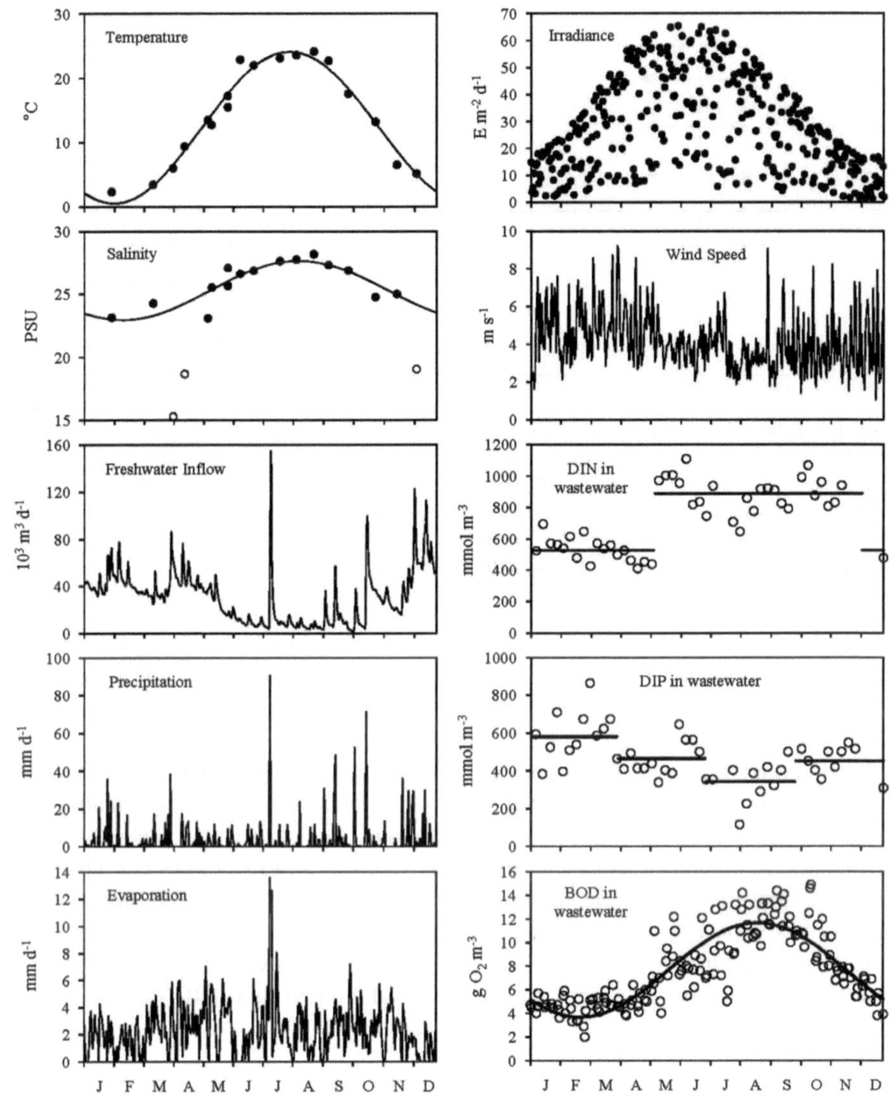

Fig. 4.3 Model forcing functions. Temperature and salinity were specified through cosine functions fit to the data of Granger et al. (2000). Daily freshwater inflow from the watershed was provided by S. Viator and R. Wright (University of Rhode Island, unpublished data). Precipitation and wind speeds were obtained from the National Climatic Data Center (NOAA) for T.F. Green Airport 5 km north of the bay; evaporation was computed from meteorological data after Erikson (1998). Solar radiation was obtained from the Eppley Laboratory in Newport, R.I., 21 km southeast of the bay and converted to PAR (Vollenweider 1974). Daily freshwater discharge from the wastewater treatment facility was provided by the plant; functions for effluent concentrations were fit to plant data

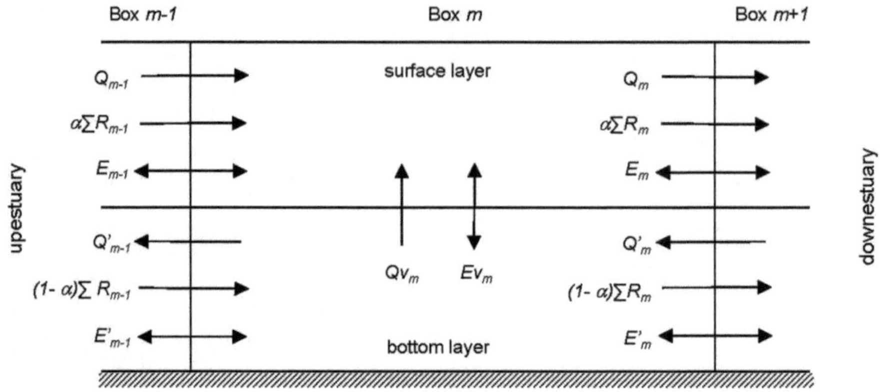

Fig. 4.4 Schematic of the hydrodynamic transport model in vertical cross section, after Officer (1980), Swanson and Jayko (1988), and Erikson (1998). Q advective flows for box m, E non-advective flows, $\sum R$ sum of all freshwater inputs upstream of and into box m, α is the fraction of freshwater inputs that enter the surface layer, subscript v refers to vertical flows, and prime (′) denotes the bottom layer. The version of the box model used here distinguishes between advective flows of freshwater ($\sum R$) and entrained ocean water (Q)

Exchanges of water and materials between the spatial elements and flushing times were computed from forced salinity distributions and freshwater inputs with an Officer-type box model (Officer 1980; Swanson and Jayko 1988; Erikson 1998) (Fig. 4.4). Freshwater inputs were averaged over a 10-day period. A version of the Officer model was used that allocates freshwater inputs between the surface and bottom layers using the parameter α, and divides total exchanges between advective (gravitational) and non-advective (tidal) flows based on the Hansen-Rattray parameter, υ, computed using the approach of Officer and Kester (1991). The mean flushing time for Greenwich Bay computed with our box model (8.8 days) closely matches estimates from a fine scale 2-D circulation model (9.2 days; Abdelrhman 2005) and is within the range of 2–20 days depending on wind speed and direction determined with the 3-D Regional Ocean Modeling System (ROMS) by Rogers (2008).

4.2.5 Calibration and Sensitivity Analysis

The model was calibrated to the data of Granger et al. (2000) by tuning parameter values within known ranges for the period May 1996–May 1997 (the standard run). Model skill was assessed by analysis of mean absolute error, mean percent error, and root-mean-squared (RMS) error for surface chlorophyll-a, surface DIN and DIP, and bottom O_2. Monitoring data were first compared to the model prediction on the same day of the measurement; however, the intent of this and many estuarine models is not to predict values to the exact day, but rather to predict general dynamics, down-estuary gradients, and approximate magnitudes of state variables. We therefore also computed skill by comparing monitoring data to the closest model prediction within 1 and 2 weeks on either side of the observation (Brush and Nixon 2010).

Sensitivity analysis was conducted sequentially on all model parameters, selected forcing functions, and initial values of all state variables. Initial values were set at ½ and 2 times the values from the standard run; parameters were changed ±20% unless approximate ranges were known. Results of the sensitivity runs were expressed as the percent difference from the standard run:

$$\%difference = \frac{|x_{standard} - x_{sensitivity}|}{x_{standard}} \cdot 100\% \tag{4.13}$$

where x is the model prediction of a given variable on a given day. Daily values of % difference were not normally distributed, so the overall model sensitivity to a given parameter was computed as the median % difference across all spatial elements for each pair of sensitivity runs.

4.3 Results and Discussion

4.3.1 Phytoplankton

Model predictions of chlorophyll-a biomass followed the observed annual cycle and captured shorter-scale bloom dynamics (Fig. 4.5, dashed line). The use of a constant slope for the BZ_pI_o regression (m_{BZI}), however, led to overestimates of phytoplankton biomass during winter and late fall. Some studies have reported different slopes between daily production and BZ_pI_o or B in summer (higher m_{BZI}) versus non-summer (lower m_{BZI}) periods, which have been attributed to changes in species composition (Pennock and Sharp 1986; Keller 1988). While there was no apparent temperature dependency of m_{BZI} in the Keller (1986) MERL data, slopes fit using a more recent dataset of ^{14}C productivity from Narragansett Bay (Oviatt et al. 2002) were related to temperature (Fig. 4.6). However, production in this study and consequently values of m_{BZI} were lower than past estimates (Brush et al. 2002). To derive a formulation for m_{BZI} in the model, we used the exponent from the Oviatt et al. (2002) dataset (0.051) and set our original value of 0.76 as the intercept at 15 °C, the mean temperature of the Keller (1986) measurements. The equation was constrained to not go above 1.1, the highest reported value of m_{BZI} (Brush et al. 2002):

$$m_{BZI} = MIN\left(0.76e^{0.051(T-15)}, 1.1\right) \tag{4.14}$$

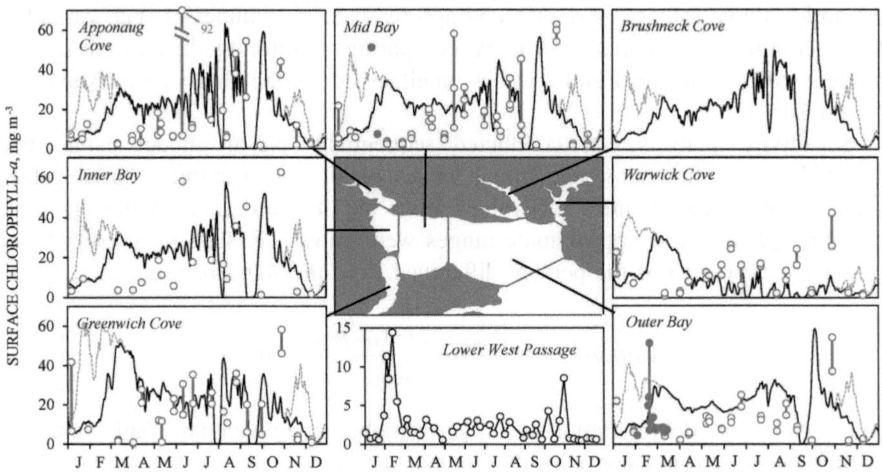

Fig. 4.5 Measured (*points*) and modeled (*lines*) chlorophyll-*a* concentrations in the surface layer of each spatial element in Greenwich Bay. Open circles are observations from the surveys of Granger et al. (2000). *Vertical lines* connect measurements from stations within the same spatial element on each sampling date. *Closed circles* are from auxiliary surveys designed to capture the winter–spring bloom. *Solid line* represents the model standard run with a temperature-dependent slope of the BZ_pI_0 regression; *dashed line* is from a run with a constant slope of 0.76. *Inset* at bottom shows MERL data collected from the lower West Passage of Narragansett Bay during the same time period

Fig. 4.6 Effect of water temperature on the slope of the regression between daily phytoplankton production and the composite parameter BZ_pI_0. Regressions were performed on the data of Oviatt et al. (2002) at 1 °C temperature intervals. *Open circles* are slopes from regressions with $r^2 < 0.50$. *Bold line* is the regression for the Oviatt et al. (2002) data. *Thin line* is the adjusted relationship used in the model

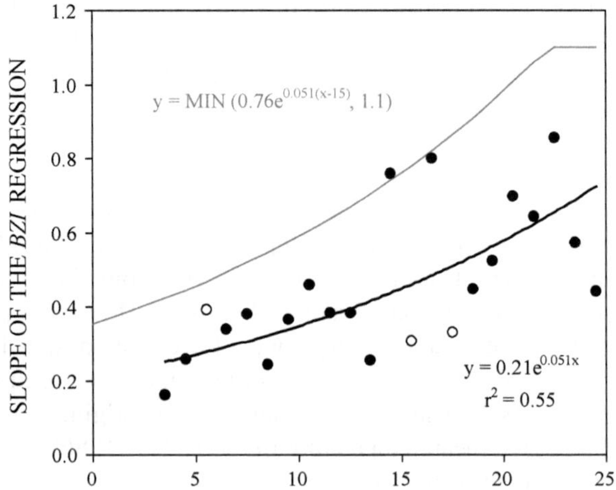

The use of this temperature-dependent relationship for m_{BZI} eliminated the winter and late fall peaks in predicted chlorophyll-a (Fig. 4.5, solid line). The model still predicted higher biomass levels in late spring than suggested by the data, which can be reduced by increasing the coefficients for water column respiration (which as formulated is a proxy representing all heterotrophic pelagic loss processes including phytoplankton respiration, mortality, and grazing). However, a winter–spring bloom was, until recently, a dominant event in the annual cycle of phytoplankton in Narragansett Bay (Pilson 1985; Li and Smayda 1998; Nixon et al. 2009). Little evidence for such a bloom exists in the Granger et al. (2000) surveys, but no data were collected between the end of January and mid-March. Auxiliary chlorophyll samples collected in the main bay during February provided evidence that a spring bloom had occurred, as did data collected by MERL in lower Narragansett Bay (Fig. 4.5). Although the timing is late, the model should indeed be predicting a winter–spring bloom for the standard year. The duration of the bloom may be artificially high due to the absence of a zooplankton state variable to graze down the bloom, which may not be adequately captured by our moving average formulation for PCR.

Following this bloom, the model reproduced both the observed levels of chlorophyll and bloom dynamics, particularly reproduction of the mid-September system-wide crash followed by a substantial fall bloom, which the model captured albeit slightly early and with a somewhat lower magnitude (Fig. 4.5). One notable exception is Warwick Cove, in which particularly high densities of clams exert a large grazing pressure; we suspect these older density estimates may be too high. The model failed to capture the unusually high chlorophyll concentrations observed in Apponaug Cove and the inner and mid bay during June. These concentrations appear related to elevated DIN (see below), but not to inputs from the surrounding watershed as evidenced by stream flow or precipitation. Such high DIN and chlorophyll concentrations were likely caused by an event external to the model formulations; when these high DIN concentrations were forced in a separate model run, predicted chlorophyll came much closer to the June observations.

4.3.2 Nutrients

Model predictions followed the typical seasonal cycle of DIN for Narragansett Bay (Pilson 1985) and tracked the lower envelope of the observations (Fig. 4.7), with periodic increases related to phytoplankton dynamics. The model failed to capture the very high (>70 mmol m^{-3}) concentrations observed in Apponaug and Warwick Coves. As for chlorophyll, these concentrations showed no consistent relationship with watershed flows or precipitation, so it is likely that they are due to processes external to the model. Further, two measurements exist for each cove on each date, one from a station near the head and one from mid-cove (Fig. 4.1); on the dates when these high concentrations were measured, only the station at the head had such high values, and only in the surface layer, perhaps due to the concentration of watershed inputs at the head of each cove. With the relatively coarse spatial resolution of the box model, and use of time-averaged flows which smooth over storm events, it is not surprising that the model misses these high concentrations.

The model should instead be capturing the element-wide mean condition, which is reflected better by the mid-cove stations which the model captures well.

The model also reproduced the seasonal pattern of DIP concentrations observed in Narragansett Bay (Pilson 1985) and reflected in the observations, although the model underestimates concentrations during summer, suggesting an additional source of DIP to the system (Fig. 4.8). This additional source may be the release of DIP from sediments due to low oxygen concentrations, or desorption of particulate-bound phosphorus in stream inputs, two processes not included in the model. As for DIN, the model does not come close to predicting the extreme (>10 μM) DIP concentrations in Greenwich Cove. Again, there was no clear relationship between these high concentrations and inputs from the watershed or

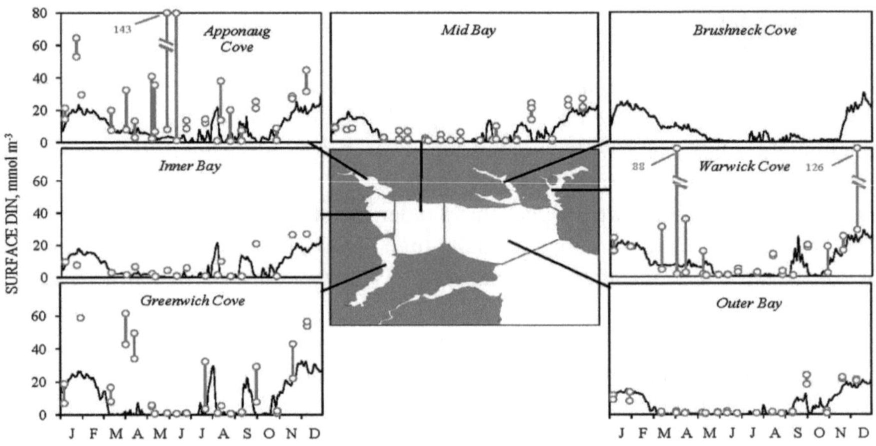

Fig. 4.7 As for Fig. 4.5, but for surface concentrations of DIN

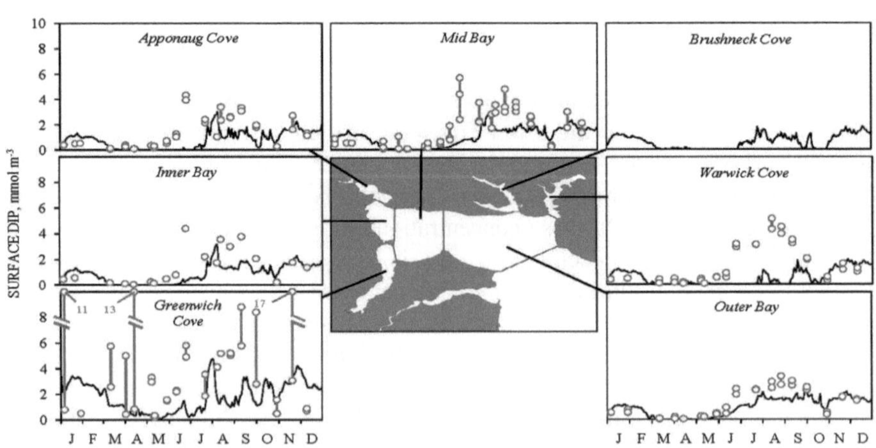

Fig. 4.8 As for Fig. 4.5, but for surface concentrations of DIP

precipitation. This cove receives effluent from the East Greenwich wastewater treatment facility which is rich in phosphorus; the plant contributes seven times the watershed contribution of DIP to Greenwich Bay, and more than twice the atmospheric load (Granger et al. 2000). These high concentrations were consistently measured at the mid-cove station, which was adjacent to the facility outfall.

4.3.3 Dissolved Oxygen

The model captured the seasonal cycle of dissolved oxygen due to the effect of temperature and salinity on solubility, along with the occurrence of bottom water hypoxia (Fig. 4.9). While the first hypoxic event during June was largely missed by the model, the other events were reproduced to varying degrees. The model does a particularly good job in Greenwich and Warwick Coves, due in part to the presence of large clam populations and their associated rates of respiration. The varying degree to which hypoxic events were captured by the model may be due to its coarse spatial resolution; alternatively, the model may simply be unable to translate phytoplankton blooms into hypoxic events due to limitations with the formulations. The latter possibility was tested by running the model while forcing the maximum observed chlorophyll biomass from the Granger et al. (2000) surveys. Model predictions of hypoxia were much improved in this scenario (Fig. 4.9), which suggests that the model kinetics are correct and the failure to reproduce all hypoxic events is largely due to the coarse spatial resolution.

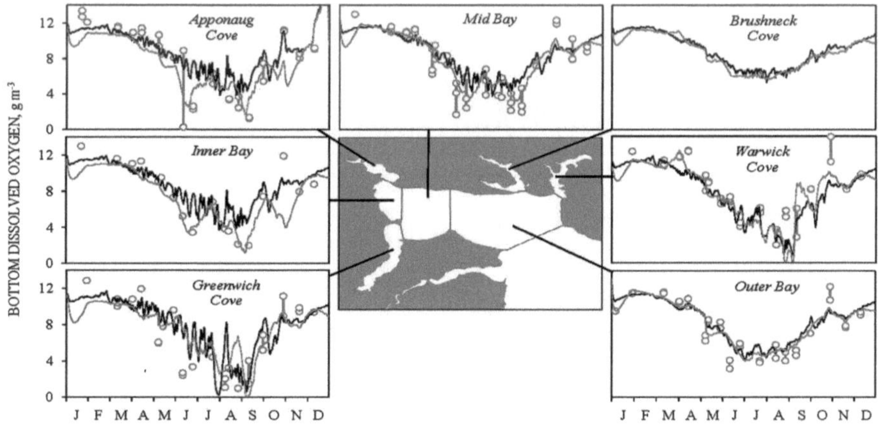

Fig. 4.9 As for Fig. 4.5, but for bottom concentrations of O_2. *Solid line* represents the model standard run; *dashed line* represents a run in which chlorophyll-*a* concentrations were forced at the maximum measured values of Granger et al. (2000)

4.3.4 Rate Processes

Predicted water column production and respiration displayed seasonal cycles with maximum values in the summer and variations on the order of a week to a month related to phytoplankton biomass dynamics (Fig. 4.10a, b). Superimposed on these cycles were high-frequency variations driven by daily fluctuations in irradiance, a product of the direct coupling of the NPP formulation to irradiance. The annual peak in production occurred during August at the peak of the temperature cycle, rather than during June when irradiance was at its peak. The only available ^{14}C production data for an annual cycle in Greenwich Bay also indicate an August peak (Oviatt et al. 2002), so it is reassuring that the model kinetics captured this despite the light-based approach. Respiration was tightly coupled to production, but did not

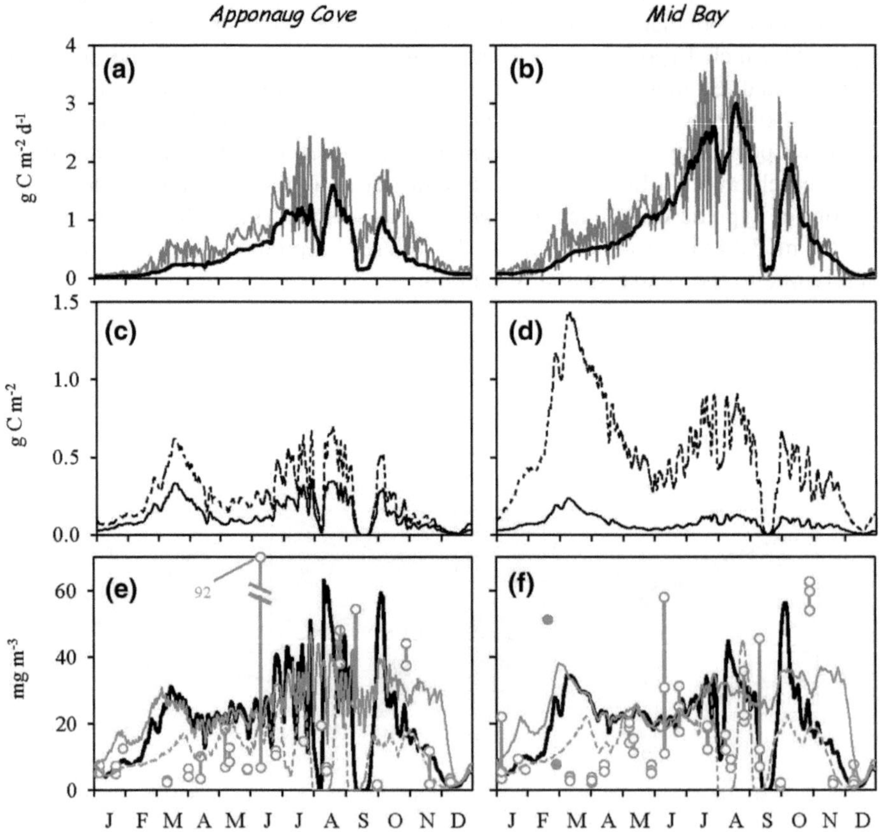

Fig. 4.10 **a**, **b** Modeled rates of daily system production (*gray*) and respiration (*black*) in representative regions of Greenwich Bay. **c**, **d** Modeled accumulation of sediment organic carbon in the surface (*solid*) and bottom (*dashed*) layers. **e**, **f** Modeled chlorophyll-*a* (surface) from the standard run (*solid bold*), Eppley run (*dashed*), and elevated Eppley run (*solid gray*)

exhibit the same high-frequency variations due to use of the 10-day moving average formulation.

The simulated pool of labile organic carbon in the sediments tracked simulated phytoplankton biomass and productivity (Fig. 4.10c, d). The bottom layer accumulated more carbon in the sediments than the surface layer because it received material deposited from the entire water column, while deposition to sediments in the surface layer was limited to nearshore regions where the sediments intersected the surface layer. As expected, the spring bloom corresponded with the largest accumulation of sediment carbon (Rudnick and Oviatt 1986; Riebesell 1989), which was then partially respired before more was deposited throughout summer and fall due to a series of phytoplankton blooms.

System-wide phytoplankton production was estimated by the model to be about 310 g C m^{-2} y^{-1} (Table 4.3). This value is nearly identical to the ^{14}C-based average value of 320 g C m^{-2} y^{-1} for all of Narragansett Bay obtained by Oviatt et al. (2002), and within the generally accepted range for the bay of 270–310 g C m^{-2} y^{-1} prior to the Oviatt et al. study (Furnas et al. 1976, Oviatt et al. 1981). However, our model estimate is higher than Oviatt et al. range of 220–255 g C m^{-2} y^{-1} for Greenwich Bay alone. That study was conducted in 1997–1998, 1 year after the model standard year and a year in which there was no winter–spring bloom. Since this bloom did occur during the modeled year, we extrapolated Oviatt et al. values to a year with a bloom by assuming that approximately 22% of annual production in Narragansett Bay occurs during the winter–spring bloom (Durbin et al. 1975; Furnas et al. 1976; Vargo 1979; Keller 1988). Our modeled estimate falls in the middle of the Oviatt et al. (2002) range when corrected to account for the winter–spring bloom (Table 4.3).

Brush et al. (2002) suggested that the traditional use of the Eppley (1972) curve between temperature and maximum daily phytoplankton growth rate in mechanistic models may result in an underestimation of phytoplankton production and that the

Table 4.3 Model estimates of annual phytoplankton production (P_y, g C m^{-2} y^{-1}) in Greenwich Bay

	P_y
Model standard run	
Main Bay[a]	320
Entire system[b]	307
Eppley curve simulations	
Eppley curve[c]	206
Increased Eppley[d]	395
Observed	
Oviatt et al. (2002)[e]	281–326

[a]Area-weighted estimate for elements 4, 6, and 7 (see Fig. 4.1)
[b]Area-weighted estimate for all seven spatial elements
[c]Using Eppley (1972) relationship of $G_{max} = 0.59e^{0.0633 \cdot Temperature}$
[d]Using Brush et al. (2002) suggested revised Eppley curve of $G_{max} = 0.97e^{0.0633 \cdot Temperature}$
[e]See text for details

BZ_pI_o formulation may produce more accurate predictions. To test this assertion, the model was run with formulations based on the traditional Eppley curve approach (Kremer and Nixon 1978). Annual phytoplankton production in the Eppley run was more than 30% lower than in the standard run (Table 4.3) and resulted in biomass predictions with equal or lower values and fewer daily fluctuations (Fig. 4.10e, f). Our model estimate is also closer to reported rates than the value of 105 g C m^{-2} y^{-1} for Greenwich Bay in the original Narragansett Bay model which was based on the Eppley curve (Kremer and Nixon 1978). The use of an elevated Eppley curve based on more recently published data (Brush et al. 2002) overestimated annual production by just under 30% and resulted in biomass predictions in the same range as or higher than the standard run, also with fewer daily variations (Table 4.3; Fig. 4.10e, f).

4.3.5 Model Skill

Absolute and RMS errors in predicted state variables were relatively low when comparing values on the same day of the observations, and errors were markedly reduced when the window for comparison was relaxed to ±1 and 2 weeks (Table 4.4). On the other hand, percent errors were high when comparisons were made on the day of the observations, but these were again markedly reduced when

Table 4.4 Computed model skill for the standard run summarized across all spatial elements and the annual cycle. Skills are reported for comparisons on the same day as the observations and within 1 and 2 weeks on either side of the observations. Units for chlorophyll-a, DIN, DIP, and O_2 are mg m^{-3}, mmol m^{-3}, mmol m^{-3}, and g m^{-3}, respectively

	Absolute error		Percent error		RMS error
	Mean	Median	Mean	Median	
Surface chlorophyll-a					
Same day	12.9	9.0	282.8	66.7	18.6
1 week	5.7	1.8	118.2	17.0	9.8
2 weeks	4.5	0.1	97.0	1.9	8.5
Surface DIN					
Same day	5.2	1.1	127.6	67.3	9.6
1 week	2.9	0.3	30.3	16.3	7.4
2 weeks	2.4	0.1	19.2	3.0	7.0
Surface DIP					
Same day	1.0	0.5	71.9	68.8	1.5
1 week	0.7	0.3	44.6	31.8	1.2
2 weeks	0.5	0.2	37.6	23.1	1.1
Bottom O_2					
Same day	1.2	0.8	29.2	11.1	1.7
1 week	0.6	0.2	13.4	2.2	1.0
2 weeks	0.4	0.0	9.8	0.3	0.8

the temporal window for comparison was relaxed. The discrepancy between mean and median errors indicates that the means are being influenced by some particularly high values, so the medians may be a better indication of model skill.

4.3.6 Sensitivity Analysis

The model was insensitive to the choice of initial conditions, as the output quickly converged to that from the standard run. Using a criterion of 10% difference from the standard run, the model was insensitive to most parameter values, with only a few meeting this 10% criterion (Table 4.5); these parameters were primarily related to our proposed alternative model formulations (i.e., Equations 4.1, 4.7, 4.8, 4.14). The choice of the carbon-to-chlorophyll (*C:Chl*) ratio had a large effect on model results, although predicted phytoplankton biomass followed the same trajectories as in the standard run (Fig. 4.11a). Values of 30 and 60 g g^{-1} were chosen as these are the first and third quartiles for light-limited cultures from the Cloern et al. (1995) database (see Brush et al. 2002). Models are often sensitive to the choice of the C: Chl ratio as it provides the conversion between carbon units, a common currency for modeled phytoplankton, to chl-*a*, the primary proxy for observed biomass used in calibration. One option for addressing this sensitivity, and potentially improving the simulation of chl-*a*, would be to consider using a time-variable model for the C: Chl ratio; various options are available and range from empirical functions of temperature, light, and nutrients (Cloern et al. 1995) or k_D (Cerco and Noel 2004b), to mechanistic approaches rooted in cellular physiology (Geider et al. 1998; Lefèvre et al. 2003).

Table 4.5 Sensitivity of water column state variables and annual system primary production (P_y) to key model parameters. Values for P_y are the average daily percent difference between the standard run and both sensitivity runs across all spatial elements. Other values are the median daily percent difference between the standard run and both sensitivity runs across all spatial elements. Results are listed only for parameters and state variables when differences were ≥ 10%

Parameter	Standard run	Sensitivity values	Chl-*a*	DIN	DIP	O_2	System P_y
C:Chl	42 g g^{-1}	30, 60	37	16	12	–	18
$m^{a}_{BZI\text{int}}$	0.76	±20%	14	12	–	–	16
R_{WC_0}	0.055 d^{-1}	±20%	11	12	–	–	–
R_{WC_k}	0.095°C^{-1}	±20%	–	12	–	–	11
t_{PHY}	10 d	±20%	10	12		–	–
$fNPP_{SED}$	0.25	0.15, 0.35	12	17	10	–	–
t^{b}_{R}	10 d	±20%	–	11		–	–
ν^{c}	Variable	0.5	12	12		–	–

[a]Intercept of the temperature – m_{BZI} formulation (Fig. 4.6)
[b]Moving average window for freshwater inputs to the hydrodynamic box model
[c]Hansen-Rattray parameter in the hydrodynamic box model

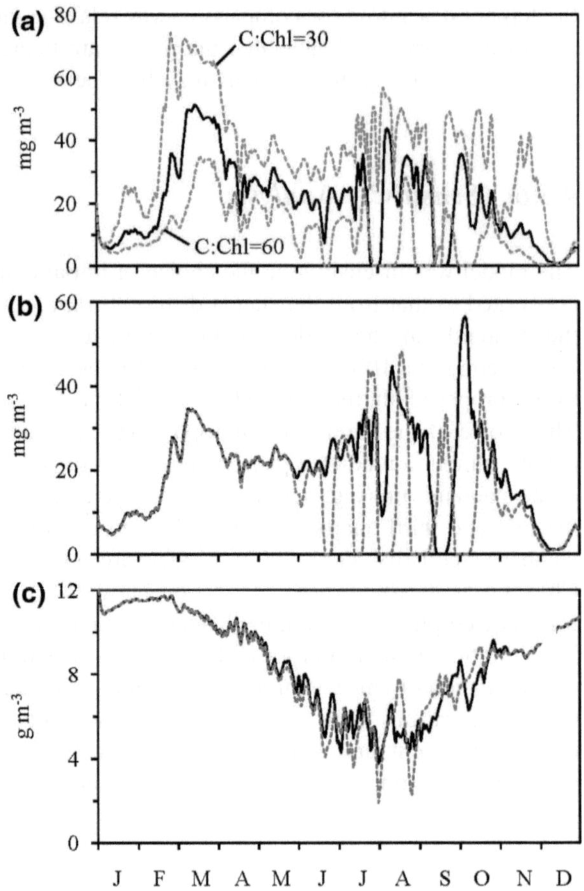

Fig. 4.11 **a** Predicted phytoplankton chlorophyll-*a* in surface waters of Greenwich Cove (Element 5) from the standard run (*solid line*) and two sensitivity runs with the C:Chl ratio set at 30 and 60 g g^{-1} (*broken lines*). **b**, **c** Predicted concentrations of **b** surface chlorophyll-*a* and **c** bottom dissolved oxygen in the mid bay (Element 6) from the standard run (*solid lines*) and a sensitivity run with increased water column respiration (*broken lines*)

The model was sensitive to the relationship between temperature and m_{BZI} (Table 4.5), which highlights the need for continued development of the *BZI* regressions for use in dynamic models, especially inclusion of the effect of temperature. The model was also sensitive to the set of parameters related to the calculation of water column respiration (R_{WC_0}, R_{WC_k}, t_{PHY}). However, increasing respiration introduced a greater and unreasonable number of oscillations in predicted phytoplankton biomass and consequently larger spikes in predicted oxygen concentrations (Fig. 4.11b, c); reducing respiration resulted in muted bloom dynamics also not supported by the data. The model was also sensitive to the fraction of daily phytoplankton carbon transferred to the sediment carbon pool ($fNPP_{SED}$), although this value was already well constrained from Nixon (1986) cross-system empirical relationship.

Another important test of the model kinetics is the degree to which predictions are driven by the formulations as opposed to forcing by boundary conditions and the choice of transport model. Freshwater inputs in the box model were smoothed with a 10-day moving average; only modeled DIN was sensitive to the length of this averaging period (t_R) (Table 4.5). Similarly, the model was only mildly sensitive to the Hansen-Rattray parameter (ν). Most importantly, when the forced concentrations of chlorophyll-*a*, DIN, DIP, and O_2 at the mouth of Greenwich Bay were replaced with constant, average values, the model continued to reproduce the dynamics of the system (Fig. 4.12), indicating that the dynamics in the model were primarily controlled by internal kinetics rather than by water exchanges and forcing at the boundary.

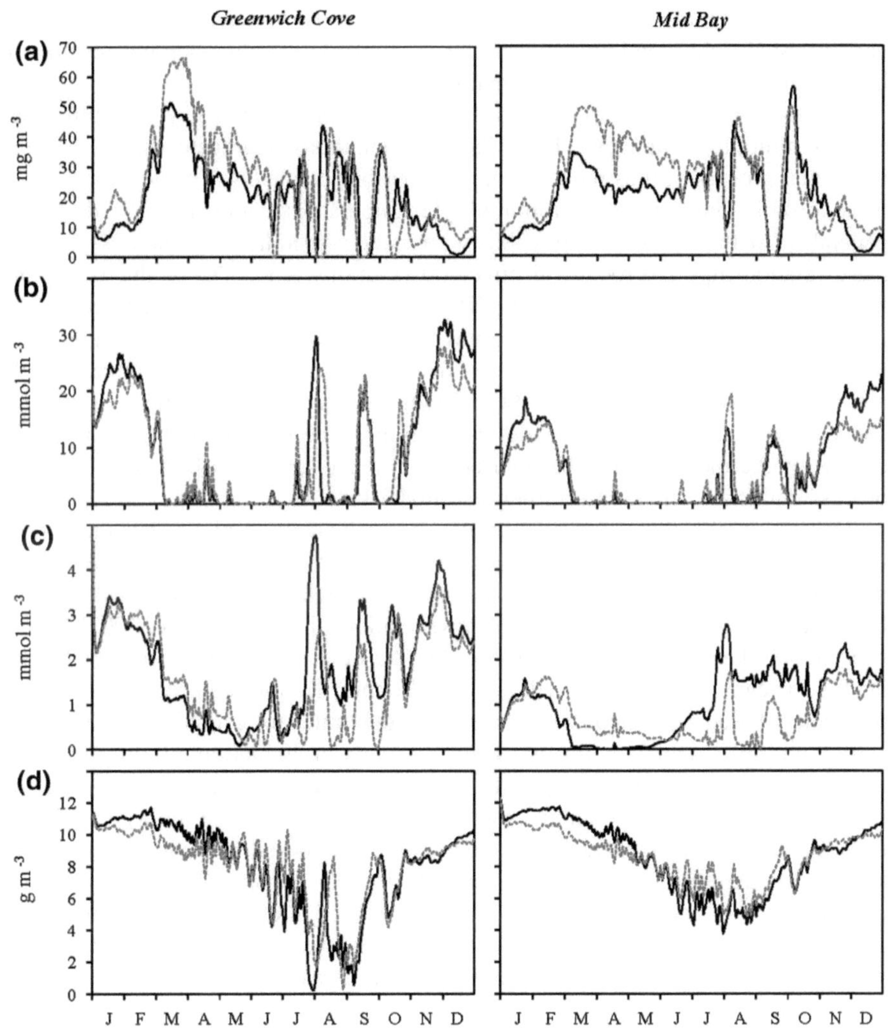

Fig. 4.12 Predicted concentrations of **a** surface chlorophyll-*a*, **b** surface DIN, **c** surface DIP, and **d** bottom dissolved oxygen from the standard run (*solid lines*) and a run in which the boundary concentrations were replaced with constant, average values (*broken lines*)

4.4 Conclusions and Future Directions

Our model combines traditional mechanistic formulations with novel empirical functions that aggregate over multiple levels of complexity to produce predictions rooted in observations across a variety of estuarine systems. The model satisfactorily reproduces observed concentrations and dynamics of phytoplankton biomass, DIN, DIP, and O_2, along with expected seasonal patterns of water column production, respiration, and accumulation of carbon in the sediments. The BZ_pI_o-based production formulation resulted in estimated annual phytoplankton production within the range of observed values. Predicted daily production and respiration demonstrated tight coupling between the phytoplankton submodel and respiration in both the water column and sediments.

The use of cross-system, empirical relationships to drive key rate processes confers multiple advantages. First, these formulations aggregate across multiple individual processes, thus reducing the number of parameters. The development of these formulations using data from multiple temperate estuaries implies a degree of generality to the model not possible with system-specific formulations. Finally, these relationships are rooted in direct observations (e.g., ^{14}C productivity) and therefore produce predictions (e.g., NPP) that can be directly compared to the observations, as opposed to growth rates for which data are rarely available. This feature helps confirm that the model is accurately simulating both concentrations and rate processes (Brush et al. 2002; Grangere et al. 2009). While we only had concentration data for use in model calibration in the present study, future applications of the model should be focused on systems for which both state and rate process data are available to further test model kinetics.

The use of the BZ_pI_0 formulation can be well justified given that these relationships are typically developed using data over one or more annual cycles and at multiple stations across a given study system. The relatively consistent slopes and high coefficients of determination for these relationships (Brush et al. 2002) further substantiate the use of this relationship. On the other hand, the use of our formulations for carbon flux to the sediments and denitrification requires application of parameters based on system-wide, annual average rates to prediction of daily rates across multiple spatial elements. This is clearly a limitation of our proposed approach. However, models typically formulate processes such as phytoplankton sinking by selecting fixed sinking rates which are known to be highly variable in time and are not well constrained in the literature (Riebesell 1989; Reckhow 1994); models for denitrification also use multiple loosely constrained parameters. We posit that use of a well-constrained annual average is defensible as an alternative approach, at least as a starting point; additional work is needed to determine whether these empirical relationships apply at seasonal and/or more spatially resolved scales.

Given the greatly reduced complexity in the present model, in terms of biology, biogeochemistry, and physics, and the use of several aggregated, empirical relationships, it is encouraging that the model nevertheless reproduces the major patterns in the observations. The limited sensitivity of the model to most parameters is also encouraging; the key sensitivities are associated with our proposed formulations and provide directions for future testing, particularly application of the model in systems where it can be constrained by available rate process data as well as concentrations. Particular focus is warranted on the relationship between temperature and slope of the BZ_pI_0 regression; Harris and Brush (2012) recently provided a temperature correction for this regression based on the Metabolic Theory of Ecology, which could be adopted. While the choice of *C:Chl* ratio in the model was justified based on the dataset of Cloern et al. (1995), the sensitivity of the model to this parameter suggests that future applications would benefit from the incorporation of a time-variable ratio based on formulations developed in the literature (e.g., Cloern et al. 1995; Geider et al. 1998). Sensitivity of the model to the parameters in our water column respiration formulation warrants future investigation and generation of data to constrain these parameters, but the choice of values is somewhat constrained by the need to generate correct dynamics without introducing unrealistic biomass fluctuations (Fig. 4.11b). While the model was sensitive to the fraction of production deposited to the sediments, this value is relatively well constrained in the literature (Nixon 1986).

Future improvements of the model can also be made to make it more generally applicable. A state variable for wastewater BOD was included given the presence of a treatment facility draining into the bay. This submodel could readily be modified to reflect the pool of non-phytoplankton associated, labile organic carbon (pooled dissolved and particulate); indeed, BOD_w in our model was immediately converted to carbon units and respired using our water column respiration formulation. To make the model applicable to a wider variety of shallow systems, it could be readily expanded to include submerged vegetation (e.g., eelgrass) and microphytobenthos; we have versions of the model with these already included.

Levins (1966) highlighted the trade-offs between precision, realism, and generality in systems models. Our goal was to develop a reduced complexity model that achieved a balance between precision (e.g., statistical/empirical models) and realism (e.g., complex mechanistic models), while at the same time being generally applicable across a range of similar coastal systems and reasonably accurate as demonstrated through calibration and skill assessment. The initial success of our proposed model highlights its potential as a reduced complexity tool useful in both heuristic and management applications. The reduced complexity approach confers the ability for rapid implementation in new study systems and fast run times on desktop computers. Additional testing is needed to demonstrate the utility of the approach, but we suggest that the model can serve as a useful tool for providing insight into ecosystem function and for informing management, both on its own and in parallel with other models.

Acknowledgements The authors wish to thank J.N. Kremer and J.W. Brawley, with whom this approach and formulations were developed while working on parallel models for Greenwich Bay and Waquoit Bay, MA. The work further benefited from frequent discussions with C.A. Oviatt, A. A. Keller, S.L. Granger, A.J. Gold, D.E. Campbell, P.V. August, L. Erikson, and M.L. Spaulding.

S.L. Granger, M. Traber, M. Richardson, and B. Buckley were responsible for much of the data collection in Greenwich Bay. Funding for this research was provided by the Rhode Island Sea Grant College Program (NOAA), Narragansett Electric, and the NOAA Center for Sponsored Coastal Ocean Research (Coastal Hypoxia Research Program, grant no. NA05NOS4781201). This is VIMS' contribution no. 3620 and NOAA Coastal Hypoxia Research Program contribution no. 213.

References

Abdelrhman MA (2005) Simplified modeling of flushing and residence times in 42 embayments in New England, USA, with special attention to Greenwich Bay, Rhode Island. Est. Coast. Shelf Sci. 62:339–351

Baird ME, Walker SJ, Wallace BB, Webster IT, Parslow JS (2003) The use of mechanistic descriptions of algal growth and zooplankton grazing in an estuarine eutrophication model. Est. Coast. Shelf Sci. 56(3–4):685–695

Beatty LL (1991) The response of benthic suspension feeders and their grazing impact on phytoplankton in eutrophied coastal ecosystems. PhD dissertation, University of Rhode Island, Kingston, RI, 351 pp

Boynton WR, Murray L, Hagy JD, Stokes C, Kemp WM, (1996) A comparative analysis of eutrophication patterns in a temperate coastal lagoon. Estuaries 19(2), 408–412

Bricker S, Longstaff B, Dennison W, Jones A, Boicourt K, Wicks C, Woerner J (2007) Effects of nutrient enrichment in the nation's estuaries: a decade of change. NOAA Coastal Ocean Program Decision Analysis Series No. 26, National Centers for Coastal Ocean Science, Silver Spring, MD, 328 pp

Brush MJ (2002) Development of a numerical model for shallow marine ecosystems with application to Greenwich Bay, RI. PhD Dissertation, University of Rhode Island, Kingston, RI, 560 pp

Brush MJ, Brawley JW (2009) Adapting the light · biomass (BZI) models of phytoplankton primary production to shallow marine ecosystems. J Marine Syst 75:227–235

Brush MJ, Brawley JW, Nixon SW, Kremer JN (2002) Modeling phytoplankton production: problems with the Eppley curve and an empirical alternative. Mar Ecol Prog Ser 238:31–45

Brush MJ, Harris LA (2010) Introduction to the special issue of *Ecological Modelling*: advances in modeling estuarine and coastal ecosystems: approaches, validation, and applications. Ecol Model 221:965–968

Brush MJ, Harris LA (2016) Ecological modeling. In: Kennish MJ (ed) Encyclopedia of Estuaries. Encyclopedia of Earth Sciences Series. Springer, The Netherlands, pp 214–223

Brush MJ, Nixon SW (2010) Modeling the role of macroalgae in a shallow sub-estuary of Narragansett Nay, RI (USA). Ecol Model 221:1065–1079

Caffrey JM (2003) Production, respiration and net ecosystem metabolism in U.S. estuaries. Environ Monit Assess 81:207–219

Caffrey JM, Cloern JE, Grenz C (1998) Changes in production and respiration during a spring phytoplankton bloom in San Francisco Bay, California, USA: implications for net ecosystem metabolism. Mar Ecol Prog Ser 172:1–12

Cerco CF, Noel MR (2004a) The 2002 Chesapeake Bay Eutrophication Model. Report 903-R-04-004, Chesapeake Bay Program Office. U.S. Environmental Protection Agency, Annapolis, MD, p 349

Cerco CF, Noel MR (2004b) Process-based primary production modeling in Chesapeake Bay. Mar Ecol Prog Ser 282:45–58

Chen C, Tian R, Beardsley RC, Qi J, Xu Q (2010) Modeling 2008 in Massachusetts Bay using an upgraded unstructured-grid Bays Eutrophication Model. Report 2010–2015, Massachusetts Water Resources Authority, Boston, MA, 118 pp

Cloern JE, Grenz C, Vidergar-Lucas L (1995) An empirical model of the phytoplankton chlorophyll:carbon ratio—the conversion factor between productivity and growth rate. Limnol Oceanogr 40(7):1313–1321
Cole BE (1989) Temporal and spatial patterns of phytoplankton production in Tomales Bay, California, USA. Est Coast Shelf Sci 28:103–115
Cole BE, Cloern JE (1987) An empirical model for estimating phytoplankton productivity in estuaries. Mar Ecol Prog Ser 36:299–305
Denman KL (2003) Modelling planktonic ecosystems: parameterizing complexity. Prog Oceanogr 57(3–4):429–452
Doering PH, Oviatt CA (1986) Application of filtration rate models to field populations of bivalves: an assessment using experimental mesocosms. Mar Ecol Prog Ser 31:265–275
Duarte CM, Amthor JS, DeAngelis DL, Joyce LA, Maranger RJ, Pace ML, Pastor J, Running SW (2003) The limits to models in ecology. In: Canham CD, Cole JJ, Lauenroth WK (eds) Models in ecosystem Science. Princeton University Press, Princeton, NJ, pp 437–451
Durbin EG, Krawiec RW, Smayda TJ (1975) Seasonal studies on the relative importance of different size fractions of phytoplankton in Narragansett Bay (USA). Mar Biol 32(3):271–287
EPA (Environmental Protection Agency) (1999) Protocol for developing nutrient TMDLs. U.S. EPA report 841-B-99-007, U.S. EPA Office of Water, Washington, DC, 135 pp
Eppley RW (1972) Temperature and phytoplankton growth in the sea. Fish Bull 70(4):1063–1085
Erikson LH (1998) Flushing times of Greenwich Bay, Rhode Island: estimates based on freshwater inputs. M.S. thesis, University of Rhode Island, Kingston, RI, 178 pp
Fourqurean JW, Webb KL, Hollibaugh JT, Smith SV (1997) Contributions of the plankton community to ecosystem respiration, Tomales Bay, California. Est Coast Shelf Sci 44(4): 493–505
Fraher J (1991) Atmospheric wet and dry deposition of fixed nitrogen to Narragansett Bay. M.S. thesis, University of Rhode Island, Kingston, RI, 165 pp
Friedrichs MAM, Hood RR, Wiggert JD (2006) Ecosystem model complexity versus physical forcing: quantification of their relative impact with assimilated Arabian Sea data. Deep-Sea Res Pt II 53:576–600
Frithsen JB, Keller AA, Pilson MEQ (1985a) Effects of inorganic nutrient additions in coastal areas: a mesocosm experiment data report, vol 1. MERL series, report no 3, University of Rhode Island, Kingston, RI, 176 pp
Frithsen JB, Keller AA, Pilson MEQ (1985b) Effects of inorganic nutrient additions in coastal areas: a mesocosm experiment data report, vol 3. MERL series, report no 5, University of Rhode Island, Kingston, RI, 244 pp
Frithsen JB, Lane PA, Keller AA, Pilson MEQ (1985c) Effects of inorganic nutrient additions in coastal areas: a mesocosm experiment data report, vol 2. MERL series, report no 4, University of Rhode Island, Kingston, RI, 330 pp
Fulton EA, Smith ADM, Johnson CR (2003) Effect of complexity on marine ecosystem models. Mar Ecol Prog Ser 253:1–16
Fulton EA, Smith ADM, Johnson CR (2004) Effects of spatial resolution on the performance and interpretation of marine ecosystem models. Ecol Model 176(1–2):27–42
Furnas MJ, Hitchcock GL, Smayda TJ (1976) Nutrient-phytoplankton relationships in Narragansett Bay during the 1974 spring bloom. In: Wiley M (ed) Estuarine Processes, vol I., Uses, stresses, and adaptation to the estuaryAcademic Press, New York, NY, pp 118–133
Ganju NK, Brush MJ, Rashleigh B, Aretxabaleta AL, del Barrio P, Forsyth M, Grear JS, Harris LA, Lake SJ, McCardell G, O'Donnell J, Ralston DK, Signell RP, Testa JM, Vaudrey JMP (2016) Progress and challenges in coupled hydrodynamic-ecological estuarine modeling. Est Coast. 39:311–332
Ganz A, Lazar N, Valliere A (1994) Quahaug management project, phase I: Greenwich Bay. Report to the Narragansett Bay Project, Rhode Island Division of Fish, Wildlife and Estuarine Resources, Coastal Fisheries Lab, Wakefield, RI, 58 pp
Geider RJ, MacIntyre HL, Kana TM (1998) A dynamic regulatory model of phytoplanktonic acclimation to light, nutrients, and temperature. Limnol Oceanogr 43(4):679–694

Giblin AE, Vallino JJ (2003) The role of models in addressing coastal eutrophication. In: Canham CD, Cole JJ, Lauenroth WK (eds) Models in ecosystem science. Princeton University Press, Princeton, NJ, pp 327–343

Goebel NL, Kremer JN (2007) Temporal and spatial variability of photosynthetic parameters and community respiration in Long Island Sound. Mar Ecol Prog Ser 329:23–42

Granger S, Brush M, Buckley B, Traber M, Richardson M, Nixon SW (2000) An assessment of eutrophication in Greenwich Bay. Paper no. 1 In: Schwartz M (ed) Restoring water quality in Greenwich Bay: a whitepaper series. Rhode Island Sea Grant College Program, Narragansett, RI, 19 pp

Grangere K, Lefebvre S, Menesguen A, Jouenne F (2009) On the interest of using field primary production data to calibrate phytoplankton rate processes in ecosystem models. Est. Coast. Shelf Sci. 81(2):169–178

Harris GP, Bigelow SW, Cole JJ, Cyr H, Janus LL, Kinzig AP, Kitchell JF, Likens GE, Reckhow KH, Scavia D, Soto D, Talbot LM, Templer PH (2003) The role of models in ecosystem management. In: Canham CD, Cole JJ, Lauenroth WK (eds) Models in ecosystem science. Princeton University Press, Princeton, NJ, pp 299–307

Harris LA, Brush MJ (2012) Bridging the gap between empirical and mechanistic models of aquatic primary production with the metabolic theory of ecology: an example from estuarine ecosystems. Ecol Model 233:83–89

Hibbert CJ (1977) Energy relations of the bivalve *Mercenaria mercenaria* on an intertidal mudflat. Mar Biol 44:77–84

Holligan PM, IeB Williams PJ, Purdie D, Harris RP (1984) Photosynthesis, respiration and nitrogen supply of plankton populations in stratified, frontal and tidally mixed shelf waters. Mar Ecol Progr Ser 17:201–213

Hopkinson CS Jr, Smith EM (2005) Estuarine respiration: an overview of benthic, pelagic, and whole system respiration. In: Giorgio PA del IeB Williams PJ (eds) Respiration in aquatic systems. Oxford University Press, Oxford, pp 122–146

Howarth RW, Marino R, Garritt R, Sherman D (1992) Ecosystem respiration and organic carbon processing in a large, tidally influenced river: the Hudson River. Biogeochemistry 16:83–102

HydroQual (1991) Water quality modeling analysis of hypoxia in Long Island Sound. Report to the Management Committee of the Long Island Sound Estuary Study and the New England Interstate Water Pollution Control Commission, HydroQual Inc., Mahwah, NJ, 280 pp

Iriarte A, Daneri G, Garcia VMT, Purdie DA, Crawford DW (1991) Plankton community respiration and its relationship to chlorophyll a concentration in marine coastal waters. Oceanol Acta 14(4):379–388

Jensen LM, Sand-Jensen K, Marcher S, Hansen M (1990) Plankton community respiration along a nutrient gradient in a shallow Danish estuary. Mar Ecol Progr Ser 61(1–2):75–85

Jiang MS, Zhou M (2008) Massachusetts Bay Eutrophication Model: 2005 simulation. Report 2008–2013 Massachusetts Water Resources Authority, Boston, MA, 85 pp

Keller AA (1986) Modeling the productivity of natural phytoplankton populations using mesocosm data along a nutrient gradient. PhD dissertation, University of Rhode Island, Kingston, RI, 240 pp

Keller AA (1988) Estimating phytoplankton productivity from light availability and biomass in the MERL mesocosms and Narragansett Bay. Mar Ecol Progr Ser 45:159–168

Kelly JR, Doering PH (1997) Monitoring and modeling primary production in coastal waters: studies in Massachusetts Bay 1992–1994. Mar Ecol Progr Ser 148:155–168

Kemp WM, Sampou PA, Garber J, Tuttle J, Boynton WR (1992) Seasonal depletion of oxygen from bottom waters of Chesapeake Bay: roles of benthic and planktonic respiration and physical exchange processes. Mar Ecol Progr Ser 85(1–2):137–152

Kremer JN, Nixon SW (1978) A coastal marine ecosystem: simulation and analysis. Springer, New York, NY 217 pp

Kremer JN, Vaudrey J, Ullman D, Bergondo D, LaSota N, Kincaid C, Codiga D, Brush MJ (2010) Simulating property exchange in estuarine ecosystem models at ecologically appropriate scales. Ecol Model 221(7):1080–1088

Lefèvre N, Taylor AH, Gilbert FJ, Geider RJ (2003) Modeling carbon to nitrogen and carbon to chlorophyll *a* ratios in the ocean at low latitudes: evaluation of the role of physiological plasticity. Limnol Oceanogr 48(5):1796–1807

Levins R (1966) The strategy of model building in population biology. Am Sci 54:421–431

Li Y, Smayda TJ (1998) Temporal variability of chlorophyll in Narragansett Bay, 1973–1990. ICES J Mar Sci 55(4):661–667

Loosanoff VL (1939) Effect of temperature upon shell movements of clams, *Venus mercenaria* (L). Biol Bull 76:171–182

Marino R, Howarth RW (1993) Atmospheric oxygen exchange in the Hudson River: dome measurements and comparison with other natural waters. Estuaries 16(3A):433–445

McGlathery KJ, Sundbäck K, Anderson IC (2007) Eutrophication in shallow coastal bays and lagoons: the role of plants in the coastal filter. Mar Ecol Progr Ser 348:1–18

Ménesguen A, Cugier P, Loyer S, Vanhoutte-Brunier A, Hoch T, Guillaud J-F, Gohin F (2007) Two- or three-layered box-models versus fine 3D models for coastal ecological modelling? A comparative study in the English Channel (Western Europe). J Marine Syst 64(1–4):47–65

Moncoiffe G, Alvarez-Salgado XA, Figueiras FG, Savidge G (2000) Seasonal and short-time-scale dynamics of microplankton community production and respiration in an inshore upwelling system. Mar Ecol Progr Ser 196:111–126

NGDC (National Geophysical Data Center) (1996) GEOphysical DAta System for hydrographic survey data (CD-ROM database). U.S. National Oceanic and Atmospheric Administration, Boulder, CO

Nixon SW (1986) Nutrient dynamics and the productivity of marine coastal waters. In: Halwagy R, Clayton D, Behbehani M (eds) Marine environment and pollution. The Alden Press, Oxford, pp 97–115

Nixon SW, Fulweiler RW, Buckley BA, Granger SL, Nowicki BL, Henry KM (2009) The impact of changing climate on phenology, productivity, and benthic-pelagic-coupling in Narragansett Bay. Est Coast Shelf Sci 82:1–18

Nixon SW, Ammerman JW, Atkinson LP, Berounsky VM, Billen G, Boicourt WC, Boynton WR, Church TM, DiToro DM, Elmgren R, Garber JH, Giblin AE, Jahnke RA, Owens NJP, Pilson MEQ, Seitzinger SP (1996) The fate of nitrogen and phosphorus at the land-sea margin of the North Atlantic Ocean. Biogeochemistry 35:141–180

Nixon S, Buckley B, Granger S, Bintz J (2001) Response of very shallow marine ecosystems to nutrient enrichment. Hum Ecol Risk Assess 7:1457–1481

Nixon SW, Granger SL, Nowicki BL (1995) An assessment of the annual mass balance of carbon, nitrogen, and phosphorus in Narragansett Bay. Biogeochemistry 31:15–61

Nixon SW, Oviatt CA (1973) Ecology of a New England salt marsh. Ecol Monogr 43:463–498

Nixon SW, Oviatt CA, Frithsen J, Sullivan B (1986) Nutrients and the productivity of estuarine and coastal marine ecosystems. J Limnol Soc South Afr 12(1/2):43–71

Nixon SW, Oviatt CA, Hale SS (1976) Nitrogen regeneration and the metabolism of coastal marine bottom communities. In: Anderson JM, Macfadyen A (eds) The role of terrestrial and aquatic organisms in decomposition processes. Blackwell Scientific, Oxford, pp 269–283

Nowicki BL (1983) Benthic community metabolism in a coastal lagoon ecosystem. M.S. thesis, University of Rhode Island, Kingston, RI, 109 pp

Nowicki BL, Oviatt CA (1990) Are estuaries traps for anthropogenic nutrients? Evidence from estuarine mesocosms. Mar Ecol Progr Ser 66:131–146

NRC (National Research Council) (2000) Clean coastal waters: understanding and reducing the effects of nutrient pollution. National Academy Press, Washington, DC, 405 pp

Obenour DR, Michalak A, Scavia D (2014) Assessing biophysical controls on Gulf of Mexico hypoxia through probabilistic modeling. Ecol Appl 25(2):492–505

Odum HT (1994) Ecological and general systems: an introduction to systems ecology, 2nd edn. University Press of Colorado, Niwot, CO 644 pp

Officer CB (1980) Box models revisited. In: Hamilton P, MacDonald KB (eds) Estuarine and wetland processes with emphasis on modeling. Plenum Press, New York, NY, pp 65–114

Officer CB, Kester DR (1991) On estimating the non-advective tidal exchanges and advective gravitational circulation exchanges in an estuary. Est Coast Shelf Sci 32:99–103

Oviatt C, Buckley B, Nixon S (1981) Annual phytoplankton metabolism in Narragansett Bay calculated from survey field measurement and micrososm observations. Estuaries 4(3): 167–175

Oviatt C, Keller A, Reed L (2002) Annual primary production in Narragansett Bay with no bay-wide winter-spring phytoplankton bloom. Est Coast Shelf Sci 54:1013–1026

Pace ML (2001) Prediction and the aquatic sciences. Can J Fish Aq Sci 58(1):63–72

Pennock JR, Sharp JH (1986) Phytoplankton production in the Delaware Estuary: temporal and spatial variability. Mar Ecol Progr Ser 34:143–155

Pilson MEQ (1985) Annual cycles of nutrients and chlorophyll in Narragansett Bay, Rhode Island. J Mar Res 43:849–873

Raick C, Soetaert K, Gregoire M (2006) Model complexity and performance: how far can we simplify? Progr Oceanogr 70(1):27–57

Reckhow KH (1994) Water quality simulation modeling and uncertainty analysis for risk assessment and decision making. Ecol Model 72(1–2):1–20

Riebesell U (1989) Comparison of sinking and sedimentation rate measurements in a diatom winter/spring bloom. Mar Ecol Prog Ser 54:109–119

Rigler FH, Peters RH (1995) Science and limnology. Book 6 In: Kinne O (ed) Excellence in ecology. International Ecology Institute, Oldendorf/Luhe, 239 pp

Rogers JM (2008) Circulation and transport in upper Narragansett Bay. M.S. thesis, University of Rhode Island, Kingston, RI, 95 pp

Rudnick D, Oviatt C (1986) Seasonal lags between organic carbon deposition and mineralization in marine sediments. J Mar Res 44(4):815–837

Sampou P, Kemp WM (1994) Factors regulating plankton community respiration in Chesapeake Bay. Mar Ecol Prog Ser 110:249–258

Scavia D, Justic D, Bierman VJ Jr (2004) Reducing hypoxia in the Gulf of Mexico: advice from three models. Estuaries 27(3):419–425

Scavia D, Kelly ELA, Hagy JD (2006) A simple model for forecasting the effects of nitrogen loads on Chesapeake Bay hypoxia. Est Coasts 29:674–684

Scavia D, Evans MA, Obenour DR (2013) A scenario and forecast model for Gulf of Mexico Hypoxia area and volume. Environ Sci Tech 47:10423–10428

Smith EM, Kemp WM (1995) Seasonal and regional variations in plankton community production and respiration for Chesapeake Bay. Mar Ecol Prog Ser 116(1–3):217–231

Stickney AP, Stringer LD (1957) A study of the invertebrate bottom fauna of Greenwich Bay, Rhode Island. Ecology 38(1):111–122

Stow CA, Roessler C, Borsuk ME, Bowen JD, Reckhow KH (2003) Comparison of estuarine water quality models for total maximum daily load development in Neuse River Estuary. J Water Resour Plann Manage 129(4):307–314

Swaney DP, Scavia D, Howarth RW, Marino RM (2008) Estuarine classification and response to nitrogen loading: insights from simple ecological models. Est. Coast. Shelf Sci. 77:253–263

Swanson JC, Jayko K (1988) A simplified estuarine box model of Narragansett Bay. Final report to the Narragansett Bay Project and U.S. Environmental Protection Agency, Applied Science Associates, Narragansett, RI 80 pp

Testa JM, Kemp WM (2008) Regional, seasonal, and inter-annual variability of biogeochemical processes and physical transport in a partially stratified estuary: a box-modeling analysis. Mar Ecol Prog Ser 356:63–79

Turner RE (1978) Community plankton respiration in a salt marsh estuary and the importance of macrophytic leachates. Limnol Oceanogr 23(3):442–451

Valiela I, Foreman K, LaMontagne M, Hersh D, Costa J, Peckol P, DeMeo-Andreson B, D'Avanzo C, Babione M, Sham C-H, Brawley J, Lajtha K (1992) Couplings of watersheds and

coastal waters: sources and consequences of nutrient enrichment in Waquoit Bay, Massachusetts. Estuaries 15(4):443–457
Van Nes EH, Scheffer M (2005) A strategy to improve the contribution of complex simulation models to ecological theory. Ecol Model 185:153–164
Vargo GA (1979) The contribution of ammonia excreted by zooplankton to phytoplankton production in Narragansett Bay. J Plankton Res 1(1):75–84
Vollenweider RA (1974) A manual on methods for measuring primary production in aquatic environments. Blackwell Scientific Publications, Oxford 225 pp

Chapter 5
Modeling Physical and Biogeochemical Controls on Dissolved Oxygen in Chesapeake Bay: Lessons Learned from Simple and Complex Approaches

Jeremy M. Testa, Yun Li, Younjoo J. Lee, Ming Li, Damian C. Brady, Dominic M. Di Toro and W. Michael Kemp

Abstract We compared multiple modeling approaches in Chesapeake Bay to understand the processes controlling dissolved oxygen (O_2) cycling and compare the advantages and disadvantages of the different models. Three numerical models were compared, including: (1) a 23-compartment biogeochemical model coupled to a regional scale, salt- and water-balance box model, (2) a simplified, four-term model formulation of O_2 uptake and consumption coupled to a 3D-hydrodynamic model, and (3) a 23-compartment biogeochemical model coupled to a 3D-hydrodynamic model. All three models reproduced reasonable spatial and temporal patterns of dissolved O_2, leading us to conclude that the model scale and approach one chooses to apply depends on the scientific questions motivating the study. From this analysis, we conclude the following: (1) Models of varying spatial and temporal scales and process resolution have a role in the scientific process.

J.M. Testa (✉)
Chesapeake Biological Laboratory, University of Maryland Center for Environmental Science, P.O. Box 38, Solomons, MD 20688, USA
e-mail: jtesta@umces.edu

Y. Li
College of Marine Science, University of South Florida, 140 7th Ave S, Saint Petersburg, FL 33701, USA
e-mail: yunli@usf.edu

Y.J. Lee
Department of Oceanography, Naval Postgraduate School, 1 University Circle, Monterey, CA 93943, USA
e-mail: ylee1@nps.edu

M. Li · W.M. Kemp
Horn Point Laboratory, University of Maryland Center for Environmental Science, 2020 Horns Point Rd, Cambridge, MD 21613, USA
e-mail: mingli@umces.edu

W.M. Kemp
e-mail: kemp@umces.edu

D. Justic et al. (eds.), *Modeling Coastal Hypoxia*,
DOI 10.1007/978-3-319-54571-4_5

(2) There is still much room for improvement in our ability to simulate dissolved O_2 dynamics in coastal ecosystems. (3) An ever-increasing diversity of models, three of which are presented here, will vastly improve our ability to discern physical versus biogeochemical controls on O_2 and hypoxia in coastal ecosystems.

Keywords Physical modeling • Biogeochemical modeling • Dissolved oxygen • Hypoxia • Coastal ecosystems • Chesapeake Bay

5.1 Introduction

Depleted dissolved oxygen (O_2) conditions have been a feature of Chesapeake Bay for at least the past nine decades (Newcombe and Horne 1938) and reflect both the Bay's high-productivity and the physical isolation of bottom waters from the atmosphere. Although Chesapeake Bay may be naturally susceptible to hypoxia development, analyses of long-term data indicate that summer hypoxic and anoxic water volumes have increased over the past several decades (Hagy et al. 2004) in response to some combination of elevated nutrient loading and large-scale climatic changes (Scully 2010a; Murphy et al. 2011). Because hypoxic conditions have many negative consequences for living resources and restrict habitat availability (Brady and Targett 2013; Buchheister et al. 2013), there is a strong emphasis in the management of this system to alleviate low-O_2 conditions in the Bay through nutrient input reductions within the watershed (Boesch et al. 2001).

Our understanding of O_2 dynamics in coastal ecosystems like Chesapeake Bay is complicated by the fact that O_2 is controlled by a diverse suite of physical and biogeochemical processes. Some of these processes covary or are linked (e.g., freshwater and nutrient inputs), and each has its own dominant time and space scales. Elevated freshwater input, for example, delivers inorganic nutrients that fuel phytoplankton production in seaward Bay regions, but also delivers inorganic particles that limit light availability in landward regions and may cause advection of phytoplankton biomass downstream, away from landward regions (e.g., Miller and Harding 2007). Freshwater input is also associated with elevated stratification during summer that reduces vertical diffusion of O_2 (Boicourt 1992), yet elevated freshwater flow also leads to higher landward advection in bottom water (Li et al. 2015), which

D.C. Brady
School of Marine Sciences, University of Maine, 193 Clark Cove Road,
Walpole, ME 04573, USA
e-mail: damian.brady@maine.edu

D.M. Di Toro
Department of Civil and Environmental Engineering, University of Delaware,
356 DuPont Hall, Newark, DE 19716, USA
e-mail: dditoro@udel.edu

may deliver O_2-rich water but also may transport relatively labile organic material (Kemp et al. 1997). Wind stress may lead to three-dimensional (e.g., along- and cross-channel) circulation features that physically replenish deep-water O_2 (Scully 2010b). Prevailing northward winds (up-estuary) in summer, for example, can reduce the along-channel exchange flow and the associated landward O_2 flux in the deep water, but strengthen the cross-channel circulation that exchanges O_2-poor water with overlying O_2-rich water (Scully 2010b; Li et al. 2015), and vice versa. The wind could also mix nutrient-rich water into surface layers and fuel elevated phytoplankton production (Malone et al. 1986; Li et al. 2009) that will drive additional O_2 depletion in the following months. Observing the biogeochemical response to these dynamics directly would involve near-continuous measurements of phytoplankton production and respiration, concentrations of O_2, organic carbon and key nutrients, current velocities, stratification, and other meteorological and hydrological variables (e.g., wind speed) over multiple seasons. Considering the impractical and high-cost nature of such efforts, numerical models can be used as tools to understand the response of O_2 to external forcing in a way that cannot reasonably be accomplished using observational studies alone.

Accordingly, much effort has been invested in building models to simulate O_2 dynamics on several time and space scales. Models encompassing a wide range of hydrodynamic resolution, biogeochemical complexity, and temporal scope have been applied in numerous ecosystems (Oguz et al. 2000; Justíc et al. 2007; Fennel et al. 2013; Hamidi et al. 2015), including Chesapeake Bay (Xu and Hood 2006; Liu and Scavia 2010; Scully 2010b; Brown et al. 2013; Cerco and Noel 2013; Lee et al. 2013; Feng et al. 2015) and the Great Lakes (Rucinski et al. 2010; Hamidi et al. 2015). Statistical models are often used to infer the major drivers of O_2 depletion and to guide management actions (e.g., nutrient load reductions) that alleviate low-O_2 zones, while coupled biophysical models (e.g., the three models presented in this chapter) are used to understand ecosystem interactions and feedbacks, where O_2 is one of many biogeochemically linked model variables. While multiple coupled, hydrodynamic–biogeochemical models currently exist for Chesapeake Bay (Xu and Hood 2006; Li et al. 2009; Cerco and Noel 2013), few studies have compared multiple models to emphasize their utility in answering different scientific and management questions (Irby et al. 2016).

In this chapter, we describe three modeling packages that include both simplified and complex biogeochemical and hydrodynamic configurations. Our objectives are to compare and contrast these models to illustrate the different types of simulations that investigators can use to answer key scientific questions. We also aim to assess quantitatively how well the applied models reproduce observations of dissolved O_2 in Chesapeake Bay, discuss their respective limitations, and suggest where and how the models might be utilized in future studies based on their spatio-temporal dimensions and levels of biogeochemical and hydrodynamic complexity.

5.2 Methods and Approach

High-resolution, coupled hydrodynamic–biogeochemical ocean models are valued for their ability to accurately represent physical and biogeochemical process in coastal ecosystems, especially with increasingly affordable computing power. Less spatially resolved models (regional scale) are still useful for research needs, as they can be executed quickly (minutes), represent biogeochemistry at the scales where it has been predominantly measured (regionally and seasonally), and can be used to compute simple and unambiguous budgets of key variables. In this chapter, we describe and analyze three different numerical models, including: (1) a course-scale mass-balance transport model ("box model") coupled to a water column and sediment model with detailed simulations of biogeochemical processes (Row-Column Aesop, or RCA); (2) a 3D, hydrodynamic model using the Regional Ocean Modeling System (ROMS) coupled to a simple, four-term O_2 biogeochemistry model; and (3) the ROMS coupled to a multi-compartment, water column and sediment biogeochemical process model (RCA).

5.2.1 *Box Model with Biogeochemistry (BM-RCA)*

A coarse spatial-scale model was developed for Chesapeake Bay, using a water column and sediment biogeochemical model and physical transport calculated using a salt- and water-balance model (box model, or BM). The spatial domain of this model includes 17 control volumes of water (9 surface-layer "boxes", 8 bottom-layer "boxes"; Figs. 5.1 and 5.2), whose vertical separation is based on mean-pycnocline depths for each region (Hagy 2002). The box model solves a series of linear algebraic equations to compute advective and non-advective exchanges between control volumes based upon freshwater inputs and salt distributions, and given the small number of regions (17), the computation requires minimal computing resources. The biogeochemical model (Row-Column Aesop, or RCA) was developed by HydroQual, Inc. and has been applied in many coastal ecosystems (e.g., Long Island Sound, Massachusetts Bay). RCA is the most recent extension of the family of water quality models that originated from the Water Quality Analysis Simulation Program (WASP) used by the United States Environmental Protection Agency (Di Toro et al. 1983). The model allows for up to three phytoplankton groups, as well as water column state variables representing the following; (1) particulate and dissolved organic carbon, nitrogen and phosphorus, (2) dissolved inorganic nitrogen, phosphorus, and silica, (3) biogenic particulate silica, and (4) O_2 (Fig. 5.3). RCA also includes a state variable that represents O_2 equivalents associated with sulfide and methane released at the sediment-water interface. Nitrification and denitrification are modeled in both the water column and sediments, where RCA includes a sediment biogeochemical model, which has two layers that represent the near-surface aerobic and underlying anaerobic

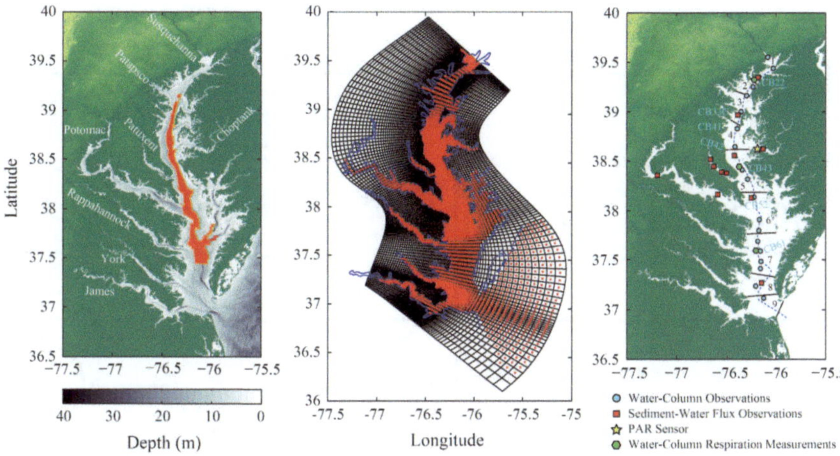

Fig. 5.1 Maps of Chesapeake Bay, including (*left panel*) bathymetry with *red* area representing the O_2 budget region (see text), (*middle panel*) the ROMS model grid with wet cells in *red*, and (*right panel*) locations of water column monitoring stations (*blue circles*), sediment-water flux observation stations (*red squares*), the location of PAR data used for simulations (*yellow star*), and water column respiration measurements (*green hexagons*). For water column monitoring stations, select stations are labeled with abbreviated station names relative to the official monitoring station names (e.g., CB22 = CB2.2 and CB33–CB43 = CB3.3C–CB4.3C). *Middle* and *right panels* reprinted from Testa et al. (2014) with permission from Elsevier

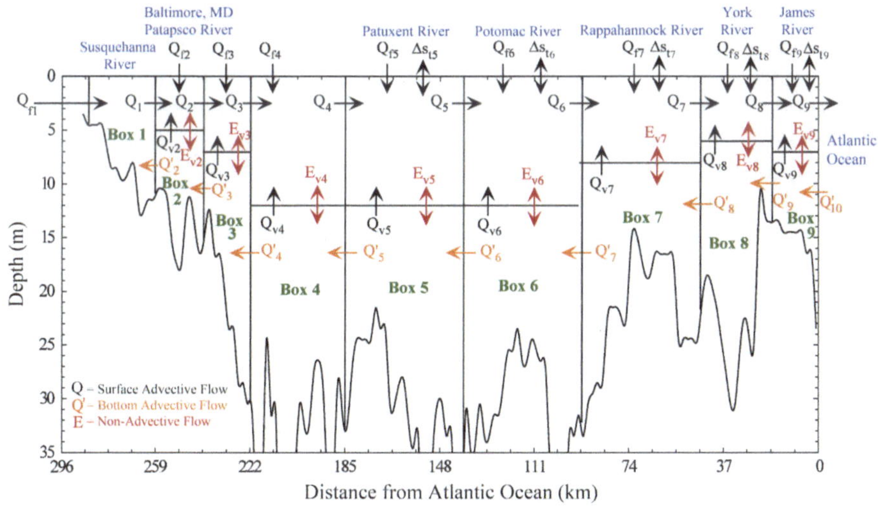

Fig. 5.2 Diagram of salt- and water-balance (box) model for Chesapeake Bay, including relevant freshwater inputs, transport coefficients (Q, Q', E), box identifiers, and key tributary estuaries. In Box 5–9, Δs represents salt exchanges between the boxes and adjacent, connected tributaries. Diagram adapted from Hagy (2002). An aerial view of this model is included in Fig. 5.1 (*right panel*)

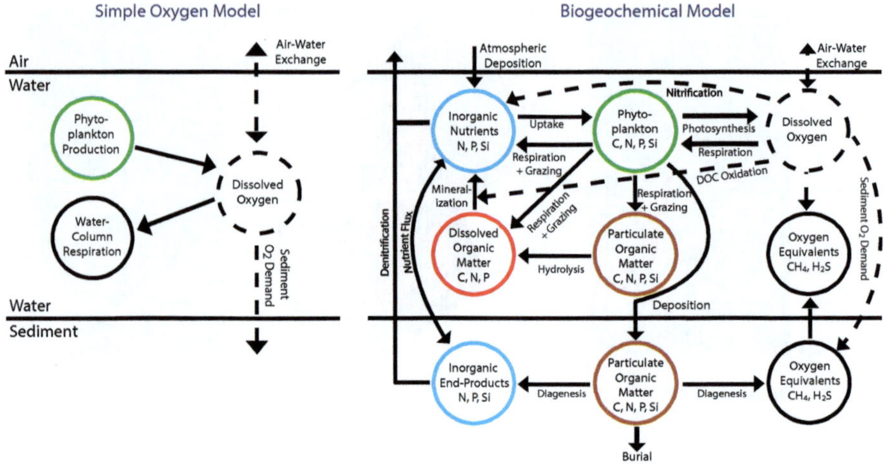

Fig. 5.3 (*left panel*) Simple O_2 model, based on empirical relationships, used to investigate physical controls on hypoxia. (*right panel*) Schematic diagram of the major state variables and transformation processes in RCA, reprinted from Testa et al. (2014) with permission from Elsevier

environments and simulates the cycling of carbon, O_2, nitrogen, phosphorus, silica, and sulfur. The sediment model predicts sediment-water fluxes of dissolved O_2, nitrate, ammonium, phosphate, dissolved methane, and sulfide, where the latter two constituents contribute to the water column state variable called "oxygen equivalents" that acts as a reservoir for non-nitrogen reduced solutes that contribute to O_2 demand. Sulfide is produced via sulfate reduction in the sediment model, where it is subsequently stored temporarily in particulate form as iron monosulfide (Cornwell and Sampou 1995), oxidized in the sediment (contributing to sediment O_2 demand), or released to the overlying water. A more extensive description of the RCA modeling package is found elsewhere (Testa et al. 2013, 2014; Xue et al. 2014). Model simulations were run on a 6-h time step over the years 1986–2006.

The key advantages of the BM-RCA approach are that it can be executed rapidly on a personal computer (20 years in ~5 min for the 17 regions) and that it captures regional patterns of O_2, carbon, and nutrient dynamics, which can be validated using available monitoring data and rate-process measurements. Its short run times make it amenable to sensitivity analysis and scenario experiments and allow it to be implemented by a wide range of users (e.g., resource managers) who may not have access to advanced computational resources. Because of its well-defined boundaries and relatively few transport terms, the model output is easily post-processed to provide regional and seasonal mass-balance budgets of key variables. The disadvantages of this modeling approach include its inability to capture lateral and vertical variability in physical and biogeochemical processes, as well as the patchy spatial patterns in plankton productivity and biomass. In particular, short temporal- and spatial-scale physical dynamics in response to wind and tidal forcing cannot be adequately represented in this model, and these processes are known to be important in Chesapeake Bay and other coastal ecosystems (Li and Zhong 2009; Scully 2010a).

5.2.2 Hydrodynamic 3D Model with Simple Oxygen (ROMS-SDO)

The second model in this analysis (ROMS-sDO) is a simple, empirical O_2 model implemented in a relatively high-resolution hydrodynamic model (the Regional Ocean Modeling System; ROMS). This model was motivated by the need to capture the key seasonal patterns in biogeochemical O_2 uptake so that experiments related to the effects of altered hydrodynamic variability due to wind stress, tidal mixing, and freshwater inflow can be achieved. This approach is designed to take into account the strong seasonal variation of both uptake and production in the water column (Fig. 5.3). Based on observed measurements of sediment O_2 demand (*SOD*), water column respiration (*WCR*), and phytoplankton community production (*PhP*), the approach empirically relates these processes to the state variables (e.g., O_2 concentration, ambient water temperature *T*), and/or environment forcing (e.g., photosynthetically available radiation *PAR*), allowing for O_2 consumption/production to vary with space (horizontal and vertical) and time to the first order (see details in Li et al. 2015).

$$
\begin{aligned}
SOD &= 9.90 \times 1.7845^{T/10^{o}\mathrm{C}} \times \left(\frac{O_2}{Ox + 59\,\mathrm{mmol\,O_2\ m^{-3}}} \right) \\
PhP &= 31.25 \times (1.0101 + 0.0314 PAR + 0.1966 T) \\
WCR &= 3.3 \times e^{0.0715T}
\end{aligned}
$$

ROMS has been implemented in Chesapeake Bay and validated against a wide range of observational data and has demonstrated considerable capability in reproducing estuarine dynamics at tidal, synoptic, and seasonal time-scales (Li et al. 2005). We use an application of this model with a 160 × 240 grid in the horizontal direction (about 500 m grid size) for ROMS-sDO and a grid 80 × 120 grid in the horizontal direction (about ~1 km grid size) for ROMS-RCA (Sect. 6.2.3). Both ROMS-sDO and ROMS-RCA include 20 layers in the vertical dimension (Fig. 5.1). The open ocean boundary consists of 10 constituents (M2, S2, N2, K2, K1, O1, P1, Q1, Mf, and Mm) for tidal forcing, de-tided observations for non-tidal water elevations, and monthly climatology for salinity and temperature. In order to provide the hindcasts of hydrodynamic fields, the model is forced by daily river discharge along with zero salinity and seasonal water temperature, and by winds, net shortwave and downward longwave radiation along with air temperature, relative humidity, and pressure. Surface water temperature was nudged to the observed SST field. Further details of the application of ROMS in Chesapeake Bay are described elsewhere (e.g., Li et al. 2005).

The ROMS-sDO approach is advantageous because it uses reasonable, but computationally meager representations of biogeochemical O_2 production and uptake to allow high temporal- and spatial-scale simulations of O_2 dynamics in coastal ecosystems. The disadvantages of this approach include the fact that the empirical O_2 uptake formulations generally cannot capture spatial and interannual variability related to nutrient loading and availability (i.e., eutrophication) and do not capture biogeochemical feedbacks, which limit the model's ability to be run for realistic multi-year experiments. Thus, we display the viability of this approach in reproducing seasonal O_2 dynamics, focusing ROMS-sDO simulations to a single year (1989) when many of the observations used to build the empirical models were made.

5.2.3 Hydrodynamic 3D Model with Biogeochemistry (ROMS-RCA)

The most complex model formulation we present in this chapter is a "soft-coupling" of ROMS to the multi-compartment, dynamic biogeochemical model RCA (as described above). The term "soft-coupling" represents the fact that ROMS hydrodynamic model simulations were performed first, where the output was saved and subsequently used to provide hydrodynamic fields (e.g., current velocity) to drive RCA simulations. This model includes high spatial and temporal resolution simulations of a wide range of hydrodynamic and biogeochemical processes. Hourly averages of ROMS temperature, salinity, and transport terms are used to force RCA along with external loads of nutrients and organic carbon based on daily freshwater inputs and monthly fortnightly nutrient concentrations (Fig. 5.1). RCA is simulated on a 10-min time step, and we utilized a multitude of available monitoring data to characterize boundary and initial conditions, as well as external flows and loadings (Testa et al. 2014; Li et al. 2016).

The key advantages of this model are that it simulates high-resolution spatial and temporal dynamics of biogeochemical and physical processes, such that even with changing nutrient loading and climatic conditions, the model can be run to reproduce reasonable estuarine dynamics over many years. This allows for experiments with altered nutrient loading, freshwater input, and wind stress to be conducted that result in improved understanding of myriad estuarine processes over multiple time and space scales. The disadvantages of this approach include high computational costs and the simulation of physical and biogeochemical processes at scales (e.g., daily, ~1 km) that are difficult to validate given limited spatial and temporal scales of observational data.

5.2.4 *Calibration and Validation Datasets*

All models require validation to ensure their ability to accurately reproduce the processes and state variables that are intended to be understood and simulated. Although many models are validated against state variables, including concentrations of nutrients, O_2, carbon, and phytoplankton biomass (chlorophyll-*a*), it is equally important to validate these models against rate processes, such as plankton community photosynthesis and respiration rates, organic matter decay rates, and nutrient transformation and sediment-water exchange rates. Validation of both "state" and "rate" observations allow one to determine that the model is predicting concentrations based on appropriate transformation rates, as opposed to situations where the model accurately simulates "state" concentrations generated by compensating but inaccurate rate processes (e.g., offsetting errors). For Chesapeake Bay, we were able to calibrate and validate model behavior for diverse variables and process for which a wide range of data exist across time and space scales. Validations of the water column state variables (e.g., chlorophyll-*a* or chl-*a*, dissolved and particulate nutrients, dissolved and particulate organic carbon, etc.) were performed using fortnightly monthly observations of these variables at several depths and stations within Chesapeake Bay (http://www.chesapeakebay.net/data). In addition, rates of water column respiration were compared to measurements of O_2-uptake in dark bottles at several stations (Sampou and Kemp 1994; Smith and Kemp 1995), while rates of photic-layer net primary production were validated with observations based on O_2 incubations and empirical model computations based on ^{14}C uptake measurements (Smith and Kemp 1995; Harding et al. 2002). Observed sediment-water fluxes of dissolved O_2 were estimated from time-course changes in solute concentrations during incubations of intact acrylic sediment cores collected at key stations in Chesapeake Bay (Fig. 5.1; Cowan and Boynton 1996).

5.3 Insights Gained from Model Simulations

In this section, we review the performance of each of the three models and highlight selected results from the simulations. We also emphasize the types of questions that can and cannot be answered with a given modeling approach and how these approaches lead to improved understanding of the dynamics of O_2 and hypoxia in Chesapeake Bay.

5.3.1 *Comparison of Model Performance*

All three models reproduced seasonal cycles of bottom-water dissolved O_2 along the central axis of Chesapeake Bay with reasonable accuracy (Fig. 5.4).

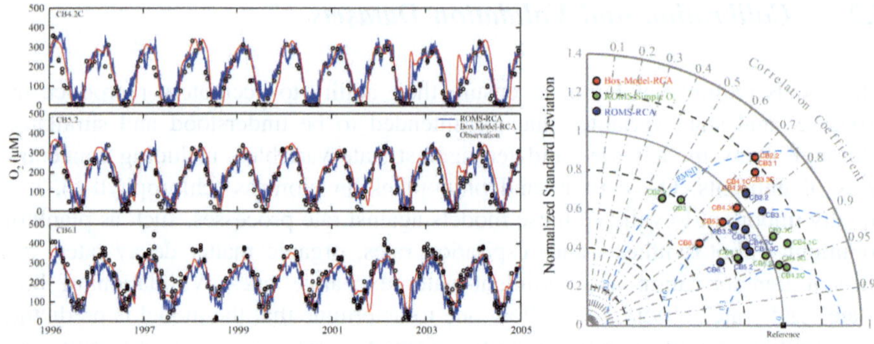

Fig. 5.4 (*left panel*) Time series of bottom-water O_2 concentrations at three stations in Chesapeake Bay, including observed concentrations (*open circles*) and those predicted by the BM-RCA (*red line*) and ROMS-RCA model (*blue lines*). (*right panel*) Taylor diagrams for bottom-water O_2 concentrations from the three models at several stations in Chesapeake Bay

Taylor diagrams (Fig. 5.4) graphically illustrate several metrics of model-data agreement (Taylor 2001), and each model was compared in such a diagram for the time period during which simulations were made, including 1996–2005 for ROMS-RCA, 1985–2006 for BM-RCA, and 1989 for ROMS-sDO. O_2 simulations from each model were highly correlated to observed values, where r-values exceeded 0.7 in all cases except the upper-Bay regions in ROMS-sDO (Fig. 5.4). These two stations (CB2.2 and CB3.1) for ROMS-sDO also tended to have the highest root-mean-squared difference values (RMSD; Fig. 5.4). Although this model-data mismatch was only based upon a single year, it reveals the potential inability for an empirical O_2 model to capture variability in respiration in the upper Bay, where observed O_2 variability is high and driven by interannual variations in the accumulation of phytoplankton biomass in bottom-waters (Testa and Kemp 2014). Somewhat over-predicted O_2 at these upper-Bay stations may also explain why ROMS-sDO tends to under-predict hypoxic volume in the early summer (see Sect. 6.3.3), as hypoxia tends to initiate in the upper Bay. In general, the ROMS-based models with high spatial resolution tended to capture variability in bottom-water O_2 better than the regionally based BM-RCA, yet BM-RCA was able to capture seasonal patterns of bottom-water O_2, as well as interannual variability associated with changes in freshwater and nutrient inputs (Fig. 5.4). That said, ROMS-RCA predicted short-term variations (daily weekly) in bottom-water O_2 that BM-RCA did not. Although the high-frequency observations necessary to validate these model simulations do not currently exist along the main channel of Chesapeake Bay, high-frequency variability in O_2 is expected to occur given tidal mixing and the passage of storm fronts.

As mentioned previously, we consider it important to validate model process rates in addition to concentration measurements. We thus compared rates of water column respiration (as O_2 uptake) and SOD measured at several stations to those rates predicted by ROMS-RCA (Fig. 5.5). Water column respiration rates, which were available from multiple studies and several stations in the main stem of Chesapeake Bay (Sampou and Kemp 1994; Smith and Kemp 1995), compared favorably with ROMS-RCA-simulated respiration over several seasons (Fig. 5.5a, c and e). Although modeled respiration rates were slightly higher than observations during March to May at three stations (10.00–14.69 (model) versus 7.39–11.17 (observed) mmol m^{-3} d^{-1}), summer (June–August) and November rates were comparable. Where measurements from multiple years were available for comparison (spring and summer at CB6.1), model estimates fell within the range of observations. At station CB4.3, anoxia during mid-summer prevented the measurement of respiration with oxygen-based techniques, but respiration was predicted by the model because anoxia was not always predicted in this region by ROMS-RCA (Figs. 5.4 and 5.5). ROMS-RCA also captured seasonal variability in sediment oxygen demand (hereafter SOD) at three stations, but tended to over-predict SOD in the middle Bay (Fig. 5.5). As with water column respiration, ROMS-RCA did not predict the true anoxic conditions that were observed at this station and SOD was allowed to persist because O_2 was available for uptake from the overlying water (Fig. 5.5). Despite this, overlying O_2 was sufficiently low and sediment pore water sulfide concentrations were high-enough for the sediment model to generate fluxes of sulfide to the water column, which would subsequently consume water column O_2. The sulfide fluxes correspond to seasonal minima in both bottom-water O_2 concentrations and the aerobic layer depth within the sediment model, which limits the storage of sulfide (Cornwell and Sampou 1995). ROMS-sDO-simulated respiration and SOD rates are plotted for comparison, to illustrate how they were similar across stations with identical seasonal variation, which contrasts with ROMS-RCA predictions that varied spatially and temporally (Fig. 5.5).

5.3.2 Insights Gained from BM-RCA

Despite the simplified physical transport model used to drive BM-RCA, it reproduces seasonal, regional, and interannual variability in bottom-water dissolved O_2 (Fig. 5.4). Given the coarse resolution in this model, these validations are restricted to the deepest stations along the main channel of central Chesapeake Bay (shallow stations flanking the channel are excluded). Thus, although BM-RCA is a useful tool to understand the effects of interannual variations of river flow and nutrient load on deep-water O_2 at regional scales, it cannot resolve episodic or spatially resolved dynamics (e.g., vertical profiles) in the Bay. However, the straightforward division of the Bay into a limited number of segments in BM-RCA allows for the computation of regional budgets of O_2, carbon, and related nutrients.

For example, several questions remain related to the timing and location of the source of organic matter fueling hypoxia in Chesapeake Bay (Kemp et al. 1997). Budget calculations for particulate organic carbon (POC) in each box during two seasons (*Spring* = March–April, *Summer* = June–July) averaged over the 1986–2006 period,

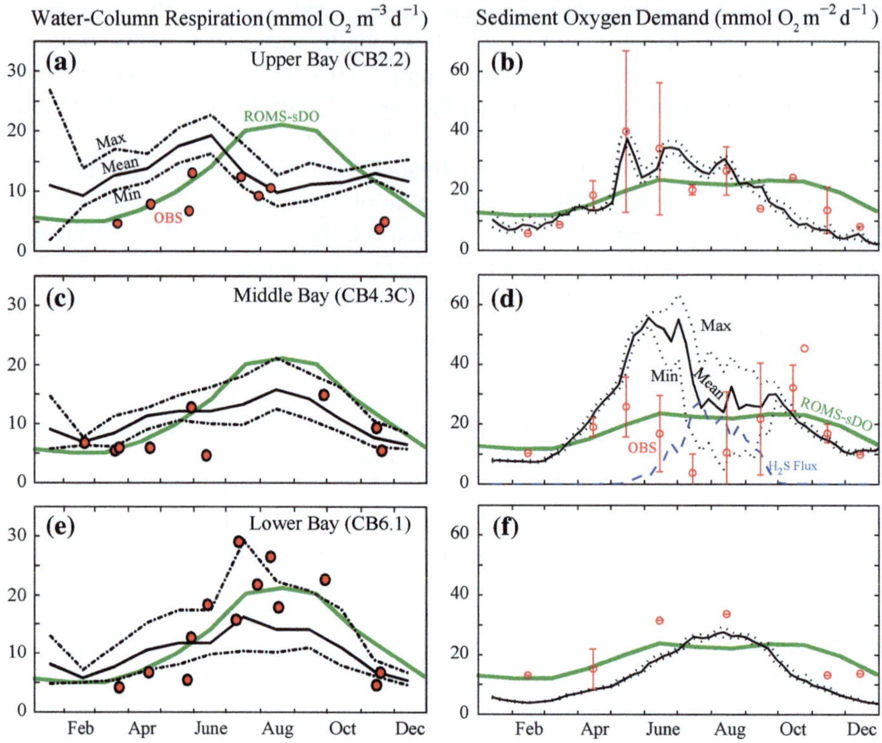

Fig. 5.5 (*left panel*; a, c, e) Comparisons of water column respiration rates modeled by ROMS-RCA (1996–2005 mean is solid *black line*, minima and maxima are *dotted lines*) and ROMS-sDO (*green lines*) with observed values indicated by *red circles*. (*right panel*; b, d, f) Comparisons of sediment oxygen demand (SOD) modeled by ROMS-RCA (1996–2005 mean is solid *black line*, minima and maxima are *dotted lines*) sediment-water sulfide (H_2S) flux modeled by ROMS-RCA (*blue lines*) and SOD modeled by ROMS-sDO (*green lines*) with observations indicated by *red circles* (±1 SD). Both *left* and *right panels* include comparisons at an *upper* Bay (a, b; CB2.2), middle Bay (c, d; CB4.3C), and *lower*-Bay (e, f; CB6.1) station

suggest the key role of *both* vertical sinking and landward longitudinal transport as mechanisms for POC delivery to bottom waters (Fig. 5.6). Landward net POC imports (the potential fuel for O_2 depletion) were greatest in lower-Bay regions, but muted in upper-Bay regions (Fig. 5.6). This is consistent with the suggestion that landward, bottom-water transport of organic carbon resulting from net surface-layer carbon production in seaward Bay regions is a key aspect of the Bay carbon budget supporting O_2 depletion (e.g., Kemp et al. 1997). Thus, although BM-RCA cannot capture small-scale variability in O_2 (which is often important) it reveals nutrient loading controls on deep-water O_2 and seasonal and regional transport of organic carbon.

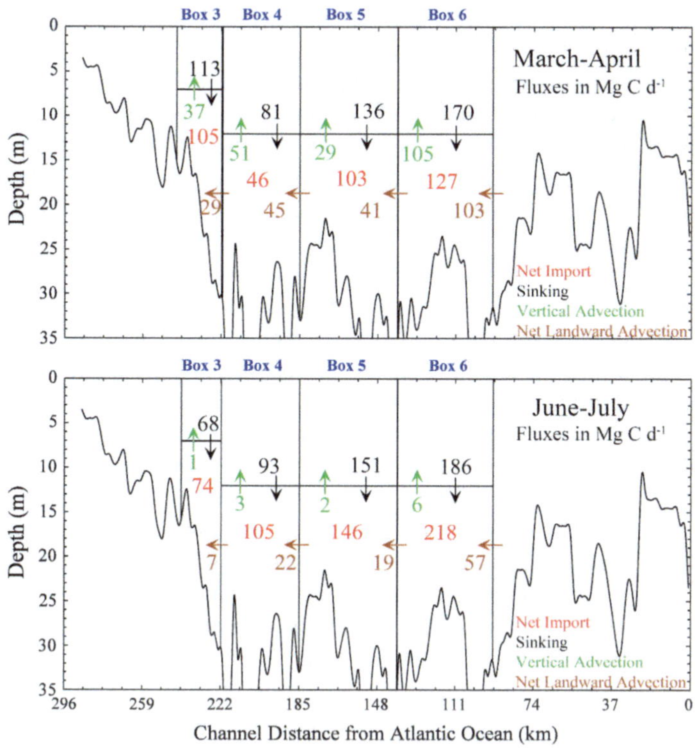

Fig. 5.6 Mean particulate organic carbon (POC) budgets for *middle*-Bay regions as computed from BM-RCA model computations in Chesapeake Bay for the March to April (*top*) and June–July (*bottom*) periods. Net Import is equivalent to the balance between the transport terms

5.3.3 Insights Gained from ROMS-SDO

Recently, state-of-the-art hydrodynamic models have been coupled to relatively simple formulations for biogeochemical O_2-uptake (Hetland and DiMarco 2008; Scully 2010a; Li et al. 2015). The motivation for these efforts has been to quantify biogeochemical effects on O_2 in a simple and computationally meager way to allow an emphasis on variations in physical controls. Such an approach for Chesapeake Bay has proven useful to understand the effects of wind speed and direction on O_2 dynamics, but these simulations also provide an opportunity to understand the balance between biogeochemical O_2 uptake and physical replenishment.

ROMS-sDO was run for Chesapeake Bay in the year 1989, where observations of primary production, water column respiration, and sediment O_2 demand were available to develop empirical formulations for these processes (Li et al. 2015). ROMS-sDO captured seasonal patterns of deep-water O_2 across many stations in Chesapeake Bay, and thus reasonably reproduced the annual cycle of hypoxic volume (Figs. 5.4 and 5.7). To understand the balance between O_2 uptake and physical replenishment, a budget was computed for O_2 by integrating model O_2-equation over a selected control volume for Chesapeake Bay, which encompasses all deep waters below 10-m depth (from the mean sea level) in the main stem to the north of York River mouth, and the northern boundary intercepts the shoaling bathymetry in the upper Bay (see Fig. 5.1 for location). The budget revealed five types of changes for deep-water O_2, including the water column respiration over the entire control volume, the sediment respiration across the seafloor, the along-channel advection of oceanic high-O_2 water across the lower-Bay section, and the vertical advection and diffusion of O_2 across the 10-m interface (Li et al. 2015). Over the April to July period, biogeochemical O_2 uptake (water column + sediment respiration) exceeded inputs via horizontal and vertical advection and diffusion, resulting in the drawdown of bottom-water O_2 and the development of hypoxia (Fig. 5.7). Over the course of spring and early summer (as in all months) this biogeochemical O_2 uptake was dominated by water column respiration (Fig. 5.7), which is consistent with cross-system analyses that suggest sediment O_2 uptake is a small fraction of total water column uptake in systems deeper than 5–8 m (Kemp et al. 1992). Interestingly, during the spring period when O_2 is drawn down, advection of O_2 is a large fraction of total input to this region due to stronger circulation resulting from buoyancy-induced along-estuary density gradient and favorable prevailing wind directions (Li et al. 2015). Although vertical diffusion is the dominant term for the physical components of the O_2 budget during mid-summer, advection is once again important during later-summer and fall, when physical imports exceed biogeochemical uptake, leading to replenishment of bottom-water O_2 (Fig. 5.4) and the decline in hypoxic volume (Fig. 5.7).

Simulations using ROMS-sDO thus make an important contribution to our understanding of O_2 dynamics in Chesapeake Bay. It became clear that even relatively simple models are useful in quantifying the seasonal and regional balance between O_2 uptake and replenishment, and discerning which processes (advection versus diffusion, water column versus sediment respiration) contribute most to variability at a given time of year. Although such models can therefore be used to investigate interannual variations in physical input and be subject to experiments that quantify the effects of freshwater input and altered wind patterns, they cannot be directly used to understand interannual variations in O_2 resulting from altered nutrient loading and other biological considerations.

5.3.4 Insights Gained from ROMS-RCA: Interannual Variation

The most complex and well-resolved model presented in this chapter is ROMS-RCA, which includes a dynamic water column and sediment biogeochemical model coupled to a relatively high-resolution hydrodynamic ocean model. This model permits investigations into short temporal- and spatial-scale dynamics as in ROMS-sDO and

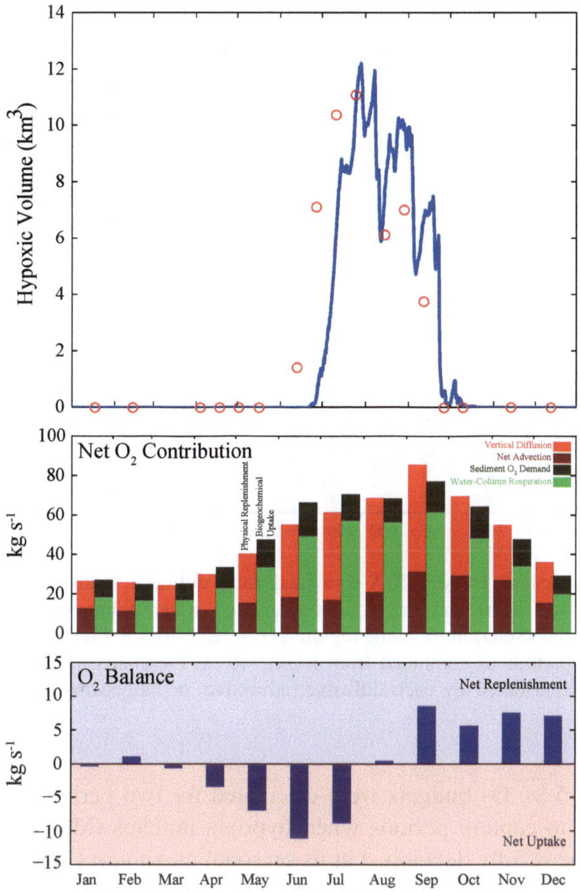

Fig. 5.7 Time series of ROMS-sDO-simulated (*top panel*) hypoxic volume (O_2 < 62.5 μM) in the main stem of Chesapeake Bay, (*middle panel*) monthly averages of the *vertical* diffusive and sum of the advective fluxes and the volume-integrated water column respiration (WCR) and sediment O_2 demand (SOD), and (*bottom panel*) monthly net O_2 flux (physical transport + biogeochemical uptake) in the control volume, (see Fig. 5.1, *left panel*) where positive value represent net replenishment and negative values represent net uptake

the analysis of individual terms in the O_2 budget, but it adds an additional value in that specific biogeochemical mechanisms can be examined, as well as interannual variability of dissolved O_2 associated with altered nutrient loading. These model simulations spanned a 10-year period in Chesapeake Bay using realistic climatic and freshwater forcing, and were also executed for a single year (2000) to examine responses specific to altered nitrogen and phosphorus loading scenarios.

A dissolved O_2 budget analysis similar to that presented for ROMS-sDO was performed on the 10-year simulation of ROMS-RCA, with the addition of separate terms for horizontal advection and diffusion, as well as vertical advection and

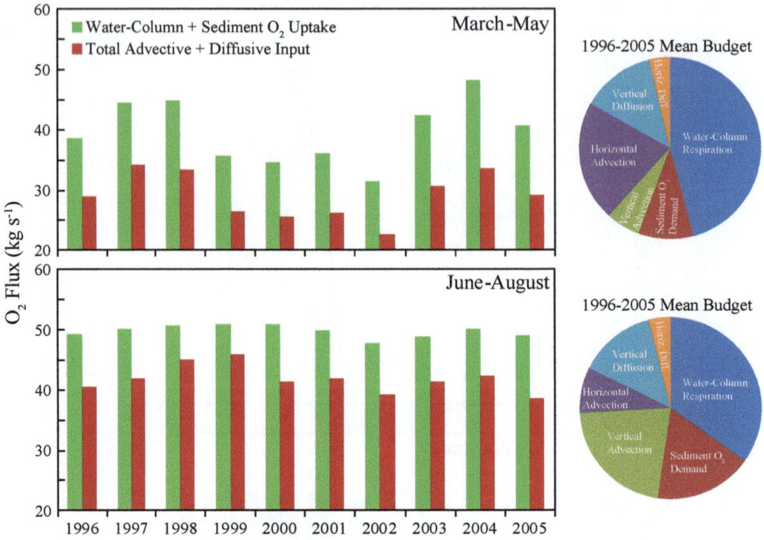

Fig. 5.8 Time series of ROMS-RCA-simulated total diffusive + advective O_2 inputs to the hypoxic region (*red bars*) and the volume-integrated water column respiration (WCR) and sediment O_2 demand (SOD) in the same region (*green bars*) for the March–May (*top*) and June to August (*bottom*) periods as computed from ROMS-RCA. Pie charts indicate the fraction of the total O_2 budget contributed by each diffusive, advective, or biogeochemical term for the given period

diffusion (Fig. 5.8). O_2 budgets were calculated for two periods (March–May and June–August), to capture periods when hypoxia initiates (March–May) and when hypoxic volume is fully developed at its seasonal maximum (June–August). In both periods, biogeochemical uptake exceeds physical replenishment, but interannual variability is much higher for the spring period (March–May) than during the summer period (June–August; Fig. 5.8). Perhaps, more importantly, physical replenishment of O_2 tends to be proportional to spring O_2 uptake, where interannual variation in spring O_2 uptake covaries with physical O_2 inputs, both of which are correlated to winter–spring nutrient loading. In addition, physical replenishment of O_2 is highest during summer, which is the period when biogeochemical O_2 uptake is at seasonal maxima (Fig. 5.8). This reveals that vertical and horizontal gradients in O_2 that are setup by biogeochemical uptake influence the physical replenishment fluxes, regardless of season. Pie charts representing the relative contribution to the total O_2 budget of the various physical and biogeochemical terms reinforce the results of ROMS-sDO, where water column respiration is the dominant uptake term and advective inputs of O_2 are comparable in magnitude to diffusive inputs in multiple seasons (Fig. 5.8).

5.3.5 *Insights Gained from ROMS-RCA: Response to Nutrient Loading*

The second set of analyses presented in this section includes a sensitivity test of Chesapeake Bay O_2 dynamics to altered loadings of nitrogen (N) and phosphorus (P). Whereas the ROMS-sDO simulations predict the same biogeochemical O_2 uptake in a given year and place to understand interannual variability in physical effects, these experiments use the same physical forcing (from the year 2000 in ROMS-RCA) to isolate variations in nutrient loading effects. These experiments reveal that hypoxic volume days (*HVD*) were consistently higher under elevated nutrient loads, but the response was stronger for N relative to P. This resulted from widespread N limitation in seaward Bay regions during summer months (Malone et al. 1996) that led to higher net primary production (NPP) and phytoplankton biomass under elevated N loads or much lower NPP and biomass under reduced loads (Fig. 5.9). Relatively lower sensitivity to P loads results from the fact that P is limiting in winter–spring in the upper and middle Bay and that phytoplankton growth during this season appears to be a less important driver of summer hypoxic volumes (e.g., Newell et al. 2007). The fact that *HVD* was more sensitive to combined NP load changes reveals the potential for alternating nutrient limitation if the load is dominated by either N or P, especially during transitional periods in Chesapeake Bay where P is limiting in spring and N is limiting in summer (Malone et al. 1996). These results highlight previously underemphasized seasonal dynamics associated with hypoxia-nutrient load relationships, as well as the interacting role of N and P loads in controlling hypoxic volume, which have been highlighted in other large coastal ecosystems (Conley et al. 2009; Greene et al. 2009; Laurent and Fennel 2014).

HVD responded nonlinearly to January to May total nitrogen (TN) loads varying over 2 orders of magnitude (e.g., Murphy et al. 2011). These simulations suggest that *HVD* would saturate (600 km^3-d) at loads approaching twice that of conditions in 2000. From a management perspective, this indicates that current nutrient reduction goals should be expected to result in observable reductions in hypoxic volume. Simple mechanistic models that simulate Chesapeake Bay hypoxia also predict similar, nonlinear relationships between TN load and hypoxic volume (Liu and Scavia 2010), although volumes at the low end of the loading range have not been observed. If *HVD* is plotted against N load for each year from the 1996–2005 simulation, the *HVD* tends to fall below the logistic curve (Fig. 5.9) for years with below-average Susquehanna Flow (lower *HVD*/load), while *HVD*/load is higher in above-average flow years. This suggests that *HVD* is sensitive to physical circulation or additional nutrient inputs under high-flow conditions.

Perhaps, the clearest conclusion of the nutrient load simulations for Chesapeake Bay is the importance of summer NPP and respiration in driving the Bay's response to N loading. This is consistent with recent historical data analyses, which have suggested that declines in late summer (July–August) hypoxic volume are associated with modest declines in January to May Susquehanna River N loads (Murphy et al. 2011).

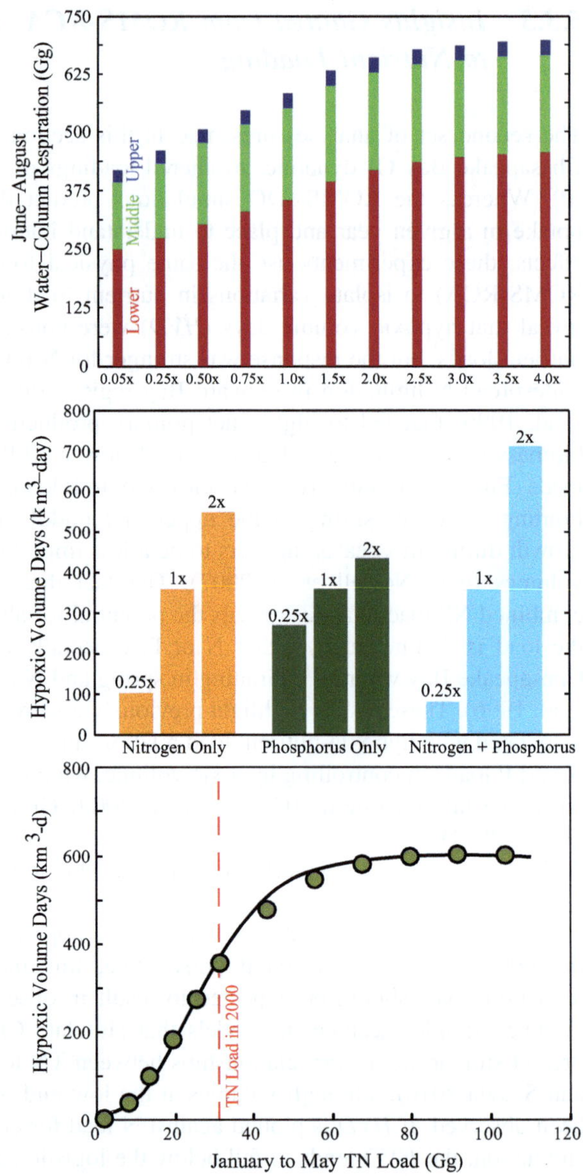

Fig. 5.9 (*top panel*) Total ROMS-RCA-simulated water column respiration in three regions (R3-*upper* Bay, R5-*middle* Bay, R7-*lower* Bay) of Chesapeake Bay (see Fig. 5.1 for regional boundaries) computed for each of 10 N load experiments during summer (June–August). (*middle panel*) Comparison of modeled hypoxic volume days (HVD) for several different nutrient loading scenarios (0.25x, 1x, 2x) for nitrogen only (*orange* bars at *left*), phosphorus only (*green* bars in *middle*), and nitrogen + phosphorus (*cyan* bars at *right*). (*bottom panel*) Relationship between January to May total nitrogen loading and HVD from model simulations (*green circles*), where the *vertical dashed line* indicates the TN load observed for the year 2000. *Middle* and *bottom panels* reprinted from Testa et al. (2014) with permission from Elsevier

Statistical analyses do not, however, provide the specific mechanisms connecting reduced winter–spring N loads to July–August hypoxic volumes. As this and other studies (Malone et al. 1996) have shown, N limitation is the primary control on phytoplankton growth during summer throughout much of Chesapeake Bay. Model simulations clearly display that NPP, phytoplankton biomass, and respiration during the summer period are more sensitive to N additions than during spring (Fig. 5.9), but

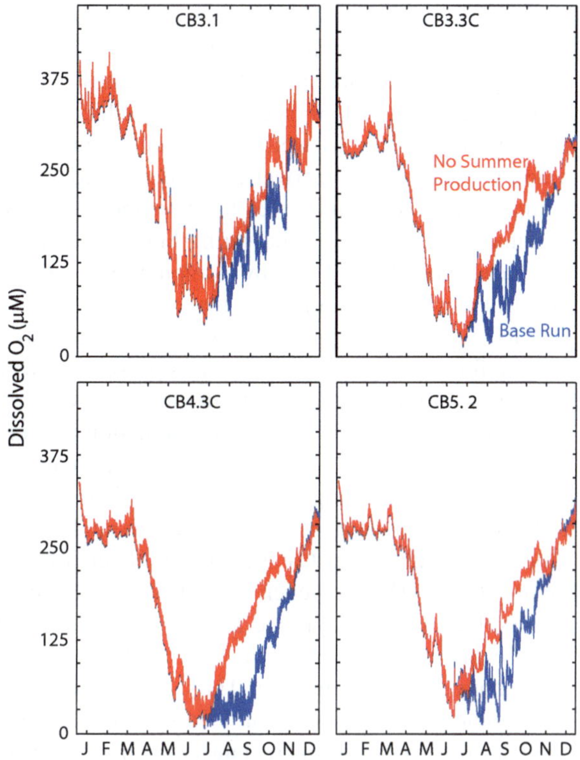

Fig. 5.10 Seasonal cycle of ROMS-RCA-simulated *bottom*-layer dissolved O_2 at four stations in Chesapeake Bay under two conditions: (1) the "Base Run," or simulation under normal conditions (*blue lines*), and (2) a simulation where summer phytoplankton growth is prevented ("No Summer Production," *red lines*)

spring (March to May) NPP and water-column respiration rates were also enhanced by elevated nutrient loads (as in Fig. 5.8). In a related ROMS-RCA model simulation, where summer phytoplankton was not allowed to grow (Fig. 5.10), bottom-water O_2 was replenished to non-hypoxic levels beginning in mid-June. The implication of this result is that summer phytoplankton growth and subsequent respiration are necessary to maintain hypoxia throughout summer. Additionally, N-loading enhancement of lower-Bay water column respiration was a primary driver of interannual variations in hypoxic volume in the 10-year simulation in Chesapeake Bay. These simulations clearly identify N load as a major driver of mid- to late summer hypoxic volume in Chesapeake Bay, and they provide mechanisms to link N load to hypoxia.

5.4 Summary and Synthesis

5.4.1 Lessons Learned from Different Models

Simulation studies for each of the three numerical models presented in this chapter provide a unique contribution to our current understanding of O_2 cycling and hypoxia in Chesapeake Bay. The modeling tool chosen to answer a particular research question should be based on a need to balance model complexity with sufficient resolution of the spatio-temporal scales and process description needed to investigate a particular suite of research questions.

BM-RCA was able to accurately reproduce seasonal and regional O_2 dynamics in Chesapeake Bay, as well as to quantify interannual variability in chlorophyll-*a* and O_2 concentrations over a two-decade-long period. Regional budgets of particulate organic carbon derived from the model revealed the importance of landward longitudinal advection in delivering labile carbon to the seasonally hypoxic region of Chesapeake Bay. The low computational cost of this model and its generic physical transport calculations make it highly portable to other coastal systems. BM-RCA, however, cannot resolve lateral patterns in O_2 uptake and transport or capture fine-scale patterns of O_2, thus preventing accurate computations of hypoxic area and volume. For example, to investigate recently recognized channel-shoal interactions associated with lateral circulation as a key replenishment process of O_2 (Scully 2010a) and associated phytoplankton responses (Malone et al. 1986), a finer-scale 3D modeling approach is required (e.g., ROMS-sDO or ROMS-RCA). ROMS-sDO can also be used to understand the nature of other hydrodynamic processes on dissolved O_2 such as the effects of tidal/wind mixing and freshwater input. ROMS-sDO, however, cannot be used to simulate the interannual variability of dissolved O_2 and hypoxic and anoxic volume in the Bay because it does not capture interannual changes in water column and sediment O_2 uptake associated with changes in nutrient load. Thus, to understand interannual variations, we need to examine variation in nutrient loading from year to year, along with the associated spatially-resolved patterns of nutrient and organic matter transport and cycling. In this case, ROMS-RCA is chosen over simplified models (BM-RCA or ROMS-sDO) because of its sufficient resolution and complexity in *both* physics and biogeochemistry.

5.4.2 Considerations for the Future

Despite the recent advances in modeling dissolved O_2 dynamics and hypoxia in Chesapeake Bay and other coastal waters worldwide, there is vast room for additional analysis and model improvement. This chapter was designed to help the coastal system modeling community by illustrating what can be learned about the effects of seasonal and interannual variability in physical forcing and nutrient loading on hypoxia in a particular system (Chesapeake Bay). The lessons learned can likely be applied elsewhere. Despite what was learned from 10 years of ROMS-RCA model simulations for the Bay, an extension of the simulations beyond a decade may be necessary to fully test the model's ability to reproduce interannual variability in hypoxia and capture trends. Biogeochemical model simulations, despite their limitations (e.g., unconstrained parameters, missing processes), are only as good as the hydrodynamic simulations used to drive them. Several applications of ROMS in Chesapeake Bay have successfully reproduced current velocities and longitudinal salinity and temperature structure (e.g., Li et al. 2005; Irby et al. 2016), but have been unable to generate accurate gradients of vertical salinity, often under-predicting these gradients in some times and places (Irby et al. 2016). To improve simulations of dissolved O_2 concentrations (vertical gradients of which tend to follow those of salinity), hydrodynamic simulations must improve this aspect of estuarine dynamics. New efforts to compare the accuracy and utility of multiple coupled hydrodynamic–biogeochemical ocean models may be a way forward in enhancing existing modeling tools.

5.4.3 Summary

In this chapter, we summarize and compare three modeling systems for simulating dissolved O_2 and hypoxia dynamics in Chesapeake Bay. We conclude that each of these modeling approaches has its advantages and disadvantages, and the choice of which to apply depends on the scientific questions that are to be addressed. For example, if one seeks a tool to do sensitivity tests or examine regional patterns in biogeochemistry, a model like BM-RCA may be sufficient. On the other hand, if the driving questions are related to climatic effects on O_2 dynamics, a model similar to ROMS-sDO may be adequate. However, if one is interested in examining inter-annual variability in biogeochemical processes, biophysical interactions and feed-backs, or small-scale processes, a model like ROMS-RCA is necessary. Although the potential levels of complexity and resolution accessible with modern coastal biophysical models continue to increase, relatively simple models still have a role for addressing broader and more general research questions to understand coastal systems dynamics.

Acknowledgements We are grateful for the constructive comments from two anonymous reviewers and to Jim Hagy for sharing his box model code that we adapted for this analysis. Support from several grants and contracts have made this chapter possible, including the US National Science Foundation grants (i) DEB1353766 (OPUS; Kemp and Boynton) and (ii) CBET1360415 (WSC; Testa and Kemp), US National Oceanic and Atmospheric Administration (NOAA) grants (iii) NAO7NOS4780191, (Coastal Hypoxia Research Program; Kemp, M. Li, Di Toro) and (iv) NA15NOS4780184 (Testa, M. Li, Kemp), and (v) National Aeronautics and Space Administration grant NNX14AM37G (Kemp). This paper is contribution #5200 of the University of Maryland Center for Environmental Science and CHRP Publication number 211.

References

Boesch DF, Brinsfield RB, Magnien RE (2001) Chesapeake Bay eutrophication: scientific understanding, ecosystem restoration, and challenges for agriculture. J Environ Qual 30:303–320

Boicourt WC (1992) Influences of circulation processes on dissolved oxygen in the Chesapeake Bay. In: Smith DE, Leffler M, Mackiernan G (eds) Oxygen dynamics in the chesapeake bay, a synthesis of recent research. Maryland Sea Grant, College Park, Maryland

Brady DC, Targett TE (2013) Movement of juvenile weakfish *Cynoscion regalis* and spot *Leiostomus xanthurus* in relations to diel-cycling hypoxia in an estuarine tidal tributary. Mar Ecol Prog Ser 491:199–219

Brown CW, Hood RR, Long W, Jacobs J, Ramers DL, Wazniak C, Wiggert JD, Wood R, Xu J (2013) Ecological forecasting in Chesapeake Bay: using a mechanistic–empirical modeling approach. J Mar Syst 125:113–125

Buchheister A, Bonzek CF, Gartland J, Latour RJ (2013) Patterns and drivers of the demersal fish community of Chesapeake Bay. Mar Ecol Prog Ser 481:161–180

Cerco CF, Noel MR (2013) Twenty-one-year simulation of Chesapeake Bay water quality using the CE-QUAL-ICM eutrophication model. J Am Water Resour Assoc. doi:10.1111/jawr.12107

Conley DJ, Paerl HW, Howarth RW, Boesch DF, Seitzinger SP, Havens KE, Lancelot C, Likens GE (2009) Controlling eutrophication: nitrogen and phosphorus. Science 323: 1014–1015

Cornwell JC, Sampou PA (1995) Environmental controls on iron sulfide mineral formation in a coastal plain estuary. In: Vairamurthy MA, Schoonen MAA (eds) Geochemical transformations of sedimentary sulfur. American Chemical Society, Washington D.C

Cowan JL, Boynton WR (1996) Sediment-water oxygen and nutrient exchanges along the longitudinal axis of Chesapeake Bay: seasonal patterns, controlling factors and ecological significance. Estuaries 19:562–580

Di Toro DM, Fitzpatrick JJ, Thomann RV (1983) Documentation for water quality analysis simulation program (WASP) and model verification program (MVP). In: United States Environmental Protection Agency, Washington D.C

Feng Y, Friedrichs MAM, Wilkin J, Tian H, Yang Q, Hofmann EE, Wiggert JD, Hood RR (2015) Chesapeake Bay nitrogen fluxes derived from a land-estuarine ocean biogeochemical modeling system: Model description, evaluation, and nitrogen budgets. J Geophys Res: Biogeosciences 120:1666–1695

Fennel K, Hu J, Laurent A, Marta-Almeida M, Hetland R (2013) Sensitivity of hypoxia predictions for the Northern Gulf of Mexico to sediment oxygen consumption and model nesting. J Geophys Res: Oceans 1–14

Greene RM, Lehrter JC, Hagy JD (2009) Multiple regression models for hindcasting and forecasting midsummer hypoxia in the Gulf of Mexico. Ecol Appl 19:1161–1175

Hagy JD (2002) Eutrophication, hypoxia, and trophic transfer efficiency in Chesapeake Bay. Ph. D., University of Maryland, College Park, College Park, Maryland

Hagy JD, Boynton WR, Keefe CW, Wood KV (2004) Hypoxia in Chesapeake Bay, 1950–2001: long-term change in relation to nutrient loading and river flow. Estuaries 27:634–658

Hamidi SA, Bravo HR, Klump JV, Waples JT (2015) The role of circulation and heat fluxes in the formation of stratification leading to hypoxia in Green Bay, Lake Michigan. J Great Lakes Res 41:1024–1036

Harding LW, Mallonee ME, Perry E (2002) Toward a predictive understanding of primary productivity in a temperate, partially stratified estuary. Estuar Coast Shelf Sci 55:437–463

Hetland RD, DiMarco SF (2008) How does the character of oxygen demand control the structure of hypoxia on the Texas-Louisiana continental shelf? J Mar Syst 70:49–62

Irby ID, Friedrichs MAM, Friedrichs CT, Bever AJ, Hood RR, Lanerolle LWJ, Li M, Linker L, Scully ME, Sellner K, Shen J, Testa J, Wang H, Wang P, Xia M (2016) Challenges associated with modeling low-oxygen waters in Chesapeake Bay: a multiple model comparison. Biogeosciences 13:2011–2028

Justíc D, Bierman VJ, Scavia D, Hetland RD (2007) Forecasting Gulf's hypoxia: the next 50 years? Estuaries Coasts 30:791–801

Kemp WM, Sampou PA, Garber J, Tuttle J, Boynton WR (1992) Seasonal depletion of oxygen from bottom waters of Chesapeake Bay: roles of benthic and planktonic respiration and physical exchange processes. Mar Ecol Prog Ser 85:137–152

Kemp WM, Smith EM, Marvin-DiPasquale M, Boynton WR (1997) Organic carbon balance and net ecosystem metabolism in Chesapeake Bay. Mar Ecol Prog Ser 150:229–248

Laurent A, Fennel K (2014) Simulated reduction of hypoxia in the Northern Gulf of Mexico due to phosphorus limitation. Elementa. doi:10.12952/journal.elementa.000022

Lee YJ, Boynton WR, Li M, Li Y (2013) Role of late winter-spring wind influencing summer hypoxia in Chesapeake Bay. Estuaries Coasts 36:683–696

Li M, Lee YJ, Testa JM, Li Y, Ni W, Kemp WM, Toro DMD (2016) What drives interannual variability of estuarine hypoxia: climate forcing versus nutrient loading? Geophys Res Lett 43:2127–2134

Li M, Zhong L (2009) Flood-ebb and spring-neap variations of mixing, stratification and circulation in Chesapeake Bay. Cont Shelf Res 29:4–14

Li M, Zhong L, Boicourt WC (2005) Simulations of Chesapeake Bay estuary: sensitivity to turbulence mixing parameterizations and comparison with observations. J Geophys Res 110: C12004

Li M, Zhong L, Harding LW (2009) Sensitivity of plankton biomass and productivity to variations in physical forcing and biological parameters in Chesapeake Bay. J Mar Res 67:667–700

Li Y, Li M, Kemp WM (2015) A budget analysis of bottom-water dissolved oxygen in Chesapeake Bay. Estuaries Coasts. doi:10.1007/s12237-12014-19928-12239

Liu Y, Scavia D (2010) Analysis of the Chesapeake Bay hypoxia regime shift: insights from two simple mechanistic models. Estuaries Coasts 33:629–639

Malone TC, Conley DJ, Fisher TR, Glibert PM, Harding LW, Sellner KG (1996) Scales of nutrient-limited phytoplankton productivity in Chesapeake Bay. Estuaries 19:371–385

Malone TC, Kemp WM, Ducklow HW, Boynton WR, Tuttle JH, Jonas RB (1986) Lateral variation in the production and fate of phytoplankton in a partially stratified estuary. Mar Ecol Prog Ser 32:149–160

Miller WD, Harding LW (2007) Climate forcing of the spring bloom in Chesapeake Bay. Mar Ecol Prog Ser 331:11–22

Murphy RR, Kemp WM, Ball WP (2011) Long-term trends in Chesapeake Bay seasonal hypoxia, stratification, and nutrient loading. Estuaries Coasts 34:1293–1309

Newcombe CL, Horne WA (1938) Oxygen-poor waters of the Chesapeake Bay. Science 88:80–81

Newell RIE, Kemp WM, Hagy JDI, Cerco CF, Testa JM, Boynton WR (2007) Top-down control of phytoplankton by oysters in Chesapeake Bay, USA: comment on Pomeroy et al. (2006). Mar Ecol Prog Ser 341:293–298

Oguz T, Ducklow HW, Malanotte-Rizzoli P (2000) Modeling distinct vertical biogeochemical structure of the Black Sea: dynamical coupling of the oxic, suboxic, and anoxic layers. Global Biogeochem Cycles 14:1331–1352

Rucinski DK, Beletsky D, DePinto JV, Schwab DJ, Scavia D (2010) A simple 1-dimensional, climate based dissolved oxygen model for the central basin of Lake Erie. J Great Lakes Res 36:465–476

Sampou P, Kemp WM (1994) Factors regulating plankton community respiration in Chesapeake Bay. Mar Ecol Prog Ser 110:249–258

Scully ME (2010a) The importance of climate variability to wind-driven modulation of hypoxia in Chesapeake Bay. J Phys Oceanogr 40:1435–1440

Scully ME (2010b) Wind modulation of dissolved oxygen in Chesapeake Bay. Estuaries Coasts 33:1164–1175

Smith EM, Kemp WM (1995) Seasonal and regional variations in plankton community production and respiration for Chesapeake Bay. Mar Ecol Prog Ser 116:217–231

Taylor KE (2001) Summarizing multiple aspects of model performance in a single diagram. J Geophys Res 106:7183–7192

Testa JM, Brady DC, Di Toro DM, Boynton WR, Cornwell JC, Kemp WM (2013) Sediment flux modeling: nitrogen, phosphorus and silica cycles. Estuar Coast Shelf Sci 131:245–263

Testa JM, Kemp WM (2014) Spatial and temporal patterns in winter-spring oxygen depletion in Chesapeake Bay bottom waters. Estuaries Coasts 37:1432–1448

Testa JM, Li Y, Lee YJ, Li M, Brady DC, Toro DMD, Kemp WM (2014) Quantifying the effects of nutrient loading on dissolved O_2 cycling and hypoxia in Chesapeake Bay using a coupled hydrodynamic-biogeochemical model. J Mar Syst 139:139–158

Xu J, Hood RR (2006) Modeling biogeochemical cycles in Chesapeake Bay with a coupled physical-biological model. Estuar Coast Shelf Sci 69:19–46

Xue P, Chen C, Qi J, Beardsley RC, Tian R, Zhao L, Lin H (2014) Mechanism studies of seasonal variability of dissolved oxygen in Mass Bay: a multi-scale FVCOM/UG-RCA application. J Mar Syst. doi:10.1016/j.jmarsys.2013.1012.1002

Chapter 6
Modeling Hypoxia and Its Ecological Consequences in Chesapeake Bay

Jerry D. Wiggert, Raleigh R. Hood and Christopher W. Brown

Abstract The Chesapeake Bay is a valuable recreational, ecological and economic resource that is subject to environmental hazards, such as harmful algal bloom (HAB) and hypoxia, which can degrade the Bay's health and jeopardize the viability of this important natural resource. As a step toward developing the capability to forecast such hazards, a biogeochemical version of the Chesapeake Bay Regional Ocean Modeling System (ChesROMS) has been developed. The model framework encompasses the physical, biogeochemical and bio-optical effects of river borne sediments, atmospheric deposition, nutrient and dissolved organic matter inputs, and benthic interactions throughout the Bay. These influences all contribute to the evolution of dissolved oxygen in the Bay's waters, in particular the consistent annual development of anoxia in the bottom waters of the mid-Bay region. Here, we report on the performance of a newly developed, mechanistic dissolved oxygen formulation that has been incorporated into the ChesROMS model with the motivation to realistically resolve seasonally developing hypoxia/anoxia in the Bay. Insights into various biophysical interactions and biogeochemical processes of the Bay gained from these numerical experiments are considered, and the application of the ChesROMS model fields in short-term ecological forecast applications is discussed.

Keywords Hypoxia · Anoxia · Numerical modeling · Ecological forecasting · ROMS · Chesapeake Bay

---

J.D. Wiggert (✉)
Division of Marine Science, School of Ocean Science and Technology,
University of Southern Mississippi, Stennis Space Center, MS 39529, USA
e-mail: jerry.wiggert@usm.edu

R.R. Hood
Horn Point Laboratory, University of Maryland Center for Environmental Science,
Cambridge, MD 21613, USA

C.W. Brown
Center for Satellite Applications and Research, National Oceanic
and Atmospheric Administration, College Park, MD 20740, USA

D. Justic et al. (eds.), *Modeling Coastal Hypoxia*,
DOI 10.1007/978-3-319-54571-4_6

6.1 Introduction

Chesapeake Bay is the largest estuary in North America and has the highest land-to-water ratio (14:1) of any coastal water body in the world. The Chesapeake Bay watershed spans more than 64,000 square miles, encompasses the District of Columbia and parts of six states, extends northward to Otsego Lake near Cooperstown, NY and westward to the foothills of the Blue Ridge Mountains, and is home to more than 17 million people. Recreationally, the Bay's sport salt-water fishing industry annually yields $1.34 billion in sales (National_Marine_Fisheries_Service 2011), and swimming and boating are supported by numerous beaches and safe harbors. Ecologically, vast wetlands surround the Bay and its tributaries and offer a haven for a rich diversity of wildlife. Economically, the Bay supports the livelihoods of many commercial fishermen. The commercial seafood industry contributed $3.39 billion in sales, $890 million in income, and nearly 34,000 jobs to the local economy in 2009; furthermore, the Bay is the largest producer of blue crabs in the world, with yearly harvests of approximately 24.9 million kilograms (National_Marine_Fisheries_Service 2011). Clearly, maintaining the ecological health of the Bay is a priority for the quality of life and economic vitality of the mid-Atlantic region.

The seasonal variability of the physical environment within the Bay regulates the biogeochemical processes that in turn provide the framework for supporting these substantial fisheries harvests. The magnitude of the Bay's estuarine circulation is primarily set by the seasonality of the Susquehanna outflow, which typically peaks in the spring and tapers off to a late summer minimum. Seasonally varying wind forcing has also been established as an important contributor to the longitudinal circulation, with wintertime northerly winds acting to reinforce the principal estuarine circulation and southerly winds associated with the summertime Bermuda High acting in opposition to this circulation (Goodrich and Blumberg 1991). Along with its influence on the estuarine circulation, the annual cycle of freshwater inflow is a primary control on the seasonal variation in water column stratification, which is of particular biogeochemical relevance due to how this affects air-sea exchange and therefore in water dissolved oxygen concentration. The annually recurring development of severe hypoxia in the bottom waters of the main stem of the Bay is clearly linked to water column stratification that develops in late spring/early summer (Murphy et al. 2011).

Human activities over the last several decades have led to significant degradation of water quality and ecosystem health in the Bay (Kemp et al. 2005). Associated detrimental impacts include nutrient pollution and the negative consequences of eutrophication such as increased turbidity of Bay waters, which inhibits the growth of submerged aquatic vegetation (SAV) (Moore and Wetzel 2000; Moore and Jarvis 2008). Eutrophication has also amplified the annually recurring manifestation of hypoxia in bottom waters of the Bay, an environmental condition that is harmful to both invertebrates and fish. All of these negative effects are subject to substantial intra-seasonal and interannual variability that arises due to variations in freshwater/nutrient loading and the above-noted physical drivers (Murphy et al. 2011; Hagy et al. 2005). This variability is so large that it has made it difficult to assess whether efforts to restore the Bay are working; that is, detecting a clear trend is problematic given the system's inherent variability. This variability also makes it difficult to predict what the ecological health of the Bay will be in the future on timescales ranging from days to months to years and under the emerging impacts of global climate change (Najjar et al. 2010).

Developing the ability to predict the timing, location, and intensity of low oxygen events will help mitigate their impacts on human and ecosystem health by providing local, state, and federal agencies with early warnings of their arrival. Furthermore, this capability can be used by managers to evaluate the outcome of different scenarios and select the best alternative in order to better preserve the coastal resources and protect human health.

As part of an effort to predict water quality and ecosystem health in Chesapeake Bay, we have developed the Chesapeake Bay Ecological Prediction System (CBEPS), a three-dimensional, coupled estuarine physical–biogeochemical–ecological modeling system that routinely generates and provides nowcasts and short-term (3-day) forecasts of a broad suite of physical, biogeochemical and ecological properties in the Bay (Brown et al. 2013). This modeling system is built upon the Chesapeake Bay Regional Ocean Modeling System (ChesROMS) (Xu et al. 2011), an implementation of the open source Regional Ocean Modeling System (ROMS) (Shchepetkin and McWilliams 2005) for Chesapeake Bay.

The ChesROMS's biogeochemical model was developed to capture the spatio-temporal variability of the Bay's phytoplankton and nutrient distributions. Capturing the annually recurring seasonal onset of hypoxia at depth in the mid-Bay region (Hagy et al. 2004) was another specific objective of the model's development. The focus on hypoxia necessitates the inclusion of an active dissolved oxygen (DO) component in the biogeochemistry. Fulfilling this need represents a challenge in terms of maintaining sufficient complexity to capture the seasonally recurring transition to hypoxic/anoxic waters in a setting where the influence of benthic–pelagic coupling (i.e., exchange of dissolved nutrients and DO) plays a critical role in the overall elemental cycling of the Chesapeake Bay system. In addition, this model must be capable of capturing interannual variability in the Bay's biogeochemical properties and DO. Finally, a streamlined approach is needed for routine application of the ChesROMS biogeochemical model to obtain the nowcasts and short-term forecasts of ecosystem function that inform CBEPS (cf., Brown et al. 2013), and to expeditiously perform longer-term runs (e.g., annual and interannual hindcasts) used for synthesis studies of the Chesapeake Bay system.

In this paper, we describe the ChesROMS biogeochemical model, focusing on the model components that have been developed and implemented to simulate nutrient cycling under seasonally developing hypoxic and anoxic conditions in both the water column and sediments of Chesapeake Bay. The solutions presented here were all obtained through application of forcing fields from 1999, which was chosen as a focus for the biogeochemical model's development since it represents a typical hydrologic flow regime. The three primary aspects of the Bay's biogeochemical variability we targeted as key features to capture in the model simulations were the phytoplankton bloom dynamics, the spatio-temporal variation of nutrient distributions, and the onset and persistence of severe hypoxia along the main stem. We found that there are inherent tradeoffs to these three modeling objectives, whereby increased skill in one aspect can be countered by a significant degradation of one or both of the other objectives. Here, we report on our success in simultaneously attaining these three objectives, review the new insights into the workings of the Chesapeake Bay system that were gained, and outline future steps for incorporating these insights into ongoing research activities and operational model development efforts.

6.2 Methods

6.2.1 ChesROMS: Physical Model and Forcing Fields

ChesROMS is set up as a three-dimensional, sigma-coordinate model, with a horizontal resolution of 1 to 5 km and 20 vertical levels; this configuration is used to simulate the circulation and physical properties (temperature, salinity, density, velocity and mixing) of the estuary (Fig. 6.1). The physical implementation of the ChesROMS model employed for the results presented here is identical to that reported by Xu et al. (2011). A synopsis of the model setup is presented here; for full details of the implementation, validation and assessment of model skill, the reader is invited to consult the Xu et al. foundational effort.

The ChesROMS physical model provided the capacity to simulate the estuarine circulation in Chesapeake Bay that is principally controlled by hydrologic inputs from nine freshwater sources distributed around the Bay (Xu et al. 2011). The principal sources into the Bay are the Susquehanna (51%), Potomac (18%) and James (14%) Rivers, with further contributions from the Patuxent, Rappahannock and York Rivers and sources on the Eastern Shore (Guo and Valle-Levinson 2007). Forcing of the ChesROMS model consists of lateral boundary conditions associated with the noted river inflows, atmospheric boundary conditions (heat and momentum fluxes) obtained from the NARR reanalysis produced by NCAR, tidal forcing at the Bay's mouth from the ADCIRC EC2001 tidal database along with tide station data from Wachapreague, Virginia and Duck, North Carolina. At the model's open boundary with the North Atlantic, salinity and temperature were nudged to climatological values in the 2001 World Ocean Atlas.

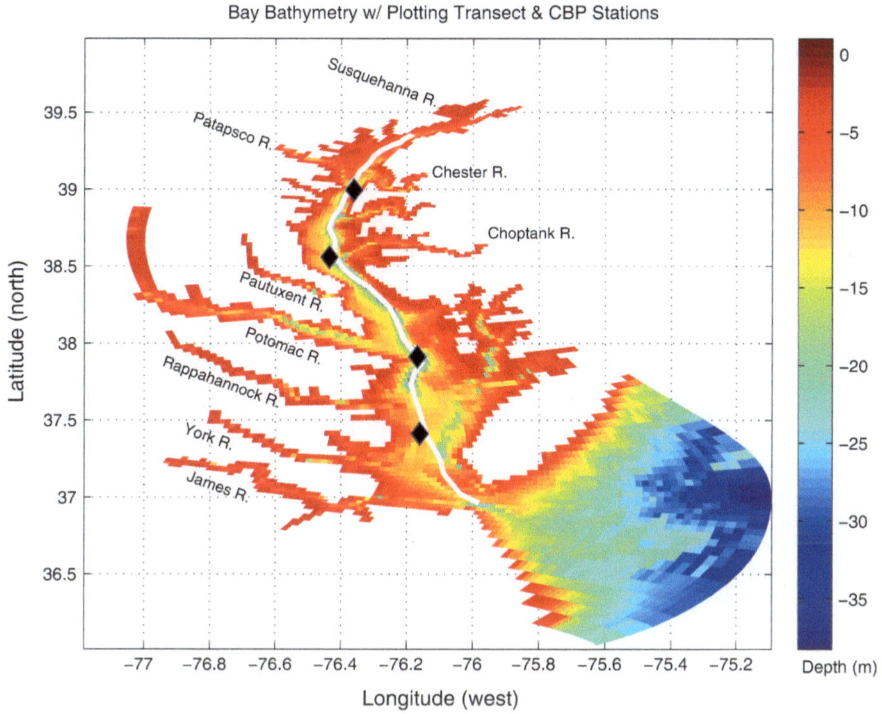

Fig. 6.1 Bathymetry for the ChesROMS model. The *white line* shows the track of data pulled in making along-bay section plots. The *black diamonds* show the CBP sites that are used in model-data comparisons of upper, mid-, and lower Bay physical and biogeochemical fields. From north to south, the four CBP sites are CB3.3C, CB4.3C, CB5.3 and CB6.3. The principal rivers flowing into Chesapeake Bay are noted

6.2.2 ChesROMS: Biogeochemical Model Configuration

A relatively simple NPZD-type ecosystem model has been implemented as a fully coupled component of ChesROMS. The Fennel et al. model (2006) that is provided as a standard component of the ROMS source (http://www.myroms.org) forms the basis for the ChesROMS ecosystem model (Fig. 6.2). Here, we describe how the model has been constructed and the capabilities that we have introduced in order to capture the elements of the Bay's biogeochemistry that were not accounted for in the standard ROMS release. Due to space considerations, the detail of the biogeochemical equations that we developed is not included herein. For the reader with interest to pursue these modifications in more depth, a slightly modified version of our biogeochemical formulations has been documented in Feng et al. (2015).

The ChesROMS ecosystem uses nitrogen as its fundamental currency but also includes a simple parameterization of phosphorus limitation. The formulation of an water light model specifically developed for the Bay by Xu et al. (2005) has been

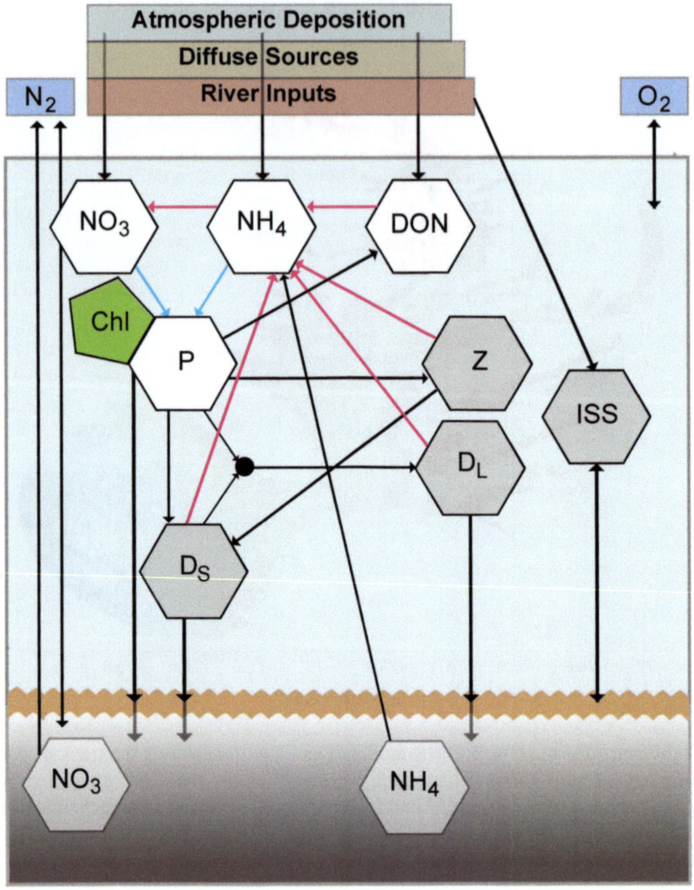

Fig. 6.2 Diagram for the ChesROMS biogeochemical model, illustrating the flows of nitrogen through the model's state variables. Benthic, terrestrial and atmospheric sources and sinks accounted for by the model are shown. The *gray* backgrounds of Z, D_S, D_L and ISS represent these constituents' role in attenuating downwelling irradiance. Similarly, the *green* background of CHL represents its role in attenuating downwelling irradiance, where CHL is a diagnostic variable of the phytoplankton state variable for which Chl:N varies with environmental conditions (Geider et al. 1997). The *black circle* represents the formation of large detritus (D_L) through the aggregation of phytoplankton (P) and small detritus (D_S). The *light blue* background of the *upper box* represents the dissolved oxygen within the water column that is subject to atmospheric exchange, variability in stratification and biotic generation and utilization. The *cyan arrows* represent production of oxygen via photosynthesis while the *magenta arrows* in the flow chart represent respiratory processes that take up oxygen. Drawdown of nitrate to fuel benthic denitrification and benthic efflux of ammonium into the water column are indicated by the pathways of NO_3 into and NH_4 out of the benthos

adopted, with model salinity (as a proxy for CDOM (colored dissolved organic matter)) contributing to attenuation of downwelling irradiance, along with chlorophyll and inorganic suspended solids (ISS). The ecosystem tracks phytoplankton biomass, and organic and inorganic components of nutrient and particulate constituents (Fig. 6.2). The phytoplankton constituent in the model has an associated chlorophyll representation that is based on a Chl:N ratio that is modulated by the light field (Geider et al. 1997). As part of the particulate organic constituents, two detritus size classes are included, where large detritus (D_L) is generated through the aggregation of small detritus (D_S) and phytoplankton (P) (black circle, Fig. 6.2). A single zooplankton size class with grazing rate modulated by temperature (Huntley and Lopez 1992) contributes to the small detritus pool via sloppy feeding and mortality. A dissolved organic nitrogen (DON) component has been added to accommodate riverine-associated DON loadings that contribute significantly to the Bay's overall nitrogen budget, which exhibits increasing DON:DIN ratio toward the Bay mouth (Bradley et al. 2010). In addition to the flows of nitrogen through the ChesROMS ecosystem noted here, the benthic, terrestrial and atmospheric sources and sinks incorporated in the ecosystem model are represented in the wire diagram (Fig. 6.2).

Dissolved oxygen (DO) concentration is mechanistically determined, with seasonal transitions toward anoxia accomplished by inclusion of explicit water column denitrification based on Oguz (2002). Within the water column, DO-based transitions between nitrifying (normoxic) and denitrifying (hypoxic) conditions are applied that modulate the remineralization of detritus by aerobic and anaerobic bacteria; zooplankton activity and metabolic losses are linked to the nitrification formulation and are thus ramped down where hypoxia is established and as it intensifies. The sources and sinks of DO associated with these biogeochemical processes in the water column are indicated on the wire diagram by the cyan (production) and magenta (uptake) arrows (Fig. 6.2). The benthic biogeochemical model of Fennel et al. (2006) that was developed for normoxic water column applications has been extended by linking benthic exchanges to the dissolved oxygen concentration of overlying bottom waters using a Michaelis–Menten-based formulation.

Through this benthic exchange linkage the model is set to: (1) mechanistically modulate the drawdown of water column DO at depth to fuel benthic denitrification and; (2) capture the observed amplification of ammonium efflux from the benthos and onset of nitrate influx to the benthos as benthic denitrification intensifies (see Fig. 6 in Middelburg et al. 1996). The Michaelis–Menten-style formulation applied to control these benthic exchanges is based on published observations from the Bay that relate ammonium efflux to bottom DO (Rysgaard et al. 1994; Cowan and Boynton 1996). A second Michaelis–Menten style mechanistic link to nitrate concentration in overlying bottom waters is implemented to prevent generation of negative concentrations resulting from nitrate drawdown; this inhibition to nitrate drawdown solely affects the benthic interaction with water column nitrate at depth in the model. That is, anaerobic respiration in the benthos that fuels ammonium efflux is assumed to shift to an alternative electron acceptor source (e.g., sulfate) that is not explicitly modeled. Finally, the sinking velocity of the large detritus pool was reduced by 40% in the bottom layer of the model to allow for advective redistribution of organic matter by the estuarine circulation and to promote oxygen demand in the water column via resuspension.

Point and diffuse source loadings of NO_3, and both organic and inorganic particulates, are based on the rates of river inflow and concentrations of NO_3, total organic nitrogen (TON) and total suspended solids (TSS) measured along the Bay's boundaries that were obtained from the Chesapeake Bay Program (CBP) data repository (http://chesapeakebay.net/). Atmospheric deposition of NO_3, NH_4 and DON over the Bay are determined from measured rates of wet atmospherically deposited nitrogen from the NADP's Wye Island Station (http://nadp.sws.uiuc.edu/) and constant annual rates of dry deposition (Meyers et al. 2001).

6.2.3 Model Assessment and Validation

The CBP data holdings are an invaluable resource that have been used here both for setting the initial state of the model and for constructing the biogeochemical boundary conditions associated with point and diffuse sources around the Bay (as noted above). These data were also critical for assessing how effective the biogeochemical model is at capturing the seasonal variability, including vertical structure, over the entire estuary. Measurements of chlorophyll, NO_3, NH_4, DO and DON were all routinely extracted and used to directly compare the seasonal variability of profiles at a targeted group of CBP stations (3.3C, 4.3C, 5.3 and 6.3) that are representative of the upper, mid- and lower Bay (Fig. 6.1). The three upper stations are also aligned with the outflows (north to south) of the Patapsco, Choptank and Potomac Rivers, while the Lower Bay site is located south of the Rappahannock River outflow.

The CBP measurements were also used to determine an along-Bay quantification of model skill (Willmott 1981) that provided a temporally integrated view of performance over 27 CBP sites along the Bay's main stem. For the results presented herein, skill values for chlorophyll, NO_3 and NH_4 are based on the full annual period while the skill values for DO are based on the summer period (May–August). The temporal and spatial alignment of the model output to the CBP profile data are as follows. The time of the data profiles is aligned to the middle of the UTC hour to associate it with the model time step. Spatially, the model values (20 layers) are interpolated to align with the vertical location of the data. In cases for which the measurement location extends beyond the model profile, the bottommost point in the model profile is used.

As a typical hydrological forcing condition, 1999 was deemed an attractive time frame for conducting the development and testing needed for establishing the biogeochemical model and gaining critical insight into its sensitivities. The results presented here will focus on the baseline model implementation and the insights it reveals regarding seasonal variation of bloom dynamics, nutrient availability and dissolved oxygen in the Bay.

6.3 Results

6.3.1 Seasonal Variability in the Physical Environment

Examining the seasonal and spatial variability in the Bay's salinity field reveals the physical setting that controls its biogeochemical processes. The comparison of modeled salinity with co-located measurements drawn from the CBP observations database at the four target stations (3.3C, 4.3C, 5.3 and 6.3) is shown in Fig. 6.3. These point-to-point model-data plots provide a direct comparison of the 15 CBP salinity profiles obtained during 1999 at the four stations highlighted. For this comparison, the smaller symbols at each profile location represent shallower sample depth.

Some generalities that can be drawn from this comparison are that at the two northern sites (3.3c and 4.3c, Fig. 6.3), bottom waters in the model are fresher than the observed condition until the fall when the model is consistent with the data. Surface waters in the model are also consistently fresher at the northernmost station

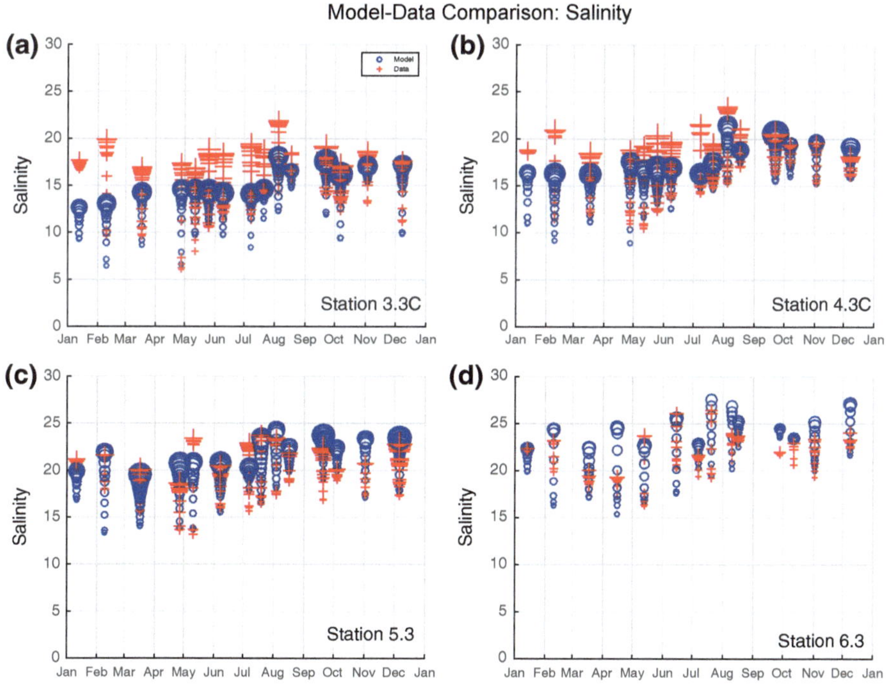

Fig. 6.3 Observed (+) and modeled (O) salinity in 1999 at four CBP stations: **a** CB3.3C; **b** CB4.3C; **c** CB5.3; and **d** CB6.3 (from upper- to lower Bay). Each vertical group of symbols represents a vertical profile at the corresponding time. Size of the symbols is coded by depth with bigger symbols for greater depth. The along-Bay distribution of these CBP stations is indicated by the *black diamonds* on Fig. 6.1

(3.3c), except during the spring freshet and in November (Fig. 6.3a). At station 4.3c (Fig. 6.3b), the surface waters are slightly fresher than observed through May; over the rest of the year, the model accurately represents surface salinity. Overall at station 5.3 (Fig. 6.3c), the salinity range for each profile over the course of 1999 is well represented by the model. The model does exhibit a tendency toward too salty conditions at depth after July. This tendency (model salinity greater than observed at depth) is more pronounced at station 6.3 (Fig. 6.3d).

Vertical sections of salinity along the Bay's main stem, for April, May, August and October, are shown in Fig. 6.4. The magenta diamonds along the distance axis represent the location of the four CBP stations featured in Fig. 6.3, while the along-Bay distance is in reference to the mouth of the Susquehanna River in Harve de Grace, MD. The extraction path of the along-Bay section from the model results follows the Bay's main stem (Fig. 6.1, white line).

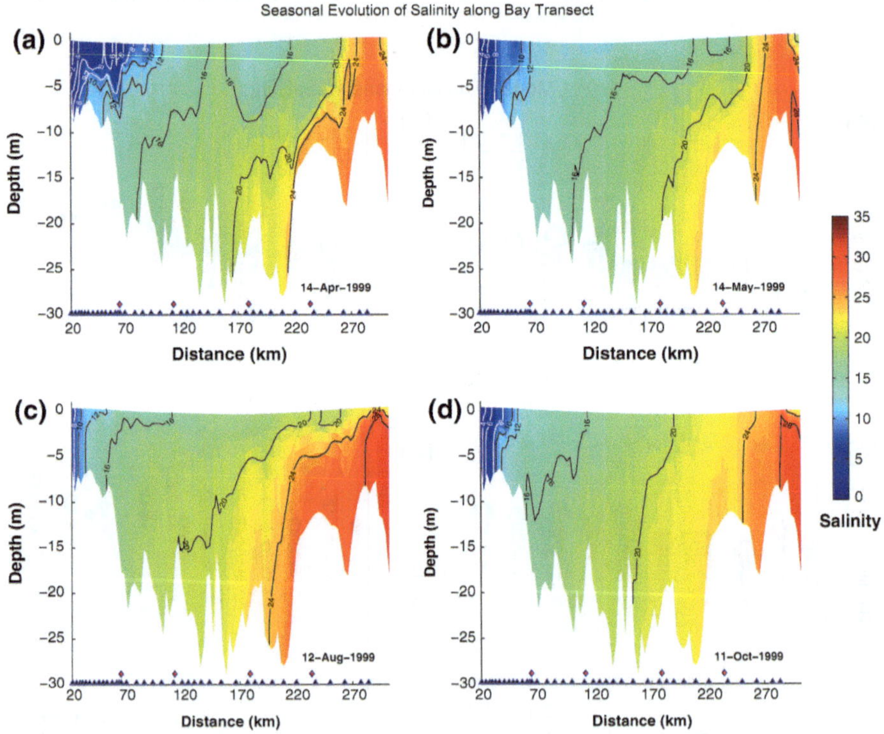

Fig. 6.4 Along-Bay vertical sections showing the seasonal progression of salinity for: **a** 14 April; **b** 14 May; **c** 12 August; and **d** 11 October. The triangles along the distance axis represent the extraction points used in forming the skill metric (Fig. 6.8). The four *diamonds* (set above the *triangles*) show the four model-data comparison sites (Figs. 6.3, 6.5 and 6.6). The northernmost point of the data extraction track (39.36 °N, 76.19 °W, Fig. 6.1), which is approximately 20 km south of Harve de Grace, MD, serves as the basis for the distance axis

The mid-April time frame of the first salinity section is chosen to align with the seasonal peak in freshwater discharge, which typically occurs late March to early April (Sanford et al. 2001; Schubel and Pritchard 1986). Over the course of the 6 months shown in Fig. 6.4, the freshwater plume associated with the inflow of Susquehanna River water exhibits a clear peak in its extent in the mid-April panel (Fig. 6.4a) with salinities of 12 or less extending ~110 km downstream of the river mouth. The profile from the end of April at station 3.3c (Fig. 6.3a) indicates that surface salinity is accurately captured in the model, whereas the deep values are fresher by ~20%. An interesting feature of the mid-April salinity section is the shoaling of the S = 16 isohaline at 155–165 km and the freshening of surface waters farther downstream, where the S = 16 isohaline again outcrops oceanward of 220 km (Fig. 6.4a). Examining an animation of model salinity sections reveals that this mid-Bay outcropping of the S = 16 isohaline occurs intermittently, and exhibits pronounced variability, from mid-February through mid-June. The fresher waters downstream of this outcrop location, which are bounded by the S = 16 isohaline when the outcrop events occur (Fig. 6.4a), are aligned with the Potomac River inflow. The magenta diamond at ~180 km downstream distance, which demarks CBP station 5.3, is adjacent to the mouth of the Potomac (Fig. 6.1).

The 6-month sequence of salinity sections provides a useful illustration of the evolution of surface salinity in the Bay (Fig. 6.4). The freshwater discharge from the Susquehanna River in May is typically 40–50% lower than the peak discharge of the late March/early April time frame. The retreat of the low salinity feature at the head of the Bay in May (note the S = 8 and S = 12 isohalines) clearly reflects this discharge reduction (Fig. 6.4a, b). The freshening of the mid-Bay region is also apparent, with the S = 16 outcrop no longer present in the mid-Bay (Fig. 6.4b). This results from the progression of the spring freshet down the Bay and the various lateral inputs (e.g., the Potomac and Patuxent Rivers).

The evolution of the estuarine circulation's return flow is represented by the S = 16 and S = 20 isohalines in the 6-month sequence (Fig. 6.4). By May, the bottom expression of both isohalines has shifted toward the Bay mouth coincident with the noted freshwater inflow. Over the summer, the two isohalines propagate 40–50 km northward (Fig. 6.4b, c). Over the summer months, model salinities at depth for the two southern stations show good agreement with the CBP measurements (Fig. 6.3c, d) while the two northern stations are consistently less salty than observed (Fig. 6.3a, b). In the fall this bias shifts, with bottom salinities at the two northern stations represented well in the model while the two southern stations tend toward being too salty. The fall salinity transect reveals the reduction in the strength of the return flow and the onset of fall breakdown of stratification with seasonal cooling (Fig. 6.4d).

As indicated above, the distinction in bottom salinity between the model and the observations at station 3.3c persists through September (Fig. 6.3a); this indicates that the return flow of the estuarine circulation in the model is slightly too weak. For the sections of model salinity, this suggests that the accumulation of elevated salinity waters at the Bay mouth is apparent as the seasonal evolution unfolds is overstated (Fig. 6.4b, c). Thus, deep salinities will tend to be too high near the Bay mouth and too low in the mid- and upper Bay, as is generally apparent in the model-data comparison time series (Fig. 6.3). This aspect of the simulated physical environment has significant implications for the functioning of the model's biogeochemical constituents, in particular the DO evolution that is attained.

6.3.2 Seasonal Variability of Biochemical Constituents

Model-data comparison for biochemical constituents of the model (chlorophyll, NO_3 and NH_4) for the two northern sites, where seasonal anoxia manifests, is shown in Fig. 6.5. The chlorophyll comparison at the northernmost site shows that the range in chlorophyll concentrations is consistent with the measurements (Fig. 6.5a). The spring bloom onset is captured, though the bloom's persistence in the model extends to early May prior to decreasing in late May and into the summer months. The observations indicate that the spring bloom begins to ramp down a couple weeks earlier than in the model; however, the maximum surface chlorophyll concentration (>40 mg m^{-3}) was measured in late May. During summer, modeled chlorophyll is within the observed range, though the model's subsurface values are consistently higher than the minimum observed concentrations. During fall and early winter, chlorophyll in the model consistently exceeds the measured values over the whole water column. At station 4.3C, the spring bloom in the model also persists longer into the early summer (Fig. 6.5b). Interestingly, during the summer and into the fall the model succeeds in capturing the minimal chlorophyll concentrations at depth that are seen in the data.

The model-data comparison of nitrate provides some insight into the functioning of the model's ecosystem. At both locations, observed nitrate concentrations during late April/early May significantly exceed the modeled values (Fig. 6.5c, d). This suggests that the termination of the spring phytoplankton bloom is due to grazer control, indicating that zooplankton activity in the model does not ramp up as quickly. Minimal values of nitrate over the summer months over the entire water column are accurately represented in the model at station 3.3c, as is the subsequent enrichment in the fall/early winter period (Fig. 6.5c). At station 4.3c, the very low summertime nitrate values are also captured; however, concentrations in the latter months are 2–3 times greater than observed (Fig. 6.5d). The model does accurately capture the observed high bottom-water nitrate concentrations.

The ammonium comparison shows that the model does a good job of simulating low surface concentrations over almost the entire seasonal cycle (Fig. 6.5e, f). Departures from the measured low surface ammonium values (<0.6 μM) in the model tend to occur in late summer (particularly mid-August) and in the early winter in the mid- to upper Bay (Fig. 6.5e, f) as well as the lower Bay (not shown). Examining animations of along-Bay sections of ammonium indicate that when they appear, these instances of elevated surface ammonium concentrations in the model result from vertical mixing that homogenizes the water column distribution (data not shown). As this mechanism implies, maximum ammonium concentrations

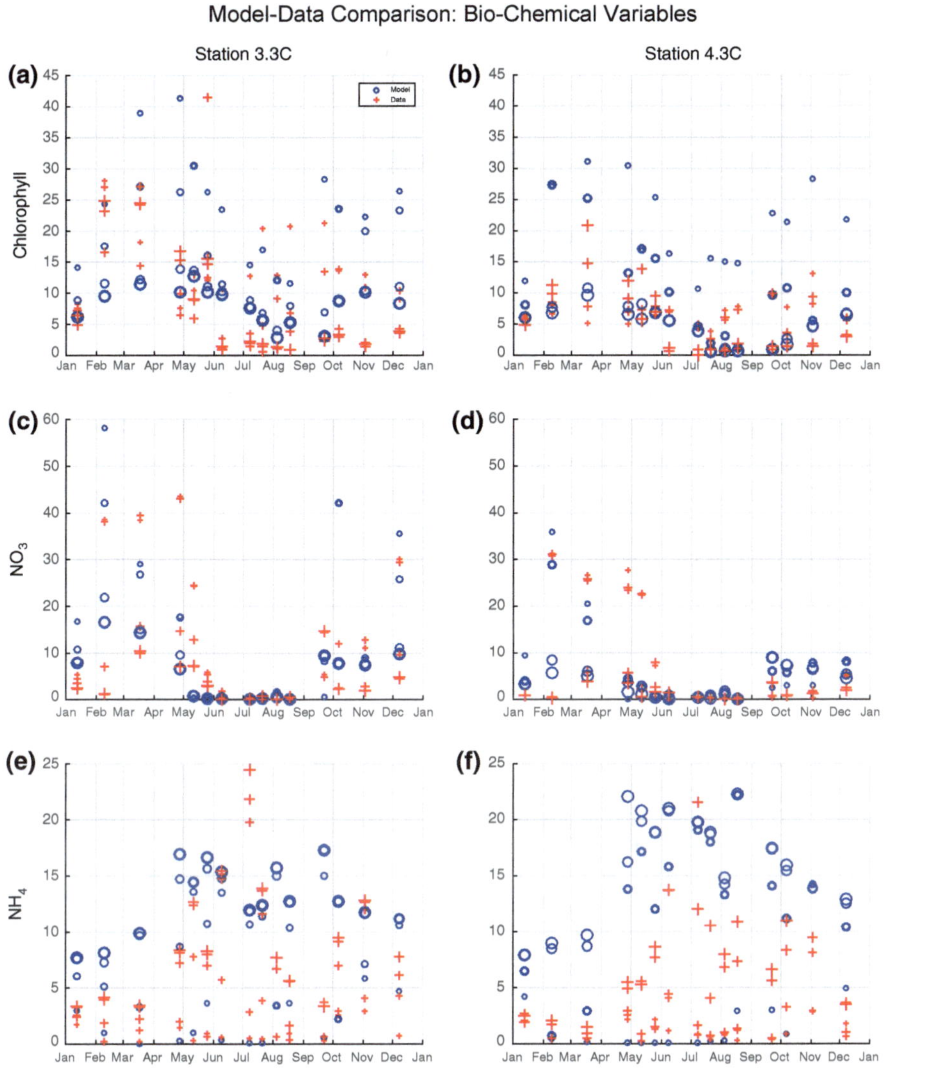

Fig. 6.5 Observed (+) and modeled (O) comparison of biochemical variables at two CBP stations from the upper (3.3C) and mid-Bay (4.3C): **a**, **b** chlorophyll (mg m^{-3}); **c**, **d** nitrate (μM); and **e**, **f** ammonium (μM). Each vertical group of symbols represents a vertical profile at the corresponding time. Size of the symbols is coded by depth with bigger symbols for greater depth. The along-Bay distribution of these CBP stations is indicated by the *black diamonds* on Fig. 6.1

occur in bottom waters that are in contact with the benthos. Peak ammonium values at depth in the model occur consistently throughout the seasonal cycle; the observed vertical distributions also follow this pattern (Fig. 6.5e, f). While the simulation's peak bottom concentration values are consistent with the highest observed values, modeled concentrations in the lower half of the water column are for the most part higher than observed and these elevated values at depth are clearly more persistent.

6.3.3 Dissolved Oxygen (DO) Results

The dissolved oxygen comparison shows that the model captures the general features of the seasonal cycle for the upper, mid and lower regions of the Bay, with peak concentrations in the spring, decreasing concentrations from late spring through the summer and re-oxygenation of the water column initiating in the fall (Fig. 6.6). A more complete representation of this seasonal evolution of dissolved oxygen in the model is revealed in the along-Bay sections (Fig. 6.7). Fully oxygenated springtime surface waters are clearly represented in the section plots; this DO condition can be seen to extend over the entire Bay (Fig. 6.7a). The late spring distribution shows reduction of dissolved oxygen concentrations over the whole water column and the

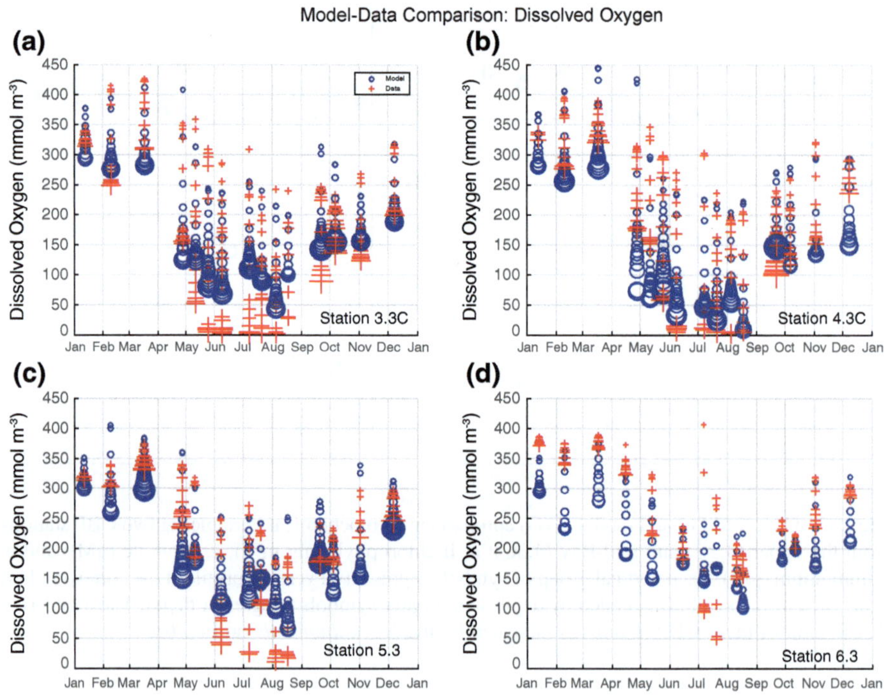

Fig. 6.6 Observed (+) and modeled (O) dissolved oxygen (mmol m^{-3}) in 1999 at four CBP stations: **a** CB3.3C; **b** CB4.3C; **c** CB5.3; and **d** CB6.3 (from upper- to lower Bay). Each vertical group of symbols represents a vertical profile at the corresponding time. Size of the symbols is coded by depth with bigger symbols for greater depth. The along-Bay distribution of these CBP stations is indicated by the *black diamonds* on Fig. 6.1

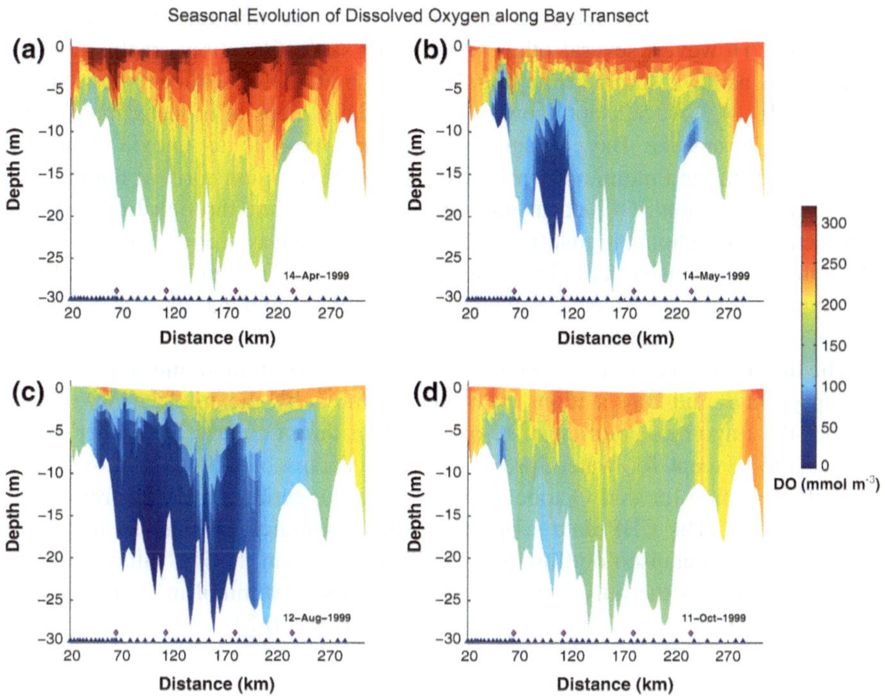

Fig. 6.7 Along-Bay vertical sections showing the seasonal progression of dissolved oxygen (mmol m^{-3}) for: **a** 14 April; **b** 14 May; **c** 12 August; and **d** 11 October. The triangles along the distance axis represent the extraction points used in forming the skill metric (Fig. 6.8). The four *diamonds* (set above the *triangles*) show the four model-data comparison sites (Figs. 6.3, 6.5 and 6.6). The northernmost point of the data extraction track (39.36 °N, 76.19 °W, Fig. 6.1), which is approximately 20 km south of Harve de Grace, MD, serves as the basis for the distance axis

appearance of very low bottom-water concentrations at several locations (Fig. 6.7b). These low DO hotspots in the late spring distribution are most pronounced in the mid- to upper Bay with two distinct features that appear just above the two northern CBP focus stations (3.3C, 4.3C); thus, they are situated just upstream of the inflows of the Patapsco and Choptank rivers. The model-data comparison for these two stations shows that the model captures the timing of this hypoxia onset in late spring quite well (Figs. 6.6a, b and 6.7b). In addition, moderate DO concentrations at this time (~180 mmol m^{-3}) are very accurately represented at station 5.3 (Fig. 6.6c). At station 6.3, the model consistently overestimates DO drawdown at depth through the late spring (mid-May) time frame (Fig. 6.6d).

The establishment of persistent summertime hypoxia is achieved at station 4.3C and intermittently realized at stations 3.3C and 5.3 (Fig. 6.6a–c). All three of these sites exhibit hypoxic conditions in the summer, with the two northern locations reaching full anoxia beginning in late May to early June that persists through the end of August (Fig. 6.6b, c). In the results shown here, full anoxia in the model's bottom waters is only achieved at station 4.3C in mid-July to mid-August (Fig. 6.6b). However, several other of the CBP station locations in this mid-Bay region also achieve summertime anoxia in the bottom DO concentrations in the model (4.1C, 4.3 W, 5.2, not shown). Further, daily maps of modeled bottom DO over the Chesapeake Bay domain reveals that hypoxic to anoxic conditions extending from the Potomac River inflow (station 5.3) to just north of the Patapsco River inflow (station 3.3C) occur intermittently in late June through July and more persistently throughout August (not shown). The along-Bay section for mid-August illustrates this extensive latitudinal range of very low DO conditions and indicates that hypoxia can range over 15 meters of the water column and extend to within 5 meters of the surface (Fig. 6.7c).

The timing of onset of re-oxygenation of the water column in mid-September for the upper and mid-Bay stations is nicely captured by the model (Fig. 6.6a–c). Further the ongoing evolution of re-oxygenation through early December is well represented. Except for waters proximal to the Patapsco outflow (Fig. 6.7d), the along-Bay section for mid-October indicates that the entire estuary has returned to oxic condition. The CBP data support the model's indication that the main stem region near the Patapsco outflow is a low DO hotspot in the late fall/early winter time frame, as it is the only site where DO is below 150 mmol m^{-3} in November and below 200 mmol m^{-3} in December (Figs. 6.6 and 6.7d).

6.3.4 Assessment of Model Skill and Parameter Sensitivities

With the goals of attaining fidelity in the model's representation of phytoplankton bloom dynamics, spatio-temporal variation of nutrient distributions and the onset/persistence of severe hypoxia in the mid- to upper Bay region, an extensive exploration of the ecosystem model's parameter space has been performed. The model-data comparisons of the biochemical fields (e.g., as featured in Figs. 6.5 and 6.6) were a prime component of the solution assessments. The other primary component of these assessments was determination of model skill (as defined in Willmott 1981), which provided an objective, overview characterization of the key state variables for each solution. An aggregate model skill over the 1999 seasonal cycle was calculated using all profile data gathered at each of 27 CBP stations and is presented as an along-Bay variation. The 27 CBP stations chosen for these along-Bay skill assessments range from station CB1.1 (39.55 °N, 76.08 °W) down to CB7.3 (37.12 °N, 76.13 °W); the linear distance of these stations from Harve de Grace, MD is represented by the black triangles on the abscissa of Figs. 6.4 and 6.7. For the most part, these stations lie along the Bay's main stem; however, several sites in the shallower waters adjacent to the main stem are included to provide full representation.

The along-Bay skill within a set of six model solutions for four model state variables (chlorophyll, ammonium, nitrate and dissolved oxygen) is shown in Fig. 6.8. The sidebar of Fig. 6.8 lists the identifiers of the 27 CBP sites included in

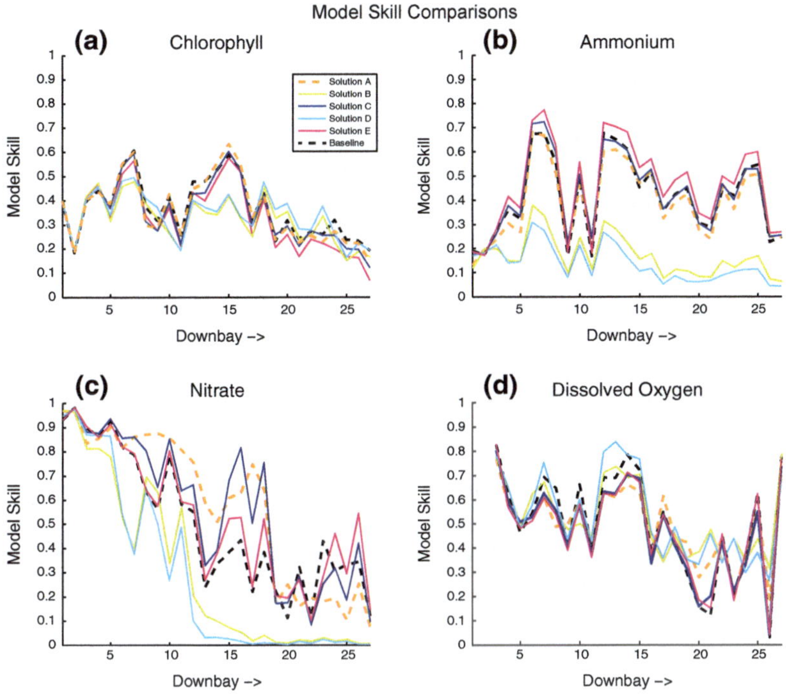

	CBP Station
1	CB1.1
2	CB2.1
3	CB2.2
4	CB3.1
5	CB3.2
6	CB3.3C
7	CB4.1C
8	CB4.1W
9	CB4.2W
10	CB4.3C
11	CB4.3W
12	CB4.4
13	CB5.1
14	CB5.2
15	CB5.3
16	CB5.4
17	CB5.5
18	CB6.1
19	CB6.2
20	CB6.3
21	CB6.4
22	CB7.1
23	CB7.1N
24	CB7.1S
25	CB7.2
26	CB7.2E
27	CB7.3

Fig. 6.8 Along-Bay skill within a set of six model solutions for four model state variables: **a** chlorophyll, **b** ammonium, **c** nitrate and **d** dissolved oxygen. An aggregate of along-Bay skill assessments is determined for 27 CBP stations that range from station CB1.1 (39.55 °N, 76.08 °W) down to CB7.3 (37.12 °N, 76.13 °W). The model-data pairs contained within each aggregate include all depths for every profile obtained by the CBP sampling in 1999 for the chlorophyll, ammonium and nitrate skill plots. For the dissolved oxygen skill plot, the May–August time frame was targeted to gain clearer insight into model performance during the summer hypoxia regime. The complete list of station IDs from which data are drawn for the skill determination is given in the included table

the skill determinations, with an integer index noting their ordering in the down Bay direction. The four CBP stations featured in the model-data comparisons (black diamonds, Fig. 6.1) are indicated by gray shading in this site listing. While skill in the model state variables throughout the Bay was desired, the skill values for DO at sites 6–7 and 12–15 were a focal point for assessing how well summertime hypoxia onset and persistence were attained. These two along-Bay foci, respectively, coincide with the Patapsco/Chester River and Potomac River inflows (Fig. 6.1). It should be noted that differences in DO skill achieved at these sites were closely examined since even minor improvement in skill at these locations was indicative of significant improvement in the summertime evolution of bottom DO concentration in the model, in particular with regard to transitioning to and maintaining hypoxic conditions.

The six featured model solutions are chosen to illustrate the tradeoffs in the fidelity of these biochemical fields that are commonly realized; that is, improvements in one of the targeted system attributes achieved through parameter adjustment are typically accompanied by some degree of degradation in one (or both) of the other attribute targets. The skill assessments shown in Fig. 6.8 represent a progression of overall improvement in model skill as parameter adjustments are adopted, albeit with tradeoffs as detailed below. In previous sensitivity analyses leading up to the group of solutions shown in Fig. 6.8, a diverse range of the model's ecosystem functionalities were explored that assessed maximum nitrification rate, temperature dependence of phytoplankton growth rate and zooplankton grazing rate, and the aggregation parameter that modulates formation of large detritus (represented by the black circle in Fig. 6.2). As a result of this sequence of parameter explorations, model solution A was achieved (dashed orange line, Fig. 6.8).

A particular target outcome that motivated the parameter exploration reported here was to improve the persistence of summertime mid-Bay hypoxia, which was difficult to achieve. Indeed, the summer evolution of mid-Bay hypoxia in Solution A nicely captures onset in late May/early June but does not maintain the severe hypoxia seen in the observations and actually transitions to non-hypoxic DO concentrations by early August (i.e., >62.5 mmol m^{-3}, Fig. 6.9a). The five subsequent solutions featured in Fig. 6.8 represent a further parameter exploration that targeted refining the formation, water column processing, and benthic delivery of organic matter in the model with the aim of improving persistence of hypoxia during the stratified, summer season. The skill curves for the baseline solution are for the standard run from which all of the model-data shown herein are taken (Figs. 6.3, 6.4, 6.5, 6.6 and 6.7). The skill curves for solutions B–E in Fig. 6.8 highlight intermediate stages from solution A (orange dashed line) that led to the baseline run (black dashed line). Table 6.1 provides a synopsis of the model parameters used for each skill assessment, how the values were adjusted and their impact on the ecosystem.

In the original water column denitrification formulation that we adopted from Oguz (2002), symmetric ramps for switching on/off nitrification and denitrification as conditions transitioned from normoxic to anoxic were employed. For solution B (Fig. 6.8, olive solid line), an asymmetry was introduced via modification of the half-saturation coefficient for the denitrification curve (KDNF, Table 6.1). This asymmetry acted to slightly retard the net remineralization by aerobic and anaerobic bacteria of both detrital size classes, thus resulting in slight amplifications of oxygen uptake within the water column and delivery of organic matter to the benthos. Relative to solution A, solution B exhibited a clear improvement in DO skill at upper and mid-Bay sites (6–7, 12–15), a mixed influence on chlorophyll and a significant degradation of ammonium and nitrate, particularly in the mid- to lower Bay (Fig. 6.8). The degradation of these DIN forms manifests as concentrations that were in excess of 6 times too high in July/August (ammonium) and October/November (nitrate) (not shown). Excessive ammonium efflux is the underlying cause, with the efflux rate being amplified by both the increased organic

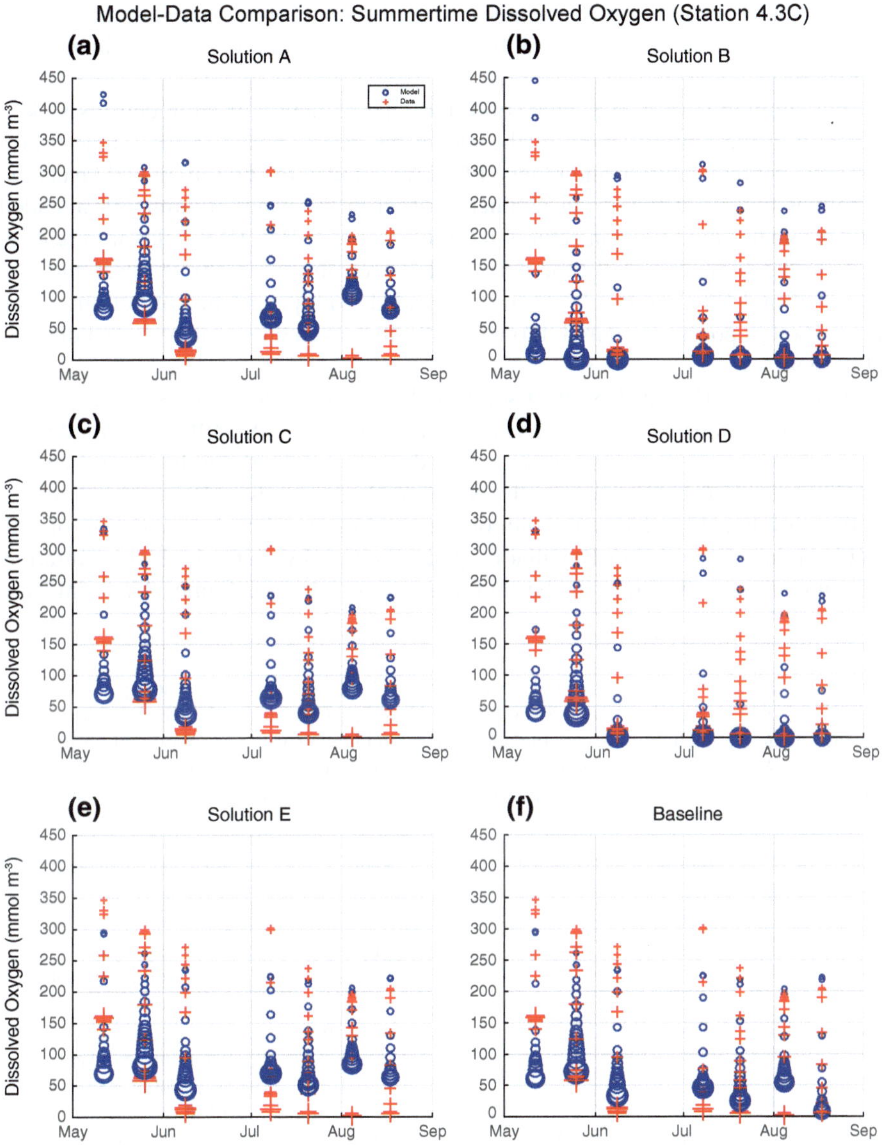

Fig. 6.9 Observed (+) and modeled (O) dissolved oxygen (mmol m^{-3}) for May–August 1999 at CBP station 4.3C for **a** Solution A, **b** Solution B, **c** Solution C, **d** Solution D, **e** Solution E and **f** Baseline solution. Each vertical group of symbols represents a vertical profile at the corresponding time. Size of the symbols is coded by depth with bigger symbols for greater depth

matter delivery to the benthos and, to a lesser degree, the lower bottom-water DO concentration. These low DO concentrations are clearly apparent in the model-data comparison (Fig. 6.9b). These show that DO values that interact with the benthos, and modulate benthic exchange rates, successfully capture the summertime persistence of bottom-water anoxia but DO values at mid-depths are lower than observed.

For solution C (Fig. 6.8, blue solid line), the reduction of large detritus sinking velocity in the bottom layers of the model has been relaxed (FwR, Table 6.1) to promote benthic delivery and relieve POM retention in the bottom model layers that amplifies oxygen demand via remineralization in the water column. As intended, this parameter modification returns the mid- to lower Bay DIN fields to a realistic state in the summer to fall time period. However, the summertime onset and evolution of mid-Bay hypoxia is severely degraded, with persistence of bottom anoxia again being poorly captured though with improvement relative to solution A (Fig. 6.9a, c).

For solution D (Fig. 6.8, cyan solid line), a more pronounced asymmetry in the nitrification/denitrification ramps is adopted with the aim of shifting the model into a solution space that can achieve the desired summertime hypoxia behavior (KDNF, Table 6.1). Similar to the solution A -> solution B impact described above, solution D (relative to solution C) exhibits significant degradation of DIN fields in the mid- to lower Bay that is again associated with too high concentrations of ammonium and nitrate in the mid-summer to fall time frame. Also similar to

Table 6.1 Summary of model parameter adjustments. Column 1 shows the line type, to reinforce the legend on Fig. 6.8. Column 2 gives the run identifier. Columns 3–5 give the details of the affected parameter including its variable name, units, and how its value was modified for the given model run. A summary of the parameter's direct impact on the model for the given experiment is given in Column 6. Over the progression from solution A to the new baseline solution, the parameter changes are cumulative

Line type	Run ID	Modified parameter	Units	Value adjustment	Modification summary
	Solution A				Original solution
	Solution B	KDNF	mmol O_2 m^{-3}	3.0 -> 2.86	Reduce 1/2 saturation coefficient of denitrification onset curve (KDNF)
	Solution C	FwR	N/A	0.9 -> 0.4	Increase D_L sinking velocity in bottom model layer. wLDet_bottom = (1 − FwR) * wLDet
	Solution D	KDNF	mmol O_2 m^{-3}	2.86 -> 2.82	Reduce 1/2 saturation coefficient of denitrification onset curve (KDNF)
	Solution E	wLDet	m d^{-1}	0.5 -> 0.95	Increase D_L sinking velocity throughout water column
	Baseline	KBO2	mmol O_2 m^{-3}	26.5 -> 20.0	Reduce 1/2 saturation coefficient applied for benthic biogeochemical exchanges

solution B, the mid-Bay DO fields better capture the onset and evolution of hypoxia in bottom waters, in particular the persistence of anoxia in bottom waters (Fig. 6.9d). However, similar to solution B, mid-depth DO is too low, particularly late July to August time frame (Fig. 6.9b, d). Solution D also realizes the highest skill in chlorophyll in the lower Bay (sites 18–23), which contrasts the poorest skill relative to the other solutions (along with solution B) at mid-Bay sites (13–17).

For solution E (Fig. 6.8, magenta solid line), the sinking velocity of large detritus was increased from 0.5 to 0.95 m/d (wLDet, Table 6.1). This parameter modification promotes flux of organic matter to the benthos and reduces particulate matter loading in the water column, with concomitant impacts on oxygen demand and denitrification. Both DIN fields are positively impacted. Solution E achieves the highest skill in ammonium throughout the Bay and the highest skill in nitrate in the lower Bay. Two pronounced tradeoffs are incurred. The first is the poorest chlorophyll skill in the lower Bay, which results from poorer fidelity of bloom dynamics during the mid-summer to early fall time frame. The second tradeoff is that bottom DO concentrations during summertime are just on the threshold of hypoxic, rather than the severe hypoxia to anoxia that is observed for the June through August time frame (Fig. 6.9e).

For the baseline solution (Fig. 6.8, black dashed line), the half-saturation coefficient in the formulation that links benthic exchanges to the DO concentration of overlying bottom waters is lowered by ~25% (KBO2, Table 6.1). In oxic to marginally hypoxic bottom waters, this modification reduces nitrate drawdown and ammonium efflux, and enhances dissolved oxygen drawdown. The reduced nitrate drawdown has an interesting, though subtle, impact that can be seen when comparing model fields for the mid-Bay stations of the standard run (Fig. 6.5b, d, and f) with those of solution E (not shown). Higher nitrate concentration in late April promotes higher plankton biomass (both P and Z) that leads to higher POM accumulation, relative to solution E, with an associated increase in water column recycling and DO uptake. This combines with the amplified benthic DO drawdown linked with export fluxes to improve hypoxic onset and persistence in the model's bottom waters during the summer (Fig. 6.9e, f).

The summertime DO time series for the six experiments demonstrate the significant variation in the modeled hypoxia evolution at one station (4.3C, Fig. 6.9). Aside from solution B, there is a consistent pattern to hypoxia onset in May. In contrast, the persistence of anoxia in bottom DO and evolution of DO concentration over the water column exhibits a range of responses across the six solutions (Fig. 6.9); this diverse model DO response to the applied parameter modifications is apparent all along the Bay in the skill values (Fig. 6.8d). At station 4.3C, the baseline solution can be seen to provide the best balance between attaining persistent hypoxic bottom DO conditions while maintaining low (non-hypoxic) conditions at mid-depths of the water column (Fig. 6.9). For additional overall comparative perspective for the six solutions in the skill assessments, mean skill values and rankings of the experiments for all four state variables have been collected in Table 6.2. The gray shading applied to some of the cells indicate results where the along-Bay skill is below one standard deviation from the mean for all

Table 6.2 Summary of impact on model skill as a result of the parameter adjustments (Table 6.1). Column 1 gives the run identifies. Columns 2–5 give the mean along-Bay skill. The temporal range for the mean skill values of chlorophyll, NH_4 and NO_3 is annual while for DO it is the summer period (May–August). The rank order over the six solutions in this comparison is given along with the mean skill. The mean and standard deviation for the four parameter skills are given in the two bottom rows. For table cells with italics, the mean along-Bay skill for a given solution is more than one σ below the mean over the six solutions

Run ID	Mean Along-Bay Skill			
	Chl	NH_4	NO_3	DO
Solution A	*0.33/5*	0.40/4	0.55/2	0.50/3
Solution B	*0.32/6*	*0.17/5*	*0.26/5*	0.53/2
Solution C	0.36/3	0.43/2	0.56/1	*0.48/6*
Solution D	0.34/4	*0.14/6*	*0.25/6*	0.54/1
Solution E	0.37/2	0.48/1	0.52/3	*0.49/5*
Baseline	0.38/1	0.43/2	0.49/4	0.50/3

experiments. This view of the results, along with the skill and summer DO comparisons (Figs. 6.8 and 6.9), summarizes the tradeoffs in the model's skill at capturing the observed spatial and temporal variability of the key biogeochemical parameters targeted in this assessment. It is apparent that each solution tended to have an attribute for which it particularly excelled, yet each was also consistently plagued by one (or more) attribute(s) for which performance was particularly poor (e.g., NH_4 and NO_3 skill in solutions B and D, Fig. 6.8b, c). Overall, it can be seen that the baseline solution achieved the best collective fidelity over the spectrum of model attribute objectives articulated in the introduction.

6.4 ChesROMS Application to Ecological Forecasting of Chesapeake Bay

The ultimate goal motivating our development of the ChesROMS biogeochemical model is its application as a means of illuminating biogeochemical processes within the Bay (described above) and providing nowcasts and short-term forecasts that can be used to inform the Chesapeake Bay Ecological Prediction System (CBEPS). The CBEPS was created with and for state and federal agencies responsible for monitoring and responding to potentially harmful biotic events and conditions in Chesapeake Bay, such as harmful algal blooms and hypoxia, to forecast these events and aid in mitigating their deleterious effects on human and ecosystem health (Brown et al. 2013).

In the application of CBEPS, the physical and biogeochemical variables are forecast mechanistically using ChesROMS, while the species predictions are generated using a novel mechanistic—empirical approach, whereby real-time output from the coupled physical—biogeochemical model drives multivariate empirical

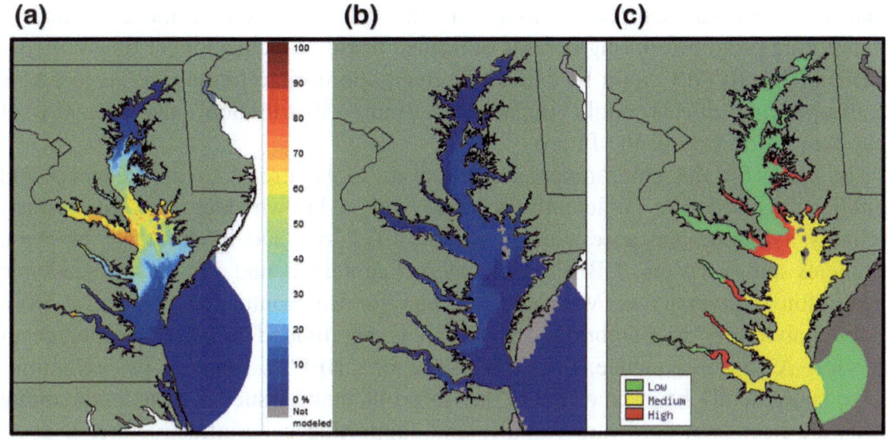

Fig. 6.10 Examples of species forecasts generated by the Chesapeake Bay Ecological Prediction System (CBEPS). **a** Probability of encountering Atlantic sea nettles, *Chrysaora quinquecirrha,* on 17 August 2007; **b** probability of encountering the pathogenic bacterium *Vibrio vulnificus* on 20 April 2011; and **c** relative abundance of the potentially toxigenic dinoflagellate *Karlodinium veneficum* on 20 April 2005. Legend: low: 0–10, med: 11–2000 cells/ml, high: >2000 cells/ml. Color bar for likelihood is the same for both A and B. Reproduced with permission from Brown et al. (2013)

habitat suitability models of the target species (see Fig. 3 in Brown et al. 2013). Environmental variables such as water temperature, water clarity, the concentrations of chlorophyll and nutrients, and the probability of encountering or (relative) abundance of several noxious species, such as the Atlantic sea nettle (*Chrysaora quinquecirrha*), a stinging jellyfish, the pathogenic bacterium *Vibrio vulnificus,* and the harmful algal species *Karlodinium veneficum* in the Bay and its tributaries are provided as forecast guidance (Fig. 6.10). Near real-time forecasts of sea nettle distribution have been widely viewed by recreational users, while predictions of *V. vulnificus* appearance are under review by state officials to assist in monitoring recreational exposure (J. Jacobs, pers. comm.) and have been deemed helpful to managers who must decide when to close beaches and shellfish beds (Pace et al. 2015). Hindcasts can be used to explore likely changes in the distribution of these organisms that might occur in the future in response to climate change (Decker et al. 2007; Jacobs et al. 2014).

The capability to predict DO and consequently hypoxic events provides these agencies with a tool that can be used to alert them of these and related potentially harmful conditions. For example, predicting the three-dimensional fields of DO and water temperature can be used by fisheries scientists to assess the volume of suitable habitat for various commercially important fish in the Bay, such as striped bass, and the stress imposed on them by hypoxia and elevated temperatures (Ludsin et al. 2009; Costantini et al. 2008). The hypoxia predictions can also be used tactically by monitoring agencies to strategically design and implement their sampling efforts in Chesapeake Bay and its tributaries. Forecasts are needed more than ever to guide sampling activities in the field and increase the efficiency of monitoring programs during periods of limited resources. On a longer time horizon, CBEPS/ChesROMS could be extended, given appropriate forcing, to project how anthropogenic effects might impact the timing, distribution, and intensity of hypoxia in the Bay in the future.

In order for the environmental predictions to be useful to the agencies and public, they must be sufficiently accurate with a known degree of uncertainty, reliably available and accessible in a timely fashion, and interpretable by these agencies and its users. CBEPS is located, maintained and run by an academic institution (University of Maryland's Center for Environmental Science at Horn Point Laboratory in Cambridge, MD) that is not funded to offer this degree of service. As a consequence, over the course of CBEPS's lifetime, its predictions have sporadically been unavailable due to problematic issues such as maintaining software licensing, hardware viability and supporting cyberinfrastructure. It does, however, offer a valuable experimental platform to test and assess new ecological forecasting algorithms and models for the Bay and to demonstrate the use of these predictions. Once validated and deemed useful by the community, the newly developed techniques and models can and should be migrated into a true operational environment within an appropriate agency, such as NOAA. This effort is one of several crucial steps in laying the foundation for generating truly operational ecological forecasts in Chesapeake Bay and serves as a roadmap for other locations.

6.5 Discussion and Conclusions

The ChesROMS physical model has considerable skill in simulating temperature and, to a lesser degree, salinity in Chesapeake Bay; hindcasts over a 15-year period (1991–2005) reveal that both temperature and salinity fields match well with observations with a correlation of approximately 0.99 and RMSE of 1–1.5 °C for temperature and a correlation of 0.95 and RMSE of 2–2.5 for salinity (Xu et al. 2011). The results presented herein reflect this lower salinity correlation, with bottom waters that are too fresh at the northern sites during summer (Fig. 6.3a, b) and too saline at the southern sites during fall (Fig. 6.3c, d). This model tendency is further demonstrated in the seasonal sections of salinity that reveal an accumulation of high salinity waters near the southern end of the Bay, which should propagate farther up the estuary. This is indicative of an estuarine circulation with a too weak return flow in the model; the core shortcoming leading to this result is an overly smoothed bathymetry with an inadequately resolved deep channel. Higher resolution ROMS simulations have been demonstrated to better resolve the estuarine return flow and more realistically capture along-Bay salinity variability (M. Scully, personal communication). However, the computational demand of such higher resolution configurations makes them impractical to use in most research efforts.

The ecosystem model results from the main solution (Figs. 6.5, 6.6, 6.7 and 6.9) indicate that rather than being solely prescribed by dynamical processes, these fields are subject, over location and season, to varying blends of physical and biochemical control; this assertion is consistent with that articulated in the synthesis of Kemp et al. (2009). The link between stratification and hypoxia in coastal and estuarine systems is well-documented (Rabalais et al. 2010) and has a clear association in the salinity and DO fields of the model (Figs. 6.3 and 6.6). Contrasting how salinity stratification relates to DO evolution in bottom waters at stations 3.3C and 4.3C underscores the importance of realistically representing stratification of bottom waters, which harkens back to the issues noted above regarding the shortcomings in the estuarine circulation. Comparing the fidelity of bottom salinity at station 3.3C to that at 4.3C in the model (Fig. 6.3a, b), it can be seen that the modeled salinity during the May–August period is notably worse at the more northern site (i.e., where the estuarine circulation is more poorly represented). It is also apparent that the greater skill at capturing summertime onset and persistence of hypoxia at these two sites aligns with how well stratification of bottom salinity is represented (Figs. 6.3a, b and 6.6a, b). Overall, examining these two sites reveals a direct linkage between the degree of mismatch in modeled bottom salinity and how well the temporal evolution of bottom hypoxia is represented.

In contrast, the spring bloom in the model is an example of a biophysical interaction subject to biochemical control where its onset is consistent with the observed timing yet its persistence is longer than is apparent in the measurements. Given that observed nutrients are non-limiting, this suggests that the key mechanism relates to establishment of top-down control of the phytoplankton population being delayed and potentially under-represented in the model. Another biophysical mechanism apparent in the results is, while the model effectively simulates the accumulation of DIN forms in bottom waters through benthic connectivity and their subsequent lateral advection, the timing of the injection of these chemical constituents into surface waters is controlled by vertical mixing (i.e., reduction of vertical stratification). Thus, it is likely that the mismatch in surface nitrate and ammonium in the late fall, where the model values are too elevated, link back to stratification shortcomings noted above.

The principal motivation of developing a mechanistic dissolved oxygen formulation applicable to an estuarine system subject to significant riverine loadings and benthic connectivity has been realized with some success. The seasonal establishment of hypoxic bottom waters over the mid- to upper Bay, and the subsequent re-oxygenation in the fall, is well represented. Further, full anoxia in late summer in the upper Bay is also realized. An interesting result appearing in the model DO field is an association between hypoxia "hot spots" in the Bay and the inflows of the Patapsco and Choptank rivers. While they may primarily relate to estuarine circulation issues in the model, it is possible that these hot spots are indicative of the flow of low DO waters into the Bay from these riverine sources, and/or augmentation of the organic matter loads remineralized within the water column and by the benthos of the main stem Bay. While the impact of the multi-decadal trend in the Susquehanna's nitrate loading on Bay eutrophication is well-established (Hagy et al. 2004), this model result suggests that the influence of the lower volume lateral riverine inputs to the main stem Bay play a significant role in the establishment and maintenance of its hypoxic waters. Elucidating the effects of these lateral inputs deserves further investigation and assessment. This is particularly crucial in light of the amplifying anthropogenic impacts, associated with evolving land use and agricultural practices (among others), that are known to afflict river dominated estuarine and coastal waters worldwide (Zhang et al. 2010).

Two aspects of our model's biogeochemical function, both relating to how the transition from normoxic to anoxic remineralization is formulated, have demonstrated sensitivity that has clear repercussions on ecosystem behavior. These are: (1) introduction of asymmetry in the nitrification/denitrification ramps applied to the water column remineralization; and (2) modification of the half-saturation coefficient applied in linking benthic exchanges of DO and DIN to the dissolved oxygen concentration of overlying bottom waters. For the ramps asymmetry, the complexity of nitrogen cycling when a nitrification—denitrification coupling can be established at the boundary of oxygen deficient waters (Codispoti and Christensen 1985) strongly suggests that mirrored onset/shutdown of these processes in the water column is improbable and requires further consideration. The formulation we have introduced to link benthic remineralization with overlying dissolved oxygen concentration was developed with the objective to forego incorporating further complexity via coupling with a detailed benthic model (e.g., Soetaert et al. 2000). However, a very limited dataset was leveraged in developing our formulation; given the clear sensitivity to adjustment of its half-saturation parameter demonstrated by the model, a larger dataset that more comprehensively establishes these exchanges is desirable.

The sensitivity studies that we have documented clearly demonstrate that modification of the biogeochemical formulation leads to notable changes in the model's biological and chemical constituents. That is, biogeochemical controls on the Chesapeake Bay system, in particular the spatio-temporal distribution of its hypoxic waters, are of primary importance to attaining realistic biogeochemical function and, in this regard, are arguably on par with the influence of physical controls that has been elegantly demonstrated previously (Scully 2010, 2013). Consequently, a well-considered, mechanism-based biogeochemical model embedded within a coupled three-dimensional model framework is essential for achieving: (1) valuable insight into how short-term and interannual variation in river inflow and nutrient loading will impact the Chesapeake Bay estuarine system; (2) successful application of ecological prediction systems to promote informed recreational and resource use; and (3) accurate prediction of the biogeochemical and ecological consequences of climate change.

Acknowledgements The authors thank Wen Long, Jiangtao Xu, and Lyon Lanerolle for their contributions to the development of the ChesROMS biogeochemical model. We would also like to thank the two reviewers of our chapter for their insightful comments, which were extremely useful in improving this contribution. Funding for this work was primarily provided by the NOAA Center for Sponsored Coastal Ocean Research's Monitoring for Event Response for Harmful Algal Bloom (MERHAB) Program (NA05NOS4781222, NA05NOS4781226, NA05NOS4781227, and NA05NOS4781229). Additional support was provided by the NOAA EcoForecasting Program, the NOAA Center for Satellite Applications and Research, and Maryland Sea Grant. This is

MERHAB publication 189 and UMCES contribution no. 5138. The views, opinions, and findings contained in this report are those of the authors and should not be construed as an official NOAA or U.S. Government position, policy, or decision.

References

Bradley PB, Lomas MW, Bronk DA (2010) Inorganic and organic nitrogen use by phytoplankton along Chesapeake Bay, measured using a flow cytometric sorting approach. Estuaries Coasts 33(4):971–984. doi:10.1007/s12237-009-9252-y

Brown CW, Hood RR, Long W, Jacobs J, Ramers DL, Wazniak C, Wiggert JD, Wood R, Xu J (2013) Ecological forecasting in Chesapeake Bay: using a mechanistic-empirical modeling approach. J Mar Syst 113–125. doi:10.1016/j.jmarsys.2012.12.007

Codispoti LA, Christensen JP (1985) Nitrification, denitrification and nitrous oxide cycling in the Eastern Tropical South Pacific Ocean. Mar Chem 16:277–300

Costantini M, Ludsin SA, Mason DM, Zhang X, Boicourt WC, Brandt SB (2008) Effect of hypoxia on habitat quality of striped bass (Morone saxatilis) in Chesapeake Bay. Can J Fish Aquat Sci 65(5):989–1002. doi:10.1139/f08-021

Cowan JLW, Boynton WR (1996) Sediment-water oxygen and nutrient exchanges along the longitudinal axis of Chesapeake Bay: seasonal patterns, controlling factors and ecological significance. Estuaries 19(3):562–580

Decker MB, Brown CW, Hood RR, Purcell JE, Gross TF, Matanoski JC, Bannon RO, Setzler-Hamilton EM (2007) Predicting the distribution of the scyphomedusa Chrysaora quinquecirrha in Chesapeake Bay. Mar Ecol Prog Ser 329:99–113. doi:10.3354/meps329099

Feng Y, Friedrichs MAM, Wilkin J, Tian HQ, Yang QC, Hofmann EE, Wiggert JD, Hood RR (2015) Chesapeake Bay nitrogen fluxes derived from a land-estuarine ocean biogeochemical modeling system: model description, evaluation, and nitrogen budgets. J Geophys Res 120 (8):1666–1695. doi:10.1002/2015jg002931

Fennel K, Wilkin J, Levin J, Moisan J, O'Reilly J, Haidvogel D (2006) Nitrogen cycling in the Middle Atlantic Bight: results from a three-dimensional model and implications for the North Atlantic nitrogen budget. Global Biogeochem Cycles 20(3). doi:10.1029/2005GB002456

Geider RJ, MacIntyre HL, Kana TM (1997) Dynamic model of phytoplankton growth and acclimation: Responses of the balanced growth rate and the chlorophyll a: carbon ratio to light, nutrient-limitation and temperature. Mar Ecol Prog Ser 148(1–3):187–200

Goodrich DM, Blumberg AF (1991) The fortnightly mean circulation of Chesapeake Bay. Estuar Coast Shelf Sci 32(5):451–462

Guo XY, Valle-Levinson A (2007) Tidal effects on estuarine circulation and outflow plume in the Chesapeake Bay. Cont Shelf Res 27(1):20–42. doi:10.1016/j.csr.2006.08.009

Hagy JD, Boynton WR, Jasinski DA (2005) Modelling phytoplankton deposition to Chesapeake Bay sediments during winter-spring: interannual variability in relation to river flow. Estuar Coast Shelf Sci 62(1–2):25–40

Hagy JD, Boynton WR, Keefe CW, Wood KV (2004) Hypoxia in Chesapeake Bay, 1950-2001: long-term change in relation to nutrient loading and river flow. Estuaries 27(4):634–658

Huntley ME, Lopez MDG (1992) Temperature-dependent production of marine copepods: a global synthesis. Am Nat 140(2):201–242

Jacobs JM, Rhodes M, Brown CW, Hood RR, Leight A, Long W, Wood R (2014) Modeling and forecasting the distribution of Vibrio vulnificus in Chesapeake Bay. J Appl Microbiol 117 (5):1312–1327. doi:10.1111/jam.12624

Kemp WM, Boynton WR, Adolf JE, Boesch DF, Boicourt WC, Brush G, Cornwell JC, Fisher TR, Glibert PM, Hagy JD, Harding LW, Houde ED, Kimmel DG, Miller WD, Newell RIE, Roman MR, Smith EM, Stevenson JC (2005) Eutrophication of Chesapeake Bay: historical trends and ecological interactions. Mar Ecol Prog Ser 303:1–29

Kemp WM, Testa JM, Conley DJ, Gilbert D, Hagy JD (2009) Temporal responses of coastal hypoxia to nutrient loading and physical controls. Biogeosciences 6(12):2985–3008

Ludsin SA, Zhang X, Brandt SB, Roman MR, Boicourt WC, Mason DM, Costantini M (2009) Hypoxia-avoidance by planktivorous fish in Chesapeake Bay: implications for food web interactions and fish recruitment. J Exp Mar Biol Ecol 381:S121–S131. doi:10.1016/j.jembe.2009.07.016

Meyers T, Sickles J, Dennis R, Russell K, Galloway J, Church T (2001) Atmospheric nitrogen deposition to coastal estuaries and their watersheds. In: Valigura RA, Alexander RB, Castro MS, Meyers TP, Paerl HW, Stacey PE, Turner RE (eds) Nitrogen loading in coastal water bodies: an atmospheric perspective. American Geophysical Union, Washington D. C., p 254

Middelburg JJ, Soetaert K, Herman PMJ, Heip CHR (1996) Denitrification in marine sediments: a model study. Global Biogeochem Cycles 10(4):661–673

Moore KA, Jarvis JC (2008) Environmental factors affecting recent summertime eelgrass diebacks in the lower Chesapeake Bay: implications for long-term persistence. J Coast Res 135–147. doi:10.2112/si55-014

Moore KA, Wetzel RL (2000) Seasonal variations in eelgrass (Zostera marina L.) responses to nutrient enrichment and reduced light availability in experimental ecosystems. J Exp Mar Biol Ecol 244(1):1–28. doi:10.1016/s0022-0981(99)00135-5

Murphy R, Kemp W, Ball W (2011) Long-term trends in Chesapeake Bay seasonal hypoxia, stratification, and nutrient loading. Estuaries Coasts 34(6):1293–1309. doi:10.1007/s12237-011-9413-7

Najjar RG, Pyke CR, Adams MB, Breitburg D, Hershner C, Kemp M, Howarth R, Mulholland MR, Paolisso M, Secor D, Sellner K, Wardrop D, Wood R (2010) Potential climate-change impacts on the Chesapeake Bay. Estuar Coast Shelf Sci 86(1):1–20

National_Marine_Fisheries_Service (2011) Fisheries economics of the United States, 2009. Economics and sociocultural status and trends series. U.S. Department of Commerce, Silver Spring, MD

Oguz T (2002) Role of physical processes controlling oxycline and suboxic layer structures in the Black Sea. Global Biogeochem Cycles 16(2). doi:10.1029/2001GB001465

Pace ML, Carpenter SR, Cole JJ (2015) With and without warning: managing ecosystems in a changing world. Front Ecol Environ 13(9):460–467. doi:10.1890/150003

Rabalais NN, Diaz RJ, Levin LA, Turner RE, Gilbert D, Zhang J (2010) Dynamics and distribution of natural and human-caused hypoxia. Biogeosciences 7(2):585–619

Rysgaard S, Risgaard-Petersen N, Sloth NP, Jensen K, Nielsen LP (1994) Oxygen regulation of nitrification and denitrification in sediments. Limnol Oceanogr 39(7):1643–1652

Sanford LP, Suttles SE, Halka JP (2001) Reconsidering the physics of the Chesapeake Bay estuarine turbidity maximum. Estuaries 24(5):655–669

Schubel J, Pritchard D (1986) Responses of upper Chesapeake Bay to variations in discharge of the Susquehanna River. Estuaries Coasts 9(4):236–249

Scully ME (2010) Wind modulation of dissolved oxygen in Chesapeake Bay. Estuaries Coasts 33 (5):1164–1175. doi:10.1007/s12237-010-9319-9

Scully ME (2013) Physical controls on hypoxia in Chesapeake Bay: a numerical modeling study. J Geophys Res 118(3):1239–1256. doi:10.1002/jgrc.20138

Shchepetkin AF, McWilliams JC (2005) The regional oceanic modeling system (ROMS): a split-explicit, free-surface, topography-following-coordinate oceanic model. Ocean Model 9 (4):347–404

Soetaert K, Middelburg JJ, Herman PMJ, Buis K (2000) On the coupling of benthic and pelagic biogeochemical models. Earth Sci Rev 51(1–4):173–201

Willmott CJ (1981) On the validation of models. Phys Geogr 2(2):184–194. doi:10.1080/02723646.1981.10642213

Xu J, Long W, Wiggert JD, Lannerolle LWJ, Brown CW, Murtugudde R, Hood RR (2011) Climate forcing and salinity variability in the Chesapeake Bay, USA. Estuaries Coasts. doi:10.1007/s12237-011-9423-5

Xu JT, Hood RR, Chao SY (2005) A simple empirical optical model for simulating light attenuation variability in a partially mixed estuary. Estuaries 28(4):572–580

Zhang J, Gilbert D, Gooday AJ, Levin L, Naqvi SWA, Middelburg JJ, Scranton M, Ekau W, Pena A, Dewitte B, Oguz T, Monteiro PMS, Urban E, Rabalais NN, Ittekkot V, Kemp WM, Ulloa O, Elmgren R, Escobar-Briones E, Van der Plas AK (2010) Natural and human-induced hypoxia and consequences for coastal areas: synthesis and future development. Biogeosciences 7(5):1443–1467. doi:10.5194/bg-7-1443-2010

Chapter 7
Modeling River-Induced Phosphorus Limitation in the Context of Coastal Hypoxia

Arnaud Laurent and Katja Fennel

Abstract The urban development of coastal areas and the increased use of chemical fertilizers over the last century have led to a worldwide expansion of coastal eutrophication and a significant increase in the occurrence and intensity of human-induced coastal hypoxia. Proportionally, nitrogen load has often increased more severely than phosphorus load and phosphorus limitation became a common seasonal phenomenon in many eutrophic coastal systems. Phosphorus limitation may alter the magnitude, timing, and location of phytoplankton production with potential effects on hypoxia. Yet, because of the difficulty in observing these effects, limited work has been carried out to assess the influence of P limitation on hypoxia. Models are thus useful tools for simulating the effects of river-induced phosphorus limitation on coastal hypoxic systems. Modeling P limitation is important to better understand the processes controlling hypoxia, to improve the predictive skill of hypoxia prediction models, and to design and evaluate nutrient management strategies for hypoxia mitigation. Here, we review the effects of phosphorus limitation on a continuum of coastal hypoxic systems, contrasting the effects of P limitation on systems that are primarily one-dimensional (or "flow-through") like the Neuse River Estuary versus more dispersive open systems like the Mississippi River plume. We discuss modeling frameworks and techniques that are relevant in this context and summarize recent modeling work that quantitatively assesses the effect of phosphorus limitation on hypoxia development in the Mississippi River plume.

Keywords Coastal eutrophication · Biogeochemical model · Phosphorus limitation · Coastal hypoxia · Nutrient load · Hypoxia mitigation

A. Laurent (✉) · K. Fennel
Department of Oceanography, Dalhousie University, Nova Scotia, Canada
e-mail: arnaud.laurent@dal.ca

D. Justic et al. (eds.), *Modeling Coastal Hypoxia*,
DOI 10.1007/978-3-319-54571-4_7

7.1 Introduction

Riverine nitrogen (N) and phosphorus (P) fluxes to the coastal oceans have both increased over the last century due to the development of coastal areas and an increasing use of chemical fertilizers (Zhang et al. 2010). N and P are essential elements for phytoplankton and, for balanced growth, are required approximately in the ratio of 16 N:1P (Redfield et al. 1963). N is typically the limiting element in marine systems because denitrification produces biologically unavailable N_2 resulting in a partial loss of N, while P has no such loss process and is rapidly recycled (Caraco et al. 1990). Increases in river nutrient loads have not necessarily been in the same proportion for N and P because their sources differ. N inputs into watersheds are mainly diffuse and associated with the use of chemical fertilizers (Boesch 2002), whereas P inputs are primarily from point sources related to urban wastewater (Harrison et al. 2010). The stoichiometry of dissolved inorganic N (DIN) and dissolved inorganic P (DIP) in river waters can thus deviate significantly from the Redfield ratio of 16 N:1P. Measures to control nutrient pollution from point sources have been put in place since the 1970s, but the use of chemical fertilizers has increased further since then. This has led to a proportionally larger increase of N loads, altered the N:P stoichiometry such that $N:P > 16$ in estuarine and river plume waters of many coastal systems (Conley 2000; Scavia and Donnelly 2007; Paerl 2009), and promoted P-limited primary production in otherwise N-limited systems (Howarth and Marino 2006).

P limitation and hypoxia often co-occur in coastal systems because they are both driven by excess nutrient load and eutrophication. A switch from N to P limitation may alter the magnitude, timing, and location of phytoplankton production with a potential effect on hypoxia (Paerl et al. 2004). Yet, limited work has been carried out to assess the influence of P limitation on hypoxia. The paucity of observational evidence for the effects of P limitation on hypoxia results in large part from the effort involved in observing coastal systems at the relevant spatial and temporal scales combined with the inability to manipulate N and P loads at the scale of the whole system. In this context, models are important for simulating the consequences of different resource limitation scenarios and for quantitatively assessing the effects of P limitation on coastal hypoxia.

The objectives of this chapter are to (1) describe the occurrence and effects of P limitation on hypoxic systems, (2) provide an overview of modeling frameworks and techniques that can be used to study river-induced P limitation in the context of coastal hypoxia, and (3) present results from a model investigation for the Mississippi River plume in the northern Gulf of Mexico as a case study of the effect of P limitation on hypoxia. First, we consider how P limitation may influence hypoxia in a one-dimensional flow-through system and then contrast this with a more open and dispersive system. Several examples representing a continuum between these two are provided. We then discuss several modeling frameworks and present recent results from the modeling investigation of Laurent and Fennel (2014) for the Mississippi River plume.

7.2 Occurrence of P Limitation in Hypoxic Systems

River-induced P limitation is most often a transient phenomenon that occurs at peak discharge in spring, when N load tends to be particularly high. When primary production is P-limited in estuaries or river plumes, the excess DIN is transported downstream into more saline, N-limited waters where it stimulates primary production (Fig. 7.1a). In other words, the location and timing of DIN uptake are shifted downstream and delayed in time. The spatial shift tends to spread eutrophication over larger regions and into downstream waters compared to a situation where P is not limiting (Paerl et al. 2004). P limitation in the spring is

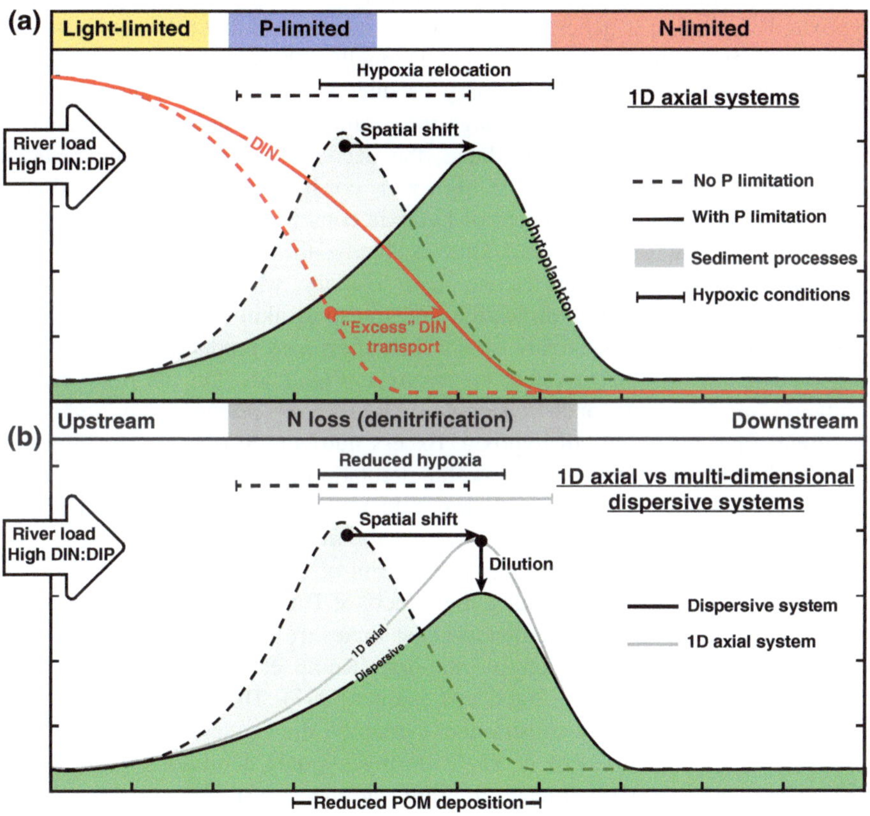

Fig. 7.1 Conceptual model showing the spatial effects of river-induced P limitation on DIN concentration, phytoplankton biomass, and hypoxia in one-dimensional flow-through systems (**A**) and open dispersive systems (**B**). Resource limitation is indicated at the *top*. For both flow-through and dispersive systems, N is partly removed by denitrification in the sediment. As an indication, spatial distributions are also represented for systems where P is not limiting (*dashed lines*)

generally followed by N limitation during the low-discharge season in summer (Conley 2000).

Sediments play an important role in the generation of coastal hypoxia. Particulate organic matter (POM) is deposited and remineralized in the sediments, which consume O_2 in the process. Sediment oxygen consumption (SOC) can be the dominant O_2 sink (Quiñones-Rivera et al. 2010) and thus a significant driver of hypoxia (Fennel et al. 2013; Yu et al. 2015a). A fraction of the deposited N is lost as biologically unavailable N_2 through sediment denitrification, an anaerobic microbial remineralization process that represents a major sink for N in coastal areas (Fennel et al. 2009). P may be adsorbed onto the sediment in oxygenated conditions, but is otherwise released back to the water column as DIP (McManus et al. 1997). This disparity in N and P recycling results in a shift back to N limitation in downstream and offshore waters (Fig. 7.1).

There is currently no consensus on whether P limitation amplifies or weakens hypoxia. P limitation is generally viewed as a mechanism that relocates or spreads hypoxia (Paerl et al. 2004; Scavia and Donnelly 2007) and thereby considered detrimental. Any relocation of primary production due to P limitation will also induce a relocation of O_2 sinks although the detailed mechanisms are nonlinear and thus hard to predict. On the one hand, the decomposition of organic matter in downstream waters should result in higher O_2 consumption rates there and could potentially lead to the development of hypoxic conditions in waters that would be normoxic without P limitation. This is likely the case in one-dimensional flow-through systems like the Neuse River Estuary, which are characterized by strong freshwater-induced stratification and transport akin to a simple translation along their upstream–downstream axis; a shift of primary production along this axis may well result in a linear effect on hypoxia (Fig. 7.1a). On the other hand, P limitation in dispersive open systems, such as river plumes, may reduce the occurrence and magnitude of hypoxia. Excess nutrients in river plumes are being diluted when plumes interact with coastal circulation forced by topography, winds, and tides. In this case, a "downstream" relocation may spread elevated primary production over a larger area while lowering the maxima of primary production in the affected area, in effect "diluting" the imprint of eutrophication (Fig. 7.1b).

Further uncertainty in predicting the effects of P limitation on hypoxia results from its differential effect on sediment-water fluxes of N and P. At low O_2, N-loss through sediment denitrification is inhibited (Kemp et al. 1990), whereas P is released from the sediments (Ingall and Jahnke 1997). The relative magnitude of these effects is often poorly constrained. Certainly, the O_2 dependence can lead to high sediment-water fluxes of N and P during hypoxic conditions (Conley et al. 2002; Kemp et al. 2005) and result in nonlinear effects of P limitation on hypoxia.

In the following sections, we discuss and compare four prominent examples of hypoxic systems where P limitation has been studied. Their locations are shown in Fig. 7.2.

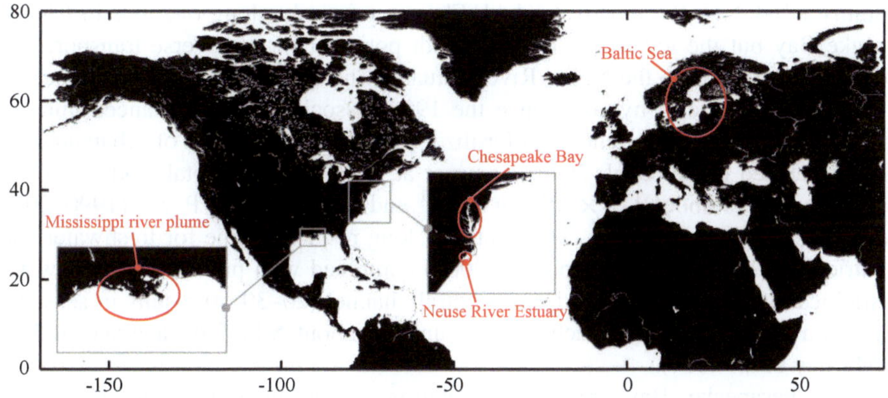

Fig. 7.2 Map showing the location of the four hypoxic systems discussed in Sect. 7.2

7.2.1 Neuse River Estuary

The Neuse River Estuary is a eutrophic to hypertrophic estuary (Paerl et al. 1998) located on the Atlantic coast of the USA (Fig. 7.2). This shallow estuary (~4 m) has a 455 km^2 surface area and drains water from a 16.1×10^3 km^2 watershed (Stow and Borsuk 2000). The estuary discharges into the Pamlico Sound, which is connected to the Atlantic Ocean. N and P loads to this estuary from urban, industrial, and agricultural sources have increased since the 1960s (Stow et al. 2001). Annual total load into the Neuse River Estuary is about 9.6×10^6 kg N y^{-1} and 9.7×10^5 kg P y^{-1} (1993–2003 average, Burkholder et al. 2006). The increase in nutrient load was followed by the emergence of cyanobacterial blooms in the upstream portion of the estuary and a concomitant increase in bottom-water hypoxia (Paerl et al. 1998, 2004) that can reach over 40% of the estuary's surface area in summer (Buzzelli et al. 2002). The decrease in P loading resulted from a ban of P detergents in the 1980s and represents a "natural" system-wide experiment on the effect of P limitation on a one-dimensional eutrophic flow-through estuary. After the ban came into effect, P limitation developed in the upstream portion of the Neuse River Estuary and led to the downstream transport of excess DIN with a subsequent increase of phytoplankton biomass in mid-estuary waters (Paerl et al. 2004). Evidence for how this affected hypoxia is limited.

7.2.2 Chesapeake Bay

Chesapeake Bay, located about 300 km north of the Neuse River Estuary (Fig. 7.2), is the largest (11.5×10^3 km^2) and most productive estuary in the USA. This relatively shallow (~8 m with a deep central channel) and partially stratified estuary drains a 164.2×10^3 km^2 watershed. Flow is primarily along one axis in Chesapeake Bay but the lower bay is broad with potential for transverse transport and dispersion. Similar to the Neuse River Estuary, Chesapeake Bay has experienced a long-term increase in hypoxia since the 1950s associated with enhanced nutrient loading due to the use of chemical fertilizers and the development of urban areas in the watershed (Hagy et al. 2004; Kemp et al. 2005). Annual total load into Chesapeake Bay is about 1.5×10^8 kg N y^{-1} and 9.6×10^6 kg P y^{-1} (1990–2012 average, Hirsch et al. 2013). The relatively long residence time for freshwater and nutrients (90–180 days, Kemp et al. 2005) combined with pronounced freshwater stratification that isolates the deeper central channel (20–30 m) results in seasonal hypoxia in the mid-bay, reaching a volume of about 8 km^3 on average in July (Murphy et al. 2011).

In Chesapeake Bay, resource limitation varies in space and follows a well-defined seasonal cycle controlled by freshwater inputs (Kemp et al. 2005). P limitation develops in the upper section of the bay which receives high N:P river inputs in spring during the peak of freshwater runoff (Fisher et al. 1992, 1999) and is alleviated by remineralized P fluxes from the sediment as waters move downstream (Fisher et al. 1999). The intrusion of N-limited marine waters and the benthic source of P modify nutrient stoichiometry in the lower section of Chesapeake Bay where primary production is typically N-limited. Using N and P load scenarios with a multi-nutrient biogeochemical model, Wang et al. (2016) showed that P load reduction in the upper bay tributaries can be an effective strategy to increase bottom-water O_2 in the upper bay due to the intensification of P limitation. This suggests a mitigation effect of P limitation on hypoxia. However, direct evidence of the spatial and temporal effects of P limitation on hypoxia is still missing for Chesapeake Bay.

7.2.3 Northern Gulf of Mexico

The Mississippi–Atchafalaya River Basin is the third largest river basin in the world; it drains a 3.2×10^6 km^2 watershed and discharges onto the Louisiana shelf through the Mississippi River delta and Atchafalaya Bay. The Mississippi River plume is a dispersive open system controlled by a complex set of interactions with wind forcing, coastal topography, and cross-shore transport associated with eddies (Schiller et al. 2011). As for the previous systems, nutrient loading increased since the 1960s, mainly due to the extensive use of chemical fertilizers, and represents currently an annual total load of about 1.2×10^9 kg N y^{-1} and 1.5×10^8 kg P y^{-1}. Excess nutrient load results in high primary production (Lohrenz et al. 1997) and recurrent bottom-water hypoxia in summer that extends over 14.9×10^3 km^2 on average (Obenour et al. 2013). This is the largest hypoxic area in North American coastal waters (Rabalais et al. 2002). Numerous physical processes affect the timing, location, and extent of hypoxia on the Louisiana shelf (Hetland and DiMarco 2008; Wang and Justic 2009; Zhang et al. 2012; Yu et al. 2015a).

On the Louisiana shelf, P limitation occurs during peak discharge in spring and early summer (Sylvan et al. 2006, 2007; Quigg et al. 2011) due to a high N:P ratio in Mississippi River loads. P limitation leads to the downstream (westward) transport of excess N and a subsequent increase of primary production on the western Louisiana shelf (Laurent et al. 2012). Due to a dilution effect, this results in a reduction of hypoxia on the Louisiana shelf (Laurent and Fennel 2014; see details in Sect. 7.4). During the low-discharge season in summer, resource limitation switches back to N limitation because of N-loss through sediment denitrification and mixing with open ocean waters.

7.2.4 Baltic Sea

The Baltic Sea is the largest brackish water system in the world. It covers a total area of 381 × 10^3 km^2 and drains a 1.6 × 10^6 km^3 watershed. This relatively deep (~55 m) semienclosed estuarine system is connected to the North Sea through the Danish Straits, via a series of sills that limit deep-water exchange with the deep and strongly stratified subbasins of the Baltic Sea. The long water residence time of the Baltic Sea (20–30 years) leads to long periods of O_2 depletion in bottom waters leading to widespread hypoxia in the deep basins (Krauss 2001).

The Baltic Sea has been increasingly affected by anthropogenic eutrophication since the 1960s with a simultaneous increase of hypoxia (Conley et al. 2002, 2011). Currently, the average total N and P loads are about 7.0 × 10^8 kg N y^{-1} and 0.3 × 10^8 kg P y^{-1}, respectively (HELCOM 2013) and are dominated by inputs from major rivers such as the Vistula, Oder, Daugava, Neman, and Neva (Stålnacke et al. 1999). P loads have decreased since the 1980s due to management of point sources (Nausch et al. 1999); hence, one would expect the system to move toward P limitation. Indeed P limitation has been observed but only in nearshore areas such as the Himmerfjärd inlet on the coast of Sweden in spring and early summer following P management measures (Granéli et al. 1990). P limitation rarely occurs in the open waters of the central Baltic Sea due to the efficient release of P from sediments under hypoxic conditions (Eilola et al. 2009). Low O_2 waters also promote N removal through denitrification when nitrate is available, due to an increased oxic/anoxic interface with increasing hypoxic volume (Vahtera et al. 2007). The resulting low N:P ratio in surface waters of the open Baltic Sea stimulates the development of N_2-fixing cyanobacterial blooms (Neumann and Schernewski 2008). These blooms in combination with the external P loads exacerbate bottom-water hypoxia and further increase the internal P load. Despite P load reduction measures, this positive feedback inhibits the recovery of the system from eutrophication (Vahtera et al. 2007).

7.2.5 *One-Dimensional Flow-Through Versus Dispersive Open Systems*

The effect of P limitation on phytoplankton is similar among the four systems presented above; P limitation displaces their biomass toward downstream waters due to the transport of excess N (Fig. 7.1a). The consequence for hypoxia depends on the system. In one-dimensional systems such as the Neuse River Estuary, hypoxia is thought to respond to P limitation with a displacement or an extension of hypoxic conditions (Fig. 7.1a). On the contrary, in open dispersive systems such as the Mississippi River plume, P limitation dilutes eutrophication and reduces hypoxia (Fig. 7.1b). In intermediate systems, such as Chesapeake Bay, both effects may occur depending on the location and timing of hypoxia. However, these contrasting effects remain speculative due to the limited number of investigations.

The Baltic Sea does not seem to fit in this continuum. It is an unusual system because permanent hypoxia appears to restrict the possibility for P limitation in open waters. This example demonstrates the difference between seasonally and permanent hypoxic systems and illustrates the importance of sediment-water fluxes and recycling in controlling P limitation in eutrophic systems.

7.3 Modeling P Limitation in Coastal Hypoxic Systems

There are primarily three reasons for simulating P limitation in coastal hypoxic systems: (1) to better understand the processes controlling hypoxia, (2) to improve the predictive skill of hypoxia prediction models, and (3) to be better able to evaluate and design nutrient management strategies for hypoxia mitigation. Hereafter, we present an overview of modeling frameworks and strategies that are used to investigate the relationship between coastal hypoxia and nutrient loading. We focus on models that explicitly consider P load and are able to represent P limitation.

7.3.1 *Statistical Regressions*

Perhaps the simplest approach for assessing the effects of P limitation on hypoxia is through statistical regressions. Statistical regression models have been used to predict hypoxia or bottom O_2 concentration as a function of nutrient loading (Turner et al. 2006; Greene et al. 2009) and physical drivers (Prasad et al. 2011; Forrest et al. 2011; Obenour et al. 2012). Although not strictly a simulation model, this strategy is worth mentioning here because it is a useful tool for assessing the effects of nutrient loading as well as other factors on the development of hypoxia. In this approach, the contribution of N and P loads to hypoxia is inferred from their statistical weight in the multiple regressions. However, P limitation is not directly represented; hence, this approach does not provide mechanistic insight.

7.3.2 Coupled Physical-Biogeochemical Models

Biogeochemical models represent the cycling of chemical elements such as C, N, P, or O in an ecosystem due to biotic and abiotic processes. Resource limitation is directly represented in these models; hence, they provide mechanistic insights into the effect of P limitation on hypoxia. The physical framework is also important because nutrient supply, primary production, and freshwater transport are three-dimensional, time-dependent processes that influence the development and location of hypoxia; in addition, vertical stratification is a key driver. Physical-biogeochemical models coupling circulation, physical properties of the water column, and biogeochemistry are therefore essential to study the processes controlling hypoxia in a spatially explicit manner. These models vary in their level of complexity in terms of both their biogeochemistry and physical framework. For instance, biogeochemical models may represent P as DIP only, or may include dissolved organic P (DOP) and the O_2-dependent P scavenging on detrital POM. Stratification and transport can be represented using coarse resolution box models that parameterize large-scale transport or high-resolution hydrodynamic models that simulate circulation explicitly.

7.3.2.1 Formulations of Limitation by Multiple Nutrients

Because P limitation is often a transient phenomenon in coastal hypoxic systems, multi-nutrient models that are able to switch between limitation by P and N (and sometimes silicate) are necessary. Several model formulations have been used to represent multi-nutrient limitation in ecosystem models. They can be divided into three general categories (Flynn 2003): (1) simple Monod-type models, (2) more complex cell quota models, and (3) complex mechanistic models that represent internal nutrient pools and feedback processes. Mechanistic models are possibly more realistic in representing the biochemistry of the cell (John and Flynn 2000), but may be over-parameterized due to more unknown parameters, are computationally less efficient, and may be unnecessarily complex for the purpose of simulating the effects of nutrient limitation switches on hypoxia. The quota model, which can include different functional forms of phytoplankton, represents internal nutrient storage. This model type is more efficient and frequently used in ecosystem studies (e.g., Roelke et al. 1999). However, measurements of intracellular nutrient concentrations are scarce, and thus, modeled internal nutrient concentrations cannot be validated against observations. Monod-type models represent nutrient limitation as a direct function of nutrient availability in surrounding waters and assume balanced growth. They are frequently used in multi-nutrient ecosystem models because of their stability and computational efficiency.

Several multi-nutrient Monod-type functional forms have been proposed (O'Neill et al. 1989). The most commonly used are the minimum and the multiplicative functional forms. The minimum functional form assumes that phytoplankton growth depends on the most limiting nutrient, whereas the multiplicative form assumes co-limitation. The minimum function is frequently used to represent P limitation in biogeochemical models, e.g., to study O_2 dynamics in the Baltic Sea (Neumann et al. 2002; Eilola et al. 2009) and on the Louisiana shelf (Justić and Wang 2014; Laurent and Fennel 2014). These formulations are based on limitation factors L_N for N ($L_N = L_{NO3} + L_{NH4}$) and L_P for P, calculated as follows:

$$L_{NO3} = \frac{NO3}{k_{NO3} + NO3} \cdot \frac{1}{1 + NH4/k_{NH4}} \tag{7.1}$$

$$L_{NH4} = \frac{NH4}{k_{NH4} + NH4} \tag{7.2}$$

$$L_P = \frac{DIP}{k_{DIP} + DIP} \tag{7.3}$$

where k_{NO3}, k_{NH4}, and k_{DIP} are the half-saturation constants for nitrate, ammonium, and DIP uptake, respectively. The second factor on the right-hand side of Eq. 7.1 represents ammonium inhibition of nitrate uptake. Quadratic versions of the limiting factors have also been used (Neumann et al. 2002).

The specific phytoplankton growth rate (μ) depends on light (E), temperature (T), and the nutrient limitation factor such that

$$\mu = \mu_{max}(E, T) \cdot L_{tot} \tag{7.4}$$

where $\mu_{max}(E, T)$ is the light- and temperature-dependent maximum growth rate of phytoplankton and L_{tot} is the nutrient limitation factor ($0 < L_{tot} < 1$).

The minimum and multiplicative forms of L_{tot} are formulated, respectively, as follows:

$$L_{tot} = \min(L_N, L_P) \tag{7.5}$$

$$L_{tot} = L_N \cdot L_P \tag{7.6}$$

7.3.2.2 Sediment-Water Fluxes

Sediment-water fluxes represent a key aspect of simulating P limitation in hypoxic environments. Nutrient fluxes control the occurrence of P limitation (e.g., Baltic Sea) and its location (e.g., Chesapeake Bay), whereas SOC can be an important sink for bottom O_2 in eutrophic systems (e.g., Fennel et al. 2013; Yu et al. 2015b). In biogeochemical models, sediment-water fluxes can either be parameterized or mechanistically simulated using a diagenetic model (Fennel et al. 2009). Perhaps the most parsimonious approach is a reflective boundary where SOC, N and P fluxes are proportional to deposited POM (Laurent and Fennel 2014). Simple parameterizations based on observations are used to represent sediment-water fluxes as a function of overlying bottom-water conditions. For example, Fennel et al. (2013) represent SOC as a function of bottom O_2 and temperature based on observed relationships. In this case, nutrient fluxes are a linear function of SOC. More complex parameterizations and representations of P cycling at the sediment-water interface are also used (e.g., Neumann and Schernewski 2008; Eilola et al. 2009; Gustafsson 2012; Justić and Wang 2014). The appropriate level of complexity for these parameterizations depends on the characteristics of the system.

Vertically resolved diagenetic models are the most realistic representation of sediment processes. Yet, they do not necessarily perform better than simple parameterizations of sediment-water fluxes (Wilson et al. 2013). Given their higher computational cost, they are typically used at the expense of physical realism (e.g., Eldridge and Roelke 2010) and simple parameterizations are often preferred. Parameterizations based on diagenetic model results have been proposed as an intermediate solution (Laurent et al. 2016).

7.3.2.3 Box Models

Box models have a coarse spatial resolution that reduces hydrodynamics to the large-scale spatial features of the system. For example, a 3-layer box model of the semienclosed Szczecin Lagoon in the southern Baltic Sea was used to describe long-term effects of eutrophication in a coastal system and its response to nutrient management (Humborg et al. 2000). A 4-box, 2-layer circulation framework of the Mississippi River plume was used to study the effects of nutrient loading in the seasonally hypoxic Louisiana shelf (Eldridge and Roelke 2010). This simple transport framework allows for the use of a more complex representation of biogeochemistry, namely the use of a mechanistic diagenetic model to represent realistic sediment-water fluxes, and representation of multiple phytoplankton groups with different nutrient requirements and growth dynamics (Eldridge and Roelke 2010).

An important drawback of the box model framework is its inability to resolve the details of vertical stratification despite its importance in hypoxia formation. For example, in the Mississippi River plume, hypoxic conditions are limited to the relatively thin bottom boundary layer (Wiseman et al. 1997; Fennel et al. 2016). The lack of vertical resolution can be alleviated to some degree by increasing the horizontal resolution (Gustafsson 2012). Nonetheless, the coarse horizontal resolution remains inadequate to represent the spatial effect of P limitation on primary production (Fig. 7.1), and therefore, box models are limited in their suitability for studying the effect of P limitation on hypoxia.

7.3.2.4 Hydrodynamic Models

High-resolution hydrodynamic models enable the more realistic spatial representation of resource limitation and hypoxia that is often essential for studying the spatial and temporal effects of P limitation on hypoxia in coastal environments. The hydrodynamic models are coupled to biogeochemical models of varying complexities. However, given the computing cost, relatively simple biogeochemical models are often used to represent pelagic processes. This type of framework has been used in large hypoxic systems such as Chesapeake Bay (see Irby et al. 2016 for a model comparison), the Baltic Sea (Eilola et al. 2009), and the northern Gulf of Mexico (Justić and Wang 2014; Laurent and Fennel 2014). Laterally averaged and three-dimensional coarse resolution coupled models have also been used in the Neuse River Estuary (Bales and Robbins 1999; Wool et al. 2003).

Hereafter, we present an overview of the recent investigations of Laurent and Fennel (2014) that used a coupled hydrodynamic-biogeochemical model of the northern Gulf of Mexico to quantitatively assess the effect of P limitation on hypoxia in a relatively open and dispersive system. Model setup and results are also relevant to other coastal systems where P limitation and hypoxia co-occur.

7.4 The Mississippi River Plume Case Study

7.4.1 Model Description

The model simulates circulation and biogeochemistry with the Regional Ocean Modeling System (ROMS, Haidvogel et al. 2008). The circulation model is described in Hetland and DiMarco (2008, 2012). The biogeochemical model aims at a parsimonious representation of the N and P cycles (Fig. 7.3). It is based on the N-cycle model of Fennel et al. (2006, 2008), which was extended to include DIP (Laurent et al. 2012) and O_2 (Fennel et al. 2013).

In the model, phytoplankton growth is limited by either DIN or DIP according to the multi-nutrient formulation described by Eqs. 7.1–7.4. Nutrient limitation factors inferred from inorganic nutrient measurements on the Louisiana shelf compare well with their corresponding nutrient addition bioassays (Fig. 7.4), indicating that the formulation of resource limitation in the model is appropriate. DIN and DIP sources are the Mississippi and Atchafalaya rivers (see locations in Fig. 7.5) and remineralization of POM in the water column and in the sediment. The sediment-water interface is represented as a reflective boundary with a preferential return of DIP to

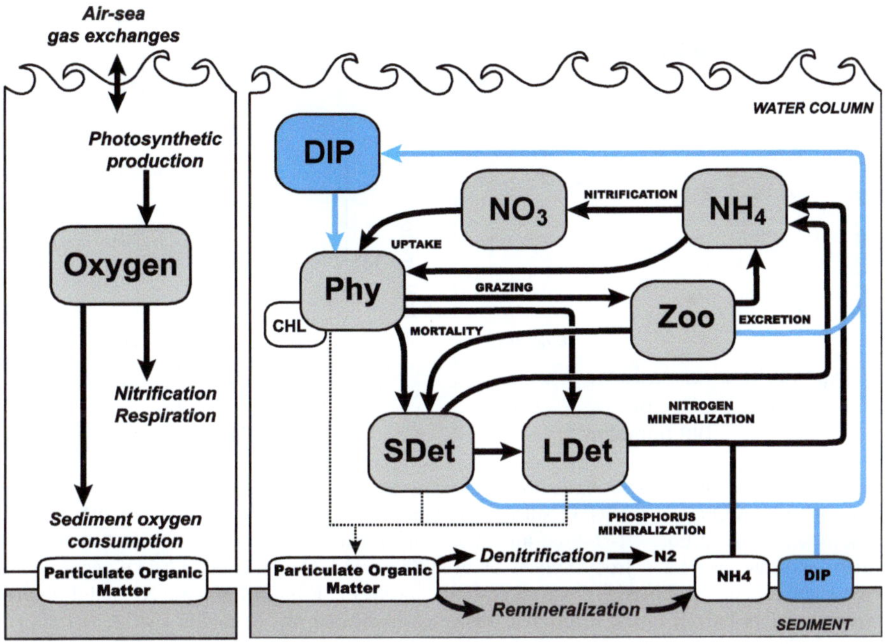

Fig. 7.3 Schematic of the biogeochemical model used in Laurent and Fennel (2014). The state variables are phytoplankton (Phy), zooplankton (Zoo), slow-sinking small detritus (SDet), fast-sinking large detritus (LDet), nitrate (NO_3), ammonium (NH_4), DIP, and O_2. The dynamics of O_2 is coupled with the production and respiration of organic matter via the sources and sinks indicated in the left panel. Figure from Laurent and Fennel (2014) subject to a CC-By 4.0 license

the water column due to denitrification in the sediment. Slowly sinking phytoplankton (Phy) and suspended detritus (SDet) coagulate into fast-sinking large detritus (LDet) as a function of their concentration according to $\tau \cdot (\text{SDet} + \text{Phy})^2$ where τ is the coagulation rate. Sources and sinks of O_2 are associated with the N-cycle (Fig. 7.3) as follows: In the water column, O_2 is produced by photosynthesis, lost through nitrification and respiration, and exchanged with the atmosphere at the air-sea interface; O_2 is lost at the sediment-water interface due to sediment oxygen consumption (Fennel et al. 2008, 2013). The model does not consider O_2 feedbacks on sediment denitrification and on the P cycle because these processes are not well constrained for the Louisiana shelf and because bottom-water O_2 is generally well-above anoxic conditions where nonlinearities are expected to become important.

Two types of simulations are used by Laurent and Fennel (2014) to quantify the effects of P limitation: (1) In a baseline simulation, the multi-nutrient version of the model with N and P cycling (as illustrated in Fig. 7.3) is used. In this simulation, either N or P can limit primary production. (2) In an additional, single-nutrient simulation, only the N-cycle version of the model is used (here, DIP is disabled, but all other processes and model forcing are identical to the baseline simulation).

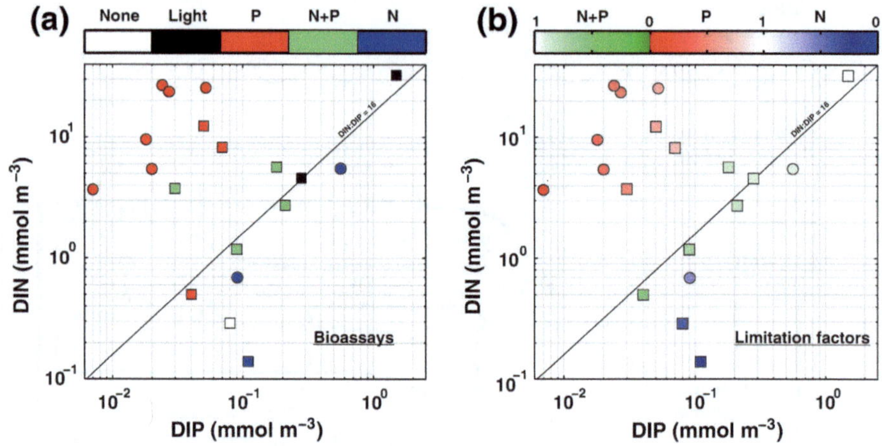

Fig. 7.4 Results of resource limitation bioassays as a function of initial (ambient) nutrient concentration. **A** Original results. **B** Nutrient limitation factors calculated from initial DIN and DIP concentrations using Eqs. 7.1–7.4 (co-limitation is assumed if $|L_N - L_P| < 0.1$). Filled squares and circles indicate bioassays from Sylvan et al. (2006) and Quigg et al. (2011), respectively. Phytoplankton in the bioassays was light-limited (*black*), P-limited (*blue*), N- and P-limited (*green*), and N-limited (*red*). In original bioassays indicating no limitation (*white*, *left panel*), phytoplankton did not respond to added nutrient or light (see references for details)

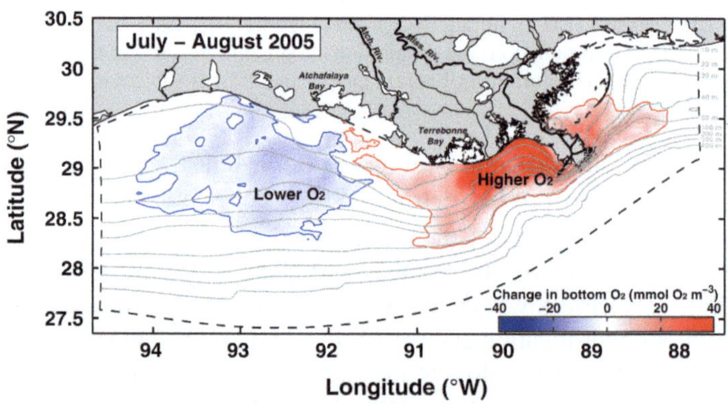

Fig. 7.5 Results from Laurent and Fennel (2014) showing average change in *bottom* O_2 concentration due to P limitation in July and August 2005 on the Louisiana shelf. Positive values (*red*) indicate an increase in O_2 concentration. *Red* and *blue* isolines indicate –3 and +3 mmol O_2 m^{-3}, respectively. The effect of P limitation on *bottom* O_2 was calculated daily in the *bottom layer* of each grid cell as the difference between O_2 concentration from the baseline and the N-only simulations. Shown is the average difference in July and August 2005

Differences between the N-only and the baseline simulations are due to the absence of P limitation in the former and illustrate the effect of P limitation on the system.

7.4.2 *Spatial/Temporal Shift in Primary Production*

The model successfully simulates resource limitation on the Louisiana shelf, namely the development of P limitation between March and July and the switch back to N limitation in September. P limitation results in excess DIN that is transported downstream (i.e., westward) between March and July, where it fuels primary production in otherwise N-limited waters (Laurent et al. 2012). This induces a time delay and a westward relocation of a fraction of primary production (Fig. 7.6a). Here, we quantify this relocation using the simulations described in Laurent and Fennel (2014), in terms of the change in horizontal transport of POM and DIN across the boundaries of three boxes representing the western, mid-, and eastern Louisiana shelf (Fig. 7.7). With P limitation, the following changes in transport occur: (1) Less primary production occurs on the eastern shelf resulting in a smaller transport of POM westward and across the shelf break into the open Gulf of Mexico, (2) Westward transport of DIN is larger, and (3) Transport of DIN across the shelf break is larger (Fig. 7.7). Offshore transport of DIN across the shelf break is larger near the Mississippi delta, but more of this DIN is recirculated onto the mid- and western shelf. On average, 34 and 26% more DIN is transported westward into the mid- and western shelf boxes, respectively. POM and DIN fluxes further west onto the Texas shelf also increase slightly (Fig. 7.7).

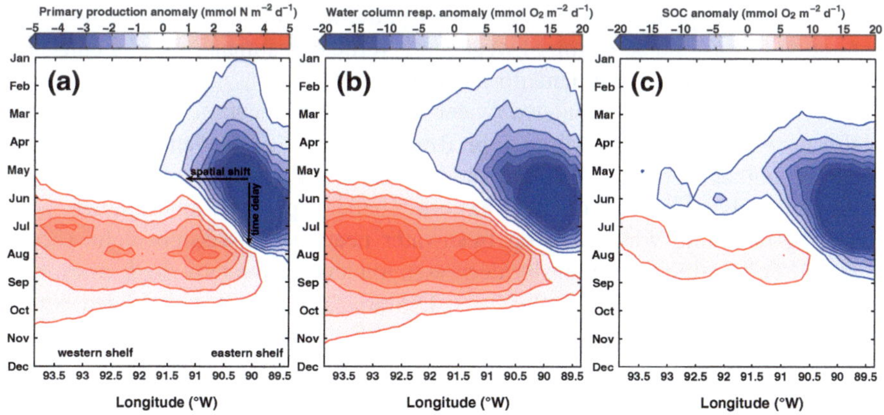

Fig. 7.6 Results from Laurent and Fennel (2014) showing spatial and temporal changes in water column integrated primary production (**A**), water column respiration (**B**), and SOC (**C**) due to P limitation on the Louisiana shelf. Values represent monthly and latitudinal averages for the Louisiana shelf ($z \leq 50$ m). Latitudinal averages were obtained by averaging over the shelf (from the coast to the 50 m isobath) for each 0.1° longitude bin. Shown is the difference between the baseline and the N-only simulations

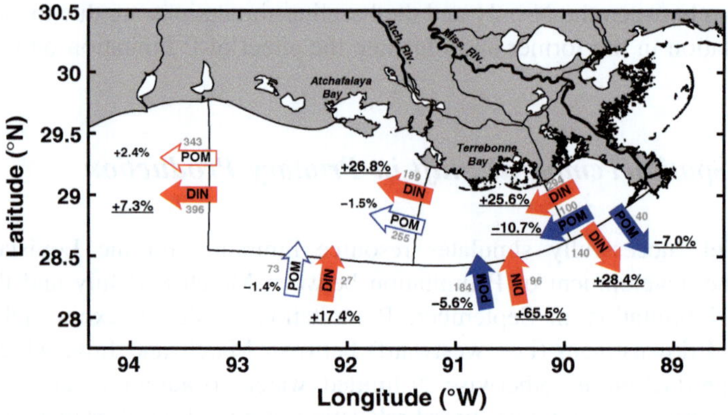

Fig. 7.7 Changes in average POM and DIN transport over the Louisiana shelf due to P limitation. Transport is calculated from the simulations of Laurent and Fennel (2014) for three boxes representative of the shelf, their boundaries shown with a *thin black line*. The southern boundaries follow the 50-m isobath. *Arrows* indicate the direction of the transport and their color an increase (*red*), a decrease (*blue*), or a small change (*white*) due to P limitation (also indicated in percent). The *gray* number beside each *arrow* indicates the magnitude of the transport (in 10^3 mmol N s^{-1})

7.4.3 The Dilution Effect

Laurent and Fennel (2014) have shown that P limitation reduces the size of the July–August hypoxic area by an average of 29%. In their simulation, changes in bottom-water O_2 are asymmetric over the Louisiana shelf with a significant increase on the eastern shelf, but only a small decrease on the western shelf (Fig. 7.5). Two additive effects explain this spatial asymmetry: the westward shift of organic matter respiration against the backdrop of weakening vertical stratification and the net shift of respiration from the sediments to the water column.

The intensity of the simulated water column stratification varies along the freshwater gradient of the Mississippi River plume (e.g., Hetland and DiMarco 2008; Zhang et al. 2012; Laurent and Fennel 2014). Simulated water column stratification is strongest on the eastern shelf, especially in summer when upwelling-favorable winds pile up freshwater near the Mississippi delta (Zhang et al. 2012). Stratification intensity decreases toward the western shelf, away from the Mississippi River delta (Hetland and DiMarco 2008). Since simulated bottom O_2 concentration is highly correlated with stratification intensity on the Louisiana shelf (Fennel et al. 2013), a westward shift in organic matter respiration results in a reduction of hypoxia (Laurent and Fennel 2014).

In addition, the simulations suggest that P limitation leads to a redistribution of respiration between sediment and water column (Fig. 7.6b, c). The westward shift of primary production occurs on a broadening shelf, thus spreading primary production over a larger area and essentially diluting phytoplankton and detritus (Fig. 7.6a). Smaller concentrations of phytoplankton and suspended detritus (even if present over a larger area) reduce coagulation into fast-sinking, larger detritus; hence, more organic matter is respired in the water column and less in the sediment when P limitation is present. This leads to an asymmetric effect of P limitation on respiration between the eastern and the western Louisiana shelf (Fig. 7.6b, c). Observations and simulations show that hypoxia mainly occurs within the bottom boundary layer (BBL) on the Louisiana shelf (i.e., within less than 5 m above the bottom; see Fennel et al. 2016). In this layer, SOC is the main O_2 sink and therefore controls hypoxia development on the Louisiana shelf (Fennel et al. 2013; Yu et al. 2015b). The net shift of respiration from sediments to the water column due to P limitation reduces respiration within the bottom boundary layer and therefore the overall extent of hypoxia (Laurent and Fennel 2014).

7.4.4 Hypoxia Remediation Strategies

Four nutrient reduction scenarios were tested in Laurent and Fennel (2014) to assess their effect on summer hypoxia: a 50% decrease of river DIN load (−N), a 50% decrease of river DIP load (−P), a 50% decrease of river DIN and DIP loads (−NP), and a 50% increase in river DIN load with a simultaneous 50% reduction of river DIP load (+N−P). The dual N and P load reduction (−NP) maximizes the decrease in hypoxia size and duration (Fig. 7.8). Decreasing N load only (−N) reduces hypoxia on the western shelf but the hypoxic area is larger than with a dual nutrient reduction. Reducing P only (−P) reduces the hypoxic area but not hypoxia duration on the western shelf. Finally, decreasing P but with a simultaneous increase of N

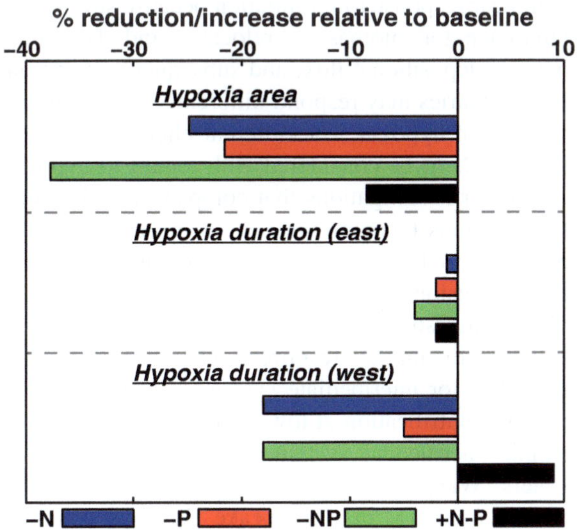

Fig. 7.8 Results from the nutrient load experiments of Laurent and Fennel (2014) showing the relative change in summer 2004 hypoxic area and hypoxia duration at an eastern and western shelf station, associated with a modification in nutrient river load as follows: −50% DIN (−N), −50% DIP (−P), −50% nutrients (−NP), and +50% DIN, −50% DIP (+N−P)

(stronger P limitation) leads to a small reduction of the hypoxic area, but also to a longer duration of hypoxic conditions on the western shelf (Fig. 7.8). A dual N and P load reduction strategy is therefore recommended to mitigate hypoxia on the Louisiana shelf.

7.5 Conclusions and Recommendations

River-induced P limitation is a common phenomenon in coastal hypoxic systems. Although consideration of P limitation is of direct relevance for hypoxia mitigation through nutrient load reduction, there is still limited direct evidence for the effect of P limitation on hypoxia. Models are invaluable tools because they allow one to simulate system-wide N and P load manipulations that are necessary to quantitatively assess the effects of resource limitation. This was illustrated by the recent modeling investigation of the Mississippi River plume (Laurent and Fennel 2014). This type of coupled hydrodynamic-biogeochemical model is most suitable to examine the effect of P limitation on hypoxia because it represents limitation by multiple resources and hypoxia within a realistic spatial and physical setting. This is essential in this context, and similar investigations should be conducted in other P-limited coastal hypoxic systems.

Based on the general understanding of P limitation and the results of the Mississippi River plume study, we were able to distinguish between two types of hypoxic systems: one-dimensional flow-through and open dispersive systems, illustrated by the conceptual model in Fig. 7.1. One-dimensional systems such as estuaries may respond linearly to P limitation with a downstream relocation of hypoxia. More open and dispersive systems such as river plumes likely have a nonlinear response to P limitation because stratification weakens downstream where phytoplankton biomass is relocated and the dilution of biomass over the shelf reduces depositional flux, and subsequently hypoxia. Intermediate systems such as large estuaries may respond differently to P limitation depending on the timing and location of hypoxia. The different effect of P limitation on this continuum of systems needs to be considered with regard to hypoxia mitigation strategies.

Further investigations that compare the effect of P limitation on hypoxia across these various types of systems are needed to confirm our conceptual model. The Neuse River Estuary is a good candidate for studying the effect of P limitation on one-dimensional systems because both resource limitation and hypoxia were previously investigated there. A relatively simple biogeochemical model should be sufficient to simulate resource limitation in this system. Chesapeake Bay is a good candidate for intermediate systems. Sediment-water feedbacks such as the inhibition of denitrification at low O_2 may need to be considered in Chesapeake Bay (or similar intermediate systems), given the relatively long residence time of the deep waters. Wang et al. (2016) simulated the effect of N and P load reduction on bottom-water O_2 in Chesapeake Bay and concluded that increased P limitation associated with P load reduction has a positive effect on bottom-water O_2 where P limitation occurs. A follow-up study is necessary to characterize the direct effect of P limitation on the timing, location, and intensity of hypoxia in this intermediate system. Model frameworks should be designed carefully in open dispersive systems and include the essential processes influencing hypoxia development. For example, coagulation and sediment remineralization were necessary to simulate the response to P limitation in the Mississippi River plume. It was also important to represent the spatial variation in stratification over the Louisiana shelf. Essential processes vary among systems, and therefore, local characteristics should be taken into account during model development.

Simulated river-induced P limitation in this continuum from one-dimensional flow-through to open dispersive systems will significantly improve our conceptual understanding of resource limitation in the context of coastal hypoxia and will help design and improve nutrient load reduction strategies to mitigate river-induced hypoxia.

Acknowledgments This work was supported by NOAA CSCOR grants NA06N0S4780198 and NA09N0S4780208. This is NOAA NGOMEX publication number 215.

References

Bales JD, Robbins JC (1999) A dynamic water-quality modeling framework for the Neuse River estuary, North Carolina. No. 99-4017. US Department of the Interior, US Geological Survey

Boesch DF (2002) Challenges and opportunities for science in reducing nutrient over-enrichment of coastal ecosystems. Estuaries 25:886–900. doi:10.1007/BF02804914

Burkholder JM, Dickey DA, Kinder CA et al (2006) Comprehensive trend analysis of nutrients and related variables in a large eutrophic estuary : a decadal study of anthropogenic and climatic influences 51:463–487. doi:10.4319/lo.2006.51.1_part_2.0463

Buzzelli C, Luettich R, Powers S et al (2002) Estimating the spatial extent of bottom-water hypoxia and habitat degradation in a shallow estuary. Mar Ecol Prog Ser 230:103–112. doi:10.3354/meps230103

Caraco N, Cole J, Likens GE (1990) A comparison of phosphorus immobilization in sediments of freshwater and coastal marine systems. Biogeochemistry 9:277–290. doi:10.1007/BF00000602

Conley DJ (2000) Biogeochemical nutrient cycles and nutrient management strategies. Hydrobiologia 410:87–96. doi:10.1023/A:1003784504005

Conley DJ, Carstensen J, Aigars J et al (2011) Hypoxia is increasing in the coastal zone of the Baltic Sea. Environ Sci Technol 45:6777–6783. doi:10.1021/es201212r

Conley DJ, Humborg C, Rahm L et al (2002) Hypoxia in the Baltic Sea and basin-scale changes in phosphorus biogeochemistry. Environ Sci Technol 36:5315–5320. doi:10.1021/es025763w

Eilola K, Meier HEM, Almroth E (2009) On the dynamics of oxygen, phosphorus and cyanobacteria in the Baltic Sea; a model study. J Mar Syst 75:163–184. doi:10.1016/j.jmarsys.2008.08.009

Eldridge PM, Roelke DL (2010) Origins and scales of hypoxia on the Louisiana shelf: importance of seasonal plankton dynamics and river nutrients and discharge. Ecol Model 221:1028–1042. doi:10.1016/j.ecolmodel.2009.04.054

Fennel K, Brady D, DiToro D et al (2009) Modeling denitrification in aquatic sediments. Biogeochemistry 93:159–178. doi:10.1007/s10533-008-9270-z

Fennel K, Hu J, Laurent A et al (2013) Sensitivity of hypoxia predictions for the Northern Gulf of Mexico to sediment oxygen consumption and model nesting. J Geophys Res-Oceans 118: 990–1002. doi:10.1002/jgrc.20077

Fennel K, Laurent A, Hetland R, Justić D, Ko DS, Lehrter J, Murrell M, Wang L, Yu L, Zhang W (2016) Effects of model physics on hypoxia simulations for the Northern Gulf of Mexico: a model intercomparison. J Geophys Res-Oceans 121. doi:10.1002/2015JC011577

Fennel K, Wilkin J, Levin J et al (2006) Nitrogen cycling in the Middle Atlantic Bight: results from a three-dimensional model and implications for the North Atlantic nitrogen budget. Glob Biogeochem Cycles 20:GB3007. doi:10.1029/2005GB002456

Fennel K, Wilkin J, Previdi M, Najjar R (2008) Denitrification effects on air-sea CO_2 flux in the coastal ocean: simulations for the Northwest North Atlantic. Geophys Res Lett 35:L24608. doi:10.1029/2008GL036147

Fisher TR, Gustafson AB, Sellner K et al (1999) Spatial and temporal variation of resource limitation in Chesapeake Bay. Mar Biol 133:763–778. doi:10.1007/s002270050518

Fisher TR, Peele ER, Ammerman JW, Harding LW (1992) Nutrient limitation of phytoplankton in Chesapeake Bay. Mar Ecol Prog Ser 90:51–63

Flynn KJ (2003) Modelling multi-nutrient interactions in phytoplankton; balancing simplicity and realism. Prog Oceanogr 56:249–279. doi:10.1016/S0079-6611(03)00006-5

Forrest DR, Hetland RD, DiMarco SF (2011) Multivariable statistical regression models of the areal extent of hypoxia over the Texas-Louisiana continental shelf. Environ Res Lett 6:045002. doi:10.1088/1748-9326/6/4/045002

Granéli E, Wallström K, Larsson U et al (1990) Nutrient limitation of primary production in the Baltic Sea area. Ambio 19:142–151

Greene RM, Lehrter JC, Hagy JD III (2009) Multiple regression models for hindcasting and forecasting midsummer hypoxia in the Gulf of Mexico. Ecol Appl 19:1161–1175. doi:10.1890/08-0035.1

Gustafsson E (2012) Modelled long-term development of Hypoxic area and nutrient pools in the Baltic Proper. J Mar Syst 94:120–134. doi:10.1016/j.jmarsys.2011.11.012

Hagy JD, Boynton WR, Keefe CW, Wood KV (2004) Hypoxia in Chesapeake Bay, 1950–2001: long-term change in relation to nutrient loading and river flow. Estuaries 27:634–658. doi:10.1007/BF02907650

Haidvogel DB, Arango H, Budgell WP et al (2008) Ocean forecasting in terrain-following coordinates: formulation and skill assessment of the regional ocean modeling system. J Comput Phys 227:3595–3624. doi:10.1016/j.jcp.2007.06.016

Harrison JA, Bouwman AF, Mayorga E, Seitzinger S (2010) Magnitudes and sources of dissolved inorganic phosphorus inputs to surface fresh waters and the coastal zone: a new global model. Glob Biogeochem Cycles 24:GB1003. doi:10.1029/2009GB003590

HELCOM (2013) Review of the fifth Baltic Sea pollution load compilation for the 2013 HELCOM Ministerial Meeting

Hetland RD, DiMarco SF (2008) How does the character of oxygen demand control the structure of hypoxia on the Texas-Louisiana continental shelf? J Mar Syst 70:49–62. doi:10.1016/j.jmarsys.2007.03.002

Hetland RD, DiMarco SF (2012) Skill assessment of a hydrodynamic model of circulation over the Texas-Louisiana continental shelf. Ocean Model 43–44:64–76. doi:10.1016/j.ocemod.2011.11.009

Hirsch RM, Moyer DL, Phillips SW (2013) Chesapeake Bay program indicator framework

Howarth RW, Marino R (2006) Nitrogen as the limiting nutrient for eutrophication in coastal marine ecosystems: evolving views over three decades. Limnol Oceanogr 51:364–376. doi:10.4319/lo.2006.51.1_part_2.0364

Humborg C, Fennel K, Pastuszak M, Fennel W (2000) A box model approach for a long-term assessment of estuarine eutrophication, Szczecin Lagoon, southern Baltic. J Mar Syst 25:387–403. doi:10.1016/S0924-7963(00)00029-4

Ingall E, Jahnke R (1997) Influence of water-column anoxia on the elemental fractionation of carbon and phosphorus during sediment diagenesis. Mar Geol 139:219–229. doi:10.1016/S0025-3227(96)00112-0

Irby I, Friedrichs MAM, Friedrichs CT, Bever AJ, Hood RR, Lanerolle LWJ, Li M, Linker L, Scully ME, Sellner K, ShenJ Testa J, Wang H, Wang P, Xia M (2016) Challenges associated with modeling low-oxygen waters in Chesapeake Bay: a multiple model comparison. Biogeosciences 13:2011–2028. doi:10.5194/bg-13-2011-2016

John EH, Flynn KJ (2000) Modelling phosphate transport and assimilation in microalgae; how much complexity is warranted? Ecol Model 125:145–157. doi:10.1016/S0304-3800(99)00178-7

Justić D, Wang L (2014) Assessing temporal and spatial variability of hypoxia over the inner Louisiana–upper Texas shelf: application of an unstructured-grid three-dimensional coupled hydrodynamic-water quality model. Cont Shelf Res. doi:10.1016/j.csr.2013.08.006

Kemp W, Boynton W, Adolf J et al (2005) Eutrophication of Chesapeake Bay: historical trends and ecological interactions. Mar Ecol Prog Ser 303:1–29. doi:10.3354/meps303001

Kemp WM, Sampou P, Cafrey J et al (1990) Ammonium recycling versus denitrification Chesapeake Bay sediments. Limnol Oceanogr 35:1545–1563. doi:10.4319/lo.1990.35.7.1545

Krauss W (2001) Baltic Sea circulation. Encycl Ocean Sci 236–244 Elsevier Ltd

Laurent A, Fennel K (2014) Simulated reduction of hypoxia in the Northern Gulf of Mexico due to phosphorus limitation. Elem Sci Anth 2:000022. doi:10.12952/journal.elementa.000022

Laurent A, Fennel K, Hu J, Hetland R (2012) Simulating the effects of phosphorus limitation in the Mississippi and Atchafalaya River plumes. Biogeosciences 9:4707–4723. doi:10.5194/bg-9-4707-2012

Laurent A, Fennel K, Wilson R, Lehrter J, Devereux R (2016) Parameterization of biogeochemical sediment–water fluxes using in situ measurements and a diagenetic model. Biogeosciences 13:77–94. doi:10.5194/bg-13-77-2016

Lohrenz SE, Fahnenstiel GL, Redalje DG et al (1997) Variations in primary production of Northern Gulf of Mexico continental shelf waters linked to nutrient inputs from the Mississippi River. Mar Ecol Prog Ser 155:45–54. doi:10.3354/meps155045

McManus J, Berelson WM, Coale KH et al (1997) Phosphorus regeneration in continental margin sediments. Geochim Cosmochim Ac 61:2891–2907. doi:10.1016/S0016-7037(97)00138-5

Murphy RR, Kemp WM, Ball WP (2011) Long-term trends in Chesapeake Bay seasonal hypoxia, stratification, and nutrient loading. Estuaries Coasts 34:1293–1309. doi:10.1007/s12237-011-9413-7

Nausch GI, Nehring D, Aertebjerg G (1999) Anthropogenic nutrient load of the Baltic Sea. Limnologica 29:233–241

Neumann T, Fennel W, Kremp C (2002) Experimental simulations with an ecosystem model of the Baltic Sea: a nutrient load reduction experiment. Glob Biogeochem Cycles 16:1033. doi:10.1029/2001GB001450

Neumann T, Schernewski G (2008) Eutrophication in the Baltic Sea and shifts in nitrogen fixation analyzed with a 3D ecosystem model. J Mar Syst 74:592–602. doi:10.1016/j.jmarsys.2008.05.003

O'Neill RV, DeAngelis DL, Pastor JJ et al (1989) Multiple nutrient limitations in ecological models. Ecol Model 46:147–163. doi:10.1016/0304-3800(89)90015-X

Obenour DR, Michalak AM, Zhou Y, Scavia D (2012) Quantifying the impacts of stratification and nutrient loading on hypoxia in the Northern Gulf of Mexico. Env Sci Technol 46:5489. doi:10.1021/es204481a

Obenour DR, Scavia D, Rabalais NN et al (2013) Retrospective analysis of midsummer Hypoxic area and volume in the Northern Gulf of Mexico, 1985–2011. Environ Sci Technol 47:9808–9815. doi:10.1021/es400983g

Paerl H (2009) Controlling eutrophication along the freshwater–marine continuum: dual nutrient (N and P) reductions are essential. Estuaries Coasts 32:593–601. doi:10.1007/s12237-009-9158-8

Paerl HW, Pinckney JL, Fear JM, Peierls BL (1998) Ecosystem responses to internal and watershed organic matter loading: consequences for hypoxia in the eutrophying Neuse River Estuary, North Carolina, USA. Mar Ecol Prog Ser 166:17–25

Paerl HW, Valdes LM, Joyner AR et al (2004) Solving problems resulting from solutions: evolution of a dual nutrient management strategy for the eutrophying Neuse River Estuary, North Carolina. Env Sci Technol 38:3068–3073. doi:10.1021/es0352350

Prasad MBK, Long W, Zhang X et al (2011) Predicting dissolved oxygen in the Chesapeake Bay: applications and implications. Aquat Sci 73:437–451. doi:10.1007/s00027-011-0191-x

Quigg A, Sylvan JB, Gustafson AB et al (2011) Going west: nutrient limitation of primary production in the Northern Gulf of Mexico and the importance of the Atchafalaya River. Aquat Geochem 17:519–544. doi:10.1007/s10498-011-9134-3

Quiñones-Rivera ZJ, Wissel B, Rabalais NN, Justic D (2010) Effects of biological and physical factors on seasonal oxygen dynamics in a stratified, eutrophic coastal ecosystem. Limnol Oceanogr 55:289–304. doi:10.4319/lo.2010.55.1.0289

Rabalais N, Turner RE, Dortch Q et al (2002) Nutrient-enhanced productivity in the Northern Gulf of Mexico: past, present and future. Hydrobiologia 475–476:39–63. doi:10.1023/A:1020388503274

Redfield AC, Ketchum BH, Richards FA (1963) The influence of organisms on the composition of sea-water. In: Hill MN (ed) The compensation of sea water comparative and descriptive oceanography. Wiley, New York, pp 26–77

Roelke DL, Eldridge PM, Cifuentes LA (1999) A model of phytoplankton competition for limiting and nonlimiting nutrients: implications for development of estuarine and nearshore management schemes. Estuaries 22:92–104

Scavia D, Donnelly KA (2007) Reassessing hypoxia forecasts for the Gulf of Mexico. Env Sci Technol 41:8111–8117. doi:10.1021/es0714235

Schiller RV, Kourafalou VH, Hogan P, Walker ND (2011) The dynamics of the Mississippi River plume: impact of topography, wind and offshore forcing on the fate of plume waters. J Geophys Res 116:C06029. doi:10.1029/2010JC006883

Stålnacke P, Grimvall A, Sundblad K, Tonderski A (1999) Estimation of riverine loads of nitrogen and phosphorus to the Baltic Sea, 1970–1993. Environ Monit Assess 58:173–200. doi:10.1023/A:1006073015871

Stow CA, Borsuk ME (2000) Neuse river estuary modeling and monitoring project stage 1: an examination of long term nutrient data in the Neuse River watershed. Duke University, Durham, North Carolina

Stow CA, Borsuk ME, Stanley DW (2001) Long-term changes in watershed nutrient inputs and riverine exports in the Neuse River, North Carolina. Water Res 35:1489–99. doi:10.1016/S0043-1354(00)00402-4

Sylvan JB, Dortch Q, Nelson DM et al (2006) Phosphorus limits phytoplankton growth on the Louisiana shelf during the period of hypoxia formation. Environ Sci Technol 40:7548–7553. doi:10.1021/es061417t

Sylvan JB, Quigg A, Tozzi S, Ammerman JW (2007) Eutrophication-induced phosphorus limitation in the Mississippi River Plume: evidence from fast repetition rate fluorometry. Limnol Ocean 52:2679–2685. doi:10.4319/lo.2007.52.6.2679

Turner RE, Rabalais NN, Justic D (2006) Predicting summer hypoxia in the Northern Gulf of Mexico: riverine N, P, and Si loading. Mar Pollut Bull 52:139–148. doi:10.1016/j.marpolbul.2005.08.012

Vahtera E, Conley D, Gustafsson BG, Kuosa H, Pitkänen H, Savchuk OP, Tamminen T, Viitasalo M, Voss M, Wasmund N, Wulff F (2007) Internal ecosystem feedbacks enhance nitrogen-fixing cyanobacteria blooms and complicate management in the Baltic Sea. Ambio 36:186–194. doi:10.1579/0044-7447(2007)36[186:IEFENC]2.0.CO;2

Wang L, Justić D (2009) A modeling study of the physical processes affecting the development of seasonal hypoxia over the inner Louisiana-Texas shelf: circulation and stratification. Cont Shelf Res 29:1464–1476. doi:10.1016/j.csr.2009.03.014

Wang P, Linker LC, Shenk GW (2016) Using geographically isolated loading scenarios to analyze nitrogen and phosphorus exchanges and explore tailored nutrient control strategies for efficient management. Environ Model Assess 21:437–454. doi:10.1007/s10666-015-9487-x

Wilson RF, Fennel K, Paul Mattern J (2013) Simulating sediment–water exchange of nutrients and oxygen: a comparative assessment of models against mesocosm observations. Cont Shelf Res 63:69–84. doi:10.1016/j.csr.2013.05.003

Wiseman WJ, Rabalais NN, Turner RE, Dinnel SP, MacNaughton A (1997) Seasonal and interannual variability within the Louisiana coastal current: stratification and hypoxia. J Mar Syst 12:237–248. doi:10.1016/S0924-7963(96)00100-5

Wool TA, Davie SR, Rodriguez HN (2003) Development of three-dimensional hydrodynamic and water quality models to support total maximum daily load decision process for the Neuse River Estuary, North Carolina. J Water Resour Plann Manage 129:295–306. doi:10.1061/(ASCE)0733-9496(2003)129:4(295)

Yu L, Fennel K, Laurent A (2015a) A modeling study of physical controls on hypoxia generation in the Northern Gulf of Mexico. J Geophys Res-Oceans 120:5019–5039. doi:10.1002/2014JC010634

Yu L, Fennel K, Laurent A, Murrell MC, Lehrter JC (2015b) Numerical analysis of the primary processes controlling oxygen dynamics on the Louisiana shelf. Biogeosciences 12:2063–2076. doi:10.5194/bg-12-2063-2015

Zhang J, Gilbert D, Gooday A et al (2010) Natural and human-induced hypoxia and consequences for coastal areas: synthesis and future development. Biogeosciences 7:1443–1467. doi:10.5194/bg-7-1443-2010

Zhang X, Hetland RD, Marta-Almeida M, DiMarco SF (2012) A numerical investigation of the Mississippi and Atchafalaya freshwater transport, filling and flushing times on the Texas-Louisiana Shelf. J Geophys Res 117:C11009. doi:10.1029/2012JC008108

Chapter 8
Predicted Effects of Climate Change on Northern Gulf of Mexico Hypoxia

John C. Lehrter, Dong S. Ko, Lisa L. Lowe and Bradley Penta

Abstract We describe the application of a coastal ocean ecosystem model to assess the effect of a future climate scenario of plus (+) 3 °C air temperature and + 10% river discharge on hypoxia ($O_2 < 63$ mmol m^{-3}) in the northern Gulf of Mexico. We applied the model to the Louisiana shelf as influenced by the runoff from the Mississippi River basin. The net effect of the future climate scenario was a mean increase in water temperature of 1.1 °C and a decrease in salinity of 0.09 for the region of the shelf where hypoxia typically occurs (<50 m depth). These changes increased the strength of water column stratification at the pycnocline and increased phytoplankton biomass. In the future scenario, the hypoxic area was only 1% larger than the present. A more significant effect was in the duration and extent of severe hypoxic areas. Severe hypoxic areas, defined as model cells having hypoxia for more than 60 days in the year, had a mean increase in hypoxia duration of 9.5 days (a 10% increase). The severely hypoxic area also increased by 1,130 km^2 (an 8% increase) in the future scenario. The results confirm that a warmer and wetter future climate will, on average, worsen the extent and duration of hypoxia in this system. Thus, it is probable that long-term Mississippi River nutrient management for hypoxia will need to be adapted for climate change.

J.C. Lehrter
Gulf Ecology Division, US EPA, Office of Research and Development,
National Health and Environmental Effects Laboratory, Gulf Breeze, FL 32561, USA

Present Address:
J.C. Lehrter (✉)
University of South Alabama and Dauphin Island Sea Lab, Dauphin Island,
AL 36528, USA
e-mail: jlehrter@southalabama.edu

D.S. Ko · B. Penta
Naval Research Laboratory, Oceanography Division, Stennis Space Center,
Bay St. Louis, MS 39529, USA

L.L. Lowe
Leidos, Research Triangle Park, Durham, NC 27709, USA

D. Justic et al. (eds.), *Modeling Coastal Hypoxia*,
DOI 10.1007/978-3-319-54571-4_8

Keywords Hypoxia • Climate change • Nutrients • Northern Gulf of Mexico • Louisiana shelf • Mississippi River • Coastal general ecosystem model (CGEM)

8.1 Introduction

Hypoxic water masses (O_2 < 63 mmol m^{-3}) are increasingly observed in the coastal ocean as a result of land-based nitrogen (N) and phosphorus (P) loading (Diaz and Rosenberg 2008). On the continental shelf of the northern Gulf of Mexico, hypoxia occurs seasonally from late spring to early fall and is correlated with peak river discharge and N and P exported from the Mississippi and Atchafalaya river basin (MARB) (Rabalais et al. 2002; Greene et al. 2009). The annual size of this hypoxic area averages 15,540 km^2 (1993–2015), and concerns about its potential effect on marine life have prompted the development of a management goal to reduce the hypoxic area to a 5-year running average of 5,000 km^2 by 2035 (Mississippi River, Gulf of Mexico Watershed Nutrient Task Force 2015). To be successful, long-term goals such as this will have to incorporate adaptive management against the backdrop of climate change and other changing stressors.

Climate change is expected to further exacerbate the hypoxia problem (Rabalais et al. 2010; Doney et al. 2012). Increased water temperatures due to global warming will impact O_2 concentrations by decreasing O_2 solubility and enhancing temperature-dependent biological metabolism. Increased sea surface temperature and freshwater discharge from rivers are expected to increase the strength of vertical stratification, which is a physical precursor necessary for bottom water hypoxia development (Wiseman et al. 1997). Other climate impacts on hypoxia could include enhanced sinking rates of organic particles due to warmer water temperature (Bach et al. 2012) that will increase the deposition of organic matter. Enhanced watershed export of nutrients, organic matter, and optically active constituents like colored dissolved organic matter (CDOM) can be expected due to their observed correlation with greater discharge and temperature (Mulholland 2002; Aulenbach et al. 2007; Spencer et al. 2009; 2012). These terrestrial exports could impact hypoxia directly through enhanced primary production and respiration or indirectly by absorbing shortwave and photosynthetically available radiation, which would be expected to further warm the surface ocean and reduce the PAR available for photosynthetic O_2 production. In sum, these changes are expected to increase the severity of hypoxia in terms of lower O_2 concentrations, longer duration of low O_2, and larger areal extent.

Various aspects of climate change impacts on Louisiana shelf hypoxia have been modeled previously. Justic et al. (1996) developed a one-dimensional, vertical model to evaluate the impacts of increased freshwater discharge and nutrient loading on stratification and hypoxia development at single point on the shelf, station C06 (Fig. 8.1). Donner and Scavia (2007) examined potential climate impacts on hypoxia using a modified Streeter–Phelps model that was scaled to the hypoxic area based on a simple one-dimensional, in the horizontal east–west

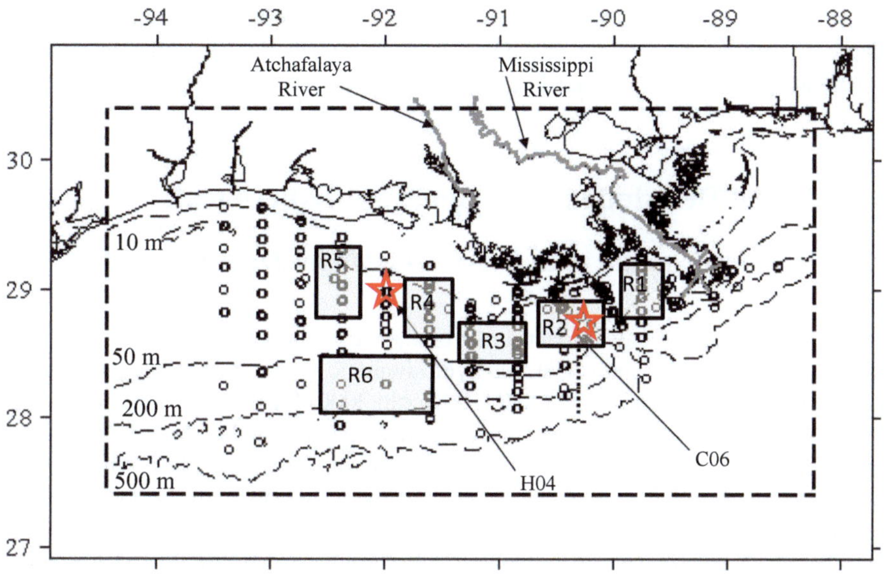

Fig. 8.1 Louisiana shelf modeling domain is shown by the box with *dashed line*. Hypoxia occurs mainly within the 50-m isobath. Sampling stations shown as *circles* were used to develop initial and boundary conditions and for model corroboration. The boxes represent regions (R1–R6) over which model results and observations were aggregated and compared. Two stations, H04 and C06, are noted (*red stars*), where modeled time-series results were compared to observations. The *dotted vertical line* at –90.4 longitude is the location of the vertical slice in Fig. 8.9

direction, physical advection model. These studies have provided first-order predictions that climate change impacts are likely to increase the severity of hypoxia.

More recent modeling studies have implemented three-dimensional (3-D) hydrodynamic and ecosystem models to explore various drivers of oxygen and primary production dynamics (Fennel et al. 2013; Justic and Wang 2014; Pauer et al. 2016). In this study, we expand upon the previous work by using a 3-D hydrodynamic and ecosystem model to simulate how a potential future climate may impact oxygen concentrations and hypoxia. We defined our future climate using a projected increase in air temperature (+3 °C, IPCC 2014) and river discharge (+10%, Sperna Weiland et al. 2012). We then compared the future scenario results with model results from the present (reference year 2006) to assess how these changes in air temperature and river discharge could impact the sea temperature, salinity, vertical stratification, phytoplankton biomass, O_2 concentration, and hypoxic area.

The manuscript is organized as follows. First, we describe the coastal ocean ecosystem model and its application to the Louisiana continental shelf. Then, the results from the numerical simulation with a potential future climate are reported with a discussion of how climate change may impact hypoxia and uncertainties in these results. The main conclusions from this numerical experiment are that the potential future climate affects the temperature and salinity of the shelf and these changes manifest in stronger stratification at the pycnocline. In turn, the stronger stratification results in increased severity of hypoxia on the shelf.

8.2 Model Description and Numerical Experiment

8.2.1 Hydrodynamic and Ecosystem Model Description

To model potential climate impacts on hypoxia, we applied a 3-D hydrodynamic and ecosystem modeling system. The 3-D hydrodynamic model is an implementation of the Navy Coastal Ocean Model (NCOM; Martin 2000) for the Louisiana continental shelf (NCOM-LCS) and was developed to provide horizontal and vertical transport and mixing, temperature, and salinity (Lehrter et al. 2013, Ko et al. 2016) for the ecosystem model. Within the Louisiana shelf model domain (Fig. 8.1), the model grid has a horizontal resolution of approximately 1.9 km. Vertical structure is represented by 20 equally spaced sigma layers on the shelf (≤ 100 m depth) that follow the ocean topography and up to 14 layers at constant depths below 100 m depth.

NCOM-LCS was run for the period October 1, 2004–December 31, 2007, at a time-step of 300 s. Hourly model outputs from January 1 to December 31, 2006 were used for this study. Land–sea forcing of the model (Table 8.1) was through river discharges to the model domain based on daily observations of the Mississippi (USACOE gauge 01100) and Atchafalaya Rivers (USACOE gauge 03045) and from 92 smaller rivers entering the domain based on USGS measurements. Aggregate discharge from the smaller rivers was calculated to be <2% of the MARB discharge. Atmospheric forcing consisted of 3 h, 9 km scale estimates of air pressure, wind stress, and air temperature from the Coupled Ocean/Atmosphere Mesoscale Prediction System (COAMPS; Hodur 1997) and solar short wave radiation and surface total heat flux from the Navy Operational Global Atmospheric Prediction System (NOGAPS; Rosmond 1992). Initial conditions and boundary conditions (Table 8.1) for currents, temperature, and salinity for the eastern, southern, and western open ocean boundaries were provided by a lower-resolution (approximately 6 km scale) regional implementation of NCOM called the Intra-Americas Sea Nowcast Forecast System (IASNFS; Ko et al. 2003; Ko and Wang 2014) covering the Gulf of Mexico, Caribbean Sea, and a portion of the North Atlantic Ocean.

The ecosystem model used for describing nutrients and O_2 dynamics was based on the biogeochemical model described by Eldridge and Roelke (2010). For this study, the one-dimensional (Eldridge and Roelke 2010) model was recoded from MATLAB to FORTRAN, made explicitly 3-D, and generalized to run on any orthogonal grid. We call this new code a coastal general ecosystem model (CGEM). The CGEM state variables used in this application (Table A.1) included one phytoplankton group to represent diatoms, which were the dominant phytoplankton taxa observed on the shelf (Murrell et al. 2014); phytoplankton cell quotas for

Table 8.1 Model boundary conditions

Model boundary	Model process	Treatment
Land–sea	Freshwater discharge, temperature, and concentrations of state variables for 94 rivers	Mississippi and Atchafalaya Rivers daily observed discharge (USACOE) and concentrations of variables at weekly to monthly sampling frequency (USGS). Other rivers had USGS gauged discharge or were assigned discharges based on gauged rivers. Nutrient concentrations (total N and P) for unmonitored rivers were obtained from the USGS SPARROW mode
Air–sea	Sea surface heat, salinity, and momentum fluxes	Temperature, solar shortwave radiation, air pressure, and winds from COAMPS model, 9 km resolution, 3 h. Precipitation from TRMM, 0.25° resolution, 3 h (Ko et al. 2008; Lehrter et al. 2013)
	O_2 and CO_2 exchange	Concentration gradients from sea to air and wind speed (Liss and Merlivat 1986; Whitfield and Turner 1986; Justic et al. 1996; Eldridge and Roelke, 2010)
Open boundaries and initial conditions	Currents, temperature, and salinity	IASNFS model, 6 km resolution, 3 h (Ko et al. 2003; Lehrter et al. 2013)
	Concentrations of state variables	Multivariate regression relationships were developed based on observed depth and salinity (independent variables) and observed concentrations reported in Lehrter et al. (2013). Regression equations were applied to IASNFS depth and salinity outputs at each i, j, k grid cell on the model boundary and each model time-step

Abbreviations: USACOE, United States Army Corps of Engineers; USGS, United States Geological Survey; IASNFS, Intra-America Seas Nowcast/Forecast System; COAMPS, Coupled Ocean/Atmosphere Mesoscale Prediction System; TRMM, Tropical Rainfall Measuring Mission; CMAQ, Community Multiscale Air Quality model

nitrogen and phosphorus; two zooplankton groups to represent large and small grazers; four classes of organic matter in particulate and dissolved forms to represent organic matter derived from phytoplankton, zooplankton, rivers, and ocean boundaries; CDOM; nutrients in the form of NO_3^-, NH_4^+, PO_4^{3-}, and Si; and O_2 (Fig. 8.2). Model equations and parameter tables are provided in Appendices A–F. The CGEM code and model output are available upon request to the authors.

CGEM was applied to the Louisiana shelf for the year 2006 with hydrodynamic forcing from NCOM-LCS. Other forcing for CGEM included (Table 8.1) photosynthetically available radiation (PAR, sum of irradiance for 400–700 nm wavelengths) delivered to the sea surface, which was calculated as a fraction (0.47) of the atmospheric short wave radiation from NOGAPS. Ocean boundary and initial conditions were derived from reported observations of nutrient and organic matter

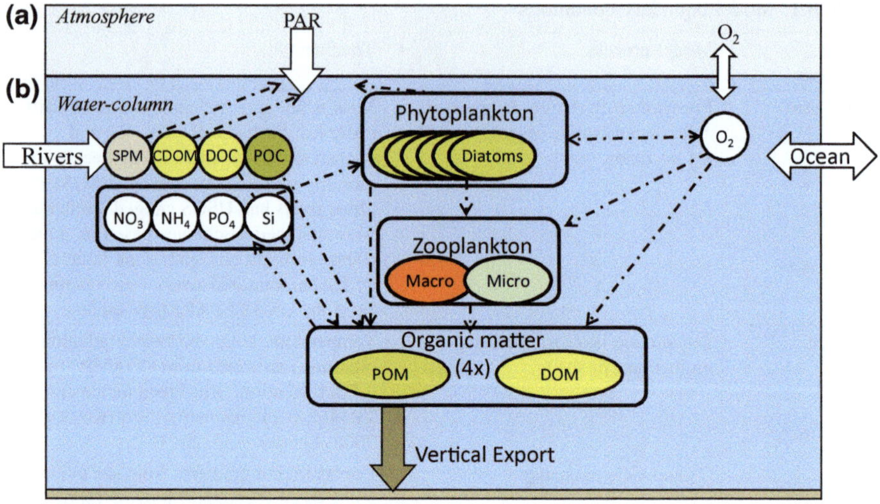

Fig. 8.2 CGEM conceptual model representing **a** atmospheric surface boundary forcing from photosynthetically available radiation (PAR) and air-sea exchange of O_2 and **b** water column horizontal exchange with river and ocean end-members and mechanisms regulating light, nutrient, phytoplankton, zooplankton, and organic matter dynamics. In the present study, only one phytoplankton group is used, but the model is flexible in representing up to 99 groups. Particulate and dissolved organic matter is represented by four types (4x) each including river, algal-derived, zooplankton-derived, and ocean boundary condition organic matter

concentrations (Lehrter et al. 2013). O_2 exchanges at the air–sea interface were calculated based on concentration gradients at the sea surface and wind speed (Liss and Merlivat 1986).

For this study, sediment-water exchanges were not modeled explicitly. As a simplification, particulate organic matter sinks to the bottom layer in the model where it continues to be remineralized through aerobic respiration and denitrification according to the availability of electron acceptors, stoichiometry of the organic matter, and parameterized decay rates.

8.2.2 *In Situ Observations Used to Assess Model Results*

Modeled state variables temperature, salinity, nitrate, chlorophyll, and O_2 were compared to observations from the Louisiana shelf (Fig. 8.1) from three cruises in April, June, and September 2006. During cruises, vertical profiles of temperature, salinity, and O_2 were collected with a CTD (Sea-Bird 911, Sea-Bird Electronics, Bellevue, WA). Chlorophyll *a* (Chla) samples were collected at discrete depths in Niskin bottles. A description of the fluorometric analysis of Chla has been reported previously (Lehrter et al. 2009; 2013; Murrell et al. 2014).

The strength and depth of the pycnocline were estimated at each station from vertical profiles of density, calculated from observed temperature and salinity profiles described above. The density profiles were used to calculate Brunt Väisälä frequencies (N^2, Pond and Pickard 1983) as $N^2 = (g/\rho 0)\ (d\rho/dz)$, where g is the gravitation parameter, ρ0 is reference sea water density, ρ is potential density, and z is water depth positive downward. The maximum Brunt Väisälä frequency (N^2_{max}) and depth were used as proxies for the stratification strength and depth of the pycnocline, respectively, at each station.

The model results were compared to observations (Murrell et al. 2014) from 2006 on both a point-to-point and regional basis. For point-to-point comparisons, model results were extracted to match station locations, depths, and times. For regional comparisons, surface and bottom median concentrations from the model results and observations were calculated for the regions shown in Fig. 8.1.

Three model assessment metrics (Stow et al. 2008; Lehman et al. 2009) were calculated for evaluating model results against observations. Bias, the mean difference between model results and observations, was calculated as

$$Bias = \frac{1}{n}\sum (M - O). \tag{8.1}$$

where O is an observed value and M the modeled value corresponding to n observations. The root-mean-square error, a measure of model agreement with observations, was calculated as

$$RMSE = \sqrt{\frac{\sum (O - M)^2}{n}} \tag{8.2}$$

The model efficiency (ME), a measure of model prediction skill, was calculated from residuals of observed and modeled values in relation to the observed sum of squares

$$ME = 1 - \frac{\sum (O - M)^2}{\sum (O - \overline{O})^2}, \tag{8.3}$$

where $\overline{O}$ is the mean of observations.

The observed hypoxic area from the July 21–27, 2006, survey by the Louisiana Universities Marine Consortium (LUMCON, http://www.gulfhypoxia.net/) was also used for comparison with modeled results. Calculated hypoxic areas from the model results did not consider model grid cells with total depths less than 7 m. These depths were excluded because they were not sampled in the LUMCON survey and because they consisted of model boundary cells at the land–sea interface.

Numerical Experiments

A future climate scenario was developed that included a +3 °C air temperature. This was based on the Intergovernmental Panel on Climate Change (IPCC) RCP6 scenario, which assumes an increase of +6 W m^{-2} of radiative forcing, that results in an approximately +3 °C change in regional air temperature in the northern Gulf of Mexico (Collins et al. 2013). The future scenario also incorporated a +10% river discharge (Sperna Weiland et al. 2012). We chose the year 2006 as the present base year because tropical storms and hurricanes were not present in 2006. Tropical systems are known to affect hypoxia by breaking down the pycnocline and mixing O_2 to depth through intensive winds and large waves. Further, there were observations (Murrell et al. 2014) available from three cruises in 2006 to compare to model results. Thus, we applied the +3 °C air temperature and +10% river discharge to 2006 to assess a future impact on hypoxia.

In the future scenario, the +3 °C air temperature was transferred to the ocean surface by latent and sensible heat fluxes. Latent heat flux occurred due to the air-sea temperature differences and the sea surface moisture pressure, and the sensible heat flux occurred due to the air-sea temperature difference. In addition, temperature enters the model domain in the freshwater discharge from the rivers for which it was assumed that the +3 °C air temperature translated to a change of +1.7 °C in freshwater temperature.

To provide appropriate boundary forcing for the future climate scenario, we first ran the IASNFS regional model using +3 °C and +10% river discharge. We then ran the NCOM-LCS model with the future climate scenario with open ocean boundary forcing from the IASNFS run with the future climate scenario. Finally, CGEM was run with the future temperature, river discharge, and transport supplied by the NCOM-LCS model. In the future climate scenario, nutrient and organic matter concentrations from the rivers and ocean boundaries were kept the same as the 2006 present case. This was done in order to focus on how changing temperature and discharge affect stratification and O_2. However, we note that a 10% increase in river discharge results in a 10% increase in nutrient load (load = discharge × concentration). In another study, we plan to address other potential future scenarios that include alterations in watershed nutrient and organic matter loading.

8.3 Results

8.3.1 Model Hindcast Comparison to Observations

Model skill metrics indicated the hydrodynamic model generally reproduced the observed spatial and temporal patterns of temperature, salinity, and density (sigma-t). A point-to-point comparison with observations of temperature, salinity, and sigma-t from three cruises in 2006 indicated small bias, small RMSE, and good model prediction skill (ME > 0.6) (Table 8.2). An ME > 0 indicates the model is a better predictor than the simple mean of all observations of a state variable, and ME = 1 is the maximum value and indicates the model has perfect agreement with observations. Thus, the model reproduced temporal and spatial patterns in temperature and salinity (Fig. 8.3).

In Fig. 8.3, the apparent bias between observed salinity and model salinity for depths <20 m is due to an under-representation of shallow depths in the

Table 8.2 Model comparison with observations on a point-by-point basis from three cruises in 2006 for temperature (T, °C), salinity (S), density (sigma-t, kg m^{-3}), nitrate (NO_3, mmol m^{-3}), chlorophyll *a* (Chla, mg m^{-3}), and oxygen (O_2, mmol m^{-3}). Shown are the number of point observations that were available for comparison to the model (n), and the estimated model bias, root-mean-square error (RMSE), and model efficiency (ME). For NO_3, Chla, and O_2, model comparison results are also shown for median surface and bottom concentrations extracted from the six regions shown in Fig. 8.1

	n	Bias	RMSE	ME
T	8,777	–0.04	0.97	0.92
S	8,777	–0.42	1.90	0.61
Sigma-T	8,777	–0.31	1.52	0.72
NO_3	723	–0.84	6.31	0.17
Chla	493	0.60	4.72	0.14
O_2	6,638	43.7	65.4	–0.89
$NO_{3regions}$	36	–0.77	3.05	0.25
$Chla_{regions}$	36	–1.02	4.21	0.15
$O_{2regions}$	36	27.2	50.7	0.22

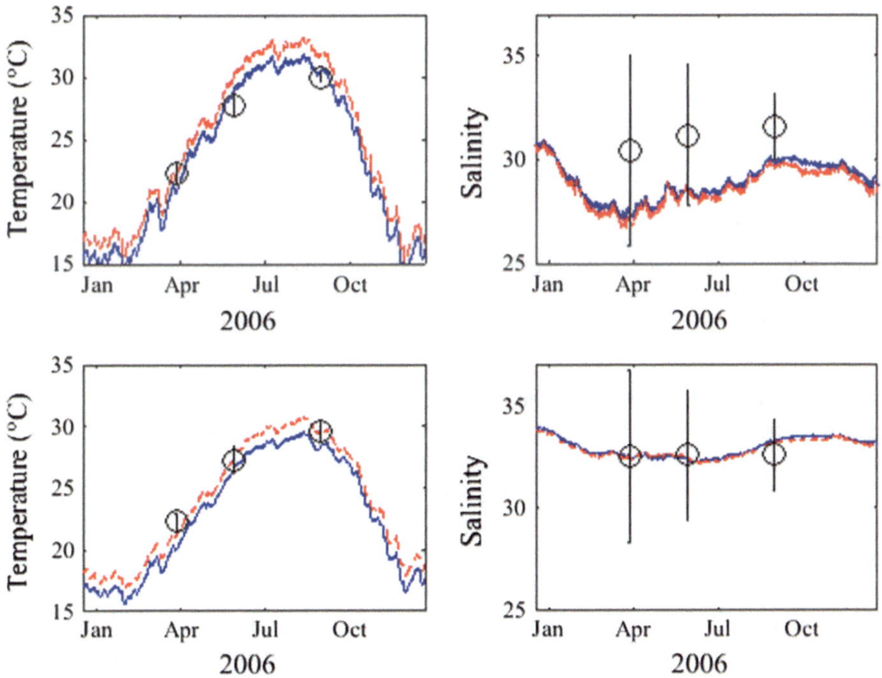

Fig. 8.3 Mean water column temperature and salinity from the model for the present case (*solid blue*) and for the future climate scenario (*dashed red*) for the area of the shelf with depths <20 m (*upper panel*) and <50 m (*lower panel*). The *circles* with error bars represent the mean ± standard deviation of observations in the same depth categories

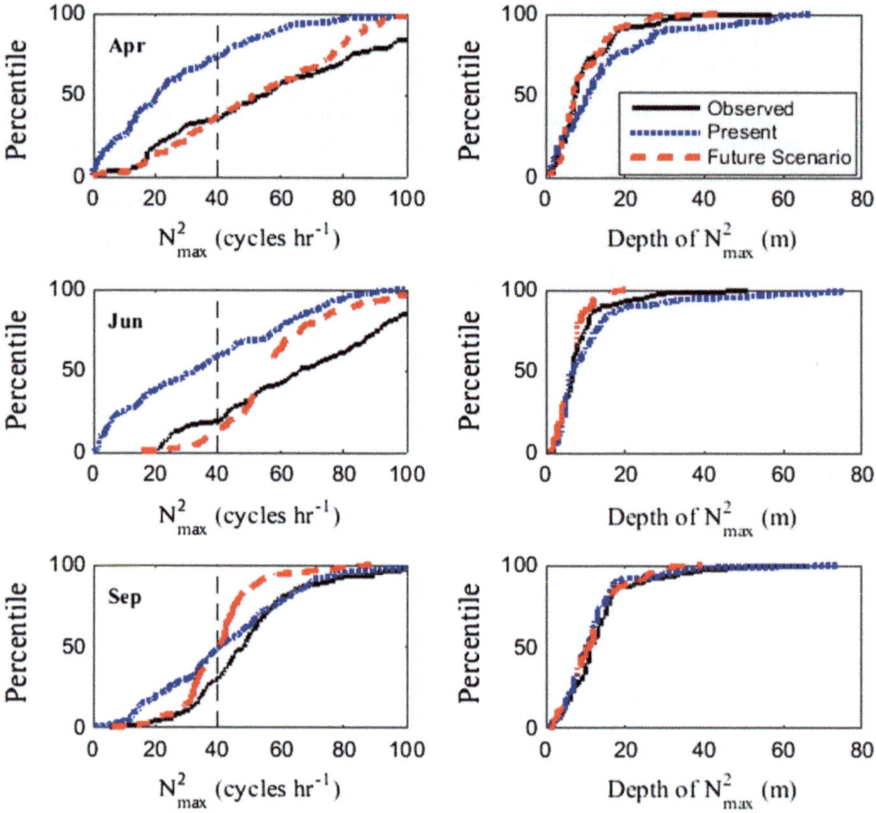

Fig. 8.4 Cumulative distributions of observed (*black line*) and modeled N^2_{max} (*left column*) and depth of N^2_{max} (*right column*). Present (*blue line*) and future scenario (*red line*) model results are shown in comparison with observations from April (*upper panel*), June (*middle panel*), and September (*lower panel*) of 2006. $N^2_{max} = 40$ is shown by the *vertical dashed line*

observational dataset. Whereas the model extends inshore into shallow coastal areas and bay, the observations are not representative of these shallow areas (Fig. 8.1), which are closer to freshwater sources and have lower salinity.

Generally, the modeled pycnocline stratification strengths were weaker than the observations (Fig. 8.4). However, both the present and future scenario models generated a large percentage of N^2 greater than 40 cycles h^{-1}, which is a value above which hypoxia occurs (Bianchi et al. 2010), especially during early summer when greater than 50% of the maximum N^2 were greater than 40. Thus, although the model is not able to reproduce the observed maximum N^2, the model does produce N^2 sufficient for the formation of hypoxia. Modeled pycnocline depths were on average 1.5 m deeper than observed pycnocline depths during April and June 2006 (Fig. 8.4). This deeper bias primarily occurred at locations less than 20 m depth. In September 2006, the modeled pycnocline depths were similar to the observed with no bias.

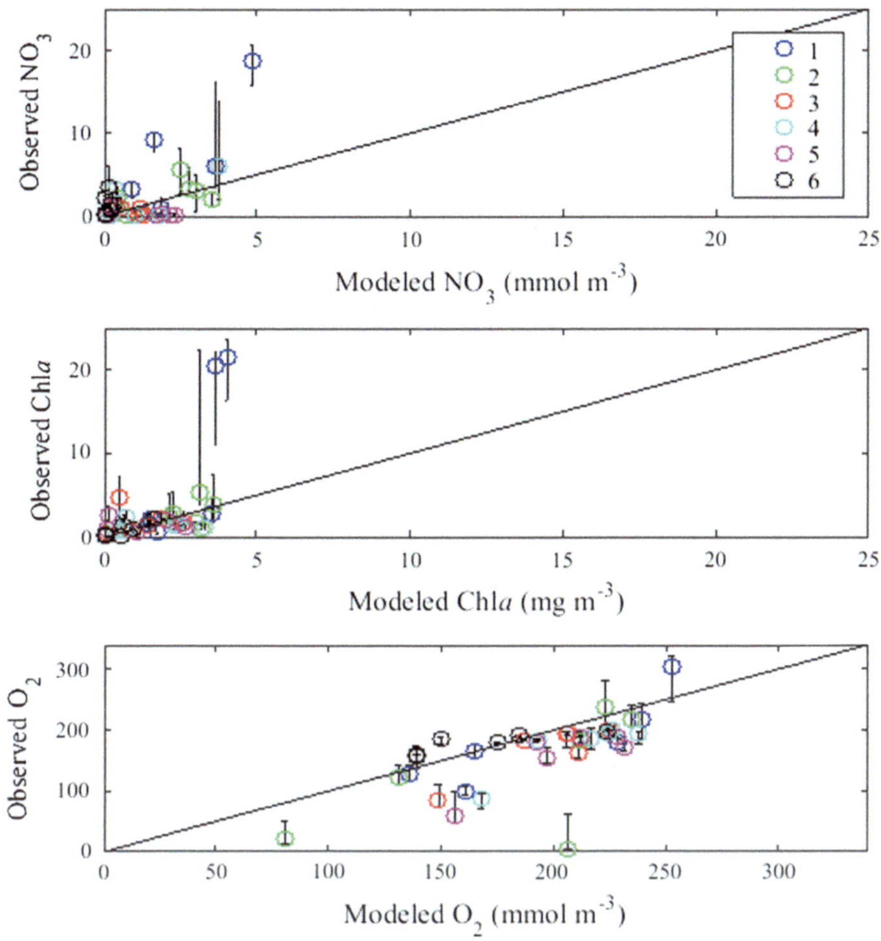

Fig. 8.5 Comparison of present modeled results and observed NO_3, Chl, and O_2. Data shown are median model and observed results from three cruises in 2006 where both median model and observed results were calculated for the six regions shown in Fig. 8.1 for surface and bottom bins (n = 36). Error bars are 25th and 75th percentiles of observations from a region. The *solid line* is the 1:1 line

Model skill metrics for point-to-point and regional comparisons with observations of NO_3^-, Chl, and O_2 are shown in Table 8.2. The model had ME > 0 for NO_3^- and Chl on a point-by-point basis and ME > 0 for all three variables on a regional basis. At a regional scale (Fig. 8.5), the model was unable to reproduce high Chl concentrations observed in Region 1 near the Southwest Pass of the lower Mississippi River, nor did it produce low O_2 concentrations observed in Region 2.

The model generally reproduced seasonal variability in observed Chl and O_2 concentrations at H04 and C06 (Fig. 8.6). The model output in Fig. 8.6 is only at noon on each day. Thus, the modeled results shown do not reproduce the diel

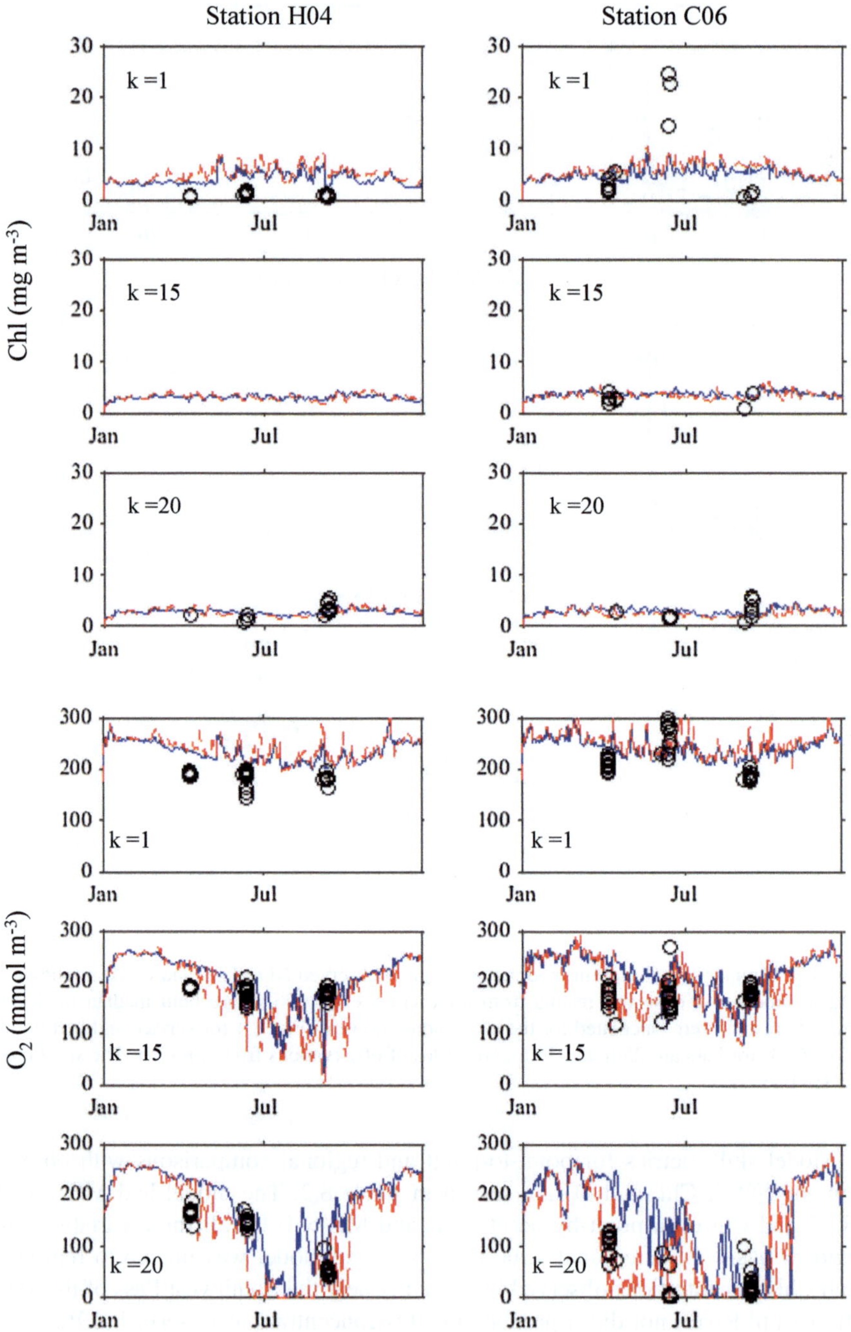
Station H04
Station C06
Chl (mg m-3)
O2 (mmol m-3)
k =1
k =15
k =20
Jan
Jul

◀**Fig. 8.6** Modeled time-series of Chl and O_2 at station H04 (*left column*) and C06 (*right column*) during 2006 for the present (*blue line*) and future (*dashed red line*) scenario. The three subplots in each column are at k levels 1, 15, and 20 (depths of approximately 1, 15, and 20 m). Observed Chl and O_2 concentrations observed in April, June, and September (*black circles*) are shown for comparison

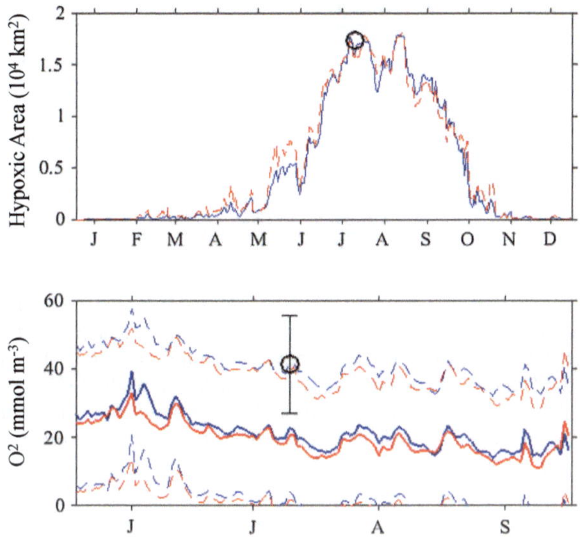

Fig. 8.7 (*Upper plot*) Annual time-series of modeled hypoxic area for the present (*blue line*) and future scenario (*dashed red line*). The observed hypoxic area of 17,280 km^2 from the LUMCON July 21–27, 2006, cruise is shown (*black circle*) for reference. (*Lower plot*) June 1–October 1 time-series of mean modeled O_2 concentration in hypoxic cells for the present (*blue line*) and future scenario (*red line*) with dashed lines representing the ± standard deviation of O_2 concentrations from hypoxic cells. The mean ± standard deviation of the observed O_2 (data from Obenour et al. 2013) in hypoxic cells is shown by the *black circle* with error bars

variability seen in the observations at these stations, which were collected every 3 h over 24–36 h of station occupation.

Modeled hypoxic area for the present scenario was similar to the observed (Fig. 8.7). In Fig. 8.7, the modeled hypoxic area starts at zero on January 1 and climbs to an August 1 value of 17,820 km^2 which is quite similar to the observed hypoxic area of 17,280 km^2. The hypoxic area then declines in mid-August before rising again to its maximum extent of 18,110 km^2 on August 26.

Despite being similar in spatial extent, the spatial distribution of modeled and observed hypoxia differed in some respects (Fig. 8.8). Modeled hypoxia tended to be larger in the offshore direction on the eastern shelf and was contiguous along the shelf. Some of this difference can be attributed to the modeled hypoxic area shown in Fig. 8.8 being from August 26 when the maximum hypoxic area occurred in the model (Fig. 8.7). The observed hypoxic area shown in Fig. 8.8 is based on one cruise from July 21–27. Modeled hypoxic areas shown along the coastline and in

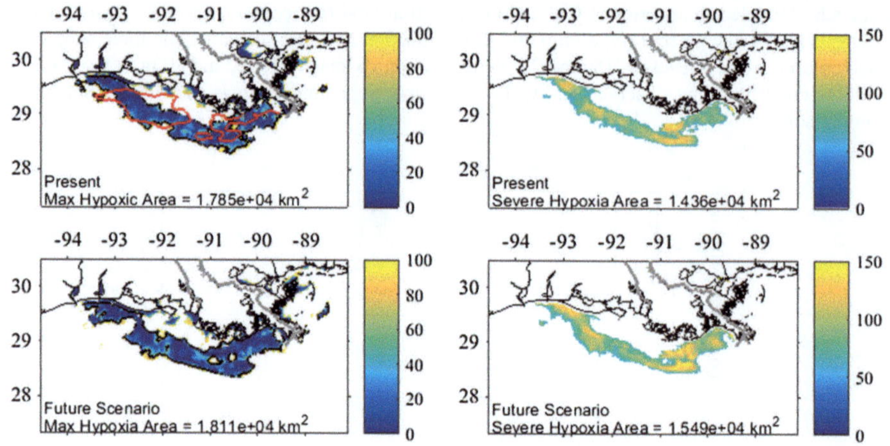

Fig. 8.8 Hypoxic area (*left column*) and duration of severe hypoxia (*right column*). The *left column* shows the present and future scenario maximum hypoxic area for 2006. The *color bar* shows the modeled bottom-water O_2 concentration (mmol m^{-3}) for O_2 ranging from 0 to 100 mmol m^{-3}. The *black solid* contour *line* is the modeled isopleth of $O_2 = 63$ mmol m^{-3}. The *red solid* contour *line* is the observed isopleth of $O_2 = 63$ mmol m^{-3} (data from Obenour et al. 2012). Note that the maximum hypoxic extent occurred on August 26 in the present model results and August 29 in the future scenario, while the observed hypoxia area was measured July 21–27 and reported to be 17,280 km^2. Annotated text is the hypoxic area on August 26 for the present and August 29 for the future scenario. The right column shows the duration of severe hypoxia. Severe hypoxia per model grid cell was defined as occurring for more than 60 days during 2006 (see Sect. 3.2). The *color map* shows the duration (days) that hypoxia occurred in each model cell that was classified as severely hypoxic. Annotated text is the area of model cells where bottom-layer hypoxia persisted longer than 60 days. Note that model results from grid cells with total depths <7 m were excluded from calculations and are not shown in the duration plots

bays in Fig. 8.8 are highly uncertain as these areas of the model are at or near the inshore boundary of the model grid.

8.3.2 Numerical Experiment with Future Climate Scenario

Application of the future climate scenario of +3 °C air temperature increased the modeled water column average temperature by +1.1 °C (Fig. 8.3). At depths shallower than 20 m, the overall temperature increase was +1.3 °C. The +10% river discharge in the future climate reduced average salinity by –0.09 for depths <50 m and –0.24 for depths <20 m (Fig. 8.3). The largest salinity changes of approximately –1 occurred in the plume region of the Atchafalaya River at depths shallower than 10 m. Salinity in the plume region of the Mississippi River was less affected by the +10% discharge (<–0.25 change) due to the deeper waters in this region of the shelf.

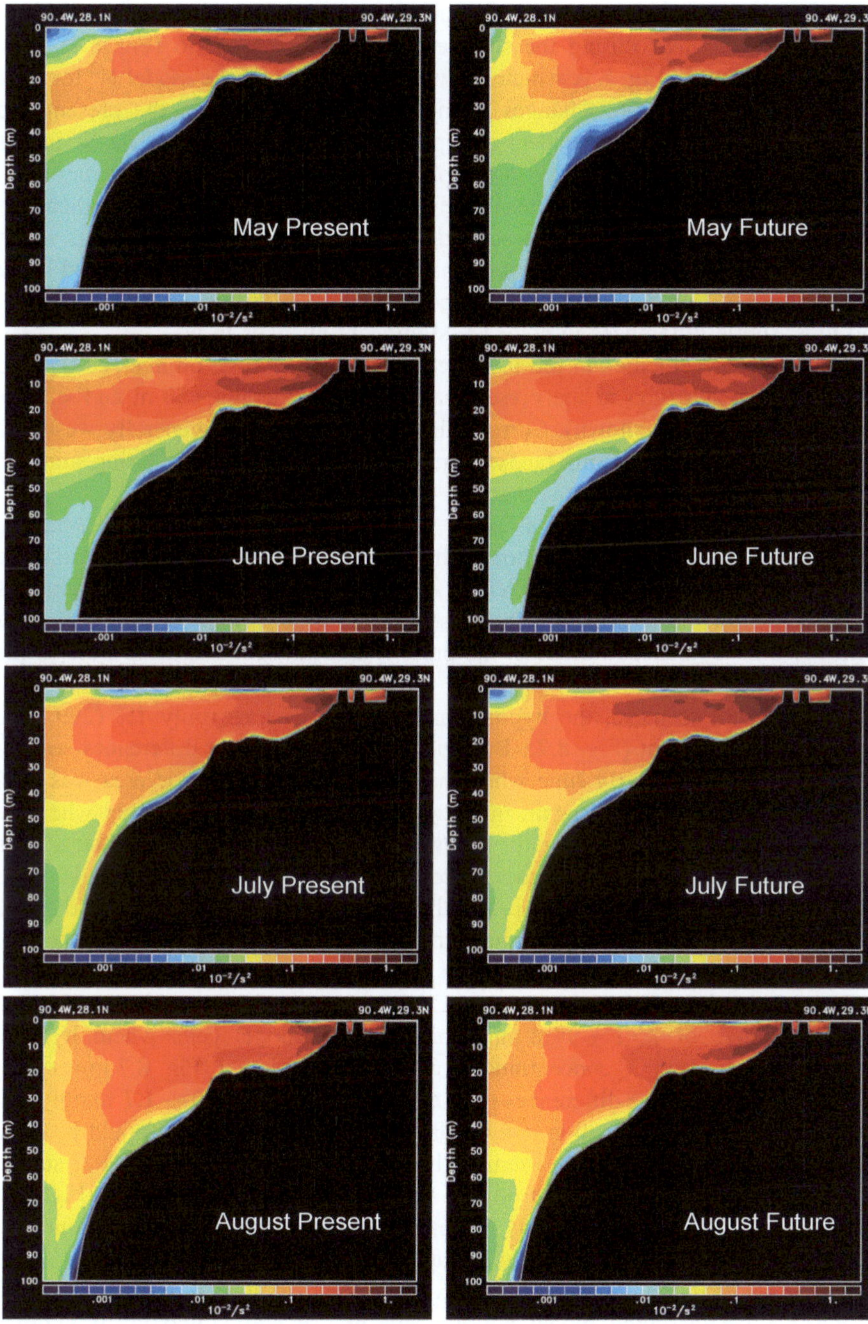

Fig. 8.9 Vertical slices of the monthly averaged N^2 (10^{-2} s^{-2}) for May through August 2006 extracted from the 20 sigma layers in the model along longitude line 90.4 °N. Model results for the present are shown in the *left column*, and future climate scenarios are shown in the *right column*. Note the *color bar* is log10 scale

As a result of the warmer and less saline water in the surface layer, stronger stratification occurred during the summer in the future scenario (Fig. 8.4). This change is evident in Fig. 8.9, which shows vertical slices of the model N^2 for May through August of 2006. Changes in the stratification between the present and future scenarios were most evident at depths between 7 and 30 m. The N^2 in this depth range were larger in May in the present model results, but thereafter from June to August the N^2 were greater in the future scenario. These results indicated a stronger pycnocline in the future scenario during June, July, and August, which coincides with the lower O_2 concentrations at C06 and H04 during this period in the future scenario (Fig. 8.6). Note that the color bar in Fig. 8.9 is N^2 on log10 scale, and thus, subtle differences in color are significant changes in N^2.

Shelf current velocities in the future climate scenario were on average 0.3 and 0.2 cm s^{-1} greater than in the present case for water depths <50 m and <20 m, respectively (data not shown). Current directions, which were generally westward inside the 50-m isobath during fall, winter, and spring and eastward during summer, were unchanged in the future scenario.

Modeled Chl, on average, was larger in the future scenario (Fig. 8.6), but the whole water column grand mean only changed from 3.8 to 4.1 mg m^{-3} from the present to the future scenario. Whole water column mean O_2 concentrations changed from 213 mmol m^{-3} in the present scenario to 207 in the future. Largest differences in O_2 concentrations in the present and future model runs were in the bottom layer (Fig. 8.6).

In the future scenario, the hypoxic area was on average only 1% (299 km^2) larger than the present (Fig. 8.7). But, there are some notable differences seasonally, for example, in early June when the future scenario hypoxic area was 23% larger than the present and again in later October and early November when the hypoxic area was approaching the fall/winter minimum. Thus, the effects of a warmer and wetter climate on hypoxic area seem to mainly occur at the tails of the hypoxia size distribution. The shapes of the hypoxic areas in the present and future scenario (Fig. 8.8) were similar, but with differences noted in the western extent near longitude 92.5° W where the future area was smaller.

The effect of the future scenario on O_2 concentration in bottom water was an average change of -6.1 ± 14.2 mmol m^{-3}. In the model cells where hypoxia occurred, the average O_2 concentration was consistently lower (Fig. 8.7 lower plot) than the present, but the differences were small on average, -2.0 ± 1.7 mmol m^{-3}. However, at a specific location, the O_2 difference in bottom waters could be quite large. For example, the future scenario bottom-water O_2 concentration at station C06 exhibited large departures from the present from mid-May to July (Fig. 8.6) and again in late October and early November, which resulted in a longer duration of hypoxia. Station H04 had a similar pattern.

To further examine the issues of hypoxia duration and areas of hypoxia severity, in Fig. 8.8 we classified model bottom cells as "severely hypoxic" if hypoxia occurred in a cell more than 60 days over the annual cycle. We then examined the change in the duration and area of severe hypoxia obtained from the present and future scenario. In the future scenario, the area of severe hypoxia increased by

1,130 km^2, a 7.9% increase. The average duration of hypoxia in the severely hypoxic cells increased by 9.5 ± 22.3 days, a 10% increase.

8.4 Discussion

8.4.1 Present Model Results and Future Scenario Implications for Hypoxia

For the first time, a 3-D hydrodynamic and ecosystem model was used to demonstrate the effects of a future climate scenario on northern Gulf of Mexico hypoxia. The use of the 3-D model allowed for comparison with observations and an examination of the spatial and temporal differences in modeled state variables as a result of the prescribed change in climate between the present and future scenario. Corroboration of the modeling results for the present included point-to-point and regional comparisons between observations and model results.

In the point-to-point comparisons, the hydrodynamic model had good skill predicting observed temperature, salinity, sigma-t (Table 8.2), and the depth of stratification (Fig. 8.4), but was unable to reproduce the maximum strength of stratification (Fig. 8.4). Accurately modeling hypoxia is contingent upon the ability of hydrodynamic models to reproduce the depth and strength of vertical stratification (Fennel et al. 2013; Fennel et al. accepted). However, hypoxia on the Louisiana shelf may occur whenever the N^2 is greater than 40 cycles per hour (Bianchi et al. 2010). This threshold was achieved at greater than 50% of the locations assessed in the model (Fig. 8.4), and stratification was sufficient for generating hypoxia (Fig. 8.7).

An obvious issue arises with the use of point-to-point comparisons. There is a mismatch in spatial scale when comparing model results with observations due to the horizontal resolution of the model (1.9 km) in comparison with point locations for sampling stations. This issue is particularly problematic for the Louisiana shelf where small-scale variability is the norm due to dynamics among many plume fronts (Marta-Almeida et al. 2013). Thus, although the model had high skill at reproducing observed temperature and salinity (Table 8.2), getting the modeled stratification to match observations requires a near-perfect representation of the observed vertical temperature and salinity profile. This is a difficult task at the coarser scales of existing models and where small-scale variability occurs. Future improvements in reproducing stratification are likely to come from data assimilation of vertical hydrographic profiles into the hydrodynamic models. This is not likely to be achieved in the near future due to a lack of sampling at the required time and space scales for data assimilation.

For the biogeochemistry, CGEM coupled to NCOM-LCS reproduced observed NO_3^-, Chl, and O_2 concentrations with decent skill (Table 8.2). In the point-to-point and regional comparisons, O_2 was biased high in the model due to an inability of the model to reproduce the lowest observed bottom-water O_2 concentrations (e.g., Fig. 8.5). In light of the discussion above about modeled stratification, this inability to reproduce the lowest O_2 concentrations is not surprising, as it has been noted previously that stratification is inversely correlated to bottom-water O_2 concentration in a model (Fennel et al. 2013). These results provide confidence that the present model was generally reproducing the temporal and spatial patterns of observations.

8.4.2 Future Scenario Implications for Hypoxia

In the future climate scenario, water temperature increased and salinity decreased (Fig. 8.3). These changes resulted in stronger stratification (Fig. 8.4), especially during summer (Fig. 8.9). In turn, O_2 concentrations were lower during late spring to fall in the future scenario. This change manifested in both the bottom layer and into the water column (Fig. 8.6). The most pronounced changes in bottom-water O_2 concentrations and hypoxic area between the present and future scenario occurred during the spring months leading to the onset of hypoxia and the fall months during the breakdown of hypoxia (Fig. 8.7).

The future scenario affected the duration and extent of severe hypoxia (Fig. 8.8). The decision to use a 60-day occurrence threshold for defining severe hypoxia was based on the continuous O_2 time-series observations at station C06 presented by Justic and Wang (2014, their Fig. 2d), which showed hypoxia occurring for approximately 60 days during late spring and summer. We repeated the analysis using different thresholds of 30 and 90 days. Using a threshold of 30 days to classify cells as severely hypoxic yielded a 0.3% increase in the area of severe hypoxia and an 8% increase in the average duration of hypoxia in severely hypoxic cells in the future. For a 90-day threshold, the future scenario had a 37% larger area and no change in the duration of hypoxia. These differences over the range of thresholds examined occur because the range in the number of hypoxia days in cells classified as exceeding the threshold becomes smaller as the threshold increases. Thus, at a low duration threshold for severe hypoxia, the areal extent of severe hypoxia cannot change much, but the overall duration can increase, while at a high duration threshold, the area can increase significantly in the future scenario but not the overall duration.

Regardless of the threshold used, the pattern that emerges is that the future scenario causes more severe hypoxia. Longer-duration hypoxia occurs due to the appearance of low O_2 conditions earlier in the spring and later in the fall in the future scenario (Fig. 8.6). The greater extent occurs due to larger areas of severe hypoxia on the eastern shelf near longitude 90.7° W and on the western shelf near longitude 92.5° W (Fig. 8.8).

The exact reasons for the difference in O_2 and hypoxia in the future scenario were not diagnosed in this study. This would require a thorough mass balance accounting of the processes controlling O_2 in bottom waters (Yu et al. 2015) and is left for another study. However, based on the patterns of O_2 responses in relation to changes in stratification we infer that physical rather than biological processes were the dominant control. We can rule out eutrophication as a driving process in the future scenario based on the muted Chl response. Though the future scenario resulted in a +10% increase in nutrient loading as a result of the +10% increase in discharge, Chl was not observed to be significantly larger in the future scenario at location where hypoxia occurs (Fig. 8.6). Further, the phytoplankton growth rate in the model (not shown) was not significantly different in the future scenario.

The inference of hypoxia change in the future scenario being driven by stratification is supported by previous work. Based on observations of O_2 and vertical hydrography, Wiseman et al. (1997) demonstrated an empirical inverse relationship between stratification strength and bottom-water O_2 concentrations. In modeling studies, simulations of future climate scenarios with 1-D models have predicted that the Louisiana shelf hypoxic area will increase as a result of increased stratification (Justic et al. 1996; Donner and Scavia 2007). More recently, 3-D modeling studies from the Louisiana shelf have demonstrated that hypoxia is sensitive to differences in water column temperature (Ko et al. 2016) and stratification (Hetland and DiMarco 2008; Fennel et al. 2013; Justic and Wang 2014). Thus, although the 3-D modeling studies were not couched in the context of climate change, their results indicate similar behavior as the future climate scenario model results presented here.

8.4.3 Climate and Modeling Uncertainties

The +3 °C air temperature used in this study is consistent with the IPCC RCP6.0 scenario (Collins et al. 2013) prediction for the southeastern USA and northern Gulf of Mexico. The range of IPCC regional predictions for this area was +2 to +4 °C, based on RCP2.6 and RCP8.5 scenarios, respectively (Romero-Lankao et al. 2014). Thus, the +3 °C increase used here may be conservative in comparison with RCP8.5, which is the business-as-usual scenario.

Climate impacts to hydrology are likely to occur, but the certainty about the amount of change is less than for air temperature (Collins et al. 2013). For example, Sperna Weiland et al. (2012) modeled global river discharge using 12 different global climate models (GCMs) as forcing and reported an ensemble mean 11% increase in discharge globally. However, at a continental scale, the change in discharge for North America ranged from –15% to +60%. Thus, the +10% increase in river discharge used in this study is consistent with the modeled global increase, but has high uncertainty.

We did not include a change in coastal winds in our scenario because the IPCC did not note a regional change in coastal winds as a future consequence of climate (Collins et al. 2013; Romero-Lankao et al. 2014). Further studies could address changes in wind with regionally downscaled climate model results with sufficient resolution to capture potential changes in regional prevailing winds. Currents on the Louisiana shelf are wind-dominated; thus, future changes in wind could have significant effects on hypoxia as has been noted in the present (Forrest et al. 2011).

Future changes in sea temperature, salinity, wind intensity, and nutrient loads could impact the food web of the shelf system through changes in the phytoplankton community structure. Phytoplankton community structure and food web dynamics are thought to play an important role in coastal hypoxia development (Falkowski et al. 1980; Eldridge and Roelke 2010), but our current understanding of these processes and how climate change may impact them is highly uncertain. Models, like CGEM, that can represent multiple phytoplankton functional types may be used to explore these climate effects. Here, however, we used only one phytoplankton type, diatoms, to simplify the model results. In future work, we plan to assess how climate impacts in combination with changing nutrient loads may affect competition among multiple phytoplankton functional types and resultant effects on trophic transfer to zooplankton and detrital organic matter pools.

The year 2006 had low river discharge for the combined Mississippi and Atchafalaya Rivers with an average discharge of 12,200 $m^3 s^{-1}$ in comparison with the long-term (1968–2015) average of 21,300 $m^3 s^{-1}$. Additional model analyses are needed to examine potential climate impacts during average and high discharge periods. Another aspect that should be examined with respect to discharge is the observed positive correlation between freshwater discharge and nutrient (Aulenbach et al. 2007) and dissolved organic matter (Mulholland 2002) concentrations and loads. In the future, if river discharge increases, we expect that elevated river nutrient and organic matter loads will enhance primary production and respiration on the shelf (Justic et al. 2003). Further increased CDOM loads may reduce light availability in the water column and potentially limit photosynthetic production of O_2 beneath the pycnocline (Lehrter et al. 2009; 2014). The accompanying increased absorption of atmospheric radiation by Chla stimulated by nutrient loads and by CDOM in surface waters could also further increase the surface water temperature, which may lead to stronger stratification (Ko et al. 2016).

Finally, there are obvious limitations in the biogeochemical model that contribute to uncertainty in the results presented here. Even a fairly complex model like CGEM is still a gross simplification of the nutrient, organic matter, and O_2 cycling processes that occur. Further work is needed to explore trade-offs between smaller model grid cells, potential improvements in modeled stratification strength and biogeochemistry, and computational requirements to run the model.

8.4.4 Conclusions

Current nutrient management strategies for mitigating hypoxia include goals with target dates that are decades into the future. Achieving these goals will require adaptive management that considers changing climate. Only the simple potential interaction between temperature and discharge has been investigated here. However, there are clearly a myriad of other interactions including changing watershed loads of nutrients and organic matter, changing ocean chemistry, and other climate impacts that should be assessed. More observation and modeling studies of these interactions are warranted to improve our conceptual model and predict the impacts on coastal hypoxia.

Acknowledgements We thank Brandon Jarvis, David Beddick, Louis Olszyk, and Barry Herchenroder for assistance with model inputs and code development. James Pauer and Steve Jordan provided helpful reviews on earlier drafts. We thank the two anonymous reviewers and the editors for their comments and suggestions to improve this manuscript. This work was supported by the USEPA Office of Research and Development. The study was reviewed and approved for publication by the USEPA National Health and Environmental Effects Research Laboratory; however, the contents are solely the views of the authors. Use of trade names of commercial products does not constitute endorsement by the USEPA.

Appendices A–F

A. State Variables

CGEM state variables (Table A.1) are represented by time-dependent differential equations. Horizontal and vertical currents, vertical mixing, and sinking are implicit, and these terms are not shown in state variable equations. Currents, mixing, and temperature are hydrodynamic model outputs that are provided to CGEM by NCOM-LCS at each model time-step and point in the grid. State variables are presented below in the order that they appear in Table A.1.

Table A.1 State variables in CGEM code

State variable	Description	Units
A	Phytoplankton abundance	cells m^{-3}
Qn and *Qp*	Nitrogen and phosphorus cell quota	mmol N $cell^{-1}$
Z_1 and Z_2	Macrozooplankton and microzooplankton	Individuals m^{-3}
OM1_A and *OM2_A*	Particulate and dissolved organic matter from phytoplankton	mmol m^{-3}
OM1_Z and *OM2_Z*	Particulate and dissolved organic matter from zooplankton	mmol m^{-3}
OM1_R and *OM2_R*	Particulate and dissolved organic matter from rivers	mmol m^{-3}
OM1_BC and *OM2_BC*	Particulate and dissolved organic matter from later boundaries	mmol m^{-3}
CDOM	Colored dissolved organic matter	ppb
NH_4	Ammonium	mmol m^{-3}
PO_4	Phosphate	mmol m^{-3}
Si	Silica	mmol m^{-3}
NO_3	Nitrate	mmol m^{-3}
O_2	Oxygen	mmol m^{-3}

A.1 Phytoplankton

For simplicity, in this implementation of CGEM, one phytoplankton functional type is modeled based on the dominant diatom on the Louisiana shelf, *Skeletonema costatum. S. costatum* accounted for 58% of the total phytoplankton abundance observed on the Louisiana shelf from 2002 to 2007 (data reported in Murrell et al. 2014). CGEM, however, is flexible in being able to represent up to 99 phytoplankton groups. Phytoplankton abundance (A, cells m^{-3}) per group (i = 1:99) is calculated as

$$\frac{d}{dt}A_i = Agrow_i - Aresp_i - ZgrazA\_tot_i - Amort_i. \tag{A1}$$

where *Agrow* is production (cells m^{-3} s^{-1}), *Aresp* (cells m^{-3} s^{-1}) is the sum of somatic and basal respiration, *ZgrazA_tot* (cells m^{-3} s^{-1}) is the total zooplankton grazing on A_i by the two zooplankton represented in the model, and *Amort* (cells m^{-3} s^{-1}) is the non-grazing mortality rate. *Agrow, Aresp, Amort,* and *ZgrazA_tot* are described in Appendix C, Eqs. (C1), (C7), (C11), and (C12), respectively.

Phytoplankton internal cell quotas (Q, mmol cell^{-1}) for nitrogen and phosphorus are calculated as

$$\frac{d}{dt}Qn_i = vN_i - Qn_i \cdot uA_i - \frac{AexudN_i}{A_i} \tag{A2}$$

$$\frac{d}{dt}Qp_i = vP_i - Qp_i \cdot uA_i - \frac{AexudP_i}{A_i} \tag{A3}$$

where vN and vP (mmol cell^{-1} d^{-1}) are phytoplankton uptake of nitrogen and phosphorus, respectively (Eqs. C15–C17), $Q \cdot uA$ (mmol cell^{-1} d^{-1}) is the utilization of Q to support the growth rate (uA, Eq. C2), and *AexudN and AexudP* are exudation (mmol cell^{-1} d^{-1}) of nitrogen and phosphorus, respectively, associated with *Aresp* (Eqs. C9 and C10).

A.2 Zooplankton

Zooplankton (Z, individuals m^{-3}) dynamics for two types (j = 1:2) are represented as

$$\frac{d}{dt}Z_j = Zgrow_j - Zresp_j - Zmort_j \tag{A4}$$

where zooplankton growth (*Zgrow*), respiration (*Zresp*), and mortality (*Zmort*) are described in Eqs. (D1), (D6), and (D9), respectively.

A.3 Organic Matter

Particulate organic matter (*OM1*, mmol C m^{-3}) from phytoplankton (*OM1_A*), zooplankton (*OM1_Z*), rivers (*OM1_R*), and boundary conditions (*OM1_BC*) are calculated as

$$\frac{d}{dt}OM1\_A = ROM1\_A + \sum_{i=1}^{6}\left(Amort_i \cdot \frac{Q\,min\,N_i}{Qn_i} \cdot Qc_i\right), \tag{A5}$$

$$\frac{d}{dt}OM1\_Z = ROM1\_Z + ZegC_1 + ZunC_1 + ZmortC_1 + ZmortC_2 + (OM1\_Ratio \cdot ZslopC\_tot) \tag{A6}$$

$$\frac{d}{dt}OM1\_R = ROM1\_R \tag{A7}$$

$$\frac{d}{dt}OM1\_BC = ROM1\_BC. \tag{A8}$$

In Eq. (A5), *ROM1* is the *OM1* remineralization rate and is a loss term for *OM1* (e.g., Eq. E39), *QminN* and *Qc* are parameters for the minimum cellular N quota and the fixed C quota per phytoplankton cell (Table A.2), respectively. In Eq. (A6), $ZegC_1$ is the Z_1 carbon in excess of growth requirements that is egested in fecal pellets (Eq. E18), $ZunC_1$ is the Z_1 ingested carbon that is unassimilated in the gut and also released in fecal pellets (Eq. D4), *ZslopC_tot* is the organic matter released during sloppy feeding by both Z_1 and Z_2 (Eq. D4), and *OM1_Ratio* (Eq. E19) is the fraction of organic matter derived from sloppy feeding that becomes *OM1_Z*.

Dissolved organic matter (*OM2*, mmol C m^{-3}) dynamics are represented by

$$\frac{d}{dt}OM2\_A = -ROM2\_A + \sum_{i=1}^{6}\left(Amort_i \cdot \frac{Qn_i - \mathrm{Qmin}N_i}{Qn_i} \cdot Qc_i\right), \tag{A9}$$

$$\frac{d}{dt}OM2\_Z = -ROM2\_Z + ZegC_2 + ZunC_2 + (OM2\_Ratio \cdot ZslopC) \tag{A10}$$

$$\frac{d}{dt}OM2\_R = -ROM2\_R \tag{A11}$$

$$\frac{d}{dt}OM2\_BC = -ROM2\_BC. \tag{A12}$$

In (A9), *ROM2* is the *OM2* remineralization rate (e.g., Eq. E40). In (A10), $ZegC_2$ is the Z_2 carbon in excess of growth requirements that is egested (Eq. E18), $ZunC_2$ is the Z_2 ingested carbon that is not assimilated (Eq. D5), and *OM2_Ratio* (Eq. E20) is the fraction of organic matter derived from sloppy feeding that becomes *OM2_Z*.

Table A.2 Phytoplankton parameters

Parameter	Description	Units	Value
volcell	Biovolume per cell	μm^3	513
Qc	Carbon per cell	10^{-7} mmol C $cell^{-1}$	0.454
umax	Maximum growth rate at 20 °C	d^{-1}	2.2
alpha	Initial slope of the photosynthesis versus irradiance curve	10^{-16} cm^2 s $quanta^{-1}d^{-1}$	3.96
beta	Photoinhibition (P1)	10^{-18} cm^2 s $quanta^{-1}d^{-1}$	1.1
respg	Growth-dependent respiration	Dimensionless	0.1
respb	Basal respiration	d^{-1}	0.02
QminN, QminP	Minimum cell quotas for N and P	10^{-9} mmol N $cell^{-1}$	0.153, 0.0107
Kn, Kp, Ksi	Half-saturation coefficients for N, P, and Si	mmol m^{-3}	1.13, 0.51, 1.13
vmaxN, vmaxP, vmaxSi	Maximum uptake rates for N, P, and Si	10^{-8} mmol $cell^{-1}d^{-1}$	0.133, 0.0407, 0.133
aN	Scaling factor	Dimensionless	1
Athresh	Phytoplankton threshold for zooplankton grazing	10^7 cells m^{-3}	7
ediblevector	Edibility of phytoplankton	Dimensionless	0.5
sink	Sinking rate	m d^{-1}	0.01
mA	Mortality of phytoplankton	d^{-1}	0.11

CDOM in concentration units (QSE ppb) is input to the model from riverine loading, and once in the model, the only biogeochemical fate is to decay. Thus, the time rate of change for *CDOM* is

$$\frac{d}{dt}CDOM = -Kcdom\_decay \cdot CDOM \tag{A13}$$

where *Kcdom_decay* is the decay rate (Table A.5).

A.4 Nutrients

NH_4 (mmol m^{-3}) sources and sinks are described by

$$\frac{d}{dt}NH_4 = RNH_4 - R\_11 - AupN\frac{NH_4}{NH_4 + NO_3} + AexudN + ZexN. \tag{A14}$$

where RNH_4 (Eq. E51) is the production of NH_4 due to remineralization of organic matter (Eq. E28), *R_11* is the nitrification rate (Eq. E43), *AupN* (Eq. C17) is the phytoplankton uptake of dissolved inorganic nitrogen (NH_4 + NO_3), *AexudN* is the phytoplankton exudation of nitrogen driven by respiration (Eq. C9), and *ZexN* is the zooplankton excretion of nitrogen driven by respiration (Eq. D7).

Change in NO_3 (mmol m^{-3}) is represented as

$$\frac{d}{dt}NO_3 = RNO_3 + R\_11 - AupN\frac{NO_3}{NO_3 + NH_4} \tag{A15}$$

where RNO_3 (Eq. E48) is negative and represents NO_3 lost to denitrification (Eq. E30) and *AupN* is modified by the fraction of NO_3 in the pool of NO_3+NH_4 to represent phytoplankton NO_3 uptake.

PO_4 (mmol m^{-3}) is calculated as

$$\frac{d}{dt}PO_4 = RPO_4 - AupP + AexudP + ZexP \tag{A16}$$

where RPO_4 (Eq. E49) is the production of PO_4 from remineralization of organic matter, *AupP* is phytoplankton uptake of PO_4 (Eq. C17), *AexudP* is phytoplankton exudation of PO_4 driven by respiration (Eq. C10), and *ZexP* is zooplankton excretion of PO_4 (Eq. D8).

Change in *Si* (mmol m^{-3}) is represented by

$$\frac{d}{dt}Si = RSi - AupSi + ZegSi + ZunSi. \tag{A17}$$

where *RSi* (Eq. E53) is *Si* produced by remineralization of organic matter, where the organic matter stoichiometry is assumed to have an Si:N = 1. *AupSi* is the phytoplankton uptake of *Si* (Eq. C17). *ZegSi* and *ZunSi* are the zooplankton egestion of *Si* and zooplankton unassimilated *Si*, respectively, and are set equal to *ZegN* (Eq. E18) and *ZunN* (Eq. D5), respectively.

A.5 Oxygen

O_2 (mmol m^{-3}) is represented by

$$\begin{aligned}\frac{d}{dt}O_2 = {} & RO_2 - 2 \cdot R\_11 + Agrow \cdot Qc - Aresp \cdot Qc - Zresp \cdot Zc \\ & \pm Air - Sea\,Exchange\end{aligned} \tag{A18}$$

where RO_2 is the aerobic oxygen consumption associated with organic matter remineralization (Eqs. E41 and E42) and air-sea exchanges are described in Appendix F.

B. Optical Equations

Irradiance is modeled using inherent optical properties (IOPs) to calculate light attenuation (k). The main advantage of this approach is that the absorption (a) and backscattering (b_b) of light are calculated, which could facilitate comparison of modeled and observed IOPs as these observations become more common. This approach could also be extended to a multi-spectral treatment. k is calculated (Penta et al. 2008; 2009) as

$$\begin{aligned} E_z &= E_0 e^{-k(z)z} \\ k(z) &= k_1 + \frac{k_2}{(1+z)^{0.5}} \\ k_1 &= \left[\chi_0 + \chi_1 (a_{490})^{0.5} + \chi_2 b_{b_{490}}\right](1 + \alpha_0 sin\, \theta_a) \\ k_2 &= \left[\zeta_0 + \zeta_1 (a_{490})^{0.5} + \zeta_2 b_{b_{490}}\right](\alpha_1 + \alpha_2 cos\, \theta_a) \end{aligned} \quad \text{(B1)}$$

where E_z is irradiance at depth z, E_0 is the irradiance at the surface layer above z, model coefficients include χ, ζ, and α (Table A.3), and θ_a is the solar zenith angle calculated as a function of latitude and time. Absorption at wavelength 490 nm (a_{490}, m^{-1}) is calculated as

$$a_{490} = a_{Chl_{490}} + a_{CDOM_{490}} + a_{SPM_{490}} + a_{w_{490}} \quad \text{(B2)}$$

where a_{Chl} is the absorption by chlorophyll (*Chl*), a_{CDOM} is the absorption by colored dissolved organic matter (*CDOM*), a_{SPM} is the absorption by suspended particulate matter (*SPM*), and a_w is the absorption by seawater and is a model parameter (Table A.3). a_{Chl} is calculated as

$$a_{Chl_{490}} = astar490 \cdot Chl \quad \text{(B3)}$$

where *astar490* (Table A.3) is *Chl*-specific absorption and total *Chl* (mg m^{-3}) is calculated as

$$Chl = \sum_{i=1}^{6} (Chl{:}\, cell) \cdot A_i \quad \text{(B4)}$$

where *Chl:cell* is the chlorophyll per cell (mg $cell^{-1}$) (see Eq. B5) and A_i is the phytoplankton cell abundance (cells m^{-3} d^{-1}). *Chl:cell* is calculated for the Louisiana shelf using an empirical equation based on observations of cell abundance and Chla (Murrell et al. 2014). A regression relating these variables ($R^2 = 0.81$), with intercept set equal to zero, has the form $Chl = 3.0 \times 10^{-9} \cdot CellAbundance$, where 3.0×10^{-9} is the slope of the observed relationship between chlorophyll a and phytoplankton abundance (i.e., Chl:cell).

Table A.3 Optical parameters

Parameter	Description	Unit	Value
chi0, chi1, chi2	Coefficients	Dimensionless	–0.057, 0.482, 4.221
zeta0, zeta1, zeta2	Coefficients	Dimensionless	0.183, 0.702, –2.567
alpha0, alpha1, alpha2	Coefficients	Dimensionless	0.090, 1.465, –0.67
astar490	*Chla* specific absorption (490 nm)	m^{-1}(mg *Chla* $m^{-3})^{-1}$	0.020
aw490	Water absorption (490 nm)	m^{-1}	0.015
astarOMA	*OM1_A* specific absorption (490 nm)	m^{-1}(g *OM1_A* $m^{-3})^{-1}$	0.01
astarOMZ	*OM1_Z* specific absorption (490 nm)	m^{-1}(g *OM1_Z* $m^{-3})^{-1}$	0.01
astarOMR	*OM1_R* specific absorption (490 nm)	m^{-1}(g *OM1_R* $m^{-3})^{-1}$	0.01
astarOMBC	*OM1_BC* specific absorption (490 nm)	m^{-1}(g *OM1_BC* $m^{-3})^{-1}$	0.01
CF_SPM	Percentage of river *SPM* that is *OM1_R*	%	1.8

a_{CDOM} is based on *CDOM* (Eq. A13) loaded to the model domain from terrestrial sources. In the case of the Louisiana shelf, *CDOM* loaded from the rivers is derived from regression equations relating observed riverine CDOM and DOC concentrations. The Mississippi River monthly time-series of observed DOC (mg l^{-1}) (USGS data) was converted to a_{CDOM} at wavelength 440 (m^{-1}) by the regression equation (Spencer et al. 2012)

$$a_{CDOM_{440}} = 1.10 \cdot DOC - 2.76. \tag{B5}$$

Next, $a_{CDOM_{440}}$ was converted to $a_{CDOM_{312}}$ with the general $a_\lambda = a_{\lambda ref} e^{-S(\lambda - \lambda ref)}$, i.e.,

$$a_{CDOM_{312}} = a_{CDOM_{440}} \cdot e^{-S(312-440)} \tag{B6}$$

where *S* is the spectral slope of *CDOM* (S = 0.016) (Spencer et al. 2012; D'Sa and Dimarco 2009). Then, $a_{CDOM_{312}}$ was converted to *CDOM* concentration using the regression equation (Conmy et al. 2004)

$$\mathrm{CDOM} = 2.933 \cdot a_{CDOM_{312}} + 0.538 \tag{B7}$$

Thus, in the model for the Louisiana shelf, Eqs. (B5–B7) are applied to riverine DOC observations. *CDOM* concentration is then input by the rivers to the shelf model domain where it is decayed by Eq. (A13) and advected and mixed. $a_{CDOM_{490}}$

is required in Eq. (B2). Thus, *CDOM* is first back-transformed to $a_{CDOM_{312}}$ using the Conmy et al. (2004) equation, and then $a_{CDOM_{312}}$ is converted to $a_{CDOM_{490}}$ by

$$a_{CDOM_{490}} = a_{CDOM_{312}} \cdot e^{-S(490-312)}. \quad \text{(B8)}$$

a_{SPM} (m^{-1}) at 490 nm is calculated as the sum of absorption by the different types of *OM1* as

$$a_{SPM_{490}} = \left(\begin{array}{c} astarOMA \cdot OM1\_A + astarOMZ \cdot OM1\_Z + \\ astarOMR \cdot \dfrac{OM1\_R}{CF\_SPM} + astarOMBC \cdot OM1\_BC \end{array} \right) \cdot \frac{12}{1000} \quad \text{(B9)}$$

where *astarOM* terms are parameters (Table A.3), *CF_SPM* is a conversion factor for adjusting *OM1_R* to river SPM. For the application of CGEM to the Louisiana shelf, *CF_SPM* = 1.8% (Table A.3) was based on the average observed POC/SPM = 1.8% in the Mississippi and Atchafalaya Rivers (USGS data). The factor 12/1000 converts from mmol m^{-3} to g m^{-3}.

Backscattering in Eq. (B1) was calculated (Penta et al. 2008; 2009) as a function of *Chl* as

$$b_{b,490} = 0.015 \cdot \left(0.3 \cdot Chl^{0.62} \cdot \left(\frac{550}{490} \right) \right). \quad \text{(B10)}$$

C. Phytoplankton Equations

C.1 Phytoplankton Growth

Growth (*Agrow*, cells m^{-3} d^{-1}) for each phytoplankton group is calculated as

$$Agrow_i = \mu_{A_i} A_i \quad \text{(C1)}$$

where μ_A is the specific growth rate (d^{-1}) and *A* is the phytoplankton abundance (cells m^{-3}). *Agrow* is converted to units of carbon (mmol C m^{-3} d^{-1}) by the product $Agrow_i \cdot Q_{C_i}$, where Q_C is a parameter specifying the carbon per cell (mmol C $cell^{-1}$) (Table A.2).

The specific growth is calculated based on Liebig's law of the minimum

$$\mu_{A_i} = u_{\max A_i} \cdot func\_T_i \cdot MIN[func\_E_i, func\_N_i, func\_P_i, func\_Si_i] \quad \text{(C2)}$$

where *func_T*, *func_E*, *func_N*, *func_P*, and *func_Si* are limiting factors due to temperature (Eldridge and Roelke 2010), PAR, nitrogen, phosphorus, and silica, respectively. PAR- and nutrient-dependent growth equations are shown below.

C.2 Phytoplankton Light–Growth Dependence

Light dependence *func_E* is represented by

$$func\_E = \left(1 - e^{\frac{-alpha_i E}{u\max_i}}\right). \tag{C3}$$

C.3 Phytoplankton Nutrient–Growth Dependence

Nitrogen and phosphorus dependence are modeled (Droop 1973) as

$$func\_N = \frac{Qn_i - Q\,min\,N_i}{Qn_i} \tag{C4}$$

$$func\_P = \frac{Qp_i - Q\,min\,P_i}{Qp_i} \tag{C5}$$

where Q_{min} is the minimum nutrient cell quota (mmol cell^{-1}) per phytoplankton group required for survival and Q is the cell quota (mmol cell^{-1}).

Silica dependence is modeled as a function of seawater silicate concentration

$$func\_Si = \frac{Si}{Si + Ksi} \tag{C6}$$

where Si is the modeled silicate concentration (mmol m^{-3}) and Ksi is the half-saturation concentration (mmol m^{-3}) of silica uptake.

C.4 Phytoplankton Losses

Phytoplankton respiration (cells m^{-3} d^{-1}) is represented as a function of growth, cell abundance, and temperature

$$Aresp_i = respg_i \cdot Agrow_i + respb_i \cdot \mathrm{A}_i \cdot func\_T \tag{C7}$$

where *respg* is a respiration coefficient that scales to growth rate and *respb* represents basal maintenance activities that scale to abundance. Phytoplankton respiration results in a loss of carbon from the cell (*ArespC*) and nutrient exudation ($Aexud_i$, mmol m^{-3} d^{-1}) as

$$Aresp\,\mathrm{C} = \sum_{i=1}^{6} Aresp_i \cdot Qc_i \tag{C8}$$

$$AexudN = \sum_{i=1}^{6} Aresp_i \cdot Qn_i \tag{C9}$$

$$AexudP = \sum_{i=1}^{6} Aresp_i \cdot Qp_i. \tag{C10}$$

Phytoplankton mortality (*Amort*, cells m^{-3} d^{-1}) is calculated by

$$Amort_i = A_i \cdot mA_i \tag{C11}$$

where mA_i is the mortality rate (d^{-1}).

Phytoplankton sinking losses (cells m^{-3} d^{-1}) are applied implicitly in the advection and mixing routines with sinking velocities prescribed by the parameter *sink* (Table A.2).

Grazing losses (*ZgrazA_tot*, cells m^{-3} d^{-1}) are calculated as

$$ZgrazA\_tot_i = \sum_{j=1}^{2} \frac{Zgrazvol_{ji}}{volcell_i}. \tag{C12}$$

The term $Zgrazvol_{ji}$ is the grazing rate by each zooplankton (j = 1:2) on each phytoplankton (i) in units of biovolume (μm3 m^{-3} d^{-1}) and is calculated by

$$Zgrazvol_{ji} = Z_j \cdot Zu\,max_j \cdot monod\,Z_{ji} \tag{C13}$$

where $Zumax_j$ is the maximum growth rate of the zooplankton in terms of volume of prey (Table A.4) and $monodZ_{ij}$ is a hyperbolic function represented by

$$monodZ_{ij} = \frac{(Abiovol_i - Athresh_i \cdot volcell_i) \cdot ediblevector_i}{ZKa_j + \sum_{i=1}^{6} (Abiovol_i \cdot ediblevector_i)}, \tag{C14}$$

where *Abiovol* (= $A_i \cdot volcell_i$) is the biovolume, *Athresh* is the threshold abundance (cells m^{-3}) below which grazing of A_i does not occur, *ediblevector* is a vector expressing prey edibility (unitless, range = 0–1), and *ZKa* is the grazing half-saturation (Table A.4).

C.5 Phytoplankton Uptake and Utilization of N, P, and Si

Nutrient uptake by the modeled phytoplankton only occurs during the day. For the rate-limiting nutrient substrate (*S*), which is determined as the min[*f_N, f_P, f_Si*], the nutrient uptake rate (*vS*, mmol cell^{-1} d^{-1}) for each phytoplankton group is calculated as

Table A.4 Phytoplankton parameters

Parameter	Description	Units	Value
Zvolcell	Volume per individual	μm^3 individual^{-1}	[2.98e+7 6.74e+5]
ZQc	Carbon per individual	mmol C individual^{-1}	[3.13e−4 7.08e−7]
ZQn	N per individual	mmol N individual^{-1}	[6.95e−5 1.57e−7]
ZQp	P per individual	mmol P individual^{-1}	[3.77e−6 8.53e−9]
Zslop	Sloppy feeding coefficient	Dimensionless	[0.25 0]
Zeffic	Assimilation efficiency as a fraction of ingestion	Dimensionless	[0.4 0.4]
ZKa	Grazing half-saturation coefficient	μm^3 m^{-3}	[1.12e+12 1.12e+12]
Zrespg	Growth-dependent respiration	Dimensionless	[0.2 0.3]
Zrespb	Biomass (basal)-dependent respiration	d^{-1}	[0.1 0.416]
Zumax	Maximum growth rate in terms of volume of prey	μm^3 ind^{-1} d^{-1}	[9.45e+8 2.98e+7]
Zm	Zooplankton mortality constant for quadratic mortality	m^6 ind^{-2} d^{-1}	[0.00072 0.00072]

$$vS_i = vmaxS_i \cdot \frac{S}{(S + K_{S_i})} \cdot Q10 \tag{C15}$$

where $v_{max}S$ is the maximum uptake rate (mmol cell $^{-1}$ d^{-1}), K_S is the half-saturation concentration (mmol m^{-3}) (Table A.2), and *Q10* is the temperature adjustment factor such that a doubling of the rate occurred for a 10 °C change in temperature.

If *S* is not the rate-limiting nutrient, the uptake is modified by an additional limitation term as

$$vS_i = vmaxS_i \cdot \frac{S}{(S + K_{S_i})} \cdot Q10 \cdot func\_Qs_i \cdot \frac{RLN}{RLN + a_S \cdot K_{RLN_i}} \tag{C16}$$

where *RLN* is the substrate concentration of the rate-limiting nutrient (*RLN*), K_{RLN} is the half-saturation concentration of the *RLN*, and a_S is a scaling factor (Roelke et al. 1999).

The total phytoplankton uptake of nutrient (shown here for nitrogen, i.e., *AupN*) is then

$$AupN = \sum_{i=1}^{6} (vN_i \cdot A)_i. \tag{C17}$$

D. Zooplankton Equations

Zooplankton growth rates (individuals m^{-3} d^{-1}) are modeled as

$$Zgrow_j = func\_T \cdot MIN\left(\frac{ZinN_j}{ZQn_j}, \frac{ZinP_j}{ZQp_j}\right). \tag{D1}$$

where *ZinN* and *ZinP* are zooplankton ingestion rates of nitrogen and phosphorus (mmol m^{-3} d^{-1}), and *ZQn* and *ZQp* are zooplankton N and P quota parameters, respectively (mmol per individual, Table A.4). *ZinN*, *ZinP,* and zooplankton ingestion of carbon (*ZinC*) are calculated similarly. For brevity, only the nitrogen equations are shown. Carbon and phosphorus equations are analogous. Thus, for *ZinN*,

$$ZinN_j = ZgrazN_j - ZslopN_j - ZunN_j \tag{D2}$$

$$ZgrazN_j = \sum_{i=1}^{6} ZgrazA_{ij} \cdot Qn_i \tag{D3}$$

$$ZslopN_j = Zslop_j \cdot ZgrazN_j \tag{D4}$$

$$ZunN_j = (1 - Zeffic_j) \cdot (ZgrazN_j - ZslopN_j) \tag{D5}$$

where *ZgrazN* (mmol N m^{-3} d^{-1}) is the grazing in units of nitrogen, *ZgrazA* is described in Eq. (C12), *ZslopN* is sloppy feeding (mmol N m^{-3} d^{-1}), *Zslop* is a sloppy feeding parameter (range = 0–1, Table A.4), and *ZunN* (mmol N m^{-3} d^{-1}) is the amount of zooplankton unassimilated nitrogen, which is a function of the assimilation efficiency (*Zeffic*, range = 0–1, Table A.4) of each zooplankton.

Zooplankton respiration loss (individuals m^{-3} d^{-1}) is represented with two terms

$$Zresp_j = Zrespg_j \cdot Zgrow_j + Zrespb_j \cdot Z_i \cdot func\_T \tag{D6}$$

where *Zrespg* and *Zrespb* are respiration coefficients (Table A.4) on zooplankton growth and basal metabolism. Zooplankton excretion of nutrients (*Zex*, mmol m^{-3} d^{-1}) is used to mass balance zooplankton respiratory loss of CO_2 by

$$ZexN = \sum_{j=1}^{2} ZrespTot_j \cdot ZQn_j \tag{D7}$$

$$ZexP = \sum_{j=1}^{2} ZrespTot_j \cdot ZQp_j. \tag{D8}$$

Zooplankton mortality (individuals m^{-3} d^{-1}) is treated as a quadratic function of zooplankton abundance (Cerco and Noel 2004) and is assumed to mainly occur by predation from other trophic levels

$$Zmort_j = Zm_j Z_j^2 \tag{D9}$$

where *Zm* is the zooplankton mortality coefficient (Table A.4)

E. Organic Matter Equations

E.1 Organic Matter Types and Stoichiometry

Eight classes of organic matter representing phytoplankton, zooplankton, river, and boundary condition *OM* are tracked in the model in both particulate (*OM1*) and dissolved (*OM2*) forms and with variable stoichiometry $C_xN_yP_z$. *OM1* and *OM2* are created (mmol m^{-3} d^{-1}) in the model by phytoplankton and zooplankton mortality, zooplankton sloppy feeding, zooplankton egestion, and unassimilated *OM* that passes through the zooplankton. We simply represent partitioning of *Amort* to *OM1_A* and *OM2_A* based on cell quotas, which assumes that *Qmin* may be used as a proxy for separating particulate and dissolved fractions

$$OM1\_CA = \sum_{i=1}^{6}\left(Amort_i \cdot \frac{Qn_i - Q\,min\,N_i}{Qn_i} \cdot Qc_i\right) \tag{E1}$$

$$OM1\_NA = \sum_{i=1}^{6}(Amort_i \cdot Q\,min\,N_i) \tag{E2}$$

$$OM1\_PA = \sum_{i=1}^{6}(Amort_i \cdot Q\,min\,P_i) \tag{E3}$$

$$OM2\_CA = \sum_{i=1}^{6}\left(Amort_i \cdot \frac{Q\,min\,N_i}{Qn_i} \cdot Qc_i\right) \tag{E4}$$

$$OM2\_NA = \sum_{i=1}^{6}(Amort_i \cdot (Qn_i - Q\,min\,N_i)) \tag{E5}$$

$$OM2\_PA = \sum_{i=1}^{6}(Amort_i \cdot (Qp_i - Q\,min\,P_i)). \tag{E6}$$

Dynamic stoichiometric ratios ($C_xN_yP_z$) of *OM1_A* and *OM2_A* are tracked as

$$stoich\_x1A = \frac{(OM1\_CA + OM1\_A)}{OM1\_PA + \frac{1}{stoich\_x1A} \cdot OM1\_A} \tag{E7}$$

$$stoich\_y1A = \frac{\left(OM1\_NA + \frac{stoich\_y1A}{stoich\_x1A} \cdot OM1\_A\right)}{OM1\_PA + \frac{1}{stoich\_x1A} \cdot OM1\_A} \tag{E8}$$

$$stoich\_x2A = \frac{(OM2\_CA + OM2\_A)}{OM2\_PA + \frac{1}{stoich\_x2A} \cdot OM2\_A} \tag{E9}$$

$$stoich\_y2A = \frac{\left(OM2\_NA + \frac{stoich\_y2A}{stoich\_x2A} \cdot OM2\_A\right)}{OM2\_PA + \frac{1}{stoich\_x2A} \cdot OM2\_A} \tag{E10}$$

$$stoich\_z1A = stoich\_z2A = 1. \tag{E11}$$

Organic matter *C*, *N*, and *P* derived from zooplankton are calculated as

$$OM1\_CZ = ZegC_1 + ZunC_1 + ZmortC_1 + ZmortC_2 + (OM1\_Ratio \cdot ZslopC) \tag{E12}$$

$$OM1\_NZ = ZegN_1 + ZunN_1 + ZmortN_1 + ZmortN_2 + (OM1\_Ratio \cdot ZslopN) \tag{E13}$$

$$OM1\_PZ = ZegP_1 + ZunP_1 + ZmortP_1 + ZmortP_2 + (OM1\_Ratio \cdot ZslopP) \tag{E14}$$

$$OM2\_CZ = ZegC_2 + ZunC_2 + (OM2\_Ratio \cdot ZslopC) \tag{E15}$$

$$OM2\_NZ = ZegN_2 + ZunN_2 + (OM2\_Ratio \cdot ZslopN) \tag{E16}$$

$$OM2\_PZ = ZegP_2 + ZunP_2 + (OM2\_Ratio \cdot ZslopP) \tag{E17}$$

where *Zun* and *Zslop* equations are presented in Appendix D and zooplankton egestion (*Zeg*) is calculated as follows. *Zeg* in the model is governed by an optimal nutrient ratio of the zooplankton, i.e., *ZQn/ZQp*, such that

$$\begin{aligned}
&if\ ZinN_j > \frac{ZQn_j}{ZQp_j} \cdot ZinP_j \\
&ZegN_j = ZinN_j - ZinP_j \cdot \frac{ZQn_j}{ZQp_j} \\
&ZegC_j = ZinC_j - \frac{ZinP_j}{ZQp_j} \cdot ZQc_j \\
&ZegP_j = 0 \\
&else \\
&ZegP_j = ZinP_j - ZinN_j \cdot \frac{ZQp_j}{ZQn_j} \\
&ZegC_j = ZinC_j - \frac{ZinN_j}{ZQn_j} \cdot ZQc_j \\
&ZegN_j = 0.
\end{aligned} \tag{E18}$$

Similar to the OM derived from mortality of phytoplankton cells being split into *OM1_A* and *OM2_A*, OM from sloppy feeding on phytoplankton cells is split into particulate and dissolved OM fractions based on the calculated *OM1_Ratio* and *OM2_Ratio* as

$$OM1\_Ratio = \frac{\sum_{i=1}^{6} A_i \cdot \frac{Qn_i - Q\,min\,N_i}{Qn_i}}{\sum_{i=1}^{6} A_i} \quad \text{(E19)}$$

$$OM2\_Ratio = \frac{\sum_{i=1}^{6} A_i \cdot \frac{Q\,min\,N_i}{Qn_i}}{\sum_{i=1}^{6} A_i}. \quad \text{(E20)}$$

C, N, and P stoichiometry of *OM1_Z* and *OM2_Z* are tracked as

$$stoich\_x1Z = \frac{(OM1\_CZ + OM1\_Z)}{OM1\_PZ + \frac{1}{stoich\_x1Z} \cdot OM1\_Z} \quad \text{(E21)}$$

$$stoich\_y1Z = \frac{\left(OM1\_NZ + \frac{stoich\_y1Z}{stoich\_x1Z} \cdot OM1\_Z\right)}{OM1\_PZ + \frac{1}{stoich\_x1Z} \cdot OM1\_Z} \quad \text{(E22)}$$

$$stoich\_x2Z = \frac{(OM2\_CZ + OM2\_Z)}{OM2\_PZ + \frac{1}{stoich\_x2Z} \cdot OM2\_Z} \quad \text{(E23)}$$

$$stoich\_y2Z = \frac{\left(OM2\_NZ + \frac{stoich\_y2Z}{stoich\_x2Z} \cdot OM2\_Z\right)}{OM2\_PZ + \frac{1}{stoich\_x2Z} \cdot OM2\_Z} \quad \text{(E24)}$$

$$stoich\_z1Z = stoich\_z2Z = 1. \quad \text{(E25)}$$

E.2 Reaction Equations

Primary production (*PrimProd)* of organic matter by phytoplankton is

$$Prim\,Prod = \sum_{i=1}^{6} Agrow_i \cdot Qc_i \quad \text{(E26)}$$

and proceeds according to the photosynthesis reaction

$$xCO_2 + yDIN + zDIP \xrightarrow{uptake} C_xN_yP_z + xO_2 \quad \text{(E27)}$$

where dissolved inorganic nitrogen (DIN) and dissolved inorganic phosphorus (DIP) are taken up to produce organic matter with C:N:P stoichiometry of *Qc:Qn:Qp*.

Organic matter oxidation by aerobic respiration is represented by

$$C_xN_yP_z + xO_2 \xrightarrow{R1} xCO_2 + yNH_4 + zPO_4 \qquad \text{(E28)}$$

where *R1* is

$$R1 = \frac{O_2}{KO_2 + O_2} \qquad \text{(E29)}$$

and KO_2 is the Monod half-saturation constant (Table A.5).

Table A.5 Organic matter parameters

Parameter	Description	Units	Value
KG1	Decay rate of OM1_A and OM1_G	y^{-1}	50
KG2	Decay rate of OM2_A and OM2_G	y^{-1}	50
KG1_R	Decay rate of OM1_R	y^{-1}	11
KG2_R	Decay rate of OM2_R	y^{-1}	
KG1_BC	Decay rate of OM1_BC	y^{-1}	1
KG2_BC	Decay rate of OM2_BC	y^{-1}	1
nitmax	Maximum rate for nitrification	mmol m^{-3} d^{-1}	0.52
KNH_4	Half-saturation constant for nitrification	mmol m^{-3}	1
KO_2	Half-saturation constant for O_2 uptake	mmol m^{-3}	10
$KstarO_2$	Inhibition constant for denitrification	mmol m^{-3}	10
KNO_3	Half-saturation constant for denitrification	mmol m^{-3}	5
stoich_x1R	Initial C:P stoichiometry of OM1_R	mol/mol	51
stoich_y1R	Initial N:P stoichiometry of OM1_R	mol/mol	4.5
stoich_x2R	Initial C:P stoichiometry of OM2_R	mol/mol	700
stoich_y2R	Initial N:P stoichiometry of OM2_R	mol/mol	50
stoich_x1BC	Initial C:P stoichiometry of OM1_BC	mol/mol	106
stoich_y1BC	Initial N:P stoichiometry of OM1_BC	mol/mol	16
stoich_x2BC	Initial C:P stoichiometry of OM2_BC	mol/mol	106
stoich_y2BC	Initial N:P stoichiometry of OM2_BC	mol/mol	16
sink_OM1_A	Sinking rate of OM1_A	m d−1	10
sink_OM2_A	Sinking rate of OM2_A	m d^{-1}	0
sink_OM1_Z	Sinking rate of OM1_Z	m d^{-1}	10
sink_OM2_Z	Sinking rate of OM2_Z	m d^{-1}	0
sink_OM1_R	Sinking rate of OM1_R	m d^{-1}	10
sink_OM2_R	Sinking rate of OM2_R	m d^{-1}	0
sink_OM1_BC	Sinking rate of OM1_BC	m d^{-1}	10
sink_OM2_BC	Sinking rate of OM2_BC	m d^{-1}	0
sink_CDOM	Sinking rate of CDOM	m d^{-1}	0
Kcdom_decay	Decay rate of CDOM	d^{-1}	0.01
K	Q10 coefficient such that a 10 °C increase results in a twofold increase in OM remineralization	Dimensionless	0.07

The organic matter reaction during denitrification uses the reaction equation of Van Cappellen and Wang (1996)

$$\begin{aligned} &C_xN_yP_z + \left(\frac{4x+3y}{5}\right) NO_3 \longrightarrow R2 \\ &\left(\frac{2x+4y}{5}\right) N_2 + \left(\frac{x-3y+10z}{5}\right) CO_2 + \left(\frac{4x+3y-10z}{5}\right) HCO3 + zPO_4 \end{aligned} \tag{E30}$$

where $R2$ is

$$R2 = \frac{NO_3}{KNO_3 + NO_3} \frac{KstarO_2}{KstarO_2 + O_2} \tag{E31}$$

and KNO_3 is a Monod half-saturation constant (Table A.5), and $KstarO_2$ is an O_2-based inhibition constant (Table A.5) that limits denitrification when O_2 concentrations approach and exceed $KstarO_2$.

Reaction rates (R) are determined for organic matter remineralization (*ROM*, mmol C m^{-3} d^{-1}), O_2 utilization (RO_2, mmol O_2 m^{-3} d^{-1}), nitrification (*R_11*, mmol N m^{-3} d^{-1}) and denitrification (RNO_3, mmol N m^{-3} d^{-1}), and remineralization of PO_4 (RPO_4, mmol P m^{-3} d^{-1}), NH_4 (*RNH4*, mmol C m^{-3} d^{-1}), and *Si* (*RSi*, mmol C m^{-3} d^{-1}). The remineralization equations are identical for the four sources of OM and, for brevity, are only shown for OM_A.

Organic matter decay coefficients are adjusted for temperature by a Q10 relation

$$KG1\_Q10 = 10^{RQ1} \tag{E32}$$

$$KG2\_Q\,10 = 10^{RQ2}. \tag{E33}$$

where *RQ1* and *RQ2* are intermediate variables calculated as

$$RQ1 = LOG10(KG1) - FACTOR \tag{E34}$$

$$RQ2 = LOG10(KG2) - FACTOR \tag{E35}$$

and

$$FACTOR = LOG10(2) \cdot 0.1 \cdot (TQ1 - TQ2) \tag{E36}$$

where *TQ1* is the reference temperature of 25 °C and *TQ2* is the temperature in the model.

The temperature-adjusted reaction (*RCT*) rates for *OM1_A* and *OM2_A* are then calculated as

$$RCT1\_A = KG1\_Q10 \cdot OM1\_A \tag{E37}$$

$$RCT2\_A = KG2\_Q10 \cdot OM2\_A. \tag{E38}$$

ROM1_A and *ROM2_A* are calculated as

$$ROM1\_A = -(RCT1\_A \cdot R1 + RCT1\_A \cdot R2) \tag{E39}$$

$$ROM2\_A = -(RCT2\_A \cdot R1 + RCT2\_A \cdot R2). \tag{E40}$$

O_2 consumption associated with remineralization (RO_2, mmol m^{-3} d^{-1}) is

$$RO_2\_A = (RCT1\_A + RCT2\_A) \cdot R1 \tag{E41}$$

and

$$RO_2 = RO_2\_A + RO_2\_Z + RO_2\_R + RO_2\_BC. \tag{E42}$$

Nitrification proceeds as a function of NH_4 and O_2 concentration (Van Cappellen and Wang 1996) as

$$NH_4^+ + 2O_2 + 2HCO_3^- \xrightarrow{R\_11} NO_3^- + 2CO_2 + 3H_2O \tag{E43}$$

where *R_11* is the reaction rate defined as

$$R\_11 = nit\,max \cdot \frac{O_2}{KO_2 + O_2} \cdot \frac{NH_4}{KNH_4 + NH_4} \cdot Q10 \tag{E44}$$

where *nitmax* is the parameterized maximum nitrification rate and KNH_4 is the half-saturation constant (Table A.5). For the denitrification reaction (Eq. E30), reaction stoichiometry is taken into account in intermediate variables (*GAM14* and *GAM24*) calculated as

$$GAM14 = \frac{4 \cdot stoich\_x1A + 3 \cdot stoich\_y1A}{\frac{5}{stoich\_x1A}} \tag{E45}$$

$$GAM24 = \frac{4 \cdot stoich\_x2A + 3 \cdot stoich\_y2A}{\frac{5}{stoich\_x2A}} \tag{E46}$$

where the coefficients 4, 3, and 5 are the coefficients to NO_3 in Eq. (E30). Denitrification (RNO_3) using *OM1_A* and *OM2_A* as electron donors is represented by

$$RNO_3\_A = -(GAM14 \cdot RCT1\_A + GAM24 \cdot RCT2\_A) \cdot R2. \tag{E47}$$

Then, RNO_3 is

$$RNO_3 = RNO_3\_A + RNO_3\_Z + RNO_3\_R + RNO_3\_BC. \tag{E48}$$

Remineralization of PO_4 during the OM oxidation reactions is calculated as

$$RPO_4 = RPO_4\_A + RPO_4\_Z + RPO_4\_R + RPO_4\_BC \tag{E49}$$

with

$$RPO_4\_A = ROM1\_A \cdot \frac{stoich\_z1A}{stoich\_x1A} + ROM2\_A \cdot \frac{stoich\_z2A}{stoich\_x2A}. \tag{E50}$$

Remineralization of NH_4 is

$$RNH_4 = RNH_4\_A + RNH_4\_Z + RNH_4\_R + RNH_4\_BC \tag{E51}$$

with

$$RNH_4\_A = \left(RCT1\_A \cdot \frac{stoich\_y1A}{stoich\_x1A} + RCT2\_A \cdot \frac{stoich\_y2A}{stoich\_x2A} \right) \cdot R1 \tag{E52}$$

The rate of Si remineralization is

$$RSi = RSi\_A + RSi\_Z + RSi\_R + RSi\_BC \tag{E53}$$

with Si stoichiometrically linked to remineralization of N as Si:N = 1 such that

$$RSi\_A = ROM1\_A \cdot \frac{stoich\_y1A}{stoich\_x1A} + ROM2\_A \cdot \frac{stoich\_y2A}{stoich\_x2A}. \tag{E54}$$

F. Air–Sea Exchange

Air–sea exchanges of O_2 were modeled based on concentration gradients and wind speed (Eldridge and Roelke 2010).

References

Aulenbach BT, Buxton HT, Battaglin WA, Coupe RH (2007) Streamflow and nutrient fluxes of the Mississippi–Atchafalaya River Basin and subbasins for the period of record through 2005: U.S. Geological Survey Open-File Report 2007-1080. http://toxics.usgs.gov/pubs/of-2007-1080/index.Html

Bach LT, Riebesell U, Sett S, Febiri S, Rzepka P, Schulz KG (2012) An approach for particle sinking velocity measurements in the 3–400 μm size range and considerations on the effect of temperature on sinking rates. Mar Biol 159:1853–1864

Bianchi TS, DiMarco SF, Cowan JH Jr, Hetland RD, Chapman P, Day JW, Allison MA (2010) The science of hypoxia in the Northern Gulf of Mexico: a review. Sci Total Environ 408:1471–1484. doi:10.1016/j.scitotenv.2009.11.047

Cerco CF, Noel MR (2004) Process-based primary production modeling in Chesapeake Bay. Mar Ecol Prog Ser 282:45–58

Collins M, Knutti R, Arblaster J, Dufresne J-L, Fichefet T, Friedlingstein P, Gao X, Gutowski WJ, Johns T, Krinner G, Shongwe M, Tebaldi C, Weaver AJ, Wehner M (2013) Long-term climate change: projections, commitments and irreversibility. In: Stocker TF, Qin D, Plattner G-K, Tignor M, Allen SK, Boschung J, Nauels A, Xia Y, Bex V, Midgley PM (eds.) Climate change 2013: the physical science basis. contribution of working group I to the fifth assessment report of the intergovernmental panel on climate change. Cambridge University Press, Cambridge, United Kingdom and New York, NY, USA

Conmy RN, Coble PG, Chen RF, Gardner GB (2004) Optical properties of colored dissolved organic matter in the Northern Gulf of Mexico. Mar Chem 89:127–144. doi:10.1016/j.marchem.2004.02.010

D'Sa EJ, DiMarco SF (2009) Seasonal variability and controls on chromophoric dissolved organic matter in a large river-dominated coastal margin. Limnol Oceanogr 54:2233–2242

Diaz RJ, Rosenberg R (2008) Spreading dead zones and consequences for marine ecosystems. Science 321:926–929. doi:10.1126/science.1156401

Doney SC, Ruckleshaus M, Duffy JE et al (2012) Climate change impacts on marine ecosystems, Annu Rev Mar Sci 4:11–37

Donner SD, Scavia D (2007) How climate controls the flux of nitrogen by the Mississippi River and the development of hypoxia in the Gulf of Mexico. Limnol Oceanogr 52:856–861

Droop MR (1973) Some thoughts on nutrient limitation in Algae. J Phycol 9:264–272. doi:10.1111/j.1529-8817.1973.tb04092.x

Eldridge PM, Roelke DL (2010) Origins and scales of hypoxia on the Louisiana shelf: Importance of seasonal plankton dynamics and river nutrients and discharge. Ecol Model 221:1028–1042. doi:10.1016/j.ecolmodel.2009.04.054

Falkowski PG, Hopkins TS, Walsh JJ (1980) An analysis of factors affecting oxygen depletion in the New York Bight. J Mar Res 38:479–506

Fennel K, Hu J, Laurent A, Marta-Almeida M, Hetland R (2013) Sensitivity of hypoxia predictions for the northern Gulf of Mexico to sediment oxygen consumption and model nesting. J Geophys Res Oceans 118:990–1002. doi:10.1002/jgrc.20077

Forrest DR, Hetland RD, DiMarco SF (2011) Multivariable statistical regression models of the areal extent of hypoxia over the Texas-Louisiana continental shelf. Environ Res Lett 6:045002. doi:10.1088/1748-9326/6/4/045002

Greene RM, Lehrter JC, Hagy JDH (2009) Multiple regression models for hindcasting and forecasting midsummer hypoxia in the Gulf of Mexico. Ecol Appl 19:1161–1175. doi:10.1890/08-0035.1

Hetland RD, DiMarco SF (2008) How does the character of oxygen demand control the structure of hypoxia on the Texas-Louisiana continental shelf. J Mar Syst 70(1–2):49–62

Hodur RM (1997) The Naval Research Laboratory's Coupled Ocean/Atmosphere Mesoscale Prediction System (COAMPS). Mon Weather Rev 125:1414–1430. doi:10.1175/1520-0493(1997)125<1414:TNRLSC>2.0.CO;2

Justic D, Rabalais NN, Turner RE (1996) Effects of climate change on hypoxia in coastal waters: a doubled CO_2 scenario for the northern Gulf of Mexico. Limnol Oceanogr 4:992–1003

Justic D, Rabalais NN, Turner RE (2003) Simulated responses of the Gulf of Mexico hypoxia to variations in climate and anthropogenic nutrient loading. J Mar Syst 42:115–126

Justic D, Wang L (2014) Assessing temporal and spatial variability of hypoxia over the inner Louisiana-upper Texas shelf: application of an unstructured-grid three-dimensional coupled hydrodynamic-water quality model. Cont Shelf Res 72:163–179

Ko DS, Gould RW, Penta B Jr, Lehrter JC (2016) Impact of satellite remote sensing data on simulations of coastal circulation and hypoxia on the Louisiana continental shelf. Remote Sens 8:435. doi:10.3390/rs8050435

Ko DS, Martin PJ, Rowley CD, Preller RH (2008) A real-time coastal ocean prediction experiment for Mississippi RiverEA04. J Mar Syst 4:17–28. doi:10.1016/j.jmarsys.2007.02.022

Ko DS, Preller RH, Martin PJ (2003) An experimental real-time intra-americas sea ocean nowcast/forecast system for coastal prediction. In: AMS 5th Conference on Coastal Atmospheric and Oceanic, NRL contribution NRL/pp/7320/03/104

Ko DS, Wang D-P (2014) Intra-Americas Sea Nowcast/Forecast system ocean reanalysis to support improvement of oil-spill risk analysis in the Gulf of Mexico by Multi-Model Approach, DOI, BOEM, Herndon, VA. BOEM 2014–1003, pp 55. http://www.data.boem.gov/PI/PDFImages/ESPIS/5/5447.pdf

Lehman MK, Fennel K, He R (2009) Statistical validation of a 3-D bio-physical model of the western North Atlantic, Biogeosciences, 6:1961–1974

Lehrter JC, Murrell MC, Kurtz JC (2009) Interactions between freshwater input, light, and phytoplankton dynamics on the Louisiana continental shelf. Cont Shelf Res 29:1861–1872

Lehrter J, Ko DS, Murrell MC, Hagy JD, Schaeffer BA, Greene RM, Gould RW, Penta B (2013) Nutrient distributions, transports, and budgets on the inner margin of a river-dominated continental shelf. J Geophys Res Oceans 118:1–17

Lehrter JC, Fry B, Murrell MC (2014) Microphytobenthos production potential and contribution to bottom layer oxygen dynamics on the inner Louisiana continental shelf. Bull Mar Sci. doi:10.5343/bms.2013.1050

Liss PS, Merlivat L (1986) Air-sea gas exchange rates: introduction and synthesis. In: Buat-Menard P, Reidel D (ed) The role of air-sea exchange in geochemical cycling, pp 113–127

Marta-Almeida M, Hetland RD, Zhang X (2013) Evaluation of model nesting performance on the Texas-Louisiana continental shelf. J Geophys Res: Oceans 118(5):2476–2491

Martin PJ (2000) Description of the navy coastal ocean model version 1.0, NRL/FR/7322-00-9962. Naval Research Lab, Stennis Space Center, Mississippi

Mississippi River/Gulf of Mexico Watershed Nutrient Task Force 2015 Report to Congress (2015). https://www.epa.gov/sites/production/files/2015-10/documents/htf_report_to_congress_final_-_10.1.15.pdf. Accessed 22 Mar 2016

Mulholland P (2002) Large-scale patterns in dissolved organic carbon concentration, flux, and sources. In: Findlay SEG, Sinsabaugh RL (eds) Aquatic ecosystems: interactivity of dissolved organic matter. Academic Press, San Diego, California

Murrell MC, Beddick DL, Devereux R, Greene RM, Hagy JD, Jarvis BM, Kurtz JC, Lehrter JC, Yates DF (2014) Gulf of Mexico hypoxia research program data report: 2002–2007. U.S. Environmental Protection Agency, Washington, D.C., EPA/600/R-13/257

Obenour DR, Michalak AM, Zhou Y, Scavia D (2012) Quantifying the impacts of stratification and nutrient loading on hypoxia in the Northern Gulf of Mexico. Environ Sci Technol 46:5489–5496

Obenour DR, Scavia D, Rabalais NN, Turner RE, Michalak AM (2013) Retrospective analysis of midsummer hypoxic area and volume in the northern Gulf of Mexico, 1985-2011. Environ Sci Technol 47:9808–9815

Pauer JJ, Feist TJ, Anstead AM, DePetro PA, Melendez W, Lehrter JC, Murrell MC, Zhang X, Ko DS (2016) A modeling study examining the impact of nutrient boundaries on primary production on the Louisiana continental shelf. Ecol Model 328:136–147

Penta B, Lee Z, Kudela RM, Palacios SL, Gray DJ, Jolliff JK, Shulman IG (2008) An underwater light attenuation scheme for marine ecosystem models. Opt Express 16:16581–16591

Penta B, Lee Z, Kudela RM, Palacios SL, Gray DJ, Jolliff JK, Shulman IG (2009) An underwater light attenuation scheme for marine ecosystem models: errata. Opt Express 17:23351

Rabalais NN, Turner RE, Jr WJW (2002) Gulf of Mexico Hypoxia, a.k.a. "The Dead Zone". Annu Rev Ecol Syst 33:235–263

Rabalais NN, Diaz RJ, Levin LA, Turner RE, Gilbert D, Zhang J (2010) Dynamics and distribution of natural and human-caused hypoxia. Biogeosciences 7:585–619

Roelke DL, Eldridge PM, Cifuentes LA (1999) A model of phytoplankton competition for limiting and nonlimiting nutrients: Implications for development of estuarine and nearshore management schemes. Estuaries 22:92–104. doi:10.2307/1352930

Romero-Lankao P, Smith JB, Davidson DJ, Diffenbaugh NS, Kinney PL, Kirshen P, Kovacs P, Villers Ruiz L (2014) North America. In Barros VR, Field CB, Dokken DJ, Mastrandrea MD, Mach KJ, Bilir TE, Chatterjee M, Ebi KL, Estrada YO, Genova RC, Girma B, Kissel ES, Levy AN, MacCracken S, Mastrandrea PR, White LL (eds) Climate change 2014: impacts, adaptation, and vulnerability. part b: regional aspects. contribution of working group II to the fifth assessment report of the intergovernmental panel on climate change. Cambridge University Press, Cambridge, United Kingdom and New York, NY, USA, pp 1439–1498

Rosmond TE (1992) The design and testing of the Navy operational global atmospheric prediction system. Weather Forecast 7:262–272. doi:10.1175/1520-0434

Spencer RGM, Aiken GR, Butler KD, Dornblaser MM, Striegl RG, Hernes PJ (2009) Utilizing chromophoric dissolved organic matter measurements to derive export and reactivity of dissolved organic carbon exported to the Artic Ocean: a case study of the Yukon River, Alaska. Geophys Res Lett 36:L06401

Spencer RGM, Butler KD, Aiken GR (2012) Dissolved organic carbon and chromophoric dissolved organic matter properties of rivers in the USA. J Geophys Res Biogeosci 117: G03001. doi:10.1029/2011JG001928

Sperna Weiland FC, van Beek LPH, Kwadijk JCJ, Bierkens MFP (2012) Global patterns of change in discharge regimes for 2100. Hydrol Earth Syst Sci 16:1047–1062

Stow CA, Jolliff J, McGillicuddy DJ Jr, Doney SC, Allen JI, Friedrichs MAM, Rose KA, Wallhead P (2008) Skill assessment for coupled biological/physical models of marine systems. J Mar Syst 76:4–15. doi:10.1016/j.jmarsys.2008.03.011

Van Cappellen P, Wang Y (1996) Cycling of iron and manganese in surface sediments; a general theory for the coupled transport and reaction of carbon, oxygen, nitrogen, sulfur, iron, and manganese. Am J Sci 296:197–243. doi:10.2475/ajs.296.3.197

Wiseman WJ, Rabalais NN, Turner RE, Dinnel SP, MacNaughton A (1997) Seasonal and interannual variablility within the Louisiana coastal current: stratification and hypoxia. J Mar Syst 12:237–248

Yu L, Fennel K, Laurent A (2015) A modeling study of physical controls on hypoxia generation in the northern Gulf of Mexico. J Geophys Res Oceans 120. doi:10.1002/2014JC010634

Chapter 9
Oregon Shelf Hypoxia Modeling

Andrey O. Koch, Yvette H. Spitz and Harold P. Batchelder

Abstract Bottom hypoxia on the shelf in the Northeast Pacific is caused by different processes than coastal hypoxia related to riverine inputs. Hypoxia off the coast of Oregon is a naturally occurring process as opposed to the anthropogenically forced hypoxia found in many coastal environments (e.g., Gulf of Mexico shelf, Chesapeake Bay). Off Oregon, bottom hypoxia occurs in summers that have large upwelling-driven near-bottom transport of high nitrate, low dissolved oxygen (DO) waters onto the shelf. The combination of low DO and high nitrate provides initially low (but not hypoxic) DO conditions near the bottom, and nitrate fertilization of shelf surface waters, leading to substantial phytoplankton production. Some production is grazed, and some of it sinks to the bottom where it decomposes consuming oxygen, creating bottom hypoxia in some years. Terrestrial runoff of nutrients into the system is small and not responsible for the development of bottom hypoxia. Similar processes contribute to natural hypoxia in other eastern boundary current upwelling regions, such as the Humboldt Current off Peru and the Benguela Current off Namibia and South Africa. We summarize the observational data on DO and illustrate the coupled bio-physical modeling of hypoxia that has been done on the Oregon shelf. We compare hypoxia development in summer of 2002 and 2006, which differed in timing, spatial extent and intensity of hypoxia. Sensitivity analysis using various initial and boundary conditions for nitrate and dissolved oxygen reveals some of the essential conditions responsible for hypoxia development on the Oregon shelf.

A.O. Koch (✉)
Department of Marine Science, University of Southern Mississippi, Hattiesburg, USA
e-mail: andreyokoch@gmail.com

Y.H. Spitz · H.P. Batchelder
College of Earth, Ocean, and Atmospheric Sciences, Oregon State University, Corvallis, USA
e-mail: yspitz@coas.oregonstate.edu

H.P. Batchelder
North Pacific Marine Science Organization, Sidney, BC, Canada
e-mail: hbatch@pices.int

D. Justic et al. (eds.), *Modeling Coastal Hypoxia*,
DOI 10.1007/978-3-319-54571-4_9

Keywords Oregon shelf • Natural hypoxia • Biophysical model of hypoxia • Observation-model comparisons • Interannual/seasonal variability

9.1 Introduction

Hypoxic regions in the world's oceans are becoming more numerous, lower in oxygen content, and expanding in area, volume and duration (Diaz and Rosenberg 2008). The concentration of dissolved oxygen (hereafter, DO) in coastal ocean regions is determined by air-sea exchange, ocean circulation, biological processes and anthropogenic activities. Hypoxia, defined by DO concentrations less than 1.43 ml L^{-1} (Grantham et al. 2004) and, in some cases, anoxia (DO < 0.5 ml L^{-1}; Chan et al. 2008) are known to occur in coastal regions, typically due to excessive nutrient input derived from agricultural runoff, or from waste management facilities incapable of handling the loads in regions of high human population density. The Louisiana–Texas shelf of the Gulf of Mexico (Chaps. 1, 3, 7, 8, 10, 13 and 14), the Chesapeake Bay (Chaps. 5, 6, 11 and 12), Narragansett Bay, RI (Chap. 4) and Green Bay, Lake Michigan (Chap. 2) are all regions that have experienced substantial anthropogenic nutrient enrichment and eutrophication as a primary contributing cause of hypoxia. Changes in DO include reductions in concentration, particularly near the bottom, seasonal extension of the duration or presence of hypoxia, and altered spatial patterns of hypoxia. Seasonal hypoxia is relatively common in semi-enclosed bays and estuaries where bottom waters have longer residence times due to reduced exchanges with more open regions. Summer is especially vulnerable to hypoxia development because stronger density stratification inhibits vertical exchange of surface and bottom waters, and high surface primary production creates organic detritus that sinks to bottom waters and remineralizes, consuming DO.

The open continental shelves of Oregon and Washington along the Pacific Northwest Coast have experienced a long-term (50 year) decline in late-summer DO (Chan et al. 2008), but without evidence of anthropogenic nutrient inputs and eutrophication. Instead, these shelves, and shelves more generally along eastern boundary currents, develop near-bottom hypoxia caused by natural oceanographic processes (Chan et al. 2008; Escribano and Schneider 2007).

The Pacific Northwest Coast experiences strong seasonally reversing winds. Winds during winter (October–April) are strongly from the SW, which causes onshore surface Ekman transport and downwelling of well-oxygenated surface waters on the inner to middle shelf. During this time, waters near the bottom on the central Oregon shelf are well oxygenated (ca. 4–7 ml L^{-1}). During spring winds change from predominantly northward to southward (spring transition). The timing of the spring transition and the duration of the summer upwelling season varies greatly from year to year, as illustrated by the along-shore wind data for the three years (2002, 2006 and 2008) considered in this study (Fig. 9.1 and Table 9.1). Strong, persistent southward winds in the summer lead to offshore Ekman transport of coastal surface waters. Coastal water is replaced by cold, upwelled, nutrient rich,

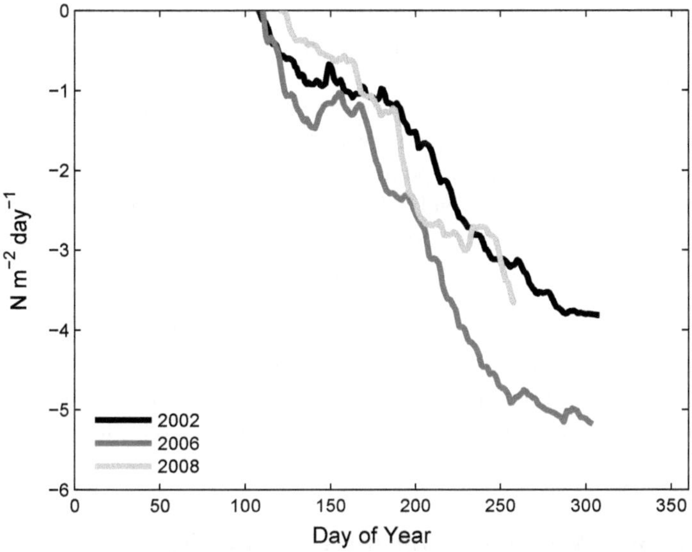

Fig. 9.1 Cumulative wind forcing (Nm^{-2} d^{-1}) for the upwelling season of 2002, 2006 and 2008. Note the later start and earlier end of upwelling in 2008, and the similar durations of upwelling in 2002 and 2006, but the 34% greater net upwelling during 2006. (Data retrieved from http://damp.coas.oregonstate.edu/windstress/index.html). April 1 is day of year 91; May 1 is day of year 121. November 1 is day of year 305

Table 9.1 Details of wind forcing for 2002, 2006 and 2008 from the National Data Buoy Center station 46050 located 20 nautical miles west of Newport, OR (station has a bottom depth of 137 m). See http://damp.coas.oregonstate.edu/windstress/index.html for summary of data

Year	2002	2006	2008
Year-day of spring transition	107	110	120
Number of upwelling days	144	143	104
Number of downwelling days	58	46	36
Length of upwelling season (days)	202	189	140
Sum of downwelling stress (N m^{-2} d^{-1})	1.03	1.13	0.72
Sum of upwelling stress (N m^{-2} d^{-1})	−4.84	−6.22	−4.38
Sum of stresses (N m^{-2} d^{-1})	−3.81	−5.09	−3.66
Proportion of days that were downwelling	0.29	0.24	0.26
Percent reduction in upwelling stress due to downwelling during upwelling period	0.21	0.18	0.16

but relatively oxygen poor water (Bakun 1990). Peterson et al. (2013) show that the source of upwelling water is from depths of 100–170 m off the shelf, where the DO is 1.5–3 ml L^{-1} (e.g., not hypoxic) in most years. The intermediate and deep water of the eastern North Pacific has been isolated from the atmosphere for ca. seven years due to global ocean circulation and has experienced substantial DO loss due to remineralization of sinking organic matter (Ueno and Yasuda 2003). The DO of the source waters may vary interannually by up to 1.0 ml L^{-1} (Grantham et al. 2004; Peterson et al. 2013).

The shelf waters in years with lower DO in offshore source waters are poised to go hypoxic earlier during the summer, reach lower minimum DO levels and possibly have more spatially extensive hypoxic regions. Years of sustained summer upwelling, such as occurred in 2002 and 2006, elevate nutrients in the inner-shelf, fueling productivity and reducing near-bottom oxygen levels due to remineralization of sinking particulate organic carbon (Grantham et al. 2004; Wheeler et al. 2003; Fig. 9.2). Thus, the development of near-bottom hypoxia on the Oregon shelf may have several contributing factors: initially low dissolved oxygen in the upwelling source waters; higher nutrient concentrations that upwell into the euphotic zone stimulate higher phytoplankton productivity; and greater sinking and remineralization of in situ produced organic matter near the bottom (Wheeler et al. 2003).

9.2 Hypoxia Variability on the Oregon Shelf

Hypoxia development on the Oregon shelf is seasonal and linked with upwelling-favorable winds. Anthropogenic eutrophication is not a significant contributor to the development of near-bottom hypoxia on the shelf of Oregon (and Washington and Southern British Columbia) in the Pacific Northwest (Chan et al. 2008; Crawford and Pena 2013). Upwelling-favorable winds may occur at any time of the year, but winds are persistently upwelling favorable during the spring and summer. Near-bottom hypoxia intensifies with duration of upwelling following the spring transition (Huyer et al. 1979) and often peaks in intensity and spatial extent on the Oregon shelf during late summer (August–September). Cumulative upwelling-favorable wind stress varies significantly interannually (Fig. 9.1). Although later retrospective studies have revealed periods of bottom hypoxia on the Oregon shelf several decades ago, the extent and ecological impacts of bottom hypoxia were not fully appreciated until the summer of 2002, when video surveys from remote operated vehicles showed dead and dying fish and invertebrates on the central Oregon shelf (Grantham et al. 2004). Prior to 2002, few hydrographic surveys included measurements of dissolved oxygen, as low oxygen was not considered a common occurrence in Oregon shelf waters.

Cross-shelf sampling of the Newport Hydrographic Line (44.65° N) during August–September of 1998–2009 showed 6 years (2000, 02, 05, 06, 07, 09) with extensive cross-shelf bottom hypoxia (Fig. 9.2). A few years (1998, 99, 2001, 03 and 08) showed little to no bottom hypoxia. What is responsible for these differences? To address this, we first examine the physical conditions leading to the 2002, 2006 and 2008 patterns of hypoxia. We then describe the results of coupled biophysical models that simulated the ecosystem and oxygen conditions on the Oregon shelf during the 2002 and 2006 upwelling period.

Spring–summer upwelling in 2002 was fairly persistent, but not exceptionally strong, whereas 2002 was remarkable in having exceptionally low oxygen

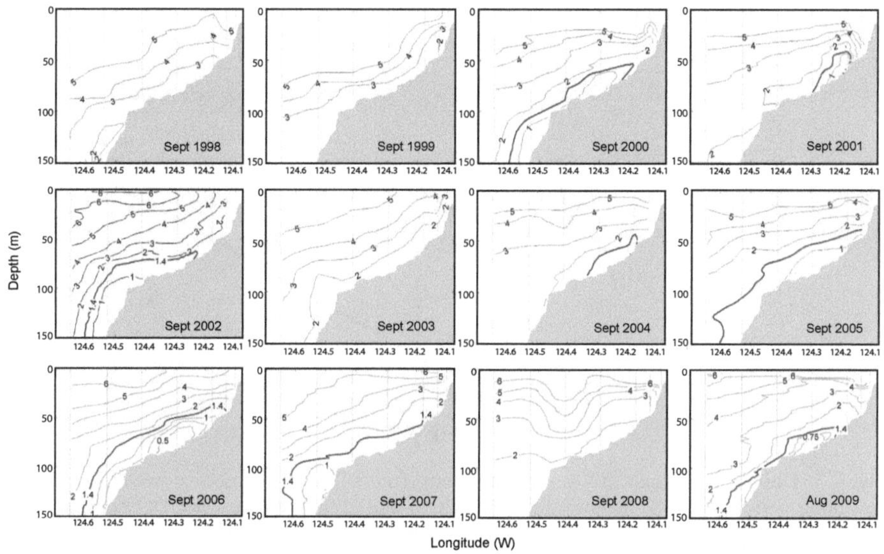

Fig. 9.2 Dissolved oxygen concentrations on the Newport Hydrographic transect in September (1998–2008) and August 2009 (lacking September data for 2009). Oxygen is contoured at 1 ml L^{-1} intervals for values of 1.0 or greater, and at 0.25 ml L^{-1} intervals where DO is less than 1 ml L^{-1}. An additional contour (*thick black line*) is shown at the hypoxic level of 1.4 ml L^{-1}. Figure provided courtesy of Bill Peterson (NWFSC, NOAA)

concentration and high nitrate concentrations in the water that upwelled from off the shelf (Fig. 9.3). In fact, the cumulative upwelling in 2002 was quite similar to 2008 (Fig. 9.1), a year in which bottom hypoxia was not prevalent. Oxygen concentrations shown by Grantham et al. (2004; their Fig. 4c) in July 2002 at 100–170 m at the shelf break off Newport, Oregon is very low, and far lower than the previously reported 95% confidence level from a decade of observations in the 1960s. Oceanic (offshore) regions of the Northeast Pacific have broad oxygen minimum zones (OMZ) at mid-depths. In 2002, the upwelled waters had very low DO (even below the hypoxic threshold in July), especially when compared to 2008 that was another "hypoxic" (though not strong) year (Fig. 9.2). The prevalence of hypoxic bottom conditions in 2002 was not recognized until the end of the summer, and there were few surveys during summer that included DO sampling in near-bottom water (Grantham et al. 2004).

During upwelling, high nutrient (especially nitrate), low oxygen waters upwell from depths of 100–200 m offshore (actual depth may vary interannually and seasonally; Peterson et al. 2013). The injection of nutrients into the photic zone enables high phytoplankton production on the shelf. In 2002, the higher nutrient concentrations resulted in shelf chlorophyll concentrations that were twice the levels of the preceding years 1998–2001 (Wheeler et al. 2003). Some of this phytoplankton production sinks as flocculates or fecal pellets to mid- and bottom waters on the shelf where it is remineralized, reducing near-bottom DO

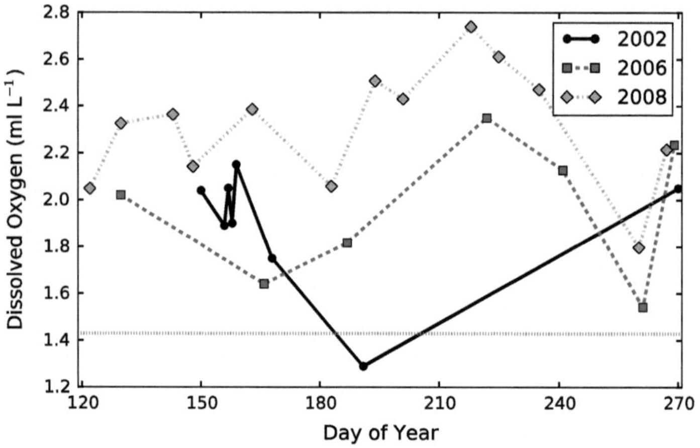

Fig. 9.3 Dissolved oxygen (DO) concentrations (ml L^{-1}) of the source water during the upwelling period (May–September) for 2002, 2006 and 2008. Source water DO concentrations were measured at the 26.44 sigma-t level at a station 25 miles off the coast of Newport Oregon (NH-25) as reported by Peterson et al. (2013). 2008 was a year of relatively little hypoxia on the Oregon shelf; 2002 and 2006 were years of widespread low DO on the shelf. (Data compiled by Jay Peterson, Oregon State University, from data collected by multiple sampling programs.)

concentrations even further from their initially low values. Gradually, the DO may be reduced below the hypoxia threshold.

Bottom DO is also strongly influenced by the local residence time of water (e.g., the near-bottom flow velocities and retention times), which may be controlled by bottom bathymetry (Castelao and Barth 2005). Wider continental shelves, complex bathymetries (bank complexes) and reduced wind stress (or intermittency of upwelling) lead to longer retention times, allowing greater localized consumption of DO (Barth et al. 2005; Hickey and Banas 2003). The region of most intense hypoxia in 2002 was associated with Heceta Bank, a wide region of the shelf, which has both sluggish flows and recirculation, and thus longer local retention times (Grantham et al. 2004; Barth et al. 2005).

The next few sections describe a physical–biological model applied to the Oregon shelf to understand the roles of physical and biological processes in modulating the development and extent of hypoxia. We focus on the summers of two hypoxic years, 2002 and 2006 that contrast in the severity and duration of their hypoxia.

9.3 Model of Oregon Shelf Hypoxia

We develop a coupled biological–physical model for the coastal ocean off Oregon (Fig. 9.4). We coupled a 3-km horizontal resolution Regional Ocean Modeling System (ROMS v3.0) physical model (Shchepetkin and McWilliams 2003, 2005) with a 5-component NAPZD (nitrate, ammonium, phytoplankton, zooplankton,

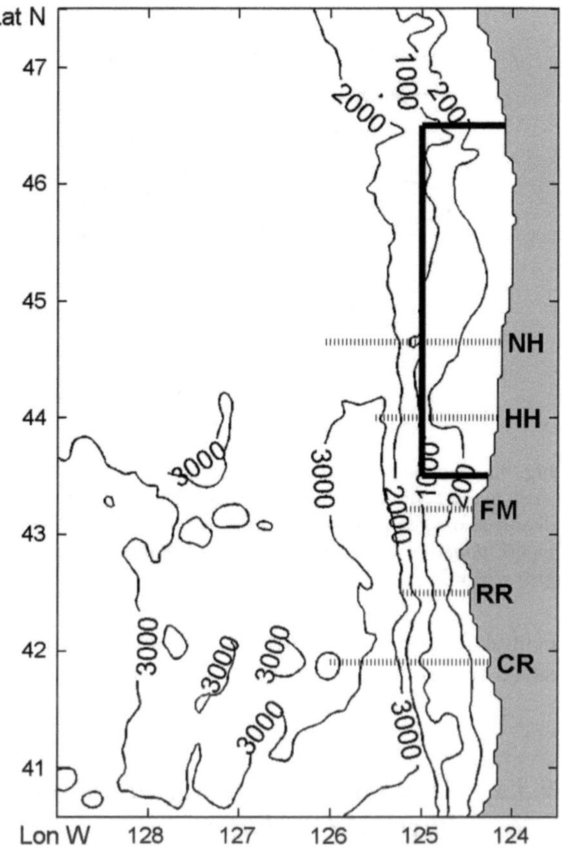

Fig. 9.4 Study location. The full region shown corresponds to the domain of the ROMS-CTZ Pacific Ocean model off Northern California (Southern end) to Northern Oregon (Northern end). The shelf region delimited by the *bold lines* (43.5–46.5° N) is the region analyzed in detail for bottom hypoxia. GLOBEC-LTOP observation locations on Crescent City (CR), Rogue River (RR), Five Miles (FM), Heceta Head (HH) and Newport Hydrographic (NH) transects are shown by *lighter dotted lines*. Bathymetry is shown by *thin numbered* (m) *lines*. The 200-m contour identifies the offshore edge of the shelf

detritus) ecosystem model (based on Spitz et al. 2005). The ROMS Coastal Transition Zone (ROMS-CTZ) computational domain spans 129–124° W meridionally and 40.5–47.5° N zonally and is identical to that used in Koch et al. (2010). Following their study, we simulate the period from April through August, which includes the transition in spring from northward winds to southward upwelling-favorable winds and summer upwelling circulation.

We add dissolved oxygen (DO) to the Spitz et al. (2005) NAPZD model to create a 6-component NAPZDO model (Fig. 9.5). DO is treated as a passive tracer with biologically mediated inputs (photosynthesis) and losses (zooplankton respiration, detritus remineralization and oxidation of ammonium), and an additional source (sink) term through air-sea exchange; the equations governing DO dynamics are shown in Appendix A. The computation of air-sea DO flux uses DO saturation concentration after Garcia and Gordon (1992) and gas transfer coefficient after Keeling et al. (1998). We do not explicitly simulate sediment-based DO consumption. Instead, we consider that settling particulate matter is retained within the bottom-most cell in the model and consumed there. We generate solutions for spring–summer of 2002 and 2006 that had extensive and severe hypoxia on the

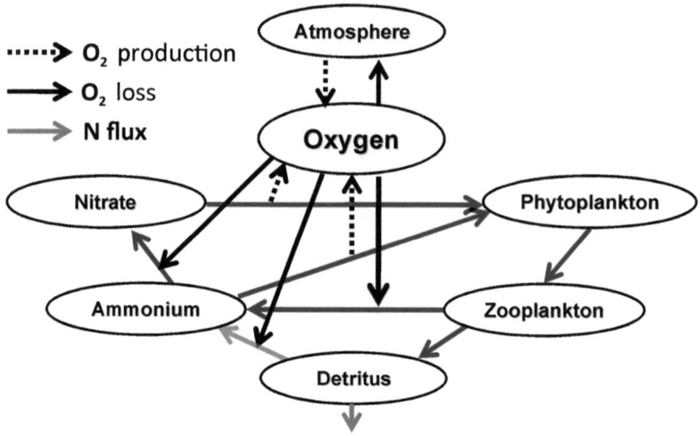

Fig. 9.5 Simple schematic of the ecological and dissolved oxygen model used for the Oregon shelf model. Model is modified from the ecosystem model of Spitz et al. (2005) to include dissolved oxygen processes as a passive tracer, with atmosphere exchange, biologically mediated inputs (*dotted black arrows*) and losses (*solid black arrows*). Nitrogen fluxes are shown with *solid gray arrows*

central Oregon shelf for most of the summer. The oxygen instantaneous values were saved twice-daily (every 12 h) for every grid cell.

9.3.1 Atmospheric Forcing, Initial and Open Boundary Physical Conditions

The coupled model simulations for 2002 are forced by 9-km horizontal resolution daily averaged COAMPS (The Coupled Ocean/Atmosphere Mesoscale Prediction System; Hodur 1997) winds and monthly averaged NCEP/NCAR (National Center for Environmental Prediction/National Center for Atmospheric Research; Kalnay et al. 1996) fields for heat-flux computation (shortwave solar radiation, air temperature, air pressure, relative humidity, precipitation) with 2.5° horizontal resolution. The simulations for 2006 were forced using 9-km atmospheric fields from the North American Mesoscale (NAM) model.

Initial conditions and open boundary conditions for velocities, temperature, salinity and sea surface height (SSH) were provided by a larger-scale NCOM-CCS (Navy Coastal Ocean Model for California Current System) model with 9-km horizontal resolution and 40 vertical levels: 20 sigma-levels (in upper 150 m) and 20 z-levels with constant depths (Shulman et al. 2004). Open boundary fields were provided daily. To suppress undesired effects of open boundary conditions as a result of merging larger-scale 9-km horizontal resolution NCOM fields with smaller-scale 3-km ROMS-CTZ fields we implemented a "sponge" layer that provided enhanced diffusivity and dissipation in the 100-km region adjacent to the open boundaries. Our analysis of biological and DO fields was restricted to the subdomain from 43.5–46.5° N that excludes the sponge layer. The ROMS-CTZ domain along with the analysis subdomain is shown in Fig. 9.4.

9.3.2 Initial and Open Boundary Ecosystem Conditions

Providing the initial and boundary conditions for model runs requires knowledge of conditions in the real ocean, which is usually collected by multiple investigators. In our case, we relied extensively on field sampling conducted during the Global Ocean Ecosystems Dynamics (GLOBEC) Northeast Pacific regional program that sampled the Oregon shelf physics and ecosystem from 1997–2004 (especially the Long-Term Observation Program; hereafter LTOP), and subsequent sampling of the mid-to-late 2000s by individual investigators.

Open boundary conditions for nitrate, ammonia, phytoplankton, zooplankton and detritus are provided by the NCOM-CCS biological solutions for April–August of 2006 and 2008. NCOM simulations for 2002 did not include an ecosystem model, so for 2002 simulations using ROMS-CTZ we use biological boundary conditions from the NCOM-CCS simulation from 2008; wind forcing in 2008 was similar to wind forcing in 2002 (Fig. 9.1). NCOM's ecosystem includes two phytoplankton types (diatoms and nanoflagellates) and two zooplankton types (microzooplankton and mesozooplankton), with most of the biomass in the diatom and mesozooplankton types. Since our model had only a single phytoplankton and single zooplankton type, we used the sum of the multiple phytoplankton and zooplankton from NCOM-CCS for boundary conditions of ROMS-CTZ.

The initial nitrate conditions for the NCOM-CCS ecosystem model are based on Levitus World Ocean climatology (Levitus 1982). During the several decades long spin-up of NCOM-CCS, the nitrate fields showed significant drift and by the 2000s both the nitrate concentration and depth of the nitrocline had become biased (I. Shulman, personal communication). We eliminated the nitrate bias by adjusting the nitrate fields from NCOM using empirical linear regression between NCOM-CCS and GLOBEC-LTOP (Strub et al. 2002; Wetz et al. 2004) nitrate values taken at the same locations of the space–time domain for eight depth layers (0–50, 50–100, 100–150, 150–250, 250–350, 350–500, 500–700, 700 m-bottom). The bias correction was done separately for NCOM-CCS data from 2006 and 2008. GLOBEC-LTOP data on NO_3 and DO were collected at standard depths spanning from surface to 1000 m depth (or bottom, if shallower) along traditional Oregon observation lines: Crescent River (CR), Rogue River (RR), Five Miles (FM), Heceta Head (HH) and Newport Hydrographic (NH, Fig. 9.4) extending from the inner-shelf offshore to 126° W. Since the NCOM model did not include DO dynamics, we use a NO_3: DO linear regression to estimate the DO field from NO_3 for both boundary and initial conditions. The NO_3: DO linear relationship was more robust than other relations between DO and density, temperature or salinity. The linear regression ratio (slope −0.16, intercept 7.23) was derived using all

GLOBEC-LTOP NO_3 and DO observations from March–April of 1997–2004. The GLOBEC-LTOP observational program ended in 2004, but we assumed that the NO_3: DO relation derived from 1997–2004 applied also to 2006.

It is critical (as shown later by "Sensitivity analysis") that the simulation begin with realistic DO and NO_3 concentrations in order to accurately reproduce the shelf ecosystem dynamics, including hypoxia. This is why we took great care to provide accurate initial (i.e., early spring) spatially explicit NO_3 and DO fields. Initial DO and NO_3 for 2002 came from 2002 in situ LTOP data. For 2006, NO_3 came from the LTOP multi-year climatology. Initial spring 2006 DO came from a shelf-wide survey (J. Peterson, Oregon State University, unpublished data). Initial conditions for phytoplankton, zooplankton and detritus were provided from NCOM-2006 fields (for 2006) or NCOM-2008 (for 2002). Figure 9.6 summarizes the vertical

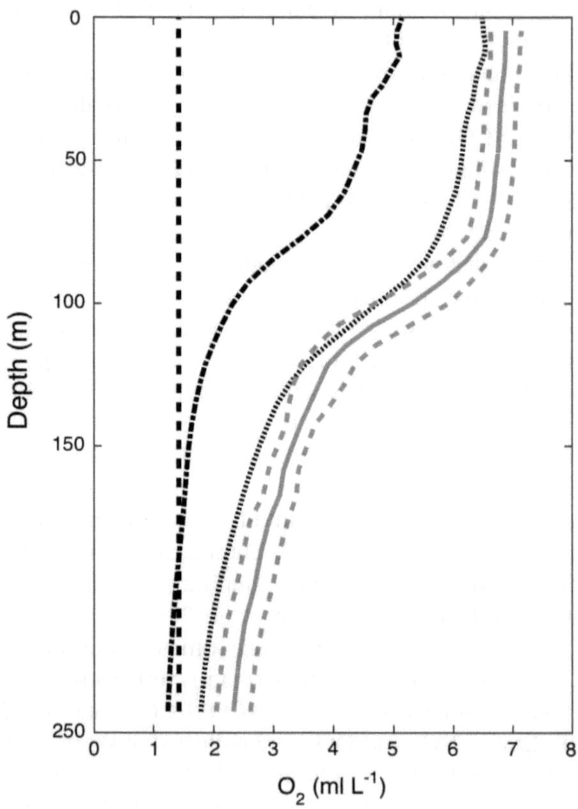

Fig. 9.6 Spring (March–April) dissolved oxygen concentration profiles at a location above the slope (25 nautical miles off the coast) along the Newport Hydrographic transect (see Fig. 9.1). The *dot-dashed black line* (LTOP2002) is the observed DO profile from the GLOBEC-LTOP cruise in 2002. The *dotted black line* (J.P. 2006) is the DO profile from the same location in 2006 (data provided by Jay Peterson, Oregon State University). The *solid gray line* is the mean climatological value from spring sampling at that station during 1998–2004 (including 2002), which is referred to as LTOP-clim in Table 9.2 and elsewhere in this chapter. The *dashed gray lines* about the climatological profile are 1 standard deviation above and below the mean. The hypoxia threshold of 1.43 ml L^{-1} is shown as the *black dashed vertical line*

profiles of dissolved oxygen from a station above the slope off Newport, OR in March–April used in computing offshore initial condition fields for various model simulations. Offshore waters in 2002 had significantly lower DO concentrations at all depths than the LTOP climatology and the 2006 observations. Not shown in this figure is the NCOM-2008 DO profile estimated from the regression analysis against NO_3, which had significantly higher concentrations than all the profiles shown, especially at depth.

9.3.3 Model–Data Comparisons

Detailed comparisons of the physical model simulations with observations are described for 2002 in Koch et al. (2010). There was good model–data agreement on the structure and seasonal development of surface and depth-averaged velocities over the shelf and in the offshore transition zone, the structure and development of the upwelling SST front, the separation and offshore intensification of the upwelling jet, and the 3-dimensional density field.

For the ecosystem components, we compare the modeled vertical profiles of NO_3 and DO to GLOBEC-LTOP July 2002 vertical profiles (e.g., 3.5 months after the start of the simulation). Figure 9.7 shows vertical profiles of NO_3 and DO along

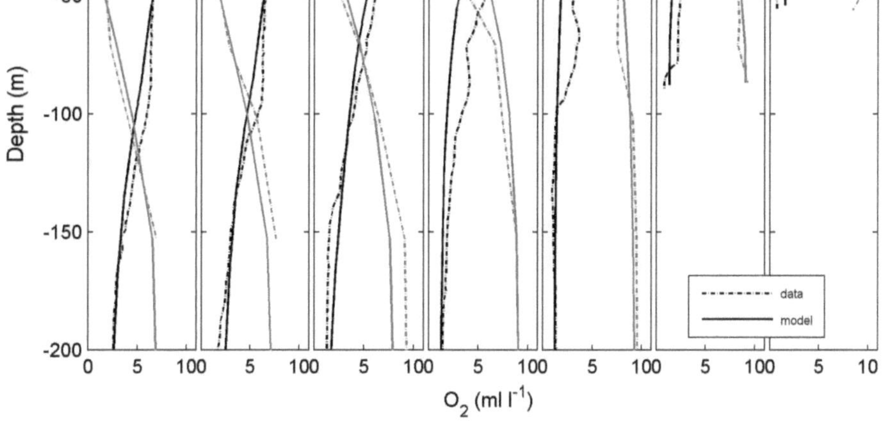

Fig. 9.7 Vertical profiles of NO_3 (mmol N m^{-3}; *gray*) and dissolved oxygen (ml L^{-1}; *black*) from the model (*solid lines*) and observations (*dotted-dashed lines*) along the Newport Hydrographic transect (Fig. 9.1, 44.65° N) during 10–12 July 2002. Leftmost panel is 85 miles offshore; *rightmost two panels* are 15 miles and 5 miles from shore

the NH line for 10–12 July 2002. There is a very good agreement between modeled and observed NO_3 and DO at offshore deep locations and on the shelf. There is a region over the continental slope (NH25, NH35), where the NO_3 and DO profiles from the model underestimate DO and overestimate NO_3 at intermediate depths, although at both shallower and deeper depths the model–data agreement is quite good.

9.4 Description of Oregon Shelf Hypoxia in 2002 and 2006

The patterns of seasonal hypoxia development in summer from the model simulations of 2002 and 2006 are illustrated by showing monthly averaged, cross-shelf vertical sections of dissolved oxygen for the Newport line (Fig. 9.8), monthly maps of the minimum dissolved oxygen concentration (Fig. 9.9), and time–latitude plots

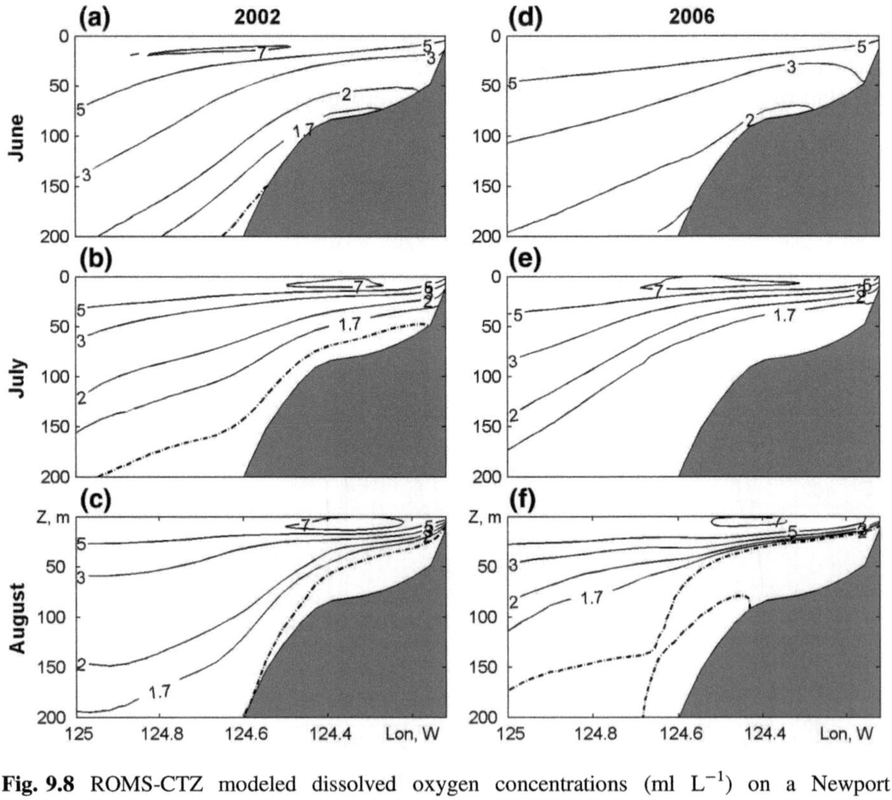

Fig. 9.8 ROMS-CTZ modeled dissolved oxygen concentrations (ml L^{-1}) on a Newport Hydrographic cross-shelf transect averaged from twice-daily values for all days in June (**a**, **d**), July (**b**, **e**) and August (**c**, **f**) of 2002 (**a–c**) and 2006 (**d–f**). The *dash-dot contour* shows the hypoxic threshold at 1.43 ml L^{-1}

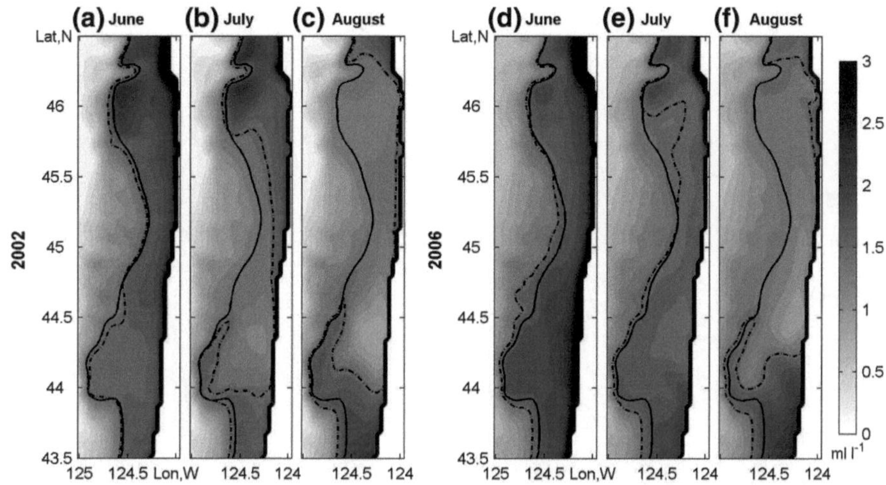

Fig. 9.9 Minimum DO concentration (ml L^{-1}) in the water column over the continental shelf for June (**a**, **d**), July (**b**, **e**), and August (**c**, **f**) of 2002 (**a**–**c**) and 2006 (**d**–**f**). *Heavy solid line* is the 200 m isobath. *Dashed line* is the location of the hypoxia threshold (1.43 ml L^{-1}) near the *bottom*. Minimum DO values were taken over vertical axis for every horizontal grid cell from daily averages of twice-daily outputs. Then, monthly averages for each horizontal grid cell were calculated

that show the seasonal cross-shelf extent of hypoxia (Fig. 9.10). Both monthly averages and minimums were calculated from daily averages of twice-daily outputs. All these results are based on the base case simulations: for 2002 simulation BC2 (Base Case-2002) and for 2006 simulation BC6 (see Table 9.2 for details).

In June 2002, the monthly averaged DO showed bottom hypoxia, but only at bottom depths exceeding 150 m (Fig. 9.8). Modeled hypoxia was severe from offshore to inshore in July and August, but temporally and spatially intermittent (Fig. 9.10), with inner- and mid-shelf bottom regions in both months having <1 ml L^{-1} DO, and the monthly averaged bottom DO being hypoxic across the entire Newport shelf (Fig. 9.8b, c). Grantham et al. (2004) reported that the few observations of inner-shelf bottom DO from mid-July to mid-September of 2002 all indicated severe bottom hypoxia. Newport transects at 44.65° N from the model reveal that inner- and mid-shelf bottom waters became hypoxic in July (Fig. 9.9b), with most of the shelf width experiencing bottom hypoxia. On average, about 20–30% of the shelf off Newport was hypoxic most days in July, gradually increasing from the first shelf hypoxia observed in mid-June (Fig. 9.10a). The model suggests that hypoxia appeared earliest at Newport and further south (near Heceta Bank) and was most severe and extended furthest off the bottom (35–40 m) in August where the bottom depths were 100 m or less (Fig. 9.8). North of 45° N, the model shows that hypoxia was not present on the inner-shelf in July (Fig. 9.9b) and was not widespread across the shelf until August (Figs. 9.9c and 9.10a).

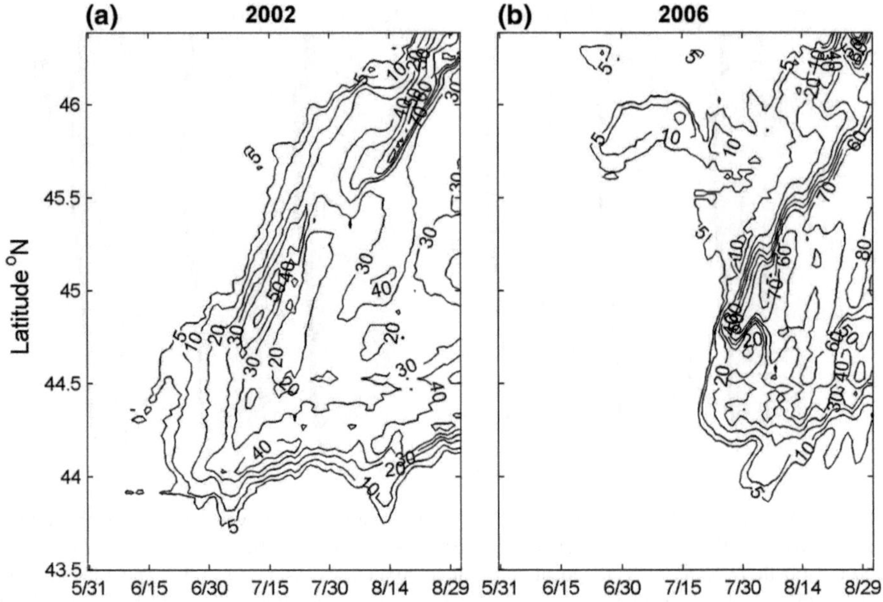

Fig. 9.10 Percentage of the latitudinal cross-shelf area (to 200 m isobath) experiencing hypoxia (DO < 1.43 ml L^{-1}) as a function of time for the base case model simulations in the sensitivity analysis (see Table 9.2): BC2 (**a** 2002) and BC6 (**b** 2006). Daily averages of twice-daily values were used

Table 9.2 Initial and open boundary conditions for dissolved oxygen (DO) and nitrate (NO_3) concentrations for four model simulations in 2002 and one in 2006 that comprised the sensitivity analysis. Two letter case prefixes BC, UI, CI and UB indicate the base case, unmodified initial conditions, climatological initial conditions and unmodified boundary conditions, respectively. One digit Case suffix 2 and 6 indicate the summer 2002 and 2006 simulations, respectively. For all four 2002 simulations, initial and boundary physical conditions are from NCOM-2002; initial and boundary biology conditions are from NCOM-2008, because NCOM-2002 did not include ecosystem variables, and wind forcing in 2008 was most similar to wind forcing in 2002; atmospheric wind forcing is from COAMPS-2002, and heat fluxes are from NCEP-2002. The single 2006 simulation (BC6) used NCOM-2006 for providing both biological and physical initial and boundary conditions, and NAM-2006 provided the wind and heat flux estimates. JP-2006 as the source for DO means in situ DO observations obtained by Jay Peterson (Oregon State University) in early spring in 2006 were used to initialize DO. See text for details on other abbreviations/methods for specifying initial and boundary conditions

Year	Case	Initial conditions		Boundary conditions	
		Dissolved oxygen (DO)	Nitrate (NO_3)	Dissolved oxygen (DO)	Nitrate (NO_3)
2002	BC2	LTOP-2002	LTOP-2002	N:O on mod NCOM-2008	NCOM:LTOP
	UI2	N:O on NCOM-2008	NCOM-2008	N:O on mod NCOM-2008	NCOM:LTOP
	CI2	LTOP-clim	LTOP-clim	N:O on mod NCOM-2008	NCOM:LTOP
	UB2	LTOP-2002	LTOP-2002	N:O on NCOM-2008	NCOM-2008
2006	BC6	JP-2006	LTOP-clim	N:O on mod NCOM-2006	NCOM:LTOP

Modeled hypoxia in 2006 exhibited a different temporal–spatial development from that shown for 2002. Although there was a short period of hypoxia locally near 46° N beginning in June (Fig. 9.10b), widespread along-shelf hypoxia did not occur until late July, and severe and spatially extensive hypoxia occurred only in August (Fig. 9.9f). Model results showed that about 60–80% of the shelf had bottom hypoxia in August, and almost 50% of the shelf volume of water had DO concentrations below the hypoxia threshold in late August, especially between Newport and 46° N.

9.5 Sensitivity Analysis Experiment in 2002

To examine the relative importance of variable NO_3 and DO initial and boundary conditions for summer hypoxia development on the Oregon shelf in 2002, we conducted a set of controlled sensitivity simulations. Table 9.2 provides details of ecosystem and physical initial and open boundary conditions, including atmospheric forcing, used in the numerical experiment. The base case simulation, BC2 (2002; BC for base case), described in the previous section, used the most realistic initial and open boundary conditions available. NO_3 and DO initial conditions for BC2 are from 2002 data collected by the US GLOBEC Northeast Pacific Long-Term Observation Program (LTOP), that sampled off the coast of Oregon from 1997–2004 (Batchelder et al. 2002). Three other model settings, which varied in either their initial conditions or the boundary conditions, were simulated to assess how much the results differ from the BC2 simulation. In simulation UI2 (2002; UI for unmodified initial conditions), the boundary conditions for NO_3 and DO are as in BC2, but initial conditions for NO_3 and DO are estimated from the unmodified NO_3 NCOM fields. For simulation CI2 (2002; CI for climatological initial conditions), a multi-year (1997–2004) climatology of DO and NO_3 derived from GLOBEC-LTOP sampling in April provided the initial conditions for DO and NO_3, and the boundary conditions were identical to those as in BC2. Simulation UB2 (2002; UB for unmodified boundary conditions) uses the same initial conditions as BC2 and boundary DO and NO_3 conditions from unmodified NCOM fields.

9.5.1 *Analysis of the Basic Simulation in the Sensitivity Experiment*

Summer hypoxia development on the shelf and its characteristics were compared among the simulations in the sensitivity experiment by computing the number of days each month (June, July and August) for each location that the bottom DO concentration was below the hypoxic threshold (Fig. 9.11).

As was noted from the analysis of BC2 in Fig. 9.10a, hypoxic waters in 2002 first appear on Heceta Bank in June. Heceta Bank is a relatively shallow shelf area surrounded by deeper bathymetry; hypoxia also occurs near 44–44.5° N, where the shelf is broad (Fig. 9.9). The along-shore coastal jet, which tends to follow the shelf edge, veers offshore near Heceta Bank (Koch et al. 2010). Circulation on the bank is slow, and recirculation and meanders are common and provide long-residence times for phytoplankton growth. Concentrations of phytoplankton (measured by chlorophyll) were high on the bank in 2002 (Grantham et al. 2004). Some of this phytoplankton is consumed by zooplankton, and some sinks to the bottom as phytodetritus. Near the bottom, DO is reduced by detritus decomposition and ammonium oxidation.

Due to the long-residence time of water on the bank, bottom waters in the BC2 simulation are more likely to become hypoxic. On the bank, bottom DO concentrations are hypoxic for 12–20 days in June (Fig. 9.11a, left panel). In July, hypoxic

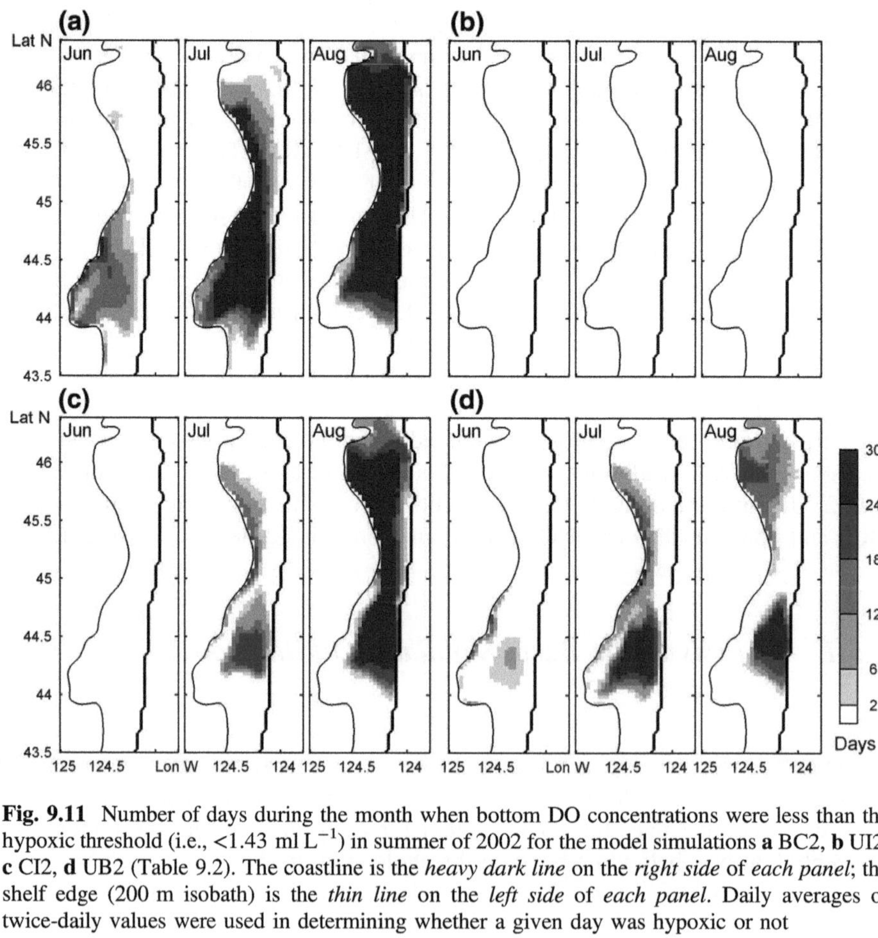

Fig. 9.11 Number of days during the month when bottom DO concentrations were less than the hypoxic threshold (i.e., <1.43 ml L^{-1}) in summer of 2002 for the model simulations **a** BC2, **b** UI2, **c** CI2, **d** UB2 (Table 9.2). The coastline is the *heavy dark line* on the *right side* of *each panel*; the shelf edge (200 m isobath) is the *thin line* on the *left side* of *each panel.* Daily averages of twice-daily values were used in determining whether a given day was hypoxic or not

conditions in the BC2 model simulation expanded northward along the shelf break and hypoxic waters occupied all of Heceta Bank, from the shelf break to the coast, with much of the bank experiencing bottom hypoxia through the entire month (Fig. 9.11a, middle panel). Inner-shelf waters are affected by hypoxia north to almost 45° N. This northward expansion of hypoxia from June to July is not due to advection, which is from north to south, but rather to delayed accumulation of phytodetritus due to weaker bloom and stronger along-shore currents where the shelf is narrower. In August, bottom hypoxia occurs nearly every day along the entire shelf from 44–46.5° N (Fig. 9.11a, right panel). The exception is the offshore and southern parts of Heceta Bank that are no longer hypoxic as a result of northward advection of higher DO concentration waters (analysis not shown).

9.5.2 Sensitivity Simulations with Modified Initial Conditions

When initial DO and NO_3 conditions are altered from the most realistic values to conditions derived from unmodified NCOM fields, the results (simulation UI2, Table 9.2) do not show development of hypoxia on the Oregon shelf (Fig. 9.11b). The reasons why hypoxia does not develop in this simulation are basically twofold. First, advection of high (overestimated) DO offshore waters onto the shelf might buffer the decline in DO due to biological processes sufficiently such that the hypoxia threshold is not exceeded. Second, the deeper than normal nitrocline from the NCOM fields may inhibit upwelling of NO_3 onto the shelf, which reduces the total phytoplankton production and the biological oxygen demand when the phytodetritus remineralizes near the bottom.

The CI2 simulation that uses NO_3 and DO from the LTOP multi-year climatology for the initial April conditions (CI2; Table 9.2) produces shelf hypoxia (Fig. 9.11c), though there are substantial spatial and temporal differences from the BC2 simulation (Fig. 9.11a). Bottom hypoxia develops along the shelf break north of 45° N and, again, in the Heceta Head transect (see Fig. 9.4) only in July, thus being delayed about one month (Fig. 9.11c, middle panel). In August, distribution of bottom hypoxic waters over the shelf is similar to BC2, although there are subtle differences in the southward extent of hypoxia along the shelf break (Fig. 9.11c, right panel).

9.5.3 Sensitivity Simulation with Modified Boundary Conditions

The UB2 simulation shows that DO and NO_3 conditions at the open boundaries influence summer hypoxia on the shelf. The shelf hypoxia for the simulation using unmodified NCOM fields at the open boundaries (UB2, Table 9.2) is shown in Fig. 9.11d. Bottom hypoxia appears first along the shelf break and on Heceta Bank in June (Fig. 9.11d, left panel), but later than in simulation BC2. The development and propagation of bottom hypoxia in July and August resemble those of BC2, but the spatial extent and duration of bottom hypoxia are considerably reduced (Fig. 9.11d, middle and right panels). Even with unrealistically high DO and low NO_3 concentrations at the open boundaries, UB2 simulated the beginning of hypoxia better than did CI2 that had more realistic boundary conditions, but used climatological initial conditions. This result suggests that having accurate initial conditions for DO and NO_3 within the model domain in early spring of 2002 was important in replicating the progression of observed summer 2002 hypoxia.

In our model domain, boundary effects might result from conditions on any of the open northern, southern or western boundaries. A simulation (not shown) where DO and NO_3 concentrations at the open western boundary were set to zero showed no difference from simulation BC in the development of bottom hypoxia on the shelf. The western boundary of the model is too distant (400 km from the coast), and the cross-shelf flow velocities are too slow to influence DO and NO_3 concentrations on the shelf within the five month simulation period. Thus, the effects of DO and NO_3 entering the shelf subdomain through the perimeter ultimately came from either the northern or southern open boundaries. This is not surprising given the much greater magnitude of along-shore flows (especially from the north) than cross-shelf flows in the California Current System.

9.6 Role of Physical and Biological Drivers

To determine the relative importance of physical and biological factors to the development of summer hypoxia on the Oregon shelf in 2002 and 2006, we integrate the DO budgets for each model process for the entire water column for the entire simulation (April 1–September 1) and for the summer (June 1–September 1) only over the shelf subdomain (Fig. 9.12). The results of base case simulations BC2 and BC6 are used. The contribution of each process to total oxygen concentration is summarized in Table 9.3. The net change of DO in modeled shelf waters for April–August 2006 is estimated as a loss of $4.7 * 10^{15}$ ml O_2 (hereafter in this paragraph we provide volumes in km^3 ($1\ km^3 = 10^{15}$ ml)), which is equivalent to a decrease of 3.0 ml L^{-1} through the entire shelf. This is 2.5 times higher than the decrease in 2002 (1.8 km^3 O_2; Table 9.3).

Although oxygen reduction due to biological sink terms (organic remineralization) is important in deeper layers in both years, the net biological effect on dissolved oxygen overall, considering the full water column, is positive and large due to the large O_2 production by phytoplankton photosynthesis in the photic zone (ca. 8.7 km^3 O_2 in both 2002 and 2006, Table 9.3). The net effect of horizontal advection is to decrease DO on the shelf through off-shelf advection during upwelling of high DO surface waters (due to photosynthesis; e.g., biological source

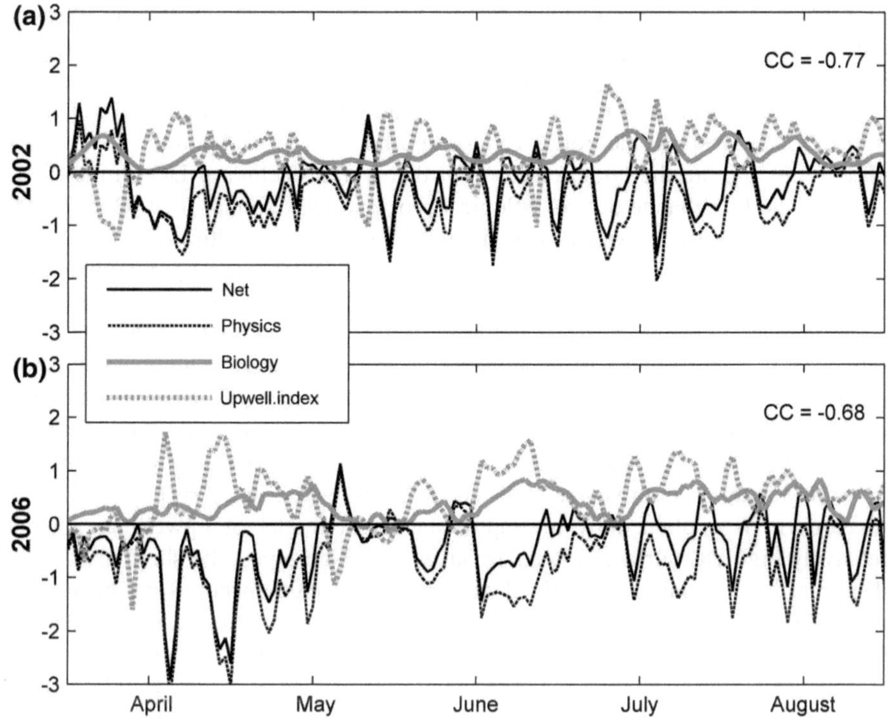

Fig. 9.12 Simulated physical (*dash black*), biological (*solid gray*) and net (*solid black*) fluxes of dissolved oxygen integrated over the shelf ($\times 10^9$ ml s^{-1}) along with the zonal integrated upwelling index (NOAA, S. Pierce; $\times 3000$ m^3 s^{-1}) for **a** 2002 and **b** 2006. Upwelling index/physical flux correlations (CC) are shown

Table 9.3 Dissolved oxygen (DO) shelf budgets (ml $O_2 \cdot 10^{15}$ (=1 km^3)) in 2002 and 2006 due to different processes: advection, diffusion, air-sea flux, biological production and loss, net physical, net biological and total net change during the entire integration period April–August and during the summer (June–August) only. Net terms are highlighted in bold

Factor/Time interval	April–August 2002	April–August 2006
Advection	−3.1	−5.7
Diffusion	−2.0	−2.9
Air-sea flux	−1.1	−1.1
Physical net	**−6.2**	**−9.7**
Biological source	8.7	8.7
Biological sink	−4.3	−3.7
Biological net	**4.4**	**5.0**
Total net	**−1.8**	**−4.7**
Factor/time interval	June–August 2002	June–August 2006
Physical net	**−4.2**	**−5.2**
Biological net	**3.0**	**3.5**
Total net	**−1.2**	**−1.7**

in Table 9.3) and deep intrusions of low DO waters onto the shelf from offshore. About 12% of the biologically produced oxygen near the surface is lost by outgassing of oxygen from supersaturated surface waters to the atmosphere. The physical mechanisms of DO reduction, especially through horizontal advection and diffusion, are responsible for losses of 6.2 km^3 O_2 in 2002 and 9.7 in 2006 (Table 9.3) and are the critical factors for the overall decline of DO on the shelf.

Since the DO change due to biological processes was similar in 2002 and 2006, the large difference between years in net DO change is ultimately due to the difference in physical forcing. In 2002, the loss of DO occurred predominantly in June–August, whereas in 2006 two-thirds of the net DO loss occurred during very strong upwelling events in April and May (Fig. 9.1). Time series of the DO fluxes (physical, biological and net) integrated over the entire shelf volume clearly show greater net biological production of DO in June–August (67–70% of total) than earlier, and the close matching of physical and net DO fluxes in both years (Fig. 9.12). There is a strong negative correlation between DO physical flux and upwelling index in both years: −0.77 in 2002 and −0.68 in 2006 (Fig. 9.12). This relationship is explained by a simple conceptual model of upwelling, wherein relatively high DO surface waters are advected offshore past the shelfbreak, while deeper waters having low dissolved oxygen are advected onshore from beyond the shelf. This circulation pattern would decrease DO concentration in shelf waters. In contrast, deeper waters advected onto the shelf would have high nutrient concentrations and with injection into the inner-shelf euphotic zone would facilitate enhanced phytoplankton growth and DO production. Considering the full depth-integrated water column on the shelf, the physical processes dominate over the biological processes in controlling the total oxygen content. These budget results are consistent with model results of Siedlecki et al. (2015) for whole water column integrations of the Oregon and Washington shelves.

Whole water integrations, as done by us and Siedlecki et al. (2015), could be supplemented by examining the water column partitioned into the euphotic DO production zone and a deeper zone dominated by remineralization processes, where each has biological and physical sources and sinks of dissolved oxygen. We did not do vertically partitioned DO budgets for this analysis.

The documented differences in hypoxia timing, spatial distribution and severity in 2002 and 2006 could be explained as follows. On the one hand, with the comparable rates of biological net oxygen production in both years, physical oxygen removal (mainly due to upwelling) rate in 2006 is about 1.5 higher resulting in net oxygen loss rate being 2.5 higher than in 2002 (Table 9.3 and Fig. 9.12). On the other hand, initial oxygen concentrations in early April of 2002 are essentially lower than in 2006 (Fig. 9.6), which provides a head start for shelf waters DO concentration to drop below hypoxic level as early as in mid-June. In 2006, despite strong (especially in April–May) upwelling, the initial DO concentration in shelf waters is relatively high, and it takes as much as 3.5–4 months to develop strong shelf hypoxia. From that point on (2nd half of July–August), hypoxia development in 2006 seems to be more rapid and ubiquitous than in 2002, which is well documented in Fig. 9.10, and is also supported by observations in September (outside of our model integration interval) on Fig. 9.2.

9.7 Discussion and Conclusions

Our results that indicate the Heceta Bank region of the Oregon shelf is a hotspot for hypoxia are consistent with the analysis of observations reported by Connolly et al. (2010) and with model simulations conducted by Siedlecki et al. (2015). There is substantial year-to-year variability in the spatial and temporal extent and intensity of hypoxia (Siedlecki et al. 2015) in the Pacific Northwest. Their model showed three regions of the Oregon–Washington coast that had persistent hypoxia in late-summer simulations of three years, which included Heceta Bank as well as two regions in Washington, one of which was the Juan de Fuca Eddy (JdFe) region offshore of the Strait of Juan de Fuca. Both Heceta Bank and JdFe are known to be regions where water residence times are relatively long because of local recirculation flows (Barth et al. 2005; Hickey and Banas 2003). Both regions had high vertically integrated water column organic matter respiration (Seidlecki et al. 2015).

The coupled biophysical model used in this study demonstrated skill in reproducing the physical and biological processes that govern Oregon shelf summer hypoxia. With realistic and controlled sensitivity simulations we were able to understand the relative contributions of ocean physics and ecological processes in creating regional hypoxia on the Oregon shelf. The sensitivity analysis demonstrated the importance of having accurate ecosystem boundary and early spring initial conditions, especially the latter, for accurate hindcasting of summer–autumn oxygen on the Oregon shelf. Unrealistically high initial DO and low NO_3 conditions in spring 2002 (from NCOM derived fields) prevented or significantly delayed bottom hypoxia development. We also showed that when lacking year-specific data for specifying the initial conditions of NO_3 and DO, a simulation initialized with a 7-year climatological average conditions was able to reveal the spatial pattern of hypoxia, although not the full temporal development (the climate scenario had weaker and delayed hypoxia). DO changes due to biological processes (photosynthesis, respiration, remineralization) are large, although physical processes, mostly horizontal advection of low DO in the bottom boundary layer associated with upwelling, are most responsible for the net reduction in DO in spring–summer and the onset of bottom hypoxia in summer on the Oregon shelf.

For realistic modeling of shelf hypoxia, it is critical to have accurate estimates of the initial and boundary conditions. These are often provided by larger-scale numerical models or, in cases where they are available, by recent ocean observations of the relevant parameters. In our modeling, initial and boundary values of NO_3 and DO concentrations are critical for evaluating hypoxia development. Many large-scale coupled physical–biological models do not provide sufficiently realistic DO and NO_3 information essential to force higher spatial resolution (local-regional) models for hindcasting (or forecasting) the intensity, duration and spatial patterns of hypoxia on the Oregon shelf. No available regional model simulations in hindcast, nowcast or forecast mode are presently able to provide the external boundary conditions and forcing required for higher resolution coastal models, like the model described here, that focus on dissolved oxygen dynamics. An alternative to downscaling from larger-scale models might be more intensive in situ sampling, which provides directly measured DO and NO_3. Often, the limited vertical resolution of the coarser resolution larger-domain models is insufficient to provide fine enough vertical information to the local model to resolve the depth of the oxygen minimum zone (OMZ) offshore, or the potential shallowing of it through recent decades. In regions where OMZ offshore has shown to be shallowing, it is important to include these changes by prescription or by modeling to enable realistic hindcasting and forecasting of hypoxia on continental shelves, which has direct effects on marine biota and livelihoods.

Appendix A. Oxygen Formulation

The dissolved oxygen dynamics in our ecosystem model is governed by the following equation:

$$\frac{\partial O_2}{\partial t} = V_m f(I)\left\{\frac{NO_3}{K_u + NO_3} e^{-\psi NH_4} r_{O_2:NO_3} + \frac{NH_4}{K_u + NH_4} r_{O_2:NH_4}\right\} P$$

$$- 2\Omega NH_4 - \Gamma r_{O_2:NH_4} Z - \varphi r_{O_2:NH_4} D + Q_{ge}(O_{2sat} - O_2)$$

where P is phytoplankton, Z is zooplankton, D is detritus, $V_m = 1.5\ \mathrm{d}^{-1}$ is phytoplankton maximum uptake rate, *f(I)* is light limitation (see Spitz et al. 2005 for details), $\psi = 1.46\ (\mathrm{mmol\ N\ m^{-3}})^{-1}$ is NH_4 inhibition parameter, $\Omega = 0.25\ \mathrm{d}^{-1}$ is NH_4 oxidation coefficient, $\Gamma = 0.1\ \mathrm{d}^{-1}$ is zooplankton specific excretion and mortality rate, $\varphi = 0.1\ \mathrm{d}^{-1}$ is detritus decomposition rate, $r_{O_2:NO_3}$ and $r_{O_2:NH_4}$ are oxygen-nitrate and oxygen-ammonium conversion parameters, respectively; $Q_{ge} = \frac{K\nu_{O_2}}{\Delta zn}$, where Δzn is the height of the top cell and $K\nu_{O_2} = 0.31u^2\sqrt{660/S_c}$ is the gas transfer coefficient with u being the average wind speed and S_c being Schmidt number calculated after Keeling et al. (1998): $S_c = 1638.0 - 81.83T + 1.483T^2 - 0.008004T^3$; $O_{2sat} = e^A 1000/22.9316$ is the saturation concentration of oxygen with $A = 2.00907 + 3.22014TS + 4.05010TS^2 + 4.94457TS^3 - 0.256847TS^4 + 3.88767TS^5 + S(-0.00624523 - 0.00737614TS - 0.0103410TS^2 - 0.00817083TS^3) - 4.88682 \cdot 10^{-7}S^2$, where $TS = \ln((298.15 - T)/T_k)$, T is in Celsius, T_k in Kelvin (after Garcia and Gordon 1992).

References

Bakun A (1990) Global climate change and intensification of coastal ocean upwelling. Science 247:198–201

Barth JA, Pierce SD, Castelao RM (2005) Time-dependent, wind-driven flow over a shallow midshelf submarine bank. J Geophys Res 110:C10S05. doi:10.1029/2004JC002761

Batchelder HP, Barth JA, Kosro PM, Strub PT, Brodeur RD, Peterson WT, Tynan CT, Ohman MD, Botsford LW, Powell TM, Schwing FB, Ainley DG, Mackas DL, Hickey BM, Ramp SR (2002) The GLOBEC Northeast Pacific California Current system program. Oceanography 15(2):36–47

Castelao RM, Barth JA (2005) Coastal ocean response to summer upwelling favorable winds in a region of alongshore bottom topography variations off Oregon. J Geophys Res 110:C10S04. doi:10.1029/2004JC002409

Chan R, Barth JA, Lubchenco J, Kirincich A, Weeks H, Peterson WT, Menge. BA (2008) Emergence of anoxia in the California Current large marine ecosystem. Science 319:920. doi:10.1126/science.1149016

Connolly TP, Hickey BM, Geier SL, Cochlan WP (2010) Processes influencing seasonal hypoxia in the northern California Current system. J Geophys Res 115:C03021. doi:10.1029/2009JC005283

Crawford WR, Pena MA (2013) Declining oxygen on the British Columbia continental shelf. Atmos Ocean 51:88–103

Diaz RJ, Rosenberg R (2008) Spreading dead zones and consequences for marine ecosystems. Science 321:926. doi:10.1126/science.1156401

Escribano R, Schneider W (2007) The structure and functioning of the coastal upwelling system off south/central Chile. Prog Oceanogr 75:343–346

Garcia HE, Gordon LI (1992) Oxygen solubility in seawater: better fitting equations. Limnol Oceanogr 37(6):1307–1312

Grantham BA, Chan F, Nielsen KJ, Fox DS, Barth JA, Huyer A, Lubchenco J, Menge BA (2004) Upwelling-driven nearshore hypoxia signals ecosystem and oceanographic changes in the Northeast Pacific. Nature 429:749–754

Hickey BM, Banas NS (2003) Oceanography of the US Pacific Northwest coastal ocean and estuaries with application to coastal ecology. Estuaries 26:1010–1031. doi:10.1007/BF02803360

Hodur RM (1997) The naval research laboratory's coupled ocean/atmosphere mesoscale prediction system (COAMPS). Mon Weather Rev 125:1414–1430

Huyer A, Sobey EJC, Smith RL (1979) The spring transition in currents over the Oregon continental shelf. J Geophys Res 84:6995–7011

Kalnay E et al (1996) The NCEP/NCAR 40-year reanalysis project. Bull Am Meteorol Soc 77:437–471

Keeling RF, Manning AC, McEvoy EM, Shertz SR (1998) Methods for measuring changes in atmospheric O_2 concentration and their application in Southern Hemisphere air. J Geophys Res 103:3381–3397

Koch AO, Kurapov AL, Allen JS (2010) Near-surface dynamics of a separated jet in the coastal transition zone off Oregon. J Geophys Res 115:C08020. doi:10.1029/2009JC005704

Levitus S (1982) Climatological atlas of the world ocean. NOAA Professional Paper 13, US Government printing office, Washington DC, 173 pp

Peterson JO, Morgan CA, Peterson WT, Di Lorenzo E (2013) Seasonal and interannual variation in the extent of hypoxia in the northern California Current from 1998-2012. Limnol Oceanogr 58:2279–2292

Shchepetkin AF, McWilliams JC (2003) A method for computing horizontal pressure-gradient force in an oceanic model with a nonaligned vertical coordinate. J Geophys Res 108(C3):3090. doi:10.1029/2001JC001047

Shchepetkin AF, McWilliams JC (2005) The regional ocean modeling system: a split–explicit, free–surface, topography–following coordinate oceanic model. Ocean Model 9:347–404. doi:10.1016/j.ocemod.2004.08.002

Shulman I, Kindle JC, deRada S, Anderson SC, Penta B, Martin PJ (2004) Development of a hierarchy of nested models to study the California Current system. Estuarine and coastal modeling. In: Malcolm L, Spaulding PE (eds) Proceedings of the 8th international conference on estuarine and coastal modeling. American Society of Civil Engineers, Reston, Va, pp 74–88

Siedlecki SA, Banas NS, Davis KA, Giddings S, Hickey BM, MacCready P, Connolly T, Geier S (2015) Seasonal and interannual oxygen variability on the Washington and Oregon continental shelves. J Geophys Res Oceans 120. doi:10.1002/2014JC010254

Spitz YH, Allen JS, Gan J (2005) Modeling of ecosystem processes on the Oregon shelf during the 2001 summer upwelling. J Geophys Res 110:C10S17. doi:10.1029/2005JC002870

Strub PT, Batchelder HP, Weingartner TJ (2002) U.S. GLOBEC Northeast Pacific program: overview. Oceanography 15:30–35

Ueno H, Yasuda O (2003) Intermediate water circulation in the North Pacific subarctic and northern subtropical regions. J Geophys Res 108(C11):3348. doi:10.1029/2002JC001372

Wetz JJ, Hill J, Corwith H, Wheeler PA (2004) Nutrient and extracted chlorophyll data from the GLOBEC long-term observation program, 1997–2004. Data report 193, COAS reference 2004-1 (Revised June 2005 [to include improved 2003 and 2004 data]). http://nepglobec.bco-dmo.org/reports/ccs_cruises/GLOBEC_nutchl_datareport_7_06_hpb.pdf

Wheeler PA, Huyer A, Fleischbein J (2003) Cold halocline, increased nutrients and higher chlorophyll off Oregon in 2002. Geophys Res Lett 30. ISSN: 0094-8276. doi:10.1029/2003GL017395

Chapter 10
Comparing Default Movement Algorithms for Individual Fish Avoidance of Hypoxia in the Gulf of Mexico

Elizabeth LaBone, Dubravko Justic, Kenneth A. Rose, Lixia Wang and Haosheng Huang

Abstract The northern Gulf of Mexico is the site of one of the largest areas of seasonal, coastal hypoxia (up to 22,000 km^2). Hypoxia can have both direct and indirect effects on fish. Atlantic croaker (*Micropogonias undulatus*) is a good model organism for studying the effects of hypoxia on fish in the Gulf of Mexico because it is a demersal species that lives in the area where hypoxia occurs and has been studied extensively. Virtual croaker movement was examined for three algorithm groups on a two-dimensional grid encompassing the Gulf hypoxia region. The model was run for seven days using four static dissolved oxygen maps reflecting progressively increasing hypoxia severity. Individual fish movement was modeled using a particle-tracking module with outputs from a three-dimensional hydrodynamic-water quality model for the 2002 hypoxia season. The three algorithm groups included the neighborhood search for hypoxia avoidance and the random walk, Cauchy correlated random walk, or kinesis for the default behavior. The results show that the default algorithms have little effect on hypoxia exposure of individual fish, but affect sinuosity (wiggle in fish path). The variables to consider when choosing between the three default algorithms are time step, dispersal, and the effects of stressors other than hypoxia. This study emphasizes the need to acquire suitable data for calibration of fish movement models that are presently not available for the northern Gulf of Mexico.

E. LaBone (✉) · D. Justic · K.A. Rose · L. Wang · H. Huang
Department of Oceanography & Coastal Sciences, College of the Coast & Environment, Louisiana State University, Baton Rouge, LA 70803, USA
e-mail: ed_labone@yahoo.com

D. Justic
e-mail: djusti1@lsu.edu

K.A. Rose
e-mail: karose@lsu.edu

L. Wang
e-mail: lxwang@lsu.edu

H. Huang
e-mail: hhuang7@lsu.edu

D. Justic et al. (eds.), *Modeling Coastal Hypoxia*,
DOI 10.1007/978-3-319-54571-4_10

Keywords Atlantic croaker · Fish movement · Avoidance behavior · Hypoxia · Numerical modeling · Gulf of Mexico

10.1 Introduction

The Louisiana–Texas (La-Tex) shelf is the location of one of the largest areas of seasonal, coastal hypoxia (up to 22,000 km^2) (Rabalais et al. 2007). For the Gulf of Mexico (GOM), hypoxia is defined as dissolved oxygen (DO) concentrations less than 2 mg/L (Rabalais et al. 2001). Hypoxia generally occurs from April through October at depths of 5–60 m and can stretch from the Birdfoot delta to the La-Tex border (Turner and Rabalais 1991). Stratification of the water column and surface primary productivity contribute to the formation of hypoxia (Justic et al. 2007). Stratification occurs during the spring and summer on the La-Tex shelf due to the spring floods of the Mississippi and Atchafalaya rivers, regional circulation, and seasonal weather patterns (Rabalais et al. 2001, 2002). Stratification can be broken down by water column mixing caused by cold fronts during late fall to early spring or by tropical cyclone activity during summer and fall. Organic matter that sinks below the pycnocline and decomposes leads to reduced oxygen levels that cannot be replenished quickly due to high stability of the water column (Justic et al. 1996; Rabalais et al. 2002). Primary production is strongly influenced by riverine nutrient loading (Justic et al. 1993).

The areal extent and severity of hypoxia have increased over the past century along with the increase in riverine nutrient concentrations (Turner and Rabalais 1991). The analyses of hypoxia proxies in sediment cores indicate that hypoxia began to appear in the 1900s and started to intensify during the 1940–1950s (Rabalais et al. 2002). Hindcasts from several hypoxia models suggest that widespread hypoxia first developed in the 1970s (Justic et al. 2002; Scavia et al. 2004). Since hypoxia monitoring program carried out by Louisiana Universities Marine Consortium (LUMCON) started in 1985, the extent of hypoxia has remained highly variable. However, a significant increase in hypoxic layer thickness was observed (Obenour et al. 2013).

Hypoxia can have both direct and indirect effects on fish. Hypoxia exposure can lead to increased mortality, decreased fecundity, decreased growth, and changes in movement (Rabalais et al. 2001). Direct mortality can occur if fish are unable to escape from hypoxic water before being asphyxiated. Indirect effects include susceptibility to predation and reduced food availability (Thomas and Rahman 2009). By avoiding hypoxic areas, fish can be more susceptible to predation and experience reduced or suboptimal habitats (Breitburg 2002). Demersal fish are more likely to be affected by a loss of habitat than pelagic fish. Benthic feeding fish can also experience reduced food availability due to changes in numbers and species composition of benthic organisms in response to hypoxia (Rabalais et al. 2002).

Fish commonly avoid hypoxic areas, though thresholds and negative effects are species-dependent. Several species of fish and invertebrates have been shown to avoid hypoxia in the GOM and in Neuse River estuary (Eby and Crowder 2002;

Craig and Bosman 2012). While most fish avoid hypoxia, thresholds can differ among species. Importantly, fish can have different DO thresholds depending on whether the individual is acclimated to hypoxic conditions or not (Brady and Targett 2013). Avoidance can cause movement out of preferred habitats (Eby and Crowder 2002) or a mismatch with prey species with a lower threshold for hypoxia avoidance (Ludsin et al. 2009).

Atlantic croaker (*Micropogonias undulatus*) is a good model organism for studying the effects of hypoxia on fish in the GOM because it is a demersal species that lives in the area where hypoxia occurs and has been studied extensively. Croaker is a dominant member of the fish fauna caught in trawls in a region where hypoxia occurs (Baustian et al. 2009). Hypoxia has been shown to reduce the growth, survival, and fecundity of croaker and other closely related species (Rose et al. 2009). There is a large amount of previous work with croaker, including field data showing the hypoxia effects on reproduction and prey availability, laboratory experiments exposing croaker to hypoxia, and models of hypoxia avoidance and effects of hypoxia on fecundity (Baustian et al. 2009; Creekmore 2011). For example, chronic exposure to hypoxia has been shown to impair oocyte maturation and sperm motility in Atlantic croaker (Thomas and Rahman 2009). This previous work allows for model parameterization and provides some knowledge of how croaker react to hypoxia.

Models of varying levels of complexity have been used to model hypoxia in the GOM, with the more complex models with higher resolution being better suited to representing smaller scale events. Simpler models include modeling the shelf with a river model (Scavia et al. 2003) or modeling the shelf as a two-layered system (Justic et al. 1996). More complex models involve modeling the shelf as a three-dimensional (3-D) system with detailed modeling of the hydrodynamics (e.g., Hetland and DiMarco 2008; Wang and Justic 2009). Two of the 3-D models adapted to model hypoxia in the GOM are ROMS (Regional Ocean Modeling System) coupled to a NPZ model (Fennel et al. 2013) and the coupled FVCOM-WASP (Finite Volume Coastal Ocean Model—Water Quality Analysis Simulation Program) (Justić and Wang 2014). These 3-D models allow for the simulation of the conditions within the hypoxic zone with high spatial and temporal resolutions and are therefore suitable for fish movement modeling. FVCOM is an unstructured grid and open-source ocean circulation model (Chen et al. 2006). WASP is an open-source water quality model with several submodels for processes such as eutrophication (Wool et al. 2006). FVCOM and FVCOM-WASP have been used to model fish movement and hypoxia in the GOM (Justić and Wang 2014; Rose et al. 2014).

Several models have been used to study the effects of hypoxia on fish and fish movement. Models looking at the effects of hypoxia include matrix models, individual-based models (IBM), and models of the internal processes of an individual fish (Rose et al. 2009). IBMs can be used to look at fish movement and hypoxia avoidance (e.g., Creekmore 2011). Even though IBMs typically require more computational power than other model types, they have the advantage of time series (such as for hypoxia exposure) being possible to create for every fish modeled. Having the exposure time series allows for a more realistic look at the direct effects of hypoxia.

There are a number of different approaches to modeling fish and other animal movements. Various representations of random walk, including simple diffusion, have been used to describe and model animal movement (Kareiva and Shigesada 1983; Marsh and Jones 1988). The distance an animal can travel with random movement decreases with the size of the time step (Landau et al. 2013). Avoidance of adverse conditions and attraction to preferred conditions can be modeled using individual movement algorithms or by using a decision-making algorithm. Some movement types include random movement components along with avoidance and attraction components, such as run and tumble (Watkins and Rose 2013) and kinesis (Humston et al. 2004; Watkins and Rose 2013). Run and tumble and kinesis both have the animal only aware of their current environmental conditions and whether those conditions match the preferred conditions. Kinesis does have fish recall their previous heading. Because these movement algorithms have a random movement component, they are still affected by the size of the time step. Other movement algorithms, such as neighborhood search, allow fish to look at the surrounding conditions and move in the direction that is most advantageous (Watkins and Rose 2013; Rose et al. 2014). These types of movements are less affected by time steps, but can allow fish to perceive environmental conditions an unrealistic distance away from their location. Decision-making algorithms, such as those based on game theory, have also been used to allow animals to choose among different movement algorithms depending on the current conditions and memory of past conditions (Goodwin et al. 2006; Rose et al. 2014). Switching among movement types allows for the movement algorithm best suited for a situation to be used, including changing the algorithm with changing conditions.

This paper explores some of the different movement algorithms and how they affect the hypoxia exposure of individual virtual croaker. Fish movement in relation to hypoxia was modeled using a combination of output from the FVCOM-WASP model and a particle-tracking module with different movement algorithms. Three algorithm groups were tested to model fish movement in four different DO scenarios. The default algorithm, i.e., the algorithm used when the fish does not perceive hypoxia, is the main difference among the algorithm groups. The hypothesis tested was that the default algorithms do not affect hypoxia exposure of individual fish if all the fish use the same avoidance algorithm. The results of the paper will give a deeper understanding of how the algorithms tested behave in different situations. A better understanding of the algorithms allows for more effective algorithm selection and ultimately better informed management decisions.

10.2 Methods

Fish movement was examined for three algorithm groups in two dimensions (2-D) for static environmental conditions (maps of bottom DO concentration and temperature). The 2-D model and static conditions were chosen to better examine fish movement for different movement algorithms. Because the model domain is much larger horizontally than vertically (674 km from Galveston to Mobile and 3000 m deep), there is a large difference in the horizontal and vertical size of model elements for any 3-D grid. Moving a particle simultaneously in the horizontal and vertical directions is more complicated than just moving horizontally or vertically. Having the model fish move in only the horizontal direction simplifies the model while allowing for the examination of the properties of different movement algorithms. Also by using static conditions, the behavior of fish using different movement algorithms can be seen more clearly.

10.2.1 FVCOM-WASP

The coupled FVCOM/WASP model (Justić and Wang 2014) was used to model DO concentrations and temperature in the GOM. The FVCOM particle-tracking module was used to model fish movement on the same FVCOM grid. The model domain covers the coastal GOM from Mobile Bay to Galveston Bay and extends offshore up to a depth of about 300m (Wang and Justic 2009). The unstructured nature of the model grid allows for a more accurate depiction of the complex GOM coastline and higher model resolution along the coast. The model has been previously calibrated to accurately represent the circulation and stratification on the La-Tex continental shelf (Wang and Justic 2009).

Fish were modeled using the FVCOM particle-tracking module with output from the FVCOM-WASP. The particle-tracking module interpolated the DO and temperature values to the location of each individual fish using the piecewise linear algorithm employed in the FVCOM. Scenarios were created by taking output from the FVCOM-WASP model for 2002, namely for May 2 (06:00), June 4 (06:00), July 23 (17:00), and September 21 (17:00). These dates and times were chosen because they depict progressively worsening hypoxic conditions in the GOM, here denoted as normoxia (May 2), mild hypoxia (June 4), intermediate hypoxia (September 21), and severe hypoxia (July 23). The May 2 scenario represented conditions before the onset of hypoxia, and the June 4 conditions represented the onset of hypoxia. During late July, the combination of primary production driven by riverine nutrients and strong water column stratification typically results in the maximum extent of hypoxia (Rabalais et al. 2007), which was represented by the July 23 scenario. Water column mixing can cause hypoxia to break up and then reform. The September 21 scenario was representative of such a breakup and reformation of hypoxia in response to Tropical Storm Hanna. All scenarios used the July 23 temperature field so that the only variation between scenarios was due to different DO concentrations. The locations for the values for plotting were listed with longitude/latitude, while the grid locations in the tracking module used Universal Transverse Mercator (UTM) projection.

10.2.2 Movement Algorithms

Movement algorithms were added to the FVCOM particle-tracking module to model fish movement in the model domain based on environmental cues. The FVCOM particle-tracking module is designed to track passive particles that move with the currents calculated by FVCOM. The module was modified as in Rose et al. (2014) to track active particles with movement behaviors such as fish. Movement algorithms were used to calculate x and y velocities.

Changes in particle position were calculated using the past positions and the velocities, as in Watkins and Rose (2013) and Rose et al. (2014). The formulas used to calculate change in position for all algorithms were as follows:

$$x(t + \Delta t) = x(t) + \mathbf{u}(t) * \Delta t \tag{10.1}$$

$$y(t + \Delta t) = y(t) + \mathbf{v}(t) * \Delta t \tag{10.2}$$

where x and y are fish positions on the x and y axes, u and v are fish velocities, and Δt is the time step for fish movement. Only the positions of fish inside the model domain were updated. For fish outside the model domain, the reflective boundary algorithm was applied as described in Sect. 10.2.2.10. Most of the algorithms, except kinesis, calculated the u and v velocities using the same equations. Velocities for kinesis will be described with the kinesis algorithm (Sect. 10.2.2.5). The velocities of fish in the x and y dimensions were calculated as follows:

$$\mathbf{u}(t) = ss * \cos(\theta(t)) \tag{10.3}$$

$$\mathbf{v}(t) = ss * \sin(\theta(t)) \tag{10.4}$$

where ss is the swimming speed and $\theta(t)$ is the swimming angle relative to due east for that time step. How the speed and angle were calculated depends on the movement algorithms.

Nine algorithms were used to model fish movement (Table 10.1). The top-level algorithm, event-based, is used to choose among the other algorithms. Seven algorithms are categorized into three behaviors for use with the event-based algorithm: default, strategic, and tactical. Default behaviors are used when there is no hypoxia. Three algorithms are classified as default: random walk, Cauchy correlated random walk, and Gaussian kinesis. Strategic behaviors are used after an individual leaves conditions that require immediate reactions, but there is still a memory of those conditions. Two strategic behaviors are used in the model: correlated random walk and logistic kinesis. Tactical behaviors are used when the fish needs to react to conditions immediately. The two tactical behaviors, neighborhood search and sprint, are triggered by encountering hypoxia for neighborhood search or spending too much time in hypoxic conditions for sprint. The ninth algorithm, reflective boundary, is used to address boundary issues with the fish and is not used in the event-based algorithm.

Table 10.1 Algorithm definitions

Algorithm	Definition	Reference	Abbr.	Event-based behavior	DO avoidance
Event-based	Switches between movement algorithms depending on environmental conditions	Watkins and Rose (2013), Rose et al. (2014)	EB	NA	Yes
Random walk	Random angle and swimming speed are selected at each time step from defined probability distributions.	Watkins and Rose (2013)	RW	Default	No
Cauchy correlated random walk	Random walk where the turning angle is taken from a non-uniform distribution (Cauchy) so that some angles are chosen more often	Wu et al. (2000)	CCRW	Default	No
Kinesis-Gaussian	"Random walk approach that continuously adjusts turning angle and swimming speed distributions based on current environmental cues"; uses a Gaussian function	Watkins and Rose (2013)	KG	Default	No
Correlated random walk	Random walk that selects a random speed but uses the previous angle with variation added	Rose et al. (2014)	CRW	Strategic	No
Kinesis-logistic	"Random walk approach that continuously adjusts turning angle and swimming speed distributions based on current environmental cues"; uses a logistic function	Modified from Watkins and Rose (2013)	KL	Strategic	Yes
Neighborhood search	Evaluate growth and mortality cues in neighboring cells, and then to move toward the cell with the highest habitat quality or away from lowest quality	Watkins and Rose (2013)	NS	Tactical	Yes
Sprint	A straight lined movement with higher swimming speeds that is triggered by a counter of consecutive hypoxic time steps		S	Tactical	Yes
Reflective boundary	A NS algorithm used to guide fish back into the model domain after they have jumped out of it; for larger time steps		RB	NA	No

10.2.2.1 Event Based

For this study, an event-based approach was used to model fish decision making. Event-based movement is based on game theory and chooses between different movement algorithms based on utilities, which depend on current and past conditions experienced by the virtual fish (Anderson 2002; Watkins and Rose 2013). Environmental conditions, such as hypoxia, are used to choose between default, strategic, and tactical behaviors. There is only one default behavior used at a time. The default behavior occurs when the main stimulus, here low DO, is not detected. Strategic and tactical behaviors come in pairs for each environmental cue or threshold. A preferred range can be defined using two thresholds, such as for temperature or salinity, resulting in two pairs of strategic and tactical behaviors. Tactical movement occurs when a threshold condition is crossed. Strategic movement occurs when fish are out of conditions that would trigger tactical movement, but a memory term in the equation still has the fish being influenced by the hypoxia exposure. Strategic is an asymmetrical behavior that occurs only after a tactical behavior has been triggered and fish are exiting the hypoxic zone.

The behavior with the highest utility is the one chosen by the event-based algorithm. In a game theory approach, utilities represent how the benefits and costs of a given behavior affect the fitness of the organism (Anderson 2002). Here, hypoxia avoidance success is used as a measure of fitness. The utility values are affected by whether or not events occur. Whether events occur was determined with:

$$e_J(t) = \begin{cases} 0 & val > th \\ 1 & val \leq th \end{cases} \tag{10.5}$$

where $e_J(t)$ is the event value at time t, a Boolean operator of whether the event J is triggered (1) or not triggered (0) at time t. The *val* is the environmental value (e.g., DO) or other value, such as a counter. An example of the threshold, *th*, is 2 mg O_2/L, which denotes the upper DO limit for hypoxia.

The utility of the different behavior options were then calculated for each fish. The behavior with the highest utility is the one chosen by the event-based algorithm. Utility is calculated by:

$$util_{J,K}(t) = utili_{J,K} * prob_{J,K}(t) \tag{10.6}$$

where *util* is the utility for that time step, *utili* is the intrinsic utility, and *prob* is the probability of a triggered event. The integers J and K in Eq. 10.6 indicate behavior groups, where the J value represents a group and the K value indicates whether the behavior is tactical (K = 1) or strategic (K = 2). The probability of an event being triggered was calculated by:

$$prob_{J,K}(t) = (1.0 - mem_{J,K}) * e_J(t) + mem_{J,K} * prob_{J,K}(t - \Delta t) \tag{10.7}$$

where $prob_{J,K}(t)$ is the probability for the behavior J, K at time t, $e_J(t)$ is the event value at time t, $prob_{J,K}(t - \Delta t)$ is the probability calculated last time step, and $mem_{J,K}$ determines what proportion of the event value and past probability make up the current probability. The probability is a running average and allows for the fish to have some memory of past events. The utilities were then compared to a minimum standard value, and the largest value determines which behavior was used. If none of the calculated utilities were larger than the standard minimum, then the default behavior was used (J=0). The event-based behavior was the same as used in Rose et al. (2014). The algorithms used for each event-based behavior are described in Table 10.1.

10.2.2.2 Random Walk

Random walk is a type of movement algorithm where random speeds and angles are chosen without considering environmental conditions. Three random walk algorithms are used in this study, two for default behaviors and one for strategic. The simplest random walk used chooses a random angle and adds variation to a baseline speed for each time step. The speed and angle are calculated using:

$$ss = ss_0 \pm 0.3 * ss_0 * ran \tag{10.8}$$

$$\theta(t) = 2\pi * ran \tag{10.9}$$

where ss_0 is the baseline swimming speed and *ran* is a uniform random number between 1 and 0. The angle is calculated with relation to a fixed axis, here due east. Equation 10.8 is used to calculate speed for all of the random walk algorithms and is slower than the speed used for the tactical behaviors. So fish slow down when switching from a tactical behavior to one of the random walk algorithms. Random walk has particles/fish move with simple diffusion.

10.2.2.3 Correlated Random Walk

A correlated random walk (CRW) is a random walk with a bias toward one direction (Kareiva and Shigesada 1983). A CRW chooses a random speed and a turning angle. Unlike the simple random walk, a CRW determines the new angle relative to the angle from the previous time step instead of a fixed axis. The variation added to the old angle to get the new angle is called the turning angle. The CRW uses the velocities from the previous time step to calculate the angle:

$$\theta(t) = \text{atan2}(\mathbf{v}(t - \Delta t), \mathbf{u}(t - \Delta t)) + 0.05 * 2\pi * (2 * ran - 1) \tag{10.10}$$

where the first half of the equation, atan2( ), is from the previous time step and the second part is a random component to add variation to the angle. If no random component was added, then when substituted into Eqs. 10.3 and 10.4, θ would produce the fish swimming velocity components u and v of the previous time step. The fish moves in mostly the same direction with some variation when a small random component is added. Speed is calculated with Eq. 10.8. CRW is the algorithm used for the strategic behavior in all runs except those with kinesis. Field data of insect movement have been described using CRW (Kareiva and Shigesada 1983).

10.2.2.4 Cauchy Correlated Random Walk

The Cauchy correlated random walk (CCRW) is a more complicated CRW where the magnitude and direction of the bias can be controlled by choosing the turning angle from a non-uniform, wrapped Cauchy distribution. The CCRW was adapted from Eq. 20 in Wu et al. (2000), which is implemented as the turning angle:

$$\theta(t) = \theta(t - \Delta t) + 2 * \mathrm{atan}\left[\frac{(1-\varepsilon)}{(1+\varepsilon)} * \tan((ran - 0.5) * \pi)\right] + \theta_m \tag{10.11}$$

where ε determines the shape of the wrapped Cauchy distribution, and θ_m determines the center of the distribution. $\theta(t - \Delta t)$ is the previous angle, and the $2 * \mathrm{atan}[\,] + \theta_m$ is the turning angle. Higher values of ε result in more correlation and less randomness to the direction of the fish. By changing θ_m, the fish can be given a bias in whether they turn left or right. The original equation from Wu et al. (2000) was changed by adding the parameter θ_m based on Batschelet (1981) and using the result of the equation as a turning angle instead of the angle of direction. Speed was calculated using Eq. 10.8. Velocities are calculated using Eqs. 10.3 and 10.4.

10.2.2.5 Kinesis

Kinesis compares the ideal condition to the current condition to determine the proportion of random versus previous velocities to use for the new velocities (Humston et al. 2004; Watkins and Rose 2013). Two forms of kinesis were used: Gaussian and logistic. As previously mentioned, kinesis uses a different method from the other algorithms to calculate velocity. Instead of calculating speed and angle at each time step, the velocities for the x and y directions are calculated using a random component and a previous velocity component. The equations are as follows:

$$\mathbf{u}(t) = \mathbf{u}(t - \Delta t) * (h1 * p1) \pm nran * (1 - h2 * p1) \tag{10.12}$$

$$\mathbf{v}(t) = \mathbf{v}(t - \Delta t) * (h1 * p1) \pm nran * (1 - h2 * p1) \tag{10.13}$$

where $p1$, $h1$, and $h2$ are weighting factors. The $p1$ weighting factor is used to determine the proportion of random versus previous velocities in the new velocities and is calculated using the comparison of the ideal and current conditions in a function. The $h1$ and $h2$ weighting factors determine the percentage of the previous velocities and random components, respectively, used in the new velocities. Fish using kinesis slow down in areas where the previous component has more weight than the random component. The previous component is smaller than the actual previous velocity because the $h1$ weighting factor is less than 1. The value *nran* is a random number from a normal distribution that is screened for extreme values ($>|3|$). The Kinesis algorithm is based on studies on a variety of organisms including wood lice and protozoa and focuses on a mechanistic approach to animal movement (Fraenkel and Gunn 1961).

10.2.2.6 Gaussian Kinesis

How the $p1$ weighting function is calculated depends on the version of kinesis. The Gaussian kinesis is the form of kinesis used in Watkins and Rose (2013) and Humston et al. (2004). It uses a Gaussian curve in determining the $p1$ weighting factor used in Eqs. 10.12 and 10.13. A Gaussian curve results in fish moving toward the mean of the curve. The $p1$ weighting factor is calculated with:

$$p1 = e^{-0.5*\left(\frac{temp(t)-\mu_{temp}}{\sigma_{temp}}\right)^2} \tag{10.14}$$

where *temp*(t) is the current temperature, μ_{temp} is the mean of the curve and ideal temperature, and σ_{temp} is the sigma value of the curve. The sensitivity of the algorithm can be altered by changing the curve width with the sigma value. Gaussian kinesis was used as a default behavior in response to temperature.

10.2.2.7 Logistic Kinesis

Logistic kinesis works better than Gaussian kinesis for avoiding poor conditions, as opposed to being attracted to good conditions. Tests were run using Gaussian kinesis to avoid hypoxia, but the algorithm did not perform well. The Gaussian curve represents an ideal condition and progressively worse conditions well, but does not represent good conditions that suddenly become poor conditions at some threshold. A logistic curve better represents the second scenario, so a logistic curve was used instead of a Gaussian curve to calculate the $p1$ weighting function:

$$p1 = \frac{e^{r*do(t)}}{e^{r*do(t)} + b} \tag{10.15}$$

where *do*(t) is the current DO value, *r* controls the slope of the curve, and *b* controls where the curve occurs. The $p1$ value was used to calculate velocities with Eqs. 10.12 and 10.13. The algorithm results in fish moving away from areas with values on the lower part of the curve and then moving mostly randomly in areas where the curve has leveled off. Logistic kinesis is used as a strategic behavior in response to DO.

10.2.2.8 Neighborhood Search

Neighborhood search is used in most of the tactical behaviors and is the main algorithm used for hypoxia avoidance. Neighborhood search works by searching the neighboring cells and then moving either away from the lowest quality cell or moving toward the highest quality cell. The quality of the cell can be calculated in different ways (e.g., growth and mortality Watkins and Rose 2013 or salinity Rose et al. 2014). Hypoxia was avoided with a neighborhood search algorithm moving away from the lowest DO value. The neighboring cells are searched for the cell with the lowest DO value and then the swimming speed and angle are calculated with:

$$\theta(t) = \mathrm{atan2}(y(t) - yl(t), x(t) - xl(t)) + 0.15 * 2\pi * (2 * ran - 1) \tag{10.16}$$

$$ss = 2 * ss_0 \pm ss_0 * ran \tag{10.17}$$

where $x(t)$ and $y(t)$ are the current x and y coordinates, $xl(t)$ and $yl(t)$ are the coordinates of the center of the cell with the lowest DO, and *ran* is a uniform random number. The first part of the equation, atan2( ), calculates the angle, and the second part calculates a random component that adds some variability to the angle. The amount of variation can be increased or decreased by replacing the 0.15 with larger or smaller values, respectively, with a maximum meaningful variation of pi. The swimming speed was faster for avoidance behaviors than for default behaviors because there was more urgency to avoid bad conditions. The variation in swimming speed is similar to Eq. 10.8, but has a higher range of variation. Velocities are calculated using Eqs. 10.3 and 10.4. Hypoxia avoidance with neighborhood search was triggered when the utility is highest for the tactical behavior for low DO avoidance.

10.2.2.9 Sprint

The sprint algorithm was created to deal with the problem of fish moving under neighborhood search getting stuck moving in a narrow region around a local DO maximum. When local maxima with values below 2 mg DO/L occur, fish will get stuck in those cells when using neighborhood search. Adding variability to the neighborhood search angle can keep some fish from getting stuck at local DO maxima, but not all. The sprint algorithm is triggered when a counter of hypoxic steps exceeds a set value. The set value determines the maximum number of time steps the fish can stay in hypoxia before drastic measures need to be used. For this paper, fish start to panic after spending two days in hypoxic conditions. The set value is the number of steps, depending on time step, that occur over two days (200 for 15 min; 25 for 2 h). The sprint is a simple algorithm that calculates the angle and speed with:

$$\theta(t) = \theta(t - \Delta t) \tag{10.18}$$

$$ss = 3 * ss_0 \tag{10.19}$$

The counter is decreased by 1 when the algorithm is triggered and is reset to zero when normoxic conditions are encountered. A decrease of 1 results in the fish sprinting until it exits the hypoxic zone. Changing the amount the counter decreased the affects how long the sprint algorithm is used, but only a decrease of 1 is used in this paper.

10.2.2.10 Reflective Boundary

Neighborhood search is also used in the reflective boundary algorithm. It is not one of the algorithms chosen by the event-based algorithm, but occurs after movement is calculated by the movement algorithms. The reflective boundary algorithm is used with kinesis, which works best at larger time steps (≥1 or 2 h). Particles in the FVCOM-tracking module tend to get stuck on the edges of the model domain at time steps large enough for kinesis to work well. The reflective boundary algorithm uses neighborhood search to calculate the cell with the fewest boundaries and move the fish toward the cell with fewest boundaries. The angle is calculated by:

$$\theta(t) = \mathrm{atan2}(yl(t) - y(t), xl(t) - x(t)) + 0.15 * 2\pi * (2 * ran - 1) \qquad (10.20)$$

where the values are the same as Eq. 10.16. The values used are from the location calculated at the last time step and not the new location calculated by the movement algorithms for the current time step. Using the previous values makes sure that the particle is in a cell so neighborhood search will work properly. The only change in the calculation of θ is the order of coordinate values in the atan2 function. Speed is calculated as in Eq. 10.17. The reflective boundary is applied only to particles that have gone outside of the model domain.

10.2.3 Algorithm Groups

Three algorithm groups were used in the model simulations (Table 10.2). Each group had a tactical, strategic, and default behavior and was named after the tactical and default behaviors. Neighborhood search was used for the tactical behavior for all groups. Neighborhood search is an efficient and effective avoidance behavior. CRW

Table 10.2 Algorithm groups: Neighborhood Search (NS), Random Walk (RW), Correlated Random Walk (CRW), Cauchy Correlated Random Walk (CCRW), Kinesis (K), Logistic Kinesis (KL), and Gaussian Kinesis (KG)

Algorithm group	Tactical	Strategic	Default	Panic response (tactical/strategic)	Time step (min)
NS/RW	NS	CRW	RW	Sprint/CRW	15
NS/CCRW	NS	CRW	CCRW	Sprint/CRW	15
NS/K	NS	KL	KG	Sprint/CRW	120

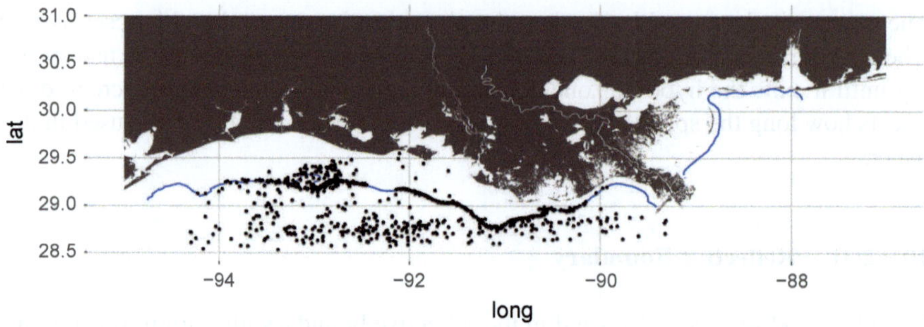

Fig. 10.1 Starting positions for particles/fish in FVCOM-tracking module in the Northern Gulf of Mexico. The *line* is the 26 °C contour

and logistic kinesis were used as the strategic behaviors. Strategic behaviors were matched to the default behaviors, where CRW was used for random walk defaults and logistic kinesis was used for the kinesis default. The default behaviors, RW, CCRW, and Gaussian kinesis (from here on, kinesis refers to Gaussian kinesis), were the major difference between algorithm groups. Each algorithm group also used a subgroup consisting of the sprint algorithm as the tactical behavior and CRW as the strategic behavior.

10.2.4 Model Runs

For simulation set 1, each of the three algorithm combinations (Table 10.2) was run for the four DO scenarios (Sect. 10.2.1). All simulations were for seven days. A time period of seven days was selected because the model fish exposure levels reach steady-state values after about five days. The starting positions of the 913 fish are the result of having fish move with Gaussian kinesis for 124 days (Fig. 10.1). The fish end up mostly gathered around the ideal temperature, which overlaps with the hypoxic zone for the severe hypoxia scenario. The time step used depends on the algorithm group. A 900-s (15 min) time step is used for the algorithm groups with random walk algorithms. A time step of 2 h is used for the kinesis algorithm group. A larger time step is used for kinesis because the algorithm is very inefficient at finding the ideal temperature at smaller time steps. The output time steps were the same as the respective time step for each algorithm group: every 15 min for the random walk groups and every 2 h for kinesis.

For simulation set 2, each of the three algorithm groups with individuals configured to be inefficient at avoidance (poor avoidance) was run on only the severe hypoxia (July 23, 2002) scenario. The effects of poor avoidance are best seen when avoidance is used a lot, so smaller hypoxic areas are less useful. Everything was the same as with the severe hypoxia scenario in set 1 except for the poor avoidance. The ability of the fish to avoid low DO was impaired by changing the 0.15 coefficient in

Eq. 10.16 to 0.5, which added more variation to the movement angle of neighborhood search used for avoidance. The sprint algorithm was also not used for these runs because it is an algorithm with good hypoxia avoidance and allows the fish to easily escape the hypoxic zone after two days. The poor avoidance runs were then compared to the runs for the severe hypoxia scenario from simulation set 1.

10.2.5 Data Analysis and Visualization

10.2.5.1 Statistics

Several statistics are used to compare model runs. The first group of three statistics looks at the conditions experienced by each fish. The first statistic, the percentage of fish in a DO or temperature range, is calculated by first taking the DO or temperature value for each fish at each time step. The number of fish in each range is then divided by the total number of fish to get the percentage. Summary statistics (minimum, mean, and maximum) for DO, the second statistic, are calculated at each time step for all 913 fish. The third statistic, exposure to hypoxia, is calculated by summing the total time spent in hypoxic conditions. For each time step, a value of one is assigned for hypoxic values and zero for non-hypoxic values. The one or zero is then multiplied by the time step, and the resulting values are summed to get the total time exposed to hypoxia. For comparing time spent in hypoxic conditions, multiplying by the time step converts to the unit of time used for the model. Data analysis, manipulation, and plotting were all performed in R (R Core Team 2013). Outliers are as defined by the R plotting method used, generally ggplot2 boxplot.

Three related statistics were used to compare the movement and distribution spread of the fish: sinuosity, net distance, and total distance. The term "distribution spread" refers to how far the fish have spread out from their initial positions. Sinuosity is the amount of wiggle in the path of the fish. It is calculated by dividing the actual path by the shortest, or net, distance between the first and last point for each fish. The actual path, or total distance travelled, is calculated by summing the distances between successive points in the path. Distances between two points use the distance formula.

A categorization test was used to compare hypoxia exposure values. It is not advised to use statistical tests with p-values for interpreting simulation models. In a simulation model, the researcher controls the degrees of freedom, which affects the p-values (White et al. 2014). The algorithm groups are also known to be different, so the questions of "are they different" is not an informative question to ask or answer with statistical tests. Whether the algorithms can be differentiated in a field sample is a more informative question to answer. A categorization test using a linear discriminant analysis (LDA) was used to determine whether the hypoxia exposures for the three algorithm groups could be told apart. The *lda* function in the R package *MASS* was used on training data sets. Each training data set was comprised of 1000 mean exposures for each algorithm group from a sample size ranging from 25 to 900 (by 25). Algorithm groups were only categorized within the same scenario and not among scenarios. The prior for the LDA was 1/3 for each algorithm group training data set, which assumes equal chance of a random sample being in any of the algorithm groups. The LDA was run for each scenario using the training data sets and the miscategorization rate was calculated as 1 minus the average categorization rate. The acceptable rate of miscategorization was 0.05. The normoxia scenario was not included because all the exposure values were zero, which caused *lda* to crash.

10.2.5.2 Growth and Vitality

The effects of hypoxia exposure on growth were calculated for the seven days of the model run and extrapolated to 30 days. The model fish are juveniles between 1 and 2 years of age. Growth was calculated by first calculating the vitality at each time step for growth using the following based on Miller Neilan and Rose (2014):

$$Gvitality = \begin{cases} 1.0 & x \geq Gne \\ 1.0 - Galpha\frac{(x-Gne)^2}{(x-Gne)^2+Gbeta^2} & x < Gne \end{cases} \tag{10.21}$$

where *Gne* (3.35), *Galpha* (110.78), and *Gbeta* (21.06) are constants. *Gne* is the threshold below which low DO has an effect on the fish, and x is the DO value experienced by the individual fish for that time step. There is also a repair term for growth vitality, *Ggamma* (0.21), which is the maximum increase in growth vitality per hour. The repair term accounts for a delay in return to normal vitality rates after returning to normoxia. The repair term was applied using:

$$Gvitality_{repair} = \min(Gvitality(t), (Gvitality(t - \Delta t) + Ggamma * \Delta t)) \tag{10.22}$$

where $Gvitality_{repair}$ is the new growth vitality term and Δt is the time step. *Ggamma* is multiplied by the time step because the data points are either 15 min or 2 h apart while the repair term is hourly. Parameters for growth vitality for Atlantic croaker in the region of the GOM hypoxic zone were provided by Sean Creekmore (personal communication, Louisiana State University).

Growth is calculated by calculating growth and new weights at each time step. The growth and new weight were calculated for each fish using:

$$Weight(t) = Weight(t - \Delta t) * e^{Gvitality_{repair} * maxGrowthM} \tag{10.23}$$

where $Weight(t)$ is the weight for the current time step, $Weight(t - \Delta t)$ is the weight of the previous time step, and *maxGrowthM* is the maximum growth under normoxic conditions. The initial weight ($Weight(0)$) and maximum growth rate were calculated using equations from Barger (1985) that determine the length from age and a weight from length. Parameters for the movement and growth equations are in Table 10.3.

Table 10.3 Parameter values for algorithm and weight calculation equations

Parameter	Value	Description
Δt	0.25, 1 h	Time step for particle-tracking module. Value used depends on default algorithm
th	2 mg O_2/L	The threshold for triggering hypoxia avoidance
Utili	2.0, 3.0, 1.0[a]	Intrinsic utility (NS, sprint, default/strategic)
Mem	0.5, 0.9[a]	Memory term (default/strategic, tactical)
Standard minimum	0.175[a]	Value utility must be greater than to affect fish movement
ss_0	0.23148 m/s[a]	Baseline swimming speed
ε	0.9[b]	Determines shape of wrapped Cauchy distribution
θ_m	0[b]	Determines what angle wrapped Cauchy distribution is biased toward
h1, h2	0.7, 0.99[a]	Kinesis weighting factors
μ_{temp}	26 °C[c]	Ideal temperature for croaker
σ_{temp}	2 °C[c]	Sigma value for Gaussian curve
r	3[d]	Controls slope of logistic curve
b	150[d]	Control where logistic curve occurs
Stuck	200, 25[E]	When the counter reaches this number of steps, the fish is considered stuck. Value depends on time step.
Galpha	100.78[f]	Constant
Gbeta	21.06[f]	Constant
Gne	3.35[f]	Threshold for hypoxia affects
Ggamma	0.21[f]	Repair term
Weight(0)	88.72 g[g]	Initial weight calculated for croaker at beginning of second year of life
MaxGrowthM	0.002 g/day[g]	Maximum growth rate in normoxic conditions for croaker from beginning to end of second year of life

[a]Based on Rose et al. (2014)
[b]Determined by testing different values affects on wrapped Cauchy distribution
[c]Based on location of temperature contour where croaker gather in the GOM
[d]Determined by testing different values so that the logistic curve lined up around 2
[e]Number of time steps that make up 2 days
[f]From Sean Creekmore (Personal communication, LSU)
[g]Calculated based on Barger (1985) and Miller Neilan and Rose (2014)

10.2.5.3 Visualization Methods

The map plots were creating using *ggplot2* layers for the coastline, interpolated values, contours, and fish generated in R. For the coastline, a shapefile for the US coast from the USGS (http://coastalmap.marine.usgs.gov/regional/contusa../gomex/gloria/data.html) was cropped down to include the Gulf coast between −95° and −87° longitude and between 27° and 31° latitude. The contour and the interpolated field layers were generated from output from FVCOM-WASP from 2002. DO values for May 2, June 4, July 23, and September 21 were combined with temperature and salinity values for July 23 to create scenarios that vary only by DO values. After converting the final fish locations to latitude and longitude, the locations were added to the plot.

10.3 Results

For simulation set 1 (good avoidance), most fish were able to escape the hypoxic zone within one day and more fish escaped more quickly with the NS/RW algorithm group than with NS/K algorithm group. In the final position maps for the fish (Fig. 10.2), the majority of fish were outside of the hypoxic zone. The fish that were in the hypoxic zone were on the very edge. For mild (d–f) and intermediate (g–i) hypoxia scenarios, it took a day or less for the percentage of fish in conditions <2 mg/L (solid orange line) to approach zero (Fig. 10.3). For the severe hypoxia scenario, it took three days for NS/RW algorithm group (j) and five days for NS/CCRW algorithm group (k) for the percentage of fish in conditions <2 mg/L to approach zero. The percentage of fish in < 2 mg/L for the NS/K algorithm group (l) in the severe hypoxia scenario leveled off at about 5% of fish in conditions < 2 mg/L.

For simulation set 2 (poor avoidance), fish took longer to escape the hypoxic zone compared to simulation set 1 fish and NS/K algorithm group fish had a greater decrease in the percentage of fish within the 1–2 mg/L range than NS/CCRW fish. There were fish in the hypoxic zone for all three algorithm groups for all seven days. All algorithm groups (m–o) had at least 5% of fish below 1 mg/L (teal line) and at least 10% of fish between 1–2 mg/L (solid orange line) during the course of the seven-day simulation (Fig. 10.3). The NS/K algorithm group (o) had the largest decrease in the percentage of fish for the 0–1 mg/L and 1–2 mg/L ranges. The NS/CCRW algorithm group (n) had the smallest decrease in the percentage for these ranges and the percentage of fish in the 1–2 mg/L range stayed constant. The other three ranges (2–3 mg/L, 3–4 mg/L, >4 mg/L) had a general increase in percentage for all three algorithm groups with NS/K algorithm group having the largest change and NS/CCRW algorithm group having the smallest. The >4 mg/L range was an exception, with NS/RW algorithm group having the smallest change in percentage of fish. For NS/CCRW algorithm group, the dispersal of fish was greater than compared to NS/RW algorithm group, resulting in NS/RW algorithm group having a smaller change in percentage for that particular range.

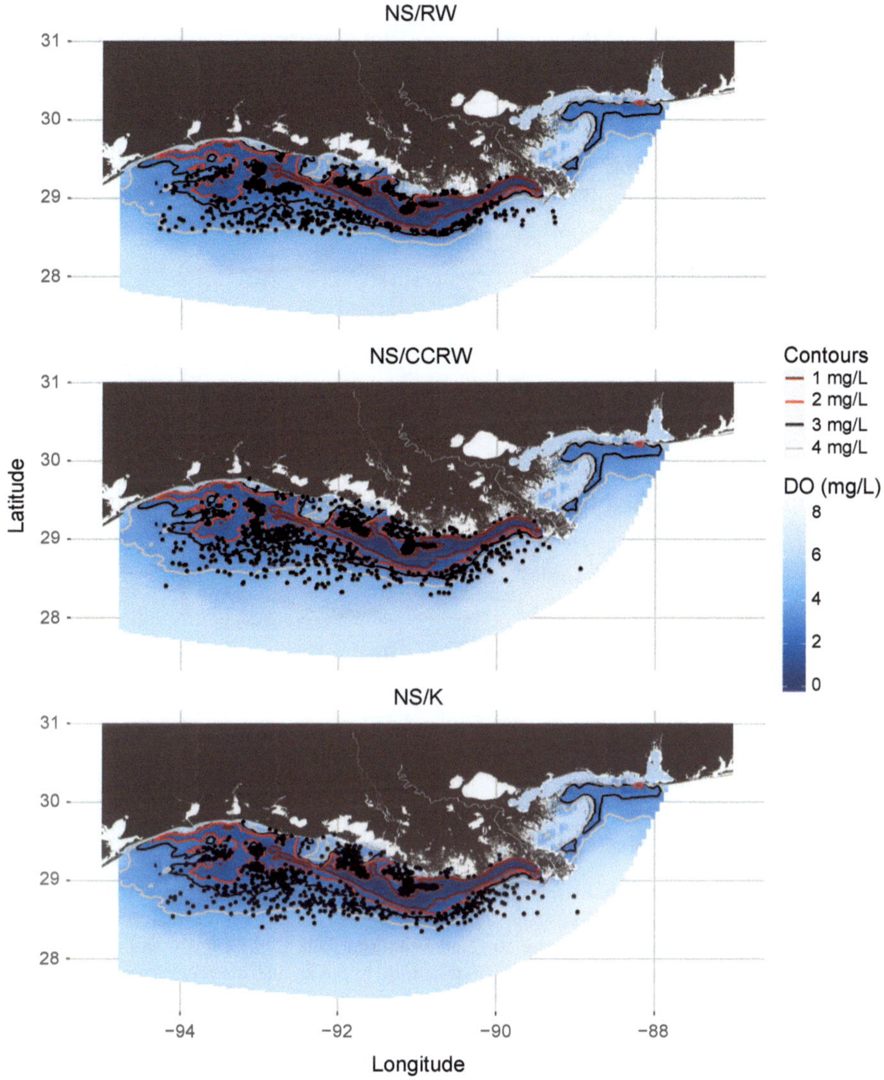

Fig. 10.2 Final fish positions for the severe hypoxia scenario along the Northern Gulf of Mexico. Algorithm groups are neighborhood search/random walk (NS/RW), neighborhood search/Cauchy correlated random walk (NS/CCRW), and Neighborhood Search/Kinesis (NS/K)

The summary statistics of DO for the algorithm groups became less similar as hypoxic area increased, with NS/K algorithm group minimum DO values never rising above 2 mg/L while NS/RW and NS/CCRW algorithm group minimums did. For normoxia (a–c), the mean (black line), minimum (dark gray line), and maximum (light gray line) DO values were about the same for the three algorithm groups

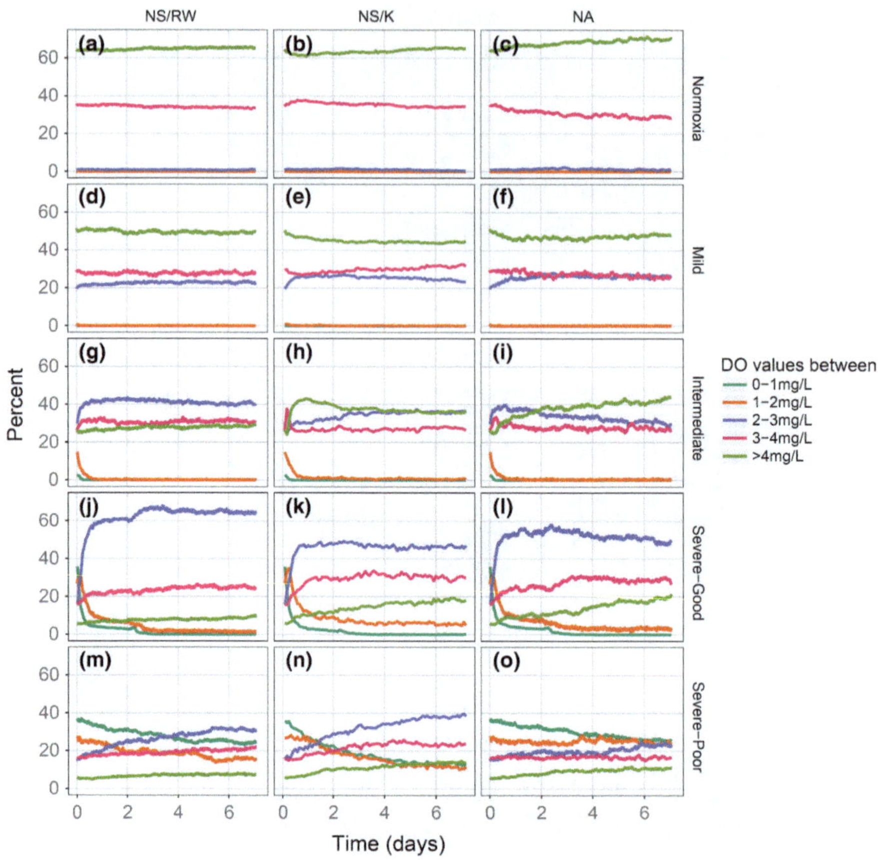

Fig. 10.3 Percentage of fish in five DO ranges (0 to <4 mg/L) over time for each algorithm group/scenario combination. Algorithm groups are neighborhood search/random walk (NS/RW), neighborhood search/Cauchy correlated random walk (NS/CCRW), and neighborhood search/kinesis (NS/K). The scenarios are normoxia, mild hypoxia, intermediate hypoxia, and severe hypoxia (with good and poor hypoxia avoidance)

(Fig. 10.4). As hypoxic area increased, the mean DO values for the fish decreased, from over 5 mg/L for normoxia to around 3 mg/L in the severe scenario (j–l). The maximum DO values for the fish also decreased with increasing hypoxic area. For NS/RW algorithm group, the maxima tended to stay around the same values for the entire model run, while the maximum values for NS/CCRW algorithm group and NS/K algorithm group tended to increase over time. The minimum values for scenarios with hypoxia tended to increase to 2 mg/L and stay around that value. The minimum values took longer to reach 2 mg/L for the intermediate (g–i) and severe scenarios (j–l), with the longest time of around 5 days for the severe scenario for NS/CCRW algorithm group (k). The intermediate (i) and severe (l) hypoxia scenarios for NS/K algorithm group are the only exceptions to the trend of the minimum

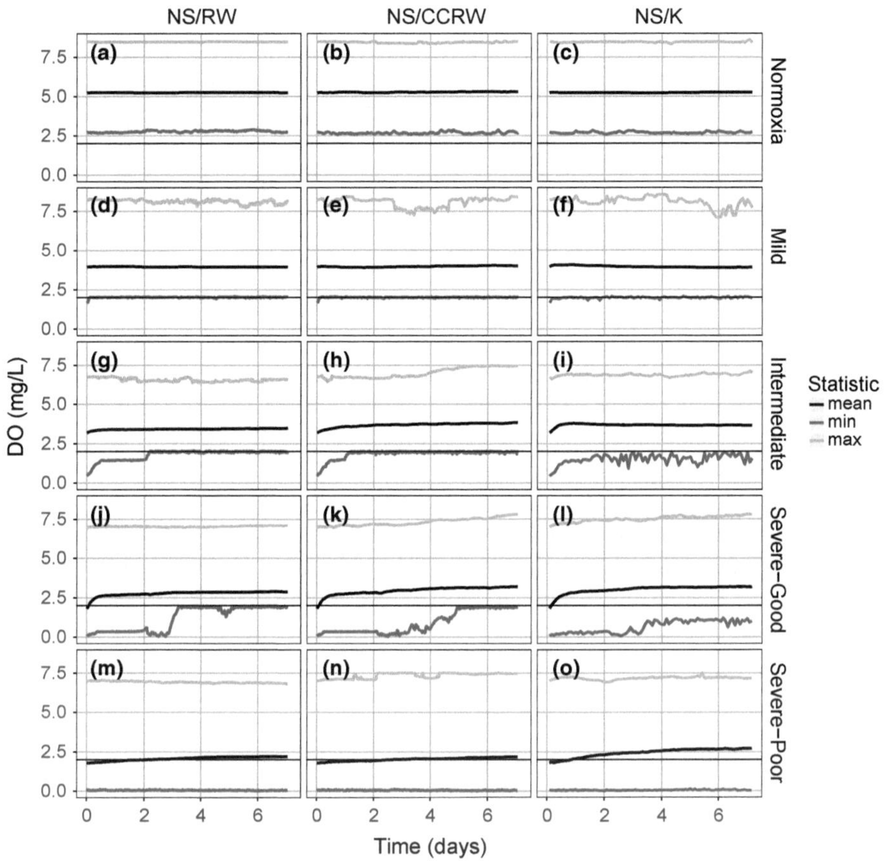

Fig. 10.4 Minimum, mean, and maximum DO values experienced by fish during the seven-day simulation period for each algorithm group/scenario combination. The *thin, black horizontal line* denotes 2 mg/L, so values below the *line* represent hypoxic conditions. Algorithm groups are neighborhood search/random walk (NS/RW), neighborhood search/Cauchy correlated random walk (NS/CCRW), and neighborhood search/kinesis (NS/K). The scenarios are normoxia, mild hypoxia, intermediate hypoxia, and severe hypoxia (with good and poor hypoxia avoidance)

leveling off at 2 mg/L. Both of these time series had more variation than the corresponding time series for NS/RW algorithm group (g, j) and NS/CCRW algorithm group (h, k). The minimum DO values experienced by fish for the intermediate scenario approach 2 mg/L and for the severe scenario stayed near 1.75 mg/L. Even though the minimum DO values experienced by fish could be below 2 mg/L during the entire simulation period, no fish spent the entire time in the hypoxic zone.

For simulation set 2, there was less variation across time and among the algorithm groups than with simulation set 1, with NS/K having the largest variation. The minimum DO value (dark gray line) for all the fish remained at or near zero for the entire seven days (m–o, Fig. 10.4). There was little variation in the three statistics for all three algorithm groups. The maximum DO value has some variation for NS/CCRW algorithm group (n) and NS/K algorithm group (o), but it was less than 0.5 mg/L. The mean DO values increased very little, less than 0.5 mg/L, over the 7 day period. The NS/K algorithm group had the largest increase in mean DO value, possibly due to the larger time step.

10.3.1 Exposure

Fish exposure to hypoxia increased with hypoxic area, and exposure was not greatly affected by the default algorithm used. As expected, there was no exposure to hypoxia in the normoxia conditions. As the area of hypoxia increased, the number of fish exposed and the exposure time increased (Fig. 10.5). NS/K algorithm group in the severe hypoxia scenario had more fish exposed for over 4 days (l, 47 fish) than did NS/RW algorithm group (j, 6 fish) or NS/CCRW algorithm group (k, 10 fish). For the severe hypoxia scenario, 83% of fish were exposed for less than a day, but there were a number of outliers with longer exposures (about 150 fish for each algorithm group). Two of the NS/K algorithm group outliers had exposures of over six days. The intermediate and severe (good avoidance) scenarios required a sample size of about 375 fish to tell the algorithm groups apart (Fig. 10.6). The mild scenario required a sample size of about 250 fish. Both of these sample sizes are very large for a study recording fish movement and hypoxia exposure. Until fish tracking technology improves, it is unlikely that the movement types described by the algorithm groups could be differentiated in field data.

The instances of longer hypoxia exposure were most likely due to spatial patterns in the hypoxic zone that occurred in the intermediate and severe hypoxia scenarios where fish were surrounded by hypoxic water on at least three sides. In the intermediate and severe hypoxia scenarios, fish tended to congregate in areas of normoxia that were either partially or entirely surrounded by hypoxic water. Fish in these areas were likely to wander back into the hypoxic zone and have higher hypoxia exposure. Because of the larger time step for the NS/K algorithm group, in narrow regions of normoxia, fish could overshoot normoxic areas multiple times and end up with large hypoxia exposures.

When avoidance is poor (simulation set 2), more fish were exposed and the three algorithm groups have similar exposures, though NS/CCRW and NS/RW algorithm groups had more fish at maximum exposure (7 days) than NS/K algorithm group. As expected, the exposures for poor avoidance did not resemble those from good avoidance (simulation set 1, Fig. 10.5). The NS/K algorithm group had fewer fish with about seven days of exposure, which was most likely due to the larger time step. Fish with a seven-day exposure were most likely stuck at local maxima that were still hypoxic and this occurred less with larger time steps. All three algorithm groups had a group of fish with no exposure, which were the fish that started outside the hypoxic area and never entered it. The NS/RW (m) and NS/CCRW (n) algorithm groups resulted in more fish being exposed for the maximum cumulative time

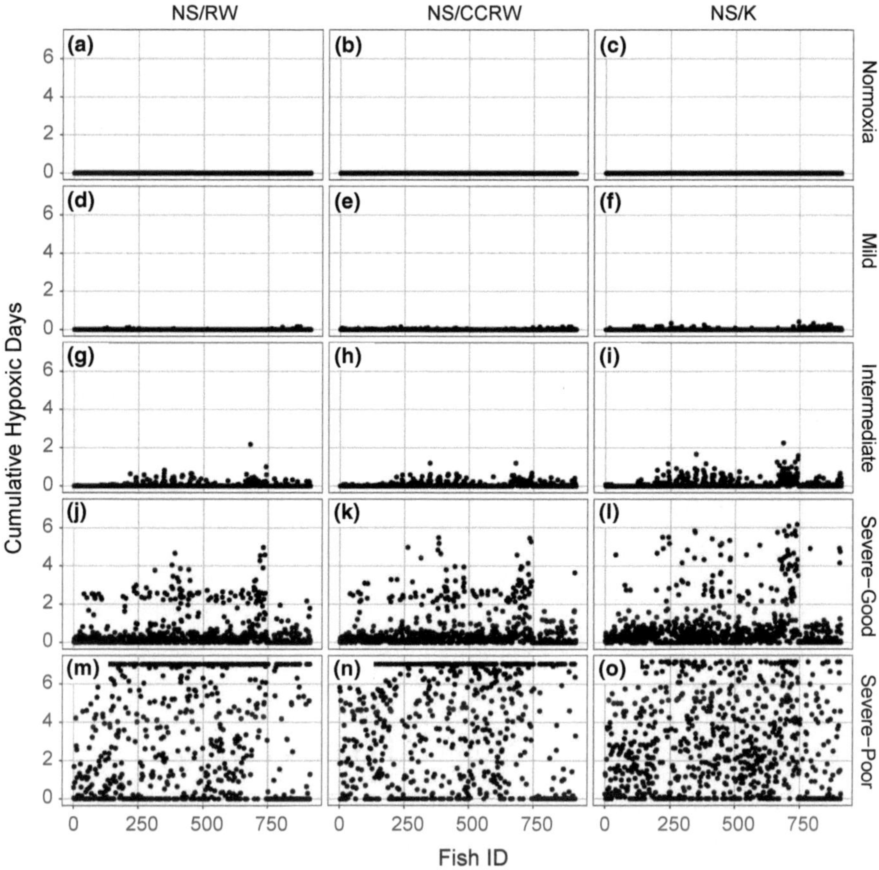

Fig. 10.5 Exposure calculated as the total number of days fish spend in hypoxic conditions plotted for each fish and each algorithm group/scenario combination. The fish are numbered by their model ID number. Algorithm groups are neighborhood search/random walk (NS/RW), neighborhood search/Cauchy correlated random walk (NS/CCRW), and neighborhood search/kinesis (NS/K). The scenarios are normoxia, mild hypoxia, intermediate hypoxia, and severe hypoxia (with good and poor hypoxia avoidance)

than NS/K algorithm group (o). The NS/CCRW algorithm group had 266 fish at maximum exposure and the NS/RW algorithm group had 273 fish, while the NS/K algorithm group had 45 fish at maximum exposure. The algorithm groups could be differentiated at a sample size of about 250 fish. This sample size was smaller than the sample size for the severe hypoxia scenario with good avoidance.

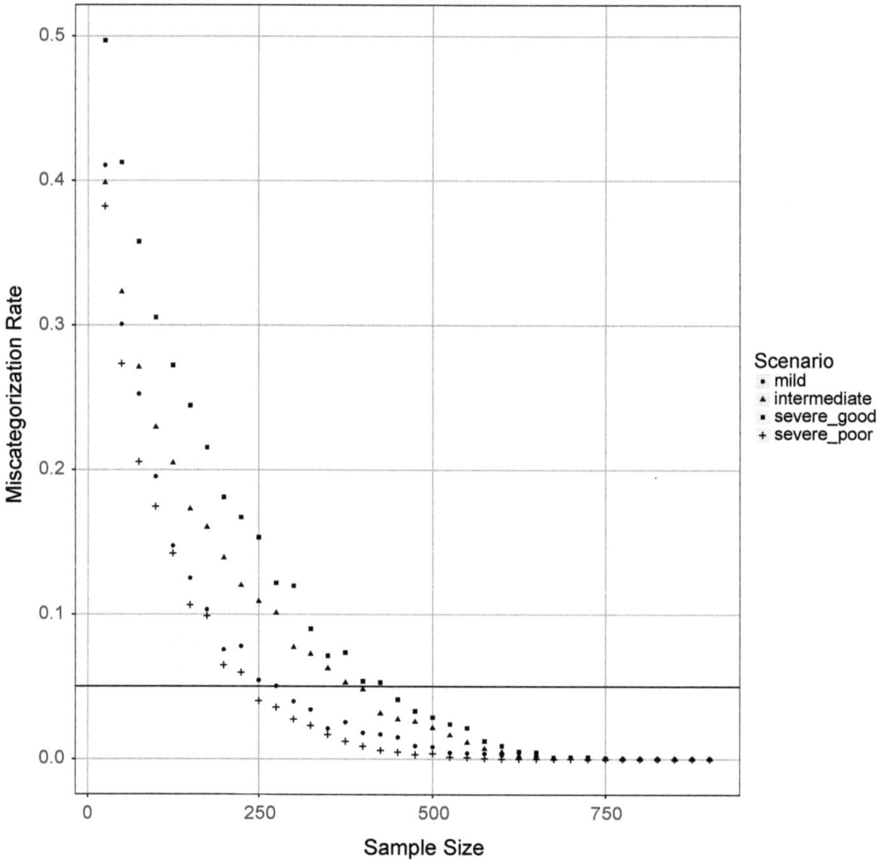

Fig. 10.6 Graph of miscategorization rate of algorithm groups plotted against sample size. The horizontal *black line* is the maximum acceptable miscategorization rate (0.05). The scenarios are mild, intermediate, and severe hypoxia. Severe hypoxia has both good and poor hypoxia avoidance

10.3.2 Distribution Spread

10.3.2.1 Maps

Fish spread further with the NS/CCRW algorithm group than with the NS/RW and NS/K algorithm groups. When looking at the final position maps for normoxia, the three algorithm groups had a different distribution. The NS/RW algorithm group fish (Fig. 10.7) diffused a short distance from the initial positions such that the initial positions are still recognizable. The NS/CCRW algorithm group fish (Fig. 10.7) spread out until the positions looked random, and the initial positions were no longer recognizable. The NS/K algorithm group fish (Fig. 10.7) looked almost exactly like the initial positions because the initial positions are the steady state for kinesis. The

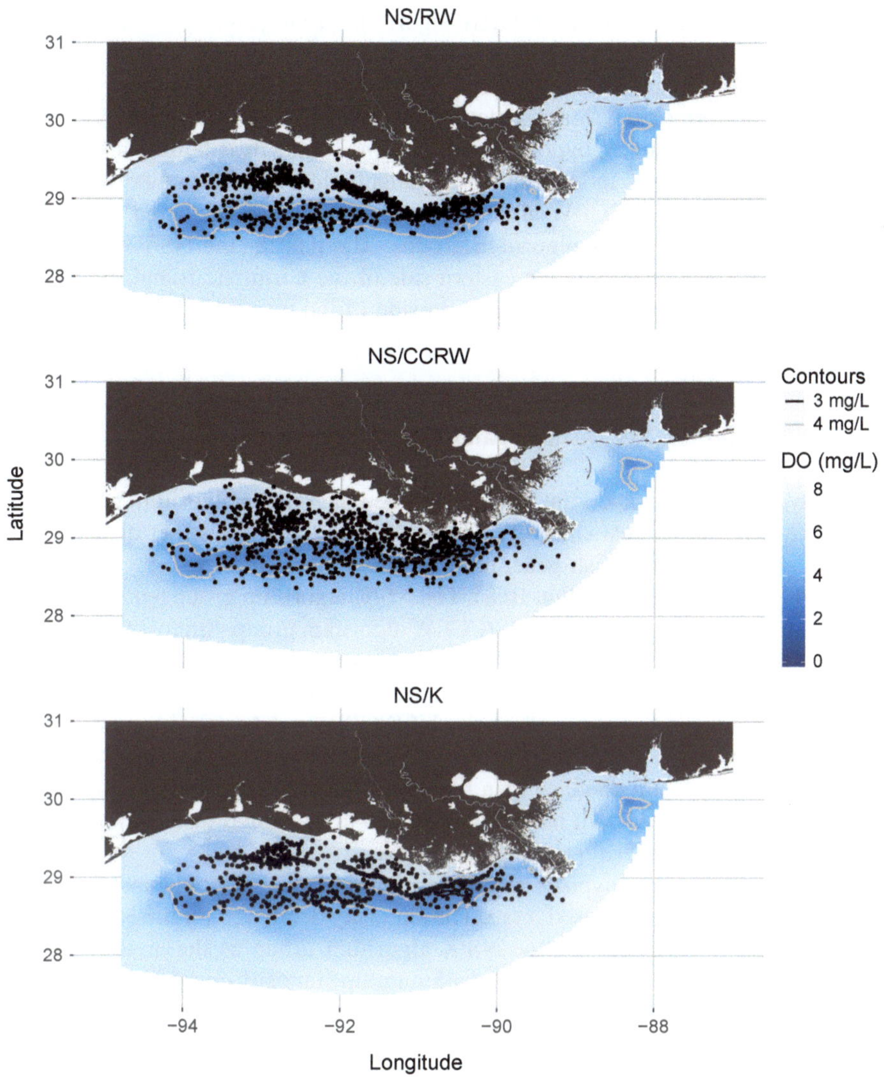

Fig. 10.7 Final fish positions for the normoxic scenario along the Northern Gulf of Mexico. Algorithm groups are neighborhood search/random walk (NS/RW), neighborhood search/Cauchy correlated random walk (NS/CCRW), and neighborhood search/kinesis (NS/K)

final position maps for mild and intermediate hypoxia had a similar distribution as the normoxia maps for the respective algorithm.

The final position maps for severe hypoxia (Fig. 10.2) show that for the NS/CCRW algorithm group fish spread more than for the other two algorithms groups and NS/K algorithm group gathered near the optimum temperature when possible. In the area outside of the hypoxic zone, fish were spread out the least with the NS/RW algorithm group and the most with the NS/CCRW algorithm group (Fig. 10.2). Fish gathered closer to the edges of the hypoxic zone with the NS/RW algorithm group and the NS/K algorithm group (Fig. 10.2) than with the NS/CCRW algorithm group. There was an area of normoxic water completely surrounded by hypoxic water (i.e., a normoxic island) for the severe hypoxia scenarios. The spatial distribution of fish within the normoxic island was influenced by the default algorithm. The fish were evenly spread out with the NS/CCRW algorithm group, while they gathered on one side with the NS/RW and NS/K algorithm groups. The fish for the NS/RW algorithm group were largely located on the eastern side of the normoxic island. In contrast, the kinesis fish gathered near the southern side of the normoxic island which was closer to their optimum temperature.

10.3.2.2 Sinuosity, Net Distance, and Total Distance

Sinuosity, net distance, and total distance differed among the algorithm groups with NS/RW having the largest sinuosity while covering the shortest net distance. The NS/RW algorithm group fish generally had larger sinuosities than the NS/CCRW and NS/K algorithm groups (Fig. 10.8). The NS/K algorithm group tended to have a slightly smaller sinuosity than the NS/CCRW algorithm group. Sinuosity was broken into its two components: the net distance from start to end and the total distance of the fish track. For net distance (Fig. 10.9), the NS/RW algorithm group fish tended to cover a smaller net distance than the NS/CCRW and NS/K algorithm groups. The opposite was true for the total distance covered, with the NS/K algorithm group covering a smaller total distance than the other two algorithm groups (Fig. 10.10). NS/CCRW and NS/RW both had similar and relatively small distributions of total distance. The NS/K algorithm group had a larger distribution.

Increasing the area of hypoxia affected sinuosity and its components because NS and sprint algorithms were used more. The sinuosity decreased for the NS/RW algorithm group and increased for the NS/K algorithm group as the area of hypoxia increased (Fig. 10.8). The NS/RW algorithm group sinuosity decreased because the net distance covered by the fish increased. Because fish travel in a mostly straight line using NS and sprint algorithms as opposed to the many turns taken by RW, increasing usage of NS and sprint algorithms increases the net distance. The sinuosity for the NS/K algorithm group increased because the total distance covered increased. The distance travelled for each time step depends on the temperature and the weighting of the random and previous components. The closer to the optimum temperature the fish are, the slower they move. By adding NS and sprint algorithms, which have set distances covered each time step, the total distance covered is increased. NS and especially sprint also contributed to the increase in the number of outliers for net and total distance as the area of hypoxia increased. There was little change for the sinuosity or its components for the NS/CCRW algorithm group as the area of hypoxia increased.

The simulation set 2 fish had larger distribution ranges for sinuosity and total distance than the simulation set 1 fish, but generally followed the same trends as

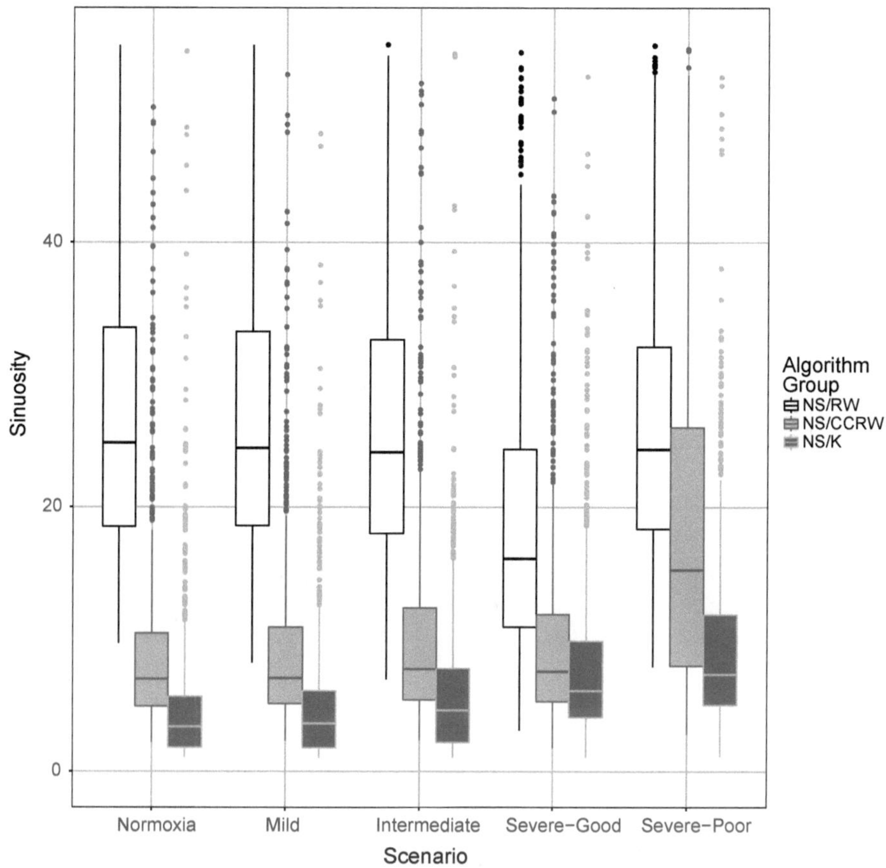

Fig. 10.8 Sinuosity (amount of wiggle) of the fish tracks. *Black dots* denote outliers. Algorithm groups are neighborhood search/random walk (NS/RW), neighborhood search/Cauchy correlated random walk (NS/CCRW), and neighborhood search/kinesis (NS/K). The scenarios are normoxia, mild hypoxia, intermediate hypoxia, and severe hypoxia (with good and poor hypoxia avoidance)

simulation set 1 fish for the respective algorithm groups. For sinuosity, the NS/RW algorithm group had the highest value and had a median and range similar to the normoxia, mild hypoxia, and intermediate hypoxia scenarios (Fig. 10.8). Because there was no sprint and NS had a large degree of randomness, the sinuosity did not decrease with a larger hypoxic area. For the NS/CCRW algorithm group, sinuosity for poor avoidance was larger than sinuosity for good avoidance. For the NS/K algorithm group, sinuosity was only a little larger, by about one. For net distance, the distribution ranges for poor avoidance were not much greater than those for good avoidance (Fig. 10.9). The poor avoidance distribution ranges were between the sizes of the good avoidance for the severe hypoxia scenario and the other three scenarios. The net distance for the NS/K algorithm group was larger than the net distance for

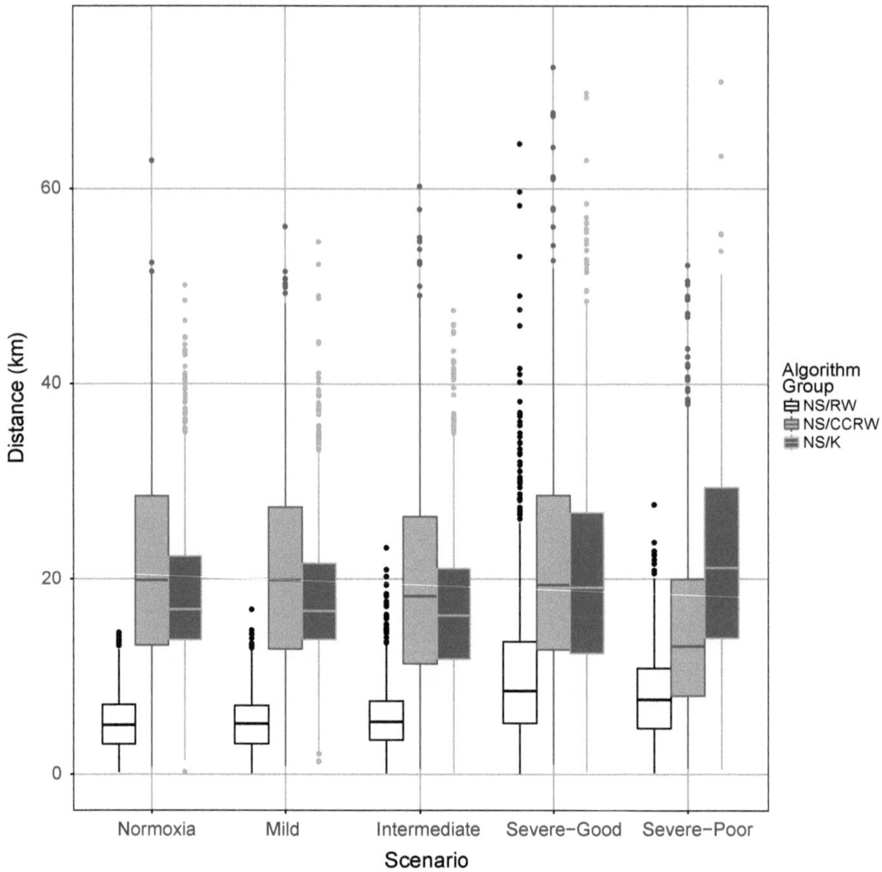

Fig. 10.9 The net, or shortest, distance between the starting and ending positions of the fish. *Black dots* denote outliers. Algorithm groups are neighborhood search/random walk (NS/RW), neighborhood search/Cauchy correlated random walk (NS/CCRW), and neighborhood search/kinesis (NS/K). The scenarios are normoxia, mild hypoxia, intermediate hypoxia, and severe hypoxia (with good and poor hypoxia avoidance)

the NS/CCRW algorithm group, while for good avoidance they were either similar or the net distance for the NS/CCRW algorithm group was greater. For total distance, the distribution ranges were much larger for the NS/RW and NS/CCRW algorithm groups (Fig. 10.10). For the NS/K algorithm group, the total distance was similar to the values for the normoxia, mild hypoxia, and intermediate hypoxia scenarios. As with good avoidance, for the NS/K algorithm group, the total distance was smaller than the total distance for the other two algorithm groups.

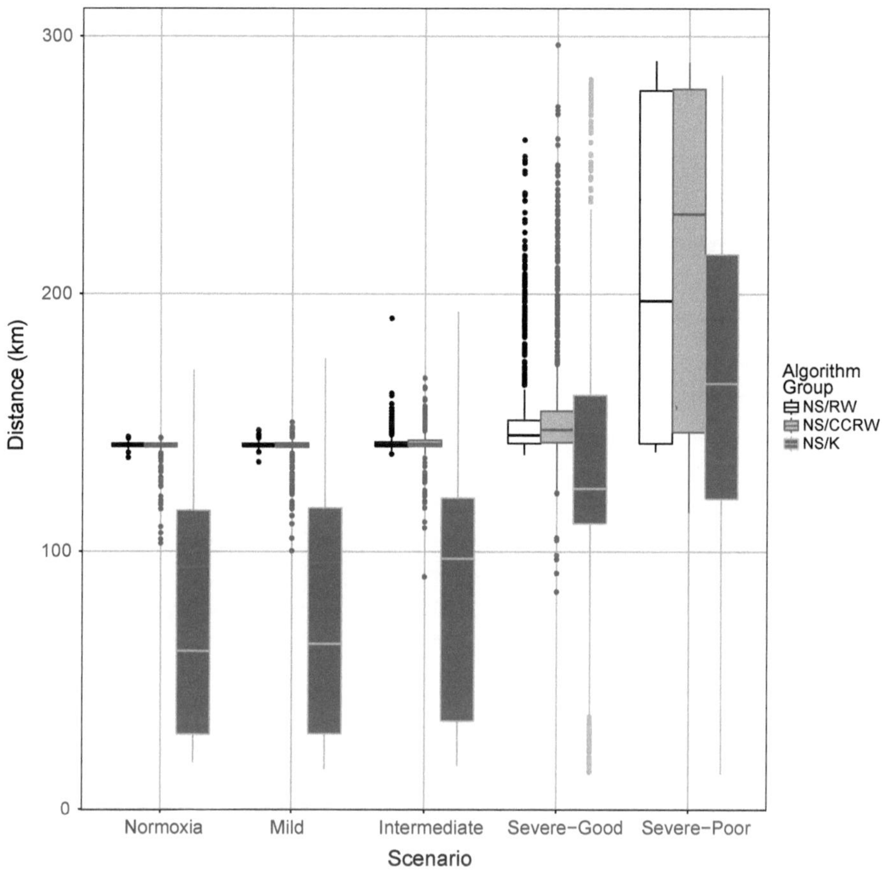

Fig. 10.10 Total distance covered by the fish. *Black dots* denote outliers. Algorithm groups are neighborhood search/random walk (NS/RW), neighborhood search/Cauchy correlated random walk (NS/CCRW), and neighborhood search/kinesis (NS/K). The scenarios are normoxia, mild hypoxia, intermediate hypoxia, and severe hypoxia (with good and poor hypoxia avoidance)

10.3.3 Temperature

The temperatures experienced by the fish depended on the default algorithm used, with a greater percentage of fish near the optimum temperature for NS/K algorithm group, except for the severe hypoxia scenario. The percentage of fish in the 25–27 °C range (solid purple line), which brackets the optimum temperature of 26 °C, decreased over the course of the seven days for NS/RW and NS/CCRW algorithm groups (Fig. 10.11). The 25–27 °C range was added to see how the percentage of fish around the optimum temperature changes. The percentage of fish in the 25–27 °C range for the NS/RW algorithm group decreased less than for the NS/CCRW algorithm group because the NS/RW algorithm group fish do not travel as far. For

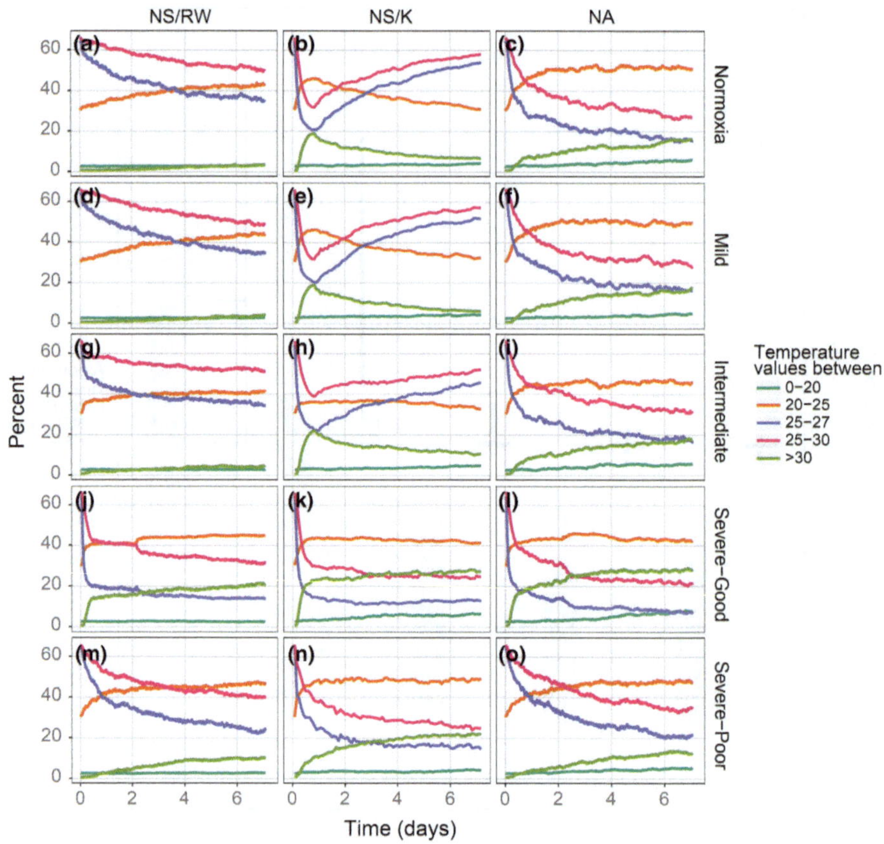

Fig. 10.11 Percentage of fish in temperature ranges over time for each algorithm group/scenario combination. The optimum temperature is 26 °C. Algorithm groups are neighborhood search/random walk (NS/RW), neighborhood search/Cauchy correlated random walk (NS/CCRW), and neighborhood search/kinesis (NS/K). The scenarios are normoxia, mild hypoxia, intermediate hypoxia, and severe hypoxia (with good and poor hypoxia avoidance)

the NS/K algorithm group, the percentage of fish in the 25–27 °C range increased for the normoxia (a–c), mild hypoxia (d–f), and intermediate hypoxia (g–i) scenarios. For the severe hypoxia scenario (j–l), the percentages for all three algorithm groups looked similar. For example, while the 25–27 °C range was over 50% by the end of the run for the less severe hypoxic scenarios, the percentage was less than 20 for the severe hypoxia scenario. The hypoxic zone overlaps with the 26 °C contour, so the NS/K algorithm group fish were not able to congregate near the contour line. The percentage of fish in the >30 °C range (dotted dark gray line) tended to increase for NS/CCRW algorithm group but not for the other algorithm groups, most likely because the NS/CCRW algorithm group fish spread out the most.

For simulation set 2, the percentage of fish in different temperature ranges was similar to the severe hypoxia scenario for simulation set 1. For all three algorithm groups (m–o), the percentage of fish in the 25–27 °C range decreased over the seven days (Fig. 10.11). The major difference between the poor and good avoidance for severe hypoxia scenarios was that there was a smaller percentage of fish in the >30 °C range for the poor avoidance (m–o) than for the good avoidance (j–l). This was probably due to the fact that many fish never left the hypoxic zone and did not have the chance to disperse away from it. The NS/K algorithm group (o) had the highest percentage of fish in the >30 °C range, but the percentage of fish was still lower than in case of good avoidance (l).

10.3.4 Growth and Vitality

There were only small difference in weights (<1 g) or growth (<0.05 g/day) due to default algorithm or scenario (Fig. 10.12). There was a small difference between different scenarios, but even the severe scenario (j–l) had less than 1 g difference in weight due to low DO exposure for most of the fish. The growth rates differed by less than 0.05 g/day among scenarios and algorithm groups. When extrapolated out to 30 days, the weights between algorithm groups covered the same numerical range. About 83% of fish were exposed for less than a day, which is probably why the effects on growth were so small.

The final weights for the poor avoidance set covered a similar range of values for all algorithm groups within that set and also a larger range compared to the good avoidance set. The poor avoidance runs had similar distributions of final weights with slight differences. The range for poor avoidance weights ranged from 90 g for the fish least affected by hypoxia to just below 87 g for the fish most affected (Fig. 10.12). For good avoidance, only less than 20 fish had weights below 89.5 g.

10.4 Discussion

10.4.1 Avoidance and Default Behaviors

The model results suggest that avoidance and default behaviors can be separated and evaluated independently. For example, avoidance can be modeled without having to consider the default algorithm because it did not affect the hypoxia exposure. The default behavior can be chosen based on its characteristics without considering any interaction between the avoidance and default behaviors. Importantly, the avoidance and default algorithms could be validated and calibrated separately. There is not a large amount of field data for validating and calibrating fish tracks from IBMs, so being able to use data even if all the environmental conditions used in a model are not recorded in the field data is essential.

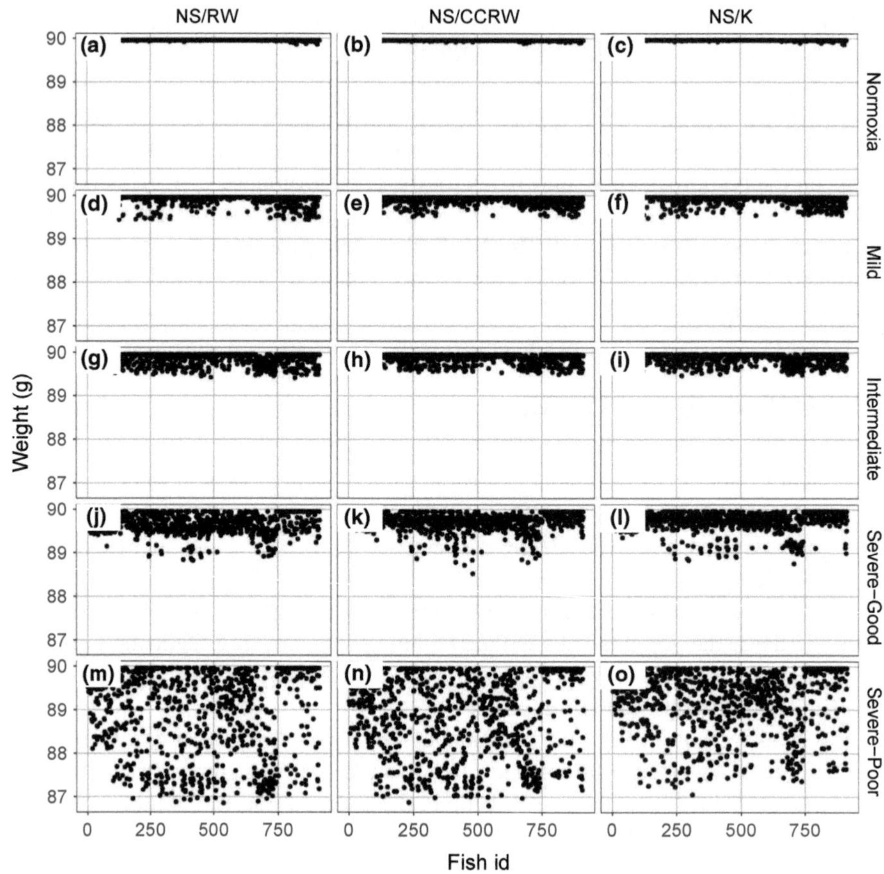

Fig. 10.12 Final weights plotted for each fish by model id number. Algorithm groups are neighborhood search/random walk (NS/RW), neighborhood search/Cauchy correlated random walk (NS/CCRW), and neighborhood search/kinesis (NS/K). The scenarios are normoxia, mild hypoxia, intermediate hypoxia, and severe hypoxia (with good and poor hypoxia avoidance)

10.4.2 Exposure

The default algorithms were found not to affect hypoxia exposure for the model fish, but the tactical behavior did affect exposure. It makes sense that the tactical behavior should have an important effect on hypoxia exposure because the tactical behavior was responsible for having the fish avoid conditions such as hypoxia. What differences there were between the default algorithms were shown when the effectiveness of the tactical behavior was changed from good to poor avoidance (Fig. 10.5). Poor avoidance magnified the differences between the default algorithms. The difference between good and poor avoidance can also be seen in the smaller sample size required to differentiate between the algorithm groups (Fig. 10.6). The difference was most likely due to the different time steps used for the random walk algorithm groups and the kinesis algorithm group. The time step used was chosen because of the characteristics of the default algorithm, but the exposure differences mostly occurred because of that time step's effect on the tactical behavior. So even when there was a difference in exposure between algorithm groups, it was generally due to the tactical behavior and not the default behavior.

The hypoxia conditions and "local" conditions had a greater effect on hypoxia exposure than the default algorithm. There was an obvious increase in exposure with increased hypoxic area. More fish were exposed in larger areas and took longer to escape the hypoxic area. There was also the increased chance of fish getting stuck and having to rely on the sprint algorithm to escape the hypoxic zone. The different scenarios also had different smaller scale or "local" features that affected hypoxia exposure. The severe scenario in particular had several features such as the normoxic island and the normoxic sliver that affected exposure. In particular, the normoxic sliver, i.e., the thin area of normoxia surrounded by hypoxia on three sides in the severe scenario, was a relatively small area for fish to inhabit. Because all the default algorithms diffuse to some extent, the fish in a small normoxic area tended to spread out into the hypoxic area. Most of these fish then used the tactical behavior and returned to the normoxic sliver. Thin hypoxic areas such as the normoxic sliver feature led to increased hypoxia exposure. Fish are known to gather at the shoal that corresponds to the normoxic sliver (Craig 2012), so whether real fish continually wander back into the hypoxic area or stay on the shoal without getting exposed would be an interesting question to explore. It would help with making the model better reflect reality and increase knowledge of the behavior of the fish.

10.4.3 Algorithm Comparison

The choice of the default algorithm did not make a significant difference in the hypoxia exposure of the fish or changes in growth due to hypoxia exposure for static conditions, so other characteristics of the algorithm must be considered when choosing which algorithm to use. As shown in the results, the temperatures experienced by the fish and the distance covered by the fish were different among the default algorithms. The time step that the algorithm can be used for is also important to consider.

The main way that the default algorithms differed was sinuosity and temperature sensitivity. There were only small differences in exposure between the default algorithms, whether avoidance was good or bad. The area covered by the fish was the major difference among the algorithms (Fig. 10.8). Fish using the RW algorithm moved only a short distance in very convoluted paths while fish using the CCRW algorithm spread out the furthest. Fish using the kinesis algorithm tended to gather along the 26 °C contour where possible, so these fish travelled farther than the RW using fish but did not spread out like the CCRW using fish. Kinesis was the only algorithm that took temperature into account and so was the only algorithm that affected temperature-related movement in the non-severe hypoxia scenarios (Fig. 10.11). The RW algorithm had more fish near the optimum temperature than the CCRW algorithm, but this was due to the RW fish not diffusing far from the starting positions along the 26 °C contour.

The time step used can limit what algorithms are effective. Kinesis is not very effective at time steps smaller than an hour on large grid cells because the fish are unable to find the preferred temperature. The CCRW and RW algorithms work at smaller time steps. While the RW algorithm works at smaller time steps, fish clearly move a shorter distance using that algorithm than with kinesis or CCRW as shown in the results. For greater diffusion, a large time step would need to be used with RW. Because CCRW is a more directional algorithm, a larger distance is covered when using that algorithm even at smaller time steps.

Selection of an appropriate time step depends on the system and the organism. Generally, a larger time step is used for slow moving particles and for large areas while a smaller time step would be used for faster moving particles and small areas. Physical variables are typically evaluated on the time frame of seconds, even for large areas. For example, the FVCOM-WASP model calculates physical and environmental data at time steps of 5 s for grid cells that can be 10 km in size (Justić and Wang 2014). For fish, the time step often depends on the grid size used. Time steps and grid sizes that have been used to model fish movement include 1-km cells with a time step of 1 hr (Creekmore 2011), 20–500 m cells with a 9-s time step (Rose et al. 2014), and <1 m with a 2-s time step (Goodwin et al. 2006). Field data could be used to determine whether croaker movement in response to hypoxia in the GOM can be modeled in sufficient detail with a 2-h time step.

The distance covered by the model fish depends on the default algorithm and the time step used. The area that a random walk algorithm can cover is limited by the time step. The average radial distance from the starting to ending point for a random walk is $\sqrt{N}$, where N is the number of steps (Landau et al. 2013). For a fixed time period, the number of steps is inversely proportional to the time step. So the number of steps is greater for the 900 s time step than for the 2-h time step. This is why for RW the fish do not travel very far and why kinesis does not work well at the shorter time step. Kinesis is the combination of the previous direction and a random walk, so the radial distance covered by kinesis is affected by the time step. The fish using kinesis do not cover a large enough area with a 900-s time step to find the preferred temperature. The CCRW algorithm is not a uniform random walk because of the use of the Cauchy distribution when choosing angles. CCRW is the only default algorithm where fish cover a large distance at the smaller time step, with the distance at 900 s for CCRW being comparable to the distance covered using kinesis for a 2-h time step. If smaller time steps are a better fit for croaker movement, the CCRW movement algorithm would be the preferred default algorithm among those compared in this paper.

In systems where multiple environmental conditions affect fish movement and distribution, being able to model fish responses to multiple conditions or stressors is useful. It is more realistic for fish to be affected by multiple stressors or conditions. Temperature and distance from shore affect the distribution of croaker in the GOM

(Craig 2012). In tidal systems, salinity and tides are important environmental conditions that can affect fish movement in conjunction with hypoxia (Brady and Targett 2013). Multiple conditions or stressors can be modeled using an event-based algorithm to choose between different algorithms depending on the condition or stressor. Event-based algorithm is useful for conditions or stressors that have clear cut offs or thresholds where the fish is either affected or not affected. For conditions or stressors that are always present and have no clear cut off or threshold, it is more useful to use a default algorithm that moves fish based on the condition or stressor. The ability to take environmental conditions, such as temperature, into account is a strength of kinesis as a default algorithm.

Determining the most appropriate default algorithm for a particular model and organism requires weighing the advantages and disadvantages of the different algorithms. For GOM croaker, more information is needed to determine which of the algorithms covered in this paper is most appropriate. Looking for additional algorithms to use as the default is also worth considering. The CCRW algorithm was adapted to work with the FVCOM particle-tracking module as an alternative to kinesis and RW. CCRW works at smaller time steps while covering a larger distance than RW. This is useful in situations where diffusion-type movement is required, but the grid cells are too large and the time step is too small for RW to sufficiently diffuse from the starting position.

10.4.4 Real Versus Model

The lack of appropriate data makes it difficult to calibrate or validate IBMs for hypoxia avoidance in the GOM. Most available data for fish locations in the GOM are from trawls or other group sampling methods that only show where fish are at one point in time. Such data sets can be compared to the general locations of fish in the model. Fish have been shown to concentrate on a shoal (Craig 2012) that corresponds to the normoxic sliver where fish gather in the severe hypoxic scenario. But for comparing fish tracks and hypoxia exposure, data sets need to have detailed information on fish movement over time and not just general locations or locations for one time step. Fish can be tracked using acoustic tags, but this does not guarantee that the horizontal movement is recorded with enough detail.

Existing data sets are either in areas that are difficult to compare to the GOM or are not detailed enough. A data set for croaker in the GOM was part of a study looking at the vertical movement of croaker in response to hypoxia. Because horizontal movement was not part of the study, it was not recorded in detail. The horizontal locations of the fish can be reconstructed after the fact from the strength of the acoustic tag signals (Grothues et al. 2013), but the resulting data have confidence intervals too large to be used to compare to the model. Two data sets that are detailed either temporally (Pepper Creek, DE, Brady and Targett 2013) or spatially (Neuse River, NC, Kevin Craig personal communication) are from shallow tidal areas with diurnal DO changes. The area, depth, and DO cycle are very different between the tidal areas and the GOM. Pepper Creek is a narrow tidal creek where the fish movement was recorded as one dimensional (Brady and Targett 2013). Neuse River is larger than Pepper Creek, but still much smaller than the GOM. Fish tracks were recorded in two dimensions at Neuse River. The tides also have a greater effect on the tidal areas, which are located on the Atlantic coast, as compared to the GOM shelf where the tides are weak. All of these differences make it difficult to compare fish behavior in the tidal areas to fish behavior in the GOM.

Despite the differences in the areas, there are still some ways that the field data can be used. The most important information that can be taken from the three existing field data sets is that fish avoid hypoxia and can survive brief exposures to hypoxic water (Brady and Targett 2013). The DO values experienced by the GOM croaker were interpolated from CTD casts near the fish location. None of the croaker were found in hypoxic conditions. This suggests that the fish were able to successfully avoid hypoxia, keeping in mind the stress due to tagging that may have affected behavior and that hypoxia was weak where the fish were tracked (Thomas Grothues personal communication). The Pepper Creek and Neuse River had fish that were exposed to hypoxia and then moved out of the hypoxic area and continued to move, indicating the exposure was sublethal. The fact that non-fatal exposure occurs means that direct effects other than mortality and indirect effects are possible. As previously mentioned, sublethal effects of low DO exposure have been found in croaker (Thomas and Rahman 2009). Even when field data cannot be directly compared to the model output, the field data can demonstrate that the processes described in the model exist in the field.

10.4.5 Impacts

The work in this paper is important because it builds on what has been done with movement algorithms and the models used. Similar movement algorithms have been used with different grids (Creekmore 2011), for different locations with different stressors (Rose et al. 2014), or in abstracted grids (Watkins and Rose 2013). This paper represents the first attempt to use the movement algorithms with a grid based on a snapshot of the FVCOM-WASP output. The CCRW algorithm was also added to the list of algorithms used before with the FVCOM-tracking module with the FVCOM-WASP output. The CCRW algorithm adds another option to the default algorithms that is not as limited by time step. There was also an increased understanding of how different algorithms behave as default algorithms in an event-based setup compared to each other. A better understanding of how algorithms behave allows for choosing the more appropriate algorithm for a given model and scenario.

Modeling movement helps to study the impacts of environmental conditions and stressors on organisms such as fish, especially when field data are hard to obtain. As mentioned in Sect. 10.4.4, there is limited field data for the interactions of croaker and the GOM hypoxic zone. There is particularly limited data for movement of individual croaker relative to the hypoxic zone. A model helps to look at the impacts of the hypoxic zone using what information there is on croaker and hypoxia. The use of a model does not negate the need for field data, but it can help to fill in gaps where field data are hard to obtain or absent.

Appropriate models can be used to help inform management decisions. Choosing the appropriate model and movement algorithms requires knowledge of the system, possible models, and possible movement algorithms. Increasing knowledge of the behaviors of different movement algorithms helps with the decision of choosing the appropriate model. Field data and a validated model can be used for management decisions. An example is a fish movement model used to model fish movement around various objects near a hydroelectric dam (Goodwin et al. 2006). If the necessary field data were collected for calibration and validation, the model in this paper could be used to inform decisions that involve the GOM hypoxic zone and fisheries.

10.5 Conclusion

This paper tested the hypothesis that the default algorithm does not affect the hypoxia exposure for model fish. The results of running three algorithm groups for four scenarios based on output from the FVCOM-WASP hydrodynamic-water quality model suggest that the default algorithm does not affect hypoxia exposure in static conditions. While exposure is not affected by the default algorithm, sinuosity and its components, total and net distance, are affected by the default algorithm used. For static conditions, the factors that should be considered when choosing a default algorithm are dispersion, time step, and environmental conditions other than DO.

For the results and conclusions of this paper to be applied to dynamic conditions or 3-D conditions and not just static 2-D conditions, the algorithms must be tested for dynamic conditions and 3-D conditions. The algorithms were tested for static 2-D conditions, which do not reflect reality very well. Consequently, the results of the paper increase the knowledge of how these algorithms perform; the results may not apply to dynamic or 3-D conditions. To better reflect reality and be more useful for management purposes, the properties of the algorithms need to be compared in dynamic conditions and in 3-D.

Validated models can help inform management decisions when field data is difficult or impossible to obtain. Models are particularly useful in situations where field data are difficult to obtain (e.g., exposure time series for large numbers of fish) or impossible to obtain (e.g., effects of future climate change). Managers can use results from a range of model scenarios to decide between different management actions. To ensure the model scenarios reflect reality as closely as possible, it is important to understand the model and to have field data for calibration and validation. By testing how the model reacts under different conditions helps the modeler to better understand the behavior of the model. Investigating how the default algorithms affect exposure helps to better understand the model for hypoxia avoidance for fish. Appropriate field data allow for calibration and validation, which help insure the model is close enough to reality. There is a need for appropriate field data for the model described in this paper. Obtaining detailed movement tracks and hypoxia time series for relatively small fish such as croaker is difficult, but worth doing to improve the model and its usefulness. Developing methods for obtaining movement and hypoxia exposure time series for croaker is also important because it establishes methods for successfully obtaining the data which could be applied to economically important species such as red snapper or shrimp.

Acknowledgements E.D. LaBone was supported by the NSF Graduate Research Fellowships Program and the Louisiana Board of Regents 8g Fellowship. The project was also funded in part by the NOAA/CSCOR Northern Gulf of Mexico Ecosystems and Hypoxia Assessment Program under award NA09NOS4780230 to Louisiana State University. This is publication number 221 of the NOAA's CSCOR NGOMEX and CHRP programs. I am grateful to Damian Brady, Kevin Craig, and Thomas Grothues for providing fish movement data for Pepper Creek, Neuse River, and the GOM, respectively. I also am thankful to Sean Creekmore for providing advice and the values for the croaker growth calculation. I would also like to thank Thomas LaBone for advice on statistics.

References

Anderson JJ (2002) An agent-based event driven foraging model. Nat Resour Model 15(1):55–82

Barger LE (1985) Age and growth of Atlantic croakers in the northern Gulf of Mexico, based on otolith sections. Trans Am Fish Soc 114(6):847–850

Batschelet E (1981) Circular statistics in biology. Academic, New York

Baustian MM, Craig JK, Rabalais NN (2009) Effects of summer 2003 hypoxia on macrobenthos and Atlantic croaker foraging selectivity in the northern Gulf of Mexico. J Exper Mar Biol Ecol 381:S31–S37

Brady DC, Targett TE (2013) Movement of juvenile weakfish Cynoscion regalis and spot Leiostomus xanthurus in relation to diel-cycling hypoxia in an estuarine tidal tributary. Mar Ecol Prog Ser 491:199–219. http://www.int-res.com/abstracts/meps/v491/p199-219/

Breitburg D (2002) Effects of hypoxia, and the balance between hypoxia and enrichment, on coastal fishes and fisheries. Estuaries Coasts 25(4):767–781. doi:10.1007/BF02804904. http://www.springerlink.com.libezp.lib.lsu.edu/content/p2306p145n47651l/abstract/

Chen C, Beardsley RC, Cowles G (2006) An unstructured grid, finite-volume coastal ocean model: FVCOM user manual, 2nd edn. Marine ecosystem dynamic modeling laboratory. http://woodshole.er.usgs.gov/staffpages/rsignell/rps/share/FVCOM_Manual.pdf

Craig JK (2012) Aggregation on the edge: effects of hypoxia avoidance on the spatial distribution of brown shrimp and demersal fishes in the Northern Gulf of Mexico. Mar Ecol Prog Ser 445:75–95. http://www.int-res.com/abstracts/meps/v445/p75-95/

Craig JK, Bosman SH (2012) Small spatial scale variation in fish assemblage structure in the vicinity of the Northwestern Gulf of Mexico Hypoxic zone. Estuaries Coasts 1–18. doi:10.1007/s12237-012-9577-9

Creekmore SB (2011) Modeling the population effects of Hypoxia on Atlantic croaker (micropogonias undulatus) in the Northwestern Gulf of Mexico. Master of Science, Louisiana State University

Eby LA, Crowder LB (2002) Hypoxia-based habitat compression in the Neuse River Estuary: context-dependent shifts in behavioral avoidance thresholds. Can J Fish Aquat Sci 59(6):952–965. doi:10.1139/f02-067

Fennel K, Hu J, Laurent A, Marta-Almeida M, Hetland R (2013) Sensitivity of hypoxia predictions for the northern Gulf of Mexico to sediment oxygen consumption and model nesting. J Geophys Res Oceans 118(2):990–1002. doi:10.1002/jgrc.20077

Fraenkel GS, Gunn DL (1961) The orientation of animals: kineses, taxes and compass reactions

Goodwin RA, Nestler JM, Anderson JJ, Weber LJ, Loucks DP (2006) Forecasting 3-D fish movement behavior using a Eulerian-Lagrangian-agent method (ELAM). Ecol Model 192(1–2):197–223

Grothues TM, Davis WC, Kraft C (2013) Sound pressure level weighting of the center of activity method to approximate sequential fish positions from acoustic telemetry. Can J Fish Aquat Sci 70(9):1359–1371. doi:10.1139/cjfas-2013-0056

Hetland RD, DiMarco SF (2008) How does the character of oxygen demand control the structure of hypoxia on the Texas–Louisiana continental shelf? J Mar Syst 70(1):49–62. http://www.sciencedirect.com/science/article/pii/S0924796307000711

Humston R, Olson DB, Ault JS (2004) Behavioral assumptions in models of fish movement and their influence on population dynamics. Trans Am Fish Soc 133(6):1304–1328

Justić D, Wang L (2014) Assessing temporal and spatial variability of hypoxia over the inner Louisiana–upper Texas shelf: application of an unstructured-grid three-dimensional coupled hydrodynamic-water quality model. Cont Shelf Res 72:163–179. http://www.sciencedirect.com/science/article/pii/S0278434313002665

Justic D, Rabalais NN, Eugene Turner R, Wiseman WJ Jr (1993) Seasonal coupling between riverborne nutrients, net productivity and hypoxia. Mar Pollut Bull 26(4):184–189

Justic D, Rabalais NN, Turner RE (1996) Effects of climate change on hypoxia in coastal waters: a doubled CO2 scenario for the northern Gulf of Mexico. Limnol Oceanogr 992–1003

Justic D, Rabalais NN, Turner RE (2002) Modeling the impacts of decadal changes in riverine nutrient fluxes on coastal eutrophication near the Mississippi River Delta. Ecol Model 152(1):33–46

Justic D, Bierman VJ, Scavia D, Hetland RD (2007) Forecasting Gulf's hypoxia: the next 50 years? Estuaries Coasts 30(5):791–801

Kareiva PM, Shigesada N (1983) Analyzing insect movement as a correlated random walk. Oecologia 56(2-3):234–238. doi:10.1007/BF00379695

Landau R, Kang N, Mulder G, Holmes R, Panoff B, Borinskaya S, Bordeianu C, Paez M (2013) Random walks. http://www.science.oregonstate.edu/~rubin/INSTANCES/WebModules/5_RandomWalk/RandomWalkFiles/Pdfs/RandomWalk.pdf

Ludsin SA, Zhang X, Brandt SB, Roman MR, Boicourt WC, Mason DM, Costantini M (2009) Hypoxia-avoidance by planktivorous fish in Chesapeake Bay: Implications for food web interactions and fish recruitment. J Exp Mar Biol Ecol 381(Supplement):S121–S131. doi:10.1016/j.jembe.2009.07.016. http://www.sciencedirect.com/science/article/pii/S0022098109002846

Marsh LM, Jones RE (1988) The form and consequences of random walk movement models. J Theoret Biol 133(1):113–131. http://www.sciencedirect.com/science/article/pii/S0022519388800286

Miller Neilan R, Rose K (2014) Simulating the effects of fluctuating dissolved oxygen on growth, reproduction, and survival of fish and shrimp. J Theoret Biol 343:54–68. doi:10.1016/j.jtbi.2013.11.004. http://www.sciencedirect.com/science/article/pii/S0022519313005298

Obenour DR, Scavia D, Rabalais NN, Turner RE, Michalak AM (2013) Retrospective analysis of midsummer hypoxic area and volume in the northern Gulf of Mexico, 1985–2011. Environ Sci Technol 47(17):9808–9815. doi:10.1021/es400983g

R Core Team (2013) R: a language and environment for statistical computing. R Foundation for Statistical Computing, Vienna, Austria. http://www.R-project.org/

Rabalais NN, Turner RE, Wiseman WJ Jr (2001) Hypoxia in the Gulf of Mexico. J Environ Qual 30(2):320–329

Rabalais NN, Turner RE, Wiseman WJ Jr (2002) Gulf of Mexico Hypoxia, aka "The Dead Zone". Ann Rev Ecol Syst 33:235–263

Rabalais NN, Turner RE, Sen Gupta BK, Boesch DF, Chapman P, Murrell MC (2007) Hypoxia in the northern Gulf of Mexico: Does the science support the plan to reduce, mitigate, and control hypoxia? Estuaries Coasts 30(5):753–772

Rose KA, Adamack AT, Murphy CA, Sable SE, Kolesar SE, Craig JK, Breitburg DL, Thomas P, Brouwer MH, Cerco CF et al (2009) Does hypoxia have population-level effects on coastal fish? Musings from the virtual world. J Exp Mar Biol Ecol 381:S188–S203

Rose KA, Huang H, Justic D, de Mutsert K (2014) Simulating fish movement responses to and potential salinity stress from large-scale river diversions. Mar Coast Fish 6(1):43–61. doi:10.1080/19425120.2013.866999

Scavia D, Rabalais NN, Turner RE, Justić D, Wiseman WJ Jr (2003) Predicting the response of Gulf of Mexico hypoxia to variations in Mississippi River nitrogen load. Limnol Oceanogr 48(3):951–956

Scavia D, Justic D, Bierman VJ (2004) Reducing hypoxia in the Gulf of Mexico: advice from three models. Estuaries Coasts 27(3):419–425

Thomas P, Rahman MS (2009) Chronic hypoxia impairs gamete maturation in Atlantic croaker induced by progestins through nongenomic mechanisms resulting in reduced reproductive success. Environ Sci Technol 43(11):4175–4180

Turner RE, Rabalais NN (1991) Changes in Mississippi River water quality this century. BioScience 140–147

Wang L, Justic D (2009) A modeling study of the physical processes affecting the development of seasonal hypoxia over the inner Louisiana-Texas shelf: Circulation and stratification. Cont Shelf Res 29(11–12):1464–1476

Watkins KS, Rose KA (2013) Evaluating the performance of individual-based animal movement models in novel environments. Ecol Model 250:214–234. http://www.sciencedirect.com/science/article/pii/S0304380012005480

White JW, Rassweiler A, Samhouri JF, Stier AC, White C (2014) Ecologists should not use statistical significance tests to interpret simulation model results. Oikos 123(4):385–388. doi:10.1111/j.1600-0706.2013.01073.x/full

Wool TA, Ambrose RB, Martin JL, Comer EA, Tech T (2006) Water quality analysis simulation program (WASP). User's manual, Version 6. http://env1.kangwon.ac.kr/CyberClass/grad/grese/2012/models/wasp/wasp6/WASP6_Manual.pdf

Wu Hi, Li BL, Springer TA, Neill WH (2000) Modelling animal movement as a persistent random walk in two dimensions: expected magnitude of net displacement. Ecol Model 132(1):115–124. http://www.sciencedirect.com/science/article/pii/S0304380000003094

Chapter 11
Hypoxia Effects Within an Intra-guild Predation Food Web of *Mnemiopsis leidyi* Ctenophores, Larval Fish, and Copepods

Sarah E. Kolesar, Kenneth A. Rose and Denise L. Breitburg

Abstract Differences in predator and prey tolerances to abiotic factors, such as seasonal low dissolved oxygen (DO) concentrations in estuarine environments, can affect planktonic food web dynamics. Summertime hypoxia in the Chesapeake Bay alters field distributions, encounter rates, and predator–prey interactions between hypoxia-tolerant ctenophores, *Mnemiopsis leidyi*, and their less tolerant ichthyoplankton and zooplankton prey. Omnivory and intra-guild predation (IGP) increase the complexity of food webs, thereby confounding the effects of predation versus competition on prey populations. Omnivorous ctenophores in temperate estuarine food webs can both eat and compete with fish larvae for copepod prey. We isolated the effects of predation and competition, and how low versus high DO, affect larval fish growth and survival, using a spatially explicit (three vertical layers) individual-based model of a ctenophore-fish larvae-copepod IGP food web. We simulated three alternative food web structures of how ctenophores affect fish larvae (full interactions, relaxed predation, relaxed competition) under normoxic and hypoxic DO scenarios. Results from laboratory experiments and field studies were

S.E. Kolesar (✉)
Chesapeake Biological Laboratory, University of Maryland Center for Environmental Science, 38 1 Williams Street, Solomons, MD 20688, USA
e-mail: sarah.kolesar@oregonstate.edu

Present Address:
S.E. Kolesar
Oregon Sea Grant, Oregon State University, 1600 SW Western Blvd., Corvallis, OR 97333, USA

K.A. Rose
Department of Oceanography and Coastal Sciences, Louisiana State University, Energy, Coast, and Environment Building, Baton Rouge, LA 70803, USA

D.L. Breitburg
Academy of Natural Sciences Estuarine Research Center, 10545 Mackall Road, St., Leonard, MD 20685, USA

Present Address:
D.L. Breitburg
Smithsonian Environmental Research Center, 647 Contees Wharf Road, Edgewater, MD 21037, USA

D. Justic et al. (eds.), *Modeling Coastal Hypoxia*,
DOI 10.1007/978-3-319-54571-4_11

used to configure and corroborate the model. Ctenophore predation had a bigger effect on survival of modeled fish larvae than did competition between ctenophores and fish larvae for shared zooplankton prey, but competition more strongly affected larval fish growth rates than did predation. Hypoxia versus normoxia did not alter the relative importance of ctenophore predation and competition, but low DO did decrease larval fish survival and increase larval growth rates. Model results suggest that consideration of the interaction strength in food webs and explicit treatment of spatial habitats to allow predator–prey overlap to emerge from movement will enhance our ability to predict hypoxia effects on fish.

Keywords Hypoxia • Bay anchovy • Fish eggs • Ichthyoplankton • Zooplankton • Predation • Trophic • Chesapeake Bay

11.1 Introduction

Hypoxia is increasing in coastal waters worldwide (Diaz and Rosenberg 2008; Gilbert et al. 2010; Rabalais et al. 2010; Zhang et al. 2010), with unknown but potentially meaningful effects on ecologically and commercially important species (Caddy 1993; Cloern 2001; Breitburg et al. 2009; Ekau et al. 2010; Levin et al. 2010). Hypoxia has well-documented effects on sessile species, and on the growth, survival, and reproduction of mobile individuals in localized areas. Furthermore, generalizations about hypoxia affecting mobile species at the population level are common (Kidwell et al. 2009 and references therein), although the quantitative evidence is mixed. Breitburg et al. (2009) did not find a strong relationship between fisheries landings and degree of hypoxia across coastal systems, but they caution that there are well-known problems with using landings data as indicators of population abundance. In a review of modeling analyses, Rose et al. (2009) determined that direct large-scale effects of hypoxia on coastal fish populations are relatively rare, but that there is potential for indirect effects of hypoxia on fish populations mediated via competitive and predation changes due to the responses of other members of the food web. Thus, examination of the effects of hypoxia within a food web context is appropriate, and may be necessary, to fully quantify hypoxia effects at the population level for key, mobile fish species.

Omnivory is common in many aquatic consumers and affects food web dynamics. Omnivory results in increased food web complexity that can dampen trophic cascades (Polis and Strong 1996; Snyder and Wise 2001) caused by strong top-down control in linear food chains. Feeding on multiple trophic levels disperses predation effects throughout the food web by creating weak trophic links (McCann et al. 1998). Trophic links are weaker when a predator is not wholly dependent upon any single resource for survival, such that the predator's actions may be more detrimental to the prey species than beneficial to the predator (Holt and Polis 1997;

Diehl and Fießel 2000). Omnivory also reduces the ability of predators to deplete any one trophic level in a system, and thus, omnivores are potentially less affected by food limitation than specialists. However, whether omnivores can limit the growth or abundance of competitors by depleting shared food resources is debatable (Polis and Strong 1996). The effect of omnivory on food webs is complicated, and it remains unclear whether the overall effect of omnivory is stabilizing or destabilizing (Fagan 1997; Vandermeer 2006).

Intraguild predation (IGP) is a specialized case of omnivory involving the consumption of one competitor by another, simultaneously conferring nutritional gain to the predator and eliminating a competitive rival (Polis et al. 1989). Intraguild predation is widespread (Ehler 1996; Arim and Marquet 2004; Vandermeer 2006; Rosenheim 2007) and is particularly ubiquitous in marine and coastal systems (Polis et al. 1989; Thompson et al. 2007). Separating the indirect effects of competition from the direct effects of predation is challenging (Wissinger and McCrady 1993; Diehl 1995; Navarette et al. 2000). Understanding the role of omnivory, and especially IGP, in food web dynamics is important for predicting how coastal food webs will respond to hypoxia.

There are a variety of conditions under which the IGP form of omnivory can promote coexistence of the predator and prey species (i.e., increase food web stability). One of the most common situations is when the prey species is more efficient than the predator at utilizing the shared resource (Polis et al. 1989; Polis and Holt 1992; Holt and Polis 1997; Rosenheim 2007). Other situations that promote coexistence include: intermediate levels of disturbance (Gurevitch et al. 2000); seasonality in environmental conditions (Polis 1984); habitat structure (Janssen et al. 2007); intermediate levels of productivity (Diehl and Feißel 2000; Heithaus 2001); spatial refuges, temporal refuges, or resource subsidies that are unique to one of the species (Polis 1984; Wissinger 1992; Navarette et al. 2000; Amaraskare 2007); and age structure in which IGP-induced competition and predation differentially affects specific age classes (Polis 1984, 1998).

Our focus here is on a specific IGP food web (Chesapeake Bay) and how an IGP food web with different degrees of competition and predation interacts with low DO to affect food web responses. In the Chesapeake Bay and its tributaries such as the Patuxent River, a major component of the open-water food web involves ctenophores (*Mnemiopsis leidyi*), planktivorous fish larvae (e.g., bay anchovy, *Anchoa mitchilli*), and calanoid copepods (e.g., *Acartia tonsa*) (Fig. 11.1). Bay anchovy is an important forage fish species and the most abundant fish in the Chesapeake Bay system (Wang and Houde 1994). *Acartia tonsa* is the dominant summer crustacean mesozooplankton species in the mesohaline Chesapeake Bay (Brownlee and Jacobs 1987; Kimmel and Roman 2004). *Acartia* is consumed by both *M. leidyi* and larval bay anchovy, and *M. leidyi* also consumes larval bay anchovy. Similar food webs, with species substitutions, are found in many temperate coastal waters (Breitburg et al. 1997).

Low dissolved oxygen (DO) during the summer is a common feature in the mainstem Chesapeake Bay and also in many of its deep tributaries (Breitburg et al. 2003;

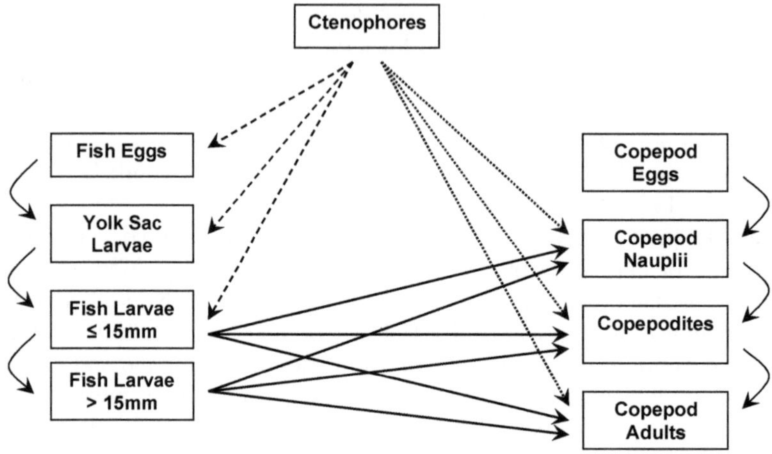

Fig. 11.1 Modeled mesohaline summertime Chesapeake Bay system food web. The food web includes intra-guild predation (IGP) with ctenophores as the intra-guild predators (IG predators) feeding on both the early life stages of fish (eggs, yolk sac larvae, and feeding larvae ≤ 15 mm) and three copepod life stages (nauplii, copepodites, and adults). Straight arrows represent influence of predator on prey. Transitions from one life stage to the next are indicated with curved arrows. The relaxed predation model scenario eliminates ctenophores feeding on the early life stages of fish (*dashed lines*), and the relaxed competition scenario reduces ctenophore feeding on the three copepod life stages (*dotted lines*). Larval fish ≤ 15 mm are the IG prey

Kemp et al. 2005) and can affect the stability of the IGP food web by differentially affecting the vertical distributions of the species. Field studies indicate that low DO concentrations can cause behavioral responses in habitat use and distribution by motile organisms such as *Mnemiopsis leidyi*, fish larvae, and zooplankton (Breitburg et al. 2003; Kolesar et al. 2010). Field data demonstrated how increased habitat overlap between ctenophores and copepods in a stratified water column led to elevated predation rates, especially near the pycnocline (Purcell et al. 2014). Indeed, increasing hypoxia has been associated with shifts in estuarine food webs to greater domination by jellyfish (Purcell et al. 2001).

In this paper, we used an individual-based simulation model to examine the roles of predation, competition, and low DO in the *M. leidyi*-fish larvae-copepod intra-guild food web. The model simulates predation by *M. leidyi* on fish larvae and zooplankton, and predation by fish larvae on zooplankton, in a three-layer water column for the summer months using information representative of the mesohaline portion of the Patuxent River. Simulations were performed to quantify the effects of low DO on food web dynamics and to isolate the effects of competition versus predation on larval fish growth and survival by *M. leidyi*. Our modeling results provide a basis for determining whether hypoxia effects in this common estuarine food web are general or are highly dependent on the relative strengths of competition and predation, which can vary over time within a system and among systems.

11.2 Model Description

11.2.1 Overview

The model follows the growth, mortality, and movement of *Mnemiopsis leidyi* ctenophores, fish larvae, and copepods every 12 h (day and night time steps) for the summer months in a three-layer water column. Temperature was assumed constant throughout the simulation, and dissolved oxygen (DO) concentrations varied over time in each of the three layers. Ctenophores and fish larvae were followed as individuals; copepods were followed as the numbers in each of three uncoupled life stages (nauplii, copepodites, and adults). Fish larvae were introduced as daily cohorts of eggs, while the net energy consumed by adult ctenophores determined the production of new ctenophores. Fish eggs and yolk sac larvae, and eggs and larvae of ctenophores, were followed using matrix projection models. The survivors to the end of the larval stages in the two matrix models (fish, ctenophores) were then treated as individuals in the simulation. Growth of individual ctenophores and fish larvae was based on similarly formulated bioenergetics models with consumption dependent on their encounters with their prey. Ctenophores ate copepods and fish larvae, and fish larvae ate copepods. Mortality of ctenophores was assumed to be constant; mortality of fish larvae and copepods included predation by other modeled individuals. Dissolved oxygen determined movement of ctenophores and fish larvae among the layers and directly affected mortality of fish eggs and growth rates of ctenophores and fish larvae. All variables used in model equations are defined in Table 11.1.

11.2.2 Water Column Structure

The simulated water column was configured to be representative of the summertime conditions typical of the deep, mesohaline region of the Patuxent River that experiences summertime hypoxia. The water column is 1 m × 1 m × 20 m deep and divided into three layers with 20% of the volume in the surface layer, 30% in the pycnocline layer, and 50% in the bottom layer. Two DO conditions were simulated: well mixed with DO concentrations of 6.0 mg L^{-1} in all three layers (high or normoxic) and stratified with surface, pycnocline, and bottom DO set to 6.0, 3.0, and 1.5 mg L^{-1} (low or hypoxic), respectively. The DO concentration of 1.5 mg L^{-1} is typical for summer conditions and is sufficiently low to alter vertical distributions of organisms (Breitburg et al. 2003) and affect predator–prey interactions (Decker et al. 2004), causing maximum overlap between ctenophores and their prey at or near the pycnocline (fish larvae and copepods avoid DO <2 mg L^{-1}). Temperature conditions were held constant at 24 °C in all layers for the duration of the simulations.

Table 11.1 Variable names, descriptions, and units used in the individual-based model

Variable	Description	Units
LvWt	Larval fish weight	mg dw
LvLn	Larval fish length	mm
CtWt	Ctenophore weight	mg dw
CtLn	Ctenophore length	mm
LvCap	Vulnerability of zooplankton prey to larval fish predators based on prey type and larval fish length	Proportion
LvCon	Larval fish consumption	mg dw 12 h^{-1}
LvAsm	Larval fish assimilation	Fraction
LvRsp	Larval fish total respiration per time step	mg dw 12 h^{-1}
j	Prey type (fish egg, yolk sac, or copepod life stage)	–
i	Water column layer (surface, pycnocline, bottom)	–
t	Time	h
LvCmax	Fish larvae maximum consumption	mg dw 12 h^{-1}
ZZ	Number of each zooplankton prey type	Number m^{-3}
T	Temperature in the water column layer	°C
KK	Half-saturation parameter for larval functional response	Number m^{-3}
ZpWt	Copepod weight	mg dw
LvRRsp	Larval fish routine respiration per time step at 24 °C	mg dw 12 h^{-1}
CtCon	Ctenophore consumption	mg dw 12 h^{-1}
CtAsm	Ctenophore assimilation	Fraction
CtRsp	Ctenophore respiration	mg dw 12 h^{-1}
CtRpr	Fraction of net energy intake used for reproduction by ctenophore	Fraction
Fpp	Ctenophore foraging rate	mm 12 h^{-1}
Fpred	Foraging rate used if ctenophore distance swum is greater	mm 12 h^{-1}
Fprey	Foraging rate used if larval fish distance swum is greater	mm 12 h^{-1}
DsPred	Distance swum by the predator	mm 12 h^{-1}
DsPrey	Distance swum by the prey	mm 12 h^{-1}
CtRd	Ctenophore reactive distance	mm
PrRd	Prey reactive distance	mm
E	Mean numbers of encounters	Number 12 h^{-1} m^{-3}
PD	Number of prey available for encounter per layer	Number m^{-3}
CtCapLv	Ctenophore capture success of individual fish larvae	Fraction
CtConCal	Ctenophore consumption	Calories
CtRprCal	Calories available for ctenophore reproduction	Calories

(continued)

Table 11.1 (continued)

Variable	Description	Units
ZProd	Copepod production rate	$12\ h^{-1}$
TotZ	Copepod maximum density	Number m^{-3}
Vol	Water column volume	m^{-3}
DO	Dissolved oxygen concentration	mg L^{-1}
SurEggDO	Fish egg survival due to DO	Number $12\ h^{-1}$
CtWorth	Modeled superindividual ctenophore	Number
LvWorth	Modeled superindividual larval fish	Number

11.2.3 Larval Fish—Energetics and Consumption

Fish eggs were introduced into the surface layer at the beginning of the nighttime step at a density of 100 m^{-3}, and the abundances of fish eggs and yolk sac larvae (total number) in each layer every 12 h were simulated using two-stage matrix projection models that were specific to each layer (Appendix A). Eggs were introduced every 3 days beginning in early June (day 150), increased to daily during July (days 190–212), and then decreased to every 3 days until August (day 220). The elements of the matrix models for each layer were determined every 12 h, and numbers of individuals in each stage in each layer were updated including the addition of newly spawned eggs and ctenophore consumption included as mortality. The number of exiting yolk sac larvae (entering the feeding larvae stage) on each day was lumped over layers and subsequently followed as individual larvae. All new larval individuals were started at 3 mm (0.0084 mg dw) and in the bottom layer. No direct egg cannibalism was assumed. Fish eggs and larvae in the model were mostly based upon information about bay anchovy.

Individual larvae grew according to a bioenergetics equation with consumption based on density-dependent encounter rate of predator and zooplankton prey. Larval fish weight was incremented each 12 h based on consumption (*LvCon*, mg dw $12\ h^{-1}$, day time step only), assimilation (*LvAsm*, fraction), and respiration (*LvRsp*, mg dw $12\ h^{-1}$):

$$LvWt_t = LvWt_{t-1} + LvCon \cdot LvAsm - LvRsp \tag{11.1}$$

Larval length (*LvLn*) was then determined from weight (*LvWt*) using a length–weight relationship. Weight was allowed to increase or decrease each time step, but length was not allowed to shrink. A new length was computed if the individual was at the weight expected for its length and if the change in weight was positive.

Larval fish consumption, assimilation, and respiration were based on larval weight, temperature, and prey densities (Adamack et al. 2012; Rose et al. 1999). Maximum consumption was dependent on larval weight, and a constant temperature of 24 °C was assumed ($LvCmax = a \bullet LvWt^b$, if *LvWt* weight <0.022, a = 27.71, b = 0.76; if *LvWt* ≥ 0.022, a = 28.87, b = 0.75) and was used with a multi-species type II functional response relationship to determine realized consumption of each of the three zooplankton types (nauplii, copepodites, adults):

$$LvCon_j = LvCmax \cdot \frac{ZZ_j}{Vol} \cdot \frac{\frac{LvCap_j}{KK_j}}{1 + \sum_i^3 \cdot \frac{LvCap_j}{KK_j} \cdot ZpWt_j} \cdot ZpWt_j \qquad (11.2)$$

where $LvCon_j$ is realized cumulative consumption rate of the jth zooplankton type (mg dw 12 h^{-1}), *LvCap* is the vulnerability of zooplankton prey to larval fish predators based on prey type and larval fish length (after Rose et al. 1999; maximum $LvCap_j = 0.9$), ZZ_j is density of zooplankton type *j* (number m^{-3}), KK_j is the half-saturation parameter of the larval fish for zooplankton type *j*, and $ZpWt_j$ is the weight per individual (mg dw) of zooplankton type *j*. Vol is the volume of the layer (m^3). The sum of the three zooplankton-type specific consumption rates is the total 12-h consumption rate for the larva in Eq. 11.1 (i.e., $LvCon = \sum_{j=1}^{3} LvCon_j$). The KK_j were calibrated to obtain realistic larval fish growth rates (Table 11.2). Feeding occurred only during daytime time steps. Assimilation efficiency (*LvAsm*) was set at 0.60 (Rose et al. 1999). Respiration (*LvRRsp*) was computed as a routine rate

Table 11.2 Values used for modeled organism weight, energy density, and length. Individual larval fish and ctenophore weight and length varied with consumption during model simulations. Fish egg and yolk sac larvae weights were based on values reported in Tucker (1989). Energy densities of all fish life stages were based on values from Hunter and Leong (1981). Ctenophore energy density values were reported in Harris et al. (2000). Energies densities for all copepod life stages were reported in Laurence (1976) and weights as reported in Tester and Turner (1988). Fish eggs, yolk sac larvae, and zooplankton lengths were estimated from live samples. Larval fish consumption was modeled using a multi-species type II functional response, and realized consumption for each prey type was estimated by applying *KK*, the half-saturation parameter, to prey densities in the model

	Weight per individual (mg dw)	Energy density (calories mg dw^{-1})	Length per individual (mm)	Calibrated *KK* values (number m^{-3})
Fish eggs	0.00842	5.525	1	NA
Yolk sac larvae	0.015	5.424	3	NA
Fish larvae	Dynamic	5.350	Dynamic	NA
Ctenophores	Dynamic	2.967	Dynamic	NA
Nauplii	0.00152	5.160	0.15	75,000
Copepodites	0.0033	5.160	0.6	3000
Adult copepods	0.011	5.160	1.2	2250

dependent upon larval weight (*LvWt*) and a constant water temperature (T = 24 °C) for nighttime and twice the routine rate for daytime. The routine rate was:

$$LvRRsp = 0.073 \cdot LvWt^{0.997} e^{\frac{\log(2.2)}{10} \cdot (T - 27)} \tag{11.3}$$

Individual fish larvae died from being eaten by ctenophores and from a constant rate representative of mortality from other sources. Mortality rate was set to 3% per 12 h to reflect predation by *Chrysaora quinquecirrha* medusae and piscivorous fish (Cowan and Houde 1993; Purcell et al. 1994a; Purcell and Arai 2001). Larger larval fish (≥ 15 mm) were no longer vulnerable to ctenophore predation, but were kept in the simulation to include their consumption effects on zooplankton prey.

11.2.4 Ctenophores—General Bioenergetics

The model assumed that ctenophores could potentially produce eggs every 12 h based on their consumption rate. As with fish eggs and yolk sac larvae, the numbers of ctenophore eggs and larvae in each layer were tracked using a two-stage matrix projection model for each layer (Appendix A). The mortality rates of ctenophore eggs and larvae used in the matrix model were determined by calibration to generate reasonable ctenophore and fish larval densities in simulations. Individuals exiting the larval stage entered an intermediate holding stage (prereproductive lobate stage) where they waited until reaching 25 mm length (another 5–7 days, depending on growth rates) and entered the model as individual reproductive ctenophores in the layer they were spawned in.

Similar bioenergetics as with fish larvae, with the addition of reproduction costs, was also used for the 25 mm and longer individual ctenophores:

$$CtWt_t = CtWt_{t-1} + (CtCon \cdot CtAsm - CtRsp) \cdot (1 - CtRpr) \tag{11.4}$$

where *CtWt* is weight of a ctenophore (mg dw), *CtCon* is the consumption rate (mg dw 12 h^{-1}), *CtAsm* is assimilation (fraction), *CtRsp* is respiration rate (mg dw 12 h^{-1}), and *CtRpr* is fraction of net energy intake used for reproduction. Ctenophore length (*CtLn;* mm) was determined from weight (*CtWt*) using a length–weight relationship (Kremer 1976).

Consumption was based on a modified version of the Gerritsen and Strickler (1977) encounter model (Cowan et al. 1999; Kolesar 2006). Ctenophores fed during both day and night time steps on fish eggs, yolk sac larvae, fish larvae (≤ 15 mm), and the three stages of copepods. Encounters were dependent on swimming speeds and encounter radii of the ctenophores and each of their prey types, both of which were dependent on body length (BL; mm). Capture success was fixed for smaller, less motile prey and varied with length for larval fish prey.

11.2.5 Ctenophores—Encounters, Consumption, and Energetics

Ctenophores and fish larvae had dynamic individual lengths that varied based on their bioenergetics and growth; fixed lengths were used for eggs, yolk sac larvae, and copepods (Table 11.2). Swimming speeds were assumed to be 0.3 BL s^{-1} for ctenophores (Kolesar 2006), and 2 BL s^{-1} for yolk sac larvae, feeding larvae, and copepods; fish eggs were assumed not to swim. Reactive distance of ctenophores (*CtRd*, mm) was based on their length (mm) and modeled as an ellipse:

$$CtRd = 0.5 \cdot \left(0.33 \cdot CtLn + \frac{0.33 \cdot CtLn}{2}\right) \tag{11.5}$$

Reactive distances for all prey types (*PrRd*, mm) were assumed to be their length in mm.

By combining swimming speeds and reactive distances with prey density, we computed the mean number of encounters (*E*) in 12 h (number 12 h^{-1} m^{-3}) between a ctenophore predator and each of its prey types:

$$E_j = \pi \cdot \left(PrRd_j + CtRd\right)^2 \cdot Fpp_j \cdot 10^{-9} \cdot PD_j \tag{11.6}$$

where $PrRd_j$ is encounter radius of prey type *j* (mm), *CtRd* is the encounter radius of the ctenophore (mm), Fpp_j is the foraging rate (mm 12 h^{-1}) of the ctenophore and prey type *j*, and PD_j is density (number m^{-3}) of prey type *j*. In Eq. 11.6, there were six possible prey types: fish eggs, yolk sac larvae, fish larvae, and the three stages of the copepods. The foraging rate depended on the distances swum by the predator (*DsPred*, mm) and prey type *j* ($DsPrey_j$, mm) in 12 h:

$$Fpp_j = \begin{cases} Fpred & \text{if } DsPred > DsPrey_j \\ Fprey & \text{if } DsPred \leq DsPrey_j \end{cases} \tag{11.7}$$

where

$$Fpred = \frac{\left(DsPrey_j^2 + 3 \cdot DsPred^2\right)}{3 \cdot DsPred}$$

$$Fprey = \frac{\left(DsPred^2 + 3 \cdot DsPrey_j^2\right)}{3 \cdot DsPrey_j}$$

Prey density (PD_j, number m^{-3}) for copepods, fish eggs, and yolk sac larvae was the total number in each layer divided by the volume of that layer. For fish larvae, which were followed as individuals, Eq. 11.6 was evaluated for each individual fish larva as a possible prey item for each ctenophore.

For each ctenophore and prey type, the number of prey of each type encountered and captured was multiplied by weight per individual, adjusted for energy density (calories per mg dw, Table 11.2), and summed to obtain biomass eaten by the ctenophore (*CtCon* in Eq. 11.4). Realized number of encounters was generated as a random deviate from a Poisson distribution with mean equal to E_j. The actual number of prey encountered and successfully captured was then determined as a deviate from a binomial distribution with the number of trials equal to the number of realized encounters (Poisson deviate generated from E_j) and the probability of success set to the probability of capture. Capture success by ctenophores was 0.62 for nauplii, 0.54 for copepodites, and 0.46 for adults (Waggett and Costello 1999), and 0.80 for fish eggs and yolk sac larvae (Cowan and Houde 1993). Capture success for ctenophore feeding on individual fish larvae (*CtCapLv*) depended on the lengths (mm) of both predator and prey and was not allowed to exceed 0.80 (Cowan and Houde 1993; Kolesar 2006):

$$CtCapLv = 1.086 - 6.99 \cdot \frac{LvLn}{CtLn} \tag{11.8}$$

Biomass eaten of prey type was computed from actual numbers eaten and the weight (mg dw) per individual prey item (Table 11.2), and then converted to calories (Table 11.2) and summed to obtain total consumption in calories for the ctenophore. This total consumption was then divided by the energy density of the ctenophore to obtain prey consumption back in units of mg dw, but now in terms of ctenophore tissue. We adjusted prey consumed by ctenophores by energy densities because energy densities of ctenophores were about half of their prey.

Based on data reported in Kremer (1976, 1979), Kremer and Reeve (1989), and Reeve et al. (1989), we fit a logistic-shaped function that relates assimilation efficiency (*CtAsm*) to prey consumption (Fig. 11.2); assimilation efficiency ranged from a maximum of 0.9 at low food densities to a minimum of 0.4 at the highest prey densities.

Respiration (*CtRsp*) depended on weight and was modified from Kremer (1976) for the 24 °C and 12 h time step used in this model:

$$CtRsp = 4.4 \cdot e^{0.15 \cdot T} \cdot 7.06 \times 10^{-4} \cdot 1.67 \cdot 0.5 \cdot CtWt \cdot 0.5 \tag{11.9}$$

where 7.06×10^{-4} converts (μM CO_2) · g dw^{-1} to g dw · (μM C)$^{-1}$, 1.67 converts g dw · (μM C)$^{-1}$ to the total daily fraction of body carbon catabolized; the first 0.5 value adjusts the rate for the warmer temperature of the Patuxent River, and the second 0.5 value converts the daily rate to a rate per 12 h.

Net energy consumed was divided between somatic growth and reproduction. Maturity occurred at 25 mm (Reeve et al. 1989). Immature individuals used all of their net energy for growth, while mature individuals allocated up to 100% of their net energy to egg production. On each nighttime time step, the proportion of net energy allocated to reproduction (*CtRpr*) was calculated (Kremer 1976):

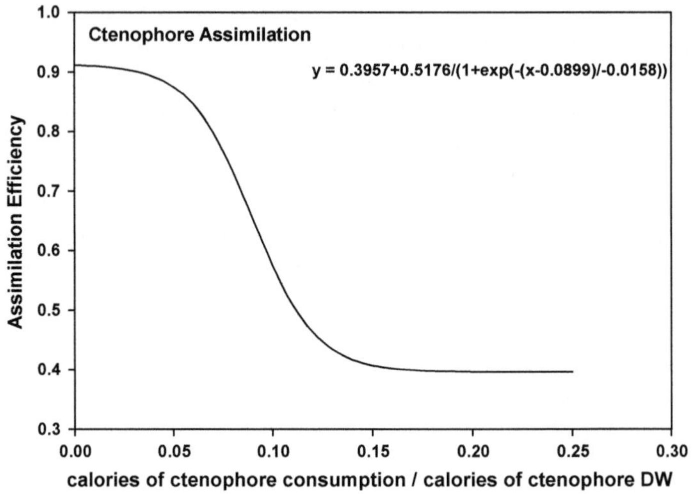

Fig. 11.2 Equation for ctenophore assimilation efficiency used in the model determined from published data on ctenophore assimilation as well as information on ctenophore bioenergetics (consumption based on biomass)

$$CtRpr = 0.01 \cdot e^{0.115 \cdot CtWt} \tag{11.10}$$

Net energy gain (i.e., $CtCon \cdot CtAsm - CtRsp$, in mg dw) was summed for the day (daytime plus nighttime; eggs were released at night) and converted to calories based on the energy densities and biomasses eaten of the prey in the diet comprising *CtCon,* and respiration rate was converted to calories ($CtRsp \cdot 2.967$). Net energy gain in calories was then multiplied by *CtRpr* to obtain energy (in calories) available for eggs, and this was done for the day and night time steps and summed to obtain a single daily value (*CtRprCal).* The daily value was used to determine the number of eggs produced by that ctenophore for that day (Grove and Breitburg 2005):

$$CtEgg = 647.5 \cdot \log(CtRprCal) + 926.75 \tag{11.11}$$

Mortality of ctenophores was 5% 12 h^{-1} for before August 1 and 15% 12 h^{-1} after August 1 (day 213). *Chrysaora quinquecirrha* and *Beroe ovata* ctenophores that consume *M. leidyi* typically peak in mid- to late summer (Kreps et al. 1997; Purcell et al. 2001).

11.2.6 Copepods

The numbers of individuals in each of the three life stages of the copepods (nauplii, copepodites, and adults) in each layer were simulated separately using a logistic production model with added mortality terms based on the summed consumption from individual ctenophores and the summed consumption from individual larvae.

$$ZZ_{j,t} = ZZ_{j,t-1} + ZProd_j \cdot \left(1 - \frac{ZZ_{j,t-1}}{TotZ_j}\right) - \sum_{i=1}^{n} \frac{LvCon_{i,t-1}}{Vol} - \sum_{k=1}^{m} \frac{CtCon_{k,t-1}}{Vol} \quad (11.12)$$

where $ZZ_{j,t}$ is the number of each copepod life stage j in the model in a layer at time t, $Zprod_j$ is the production rate (12 h^{-1}) of zooplankton type j, $TotZ_j$ is the maximum density (number m^{-3}) of type j, n is the number of larvae in the layer, and m is the number of ctenophores in the layer. *Zprodj* was set to 0.6, 0.5, and 0.4, and *TotZ* was set to 300,000 nauplii, 15,000 copepodites, and 10,000 adults.

11.2.7 Vertical Movement of Fish, Ctenophores, and Copepods

Movement of fish egg densities, yolk sac larval densities, copepod densities, and individual model fish larvae and ctenophores occurred every time step in the simulation. Modeled movement among the three water column layers was based on the proportional densities of organisms, dependent on bottom layer DO concentration (similar calculations found in Breitburg et al. 2003; Keister et al. 2000; Kolesar et al. 2010). Proportional densities of organisms were calculated for a water column, assuming equal volumes of water in all three layers, and for discrete intervals of DO (two interval scenarios are shown in Fig. 11.3). We linearly interpolated from proportional densities by DO interval to the modeled proportional densities for high DO conditions (6.0 mg L^{-1} in all three layers) and for low DO conditions (6.0 mg L^{-1} in the surface layer, 3.0 mg L^{-1} in the pycnocline layer, and 1.5 mg L^{-1} in the bottom layer), and adjusted the proportional densities for the unequal volumes of the three layers by multiplying by the volume of each layer. We then formed the cumulative distribution of these interpolated, volume-adjusted proportions and generated a random number between 0 and 1 every 12 h to determine the fraction of the individuals (eggs, yolk sac, copepods) or the probability an individual (larvae and ctenophores) would move for the next time step. Separate proportional densities by bottom DO were used for fish eggs and for fish larvae and the proportional densities of larval fish were also used for yolk sac larvae.

11.2.8 Dissolved Oxygen Effects

In addition to vertical movement, low DO directly affected larval fish growth, ctenophore growth, and fish egg survival. We combined laboratory results for

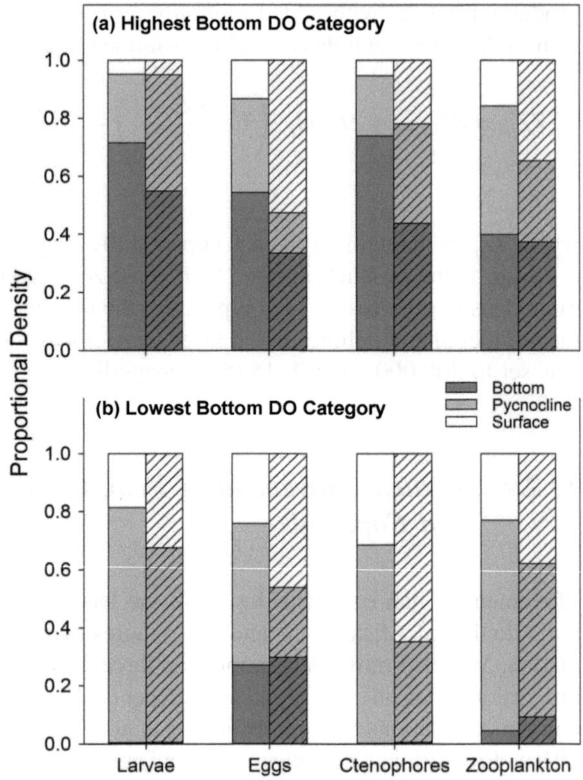

Fig. 11.3 Proportional densities of fish larvae, eggs, ctenophores, and zooplankton. Distribution of each organism type was calculated for an idealized water column with equal water volume in surface (*white*), pycnocline (*light gray*), and bottom (*dark gray*) layers during the day (unhatched bars) and night (hatched bars) for two bottom dissolved oxygen (DO, mg L^{-1}) categories, **a** highest bottom layer DO category (6–6.99 mg L^{-1}) and **b** lowest bottom layer DO category (1–1.99 mg L^{-1})

anchovy and naked goby (Zastrow et al. unpubl.) to develop a multiplier of growth rate due to low DO:

$$GrLvDO = -0.00397 + 0.482 \cdot DO^{0.389} \tag{11.13}$$

On each time step, the DO in the layer was used to determine *GrLvDO*, which was then multiplied by the growth rate according to bioenergetics (Eq. 11.1) and the adjusted growth rate was used to increment larval fish weight. The same approach was used to modify ctenophore growth (Grove and Breitburg 2005):

$$GrCtDO = \frac{0.1173 + 0.0104 \cdot DO}{0.1797} \tag{11.14}$$

Fish egg survival under low DO was based on Dorsey et al. (1996):

$$SurEggDO = \frac{95.77}{1 + e^{\frac{2.35 - DO}{0.95}}} \tag{11.15}$$

Fraction of eggs surviving 12 h used in the stage-based matrix model for fish eggs and yolk sac larvae were multiplied by *SurEggDO* based on the DO in the layer (Appendix A).

11.2.9 Numerical Considerations

We used a superindividual approach for representing ctenophore and larval fish model individuals. The superindividual approach allows for a predetermined number of model individuals to be in a simulation, thereby preventing numerical coding problems associated with following thousands or millions of model individuals (Scheffer et al. 1995). In our model, ctenophore reproduction was simulated based on energy consumed. However, adding a new model individual for every new ctenophore introduced into the model could result in the computer code exceeding memory limitations. The superindividual approach addresses this by making each model individual worth some number of identical population individuals. Thus, a known number of model individuals can be added and their worth adjusted to reflect the population number added. Mortality is then simulated by decrementing the worth of the superindividual to reflect the loss of population individuals represented by the superindividual. In all model simulations, five ctenophore superindividuals and five larval fish superindividuals were introduced into each layer at the start of every time step. The worth of ctenophore superindividuals (*CtWorth*) and larval fish superindividuals (*LvWorth*) was calculated by dividing the number of population individuals of one type introduced into each layer at each time step by 5, and assigning that same worth to each superindividual.

Mortality and predation were imposed on ctenophore and fish larvae superindividuals by adjusting the population worth of each superindividual. Mortality, either as a fixed mortality rate on either ctenophore or larval fish, or by ctenophore predation on a larval fish, resulted in a reduction of the population worth of the superindividual. Because the ctenophores (predator) and larval fish (prey) were both superindividuals, when a model ctenophore ate one or more of the population individuals of a model fish larva, we had to make adjustments to ensure mass balance. If the ctenophore worth times the number of population larvae it ate was less than the worth of the larval superindividual, then the ctenophore individual consumed the entire weight of the larval individual. The model ctenophore grew accordingly, and the larval worth was reduced by the worth of the ctenophore superindividual times the number of larvae eaten. If the ctenophore worth times the number of population larvae it ate was greater than the worth of the larval superindividual, then the ctenophore actually consumed the weight of the larval fish times the ratio of larval worth to ctenophore worth and the worth of the larva superindividual was set to 0. In this way, mass balance of the biomasses of ctenophores and fish larvae was maintained.

Ctenophore predation on larval fish and copepods, and larval fish predation on copepods, was updated after each ctenophore and larval fish evaluation as the predator. This was done to minimize the possibility of summed predation pressure over ctenophores or fish larvae exceeding the prey abundance in a layer on a time step. Because we updated the larval fish worths and copepod densities for predation after every predator, each time step the ctenophores and larval fish individuals were evaluated for growth and mortality in random order. Otherwise, modeled individuals evaluated first would always see higher prey densities.

Predation by ctenophores on fish eggs and yolk sac larvae was accounted for by the dynamic mortality term included in the estimation of the diagonal and subdiagonal terms of their stage-based matrix projection model. Predation by ctenophores and larval fish on each of the three copepods stages was accounted for by inclusion of the mortality rate in each logistic production equation.

11.3 Design of Model Simulations

All model simulations were for 100 d from May 25 (day 145) to September 2 (day 245). Five replicate simulations were performed for each condition using different random number sequences that affected encounter rates, capture success, and movement. The model was first calibrated and corroborated under baseline conditions. Baseline conditions included the full IGP food web: ctenophore consumption caused mortality of fish larvae, and ctenophore and fish larval consumption caused mortality of copepods. We then used the calibrated model to explore how predation, competition, and hypoxia effects interact to affect food web dynamics. Because model predictions were very similar among replicates (see minimum and maximum values in Table 11.3), we focus on the output variables averaged over the five replicates and graph results (e.g., time series plots, diets) from one replicate simulation.

11.3.1 Calibration and Corroboration

We adjusted several key parameters in the baseline model under both high and low DO conditions to obtain realistic model behavior compared to field data. The KK_j in Eq. (11.2) were adjusted to obtain realistic larval fish growth rates, and then the fixed mortality rates of fish and ctenophore eggs and larvae were adjusted to generate reasonable summertime averaged fish densities and ctenophore densities. The high DO (normoxic, 6.0 mg L^{-1} in all layers) and low DO (hypoxic, 6.0, 3.0, and 1.5 mg L^{-1}) simulations were used because the field data reflect a range of DO conditions. We first crudely compared larval fish growth rates and ctenophore

Table 11.3 Comparison of simulated densities of ctenophores, fish eggs, yolk sac larvae, post-yolk sac (feeding) larvae, copepod nauplii, copepodites, and adult copepods with reported field densities (all m^{-3}). Model values presented are the mean ± SE, and minimum and maximum, of the 100 daily densities entering each life stage. Each value for the field data was a summertime mean based on the available stations and sampling dates. There were generally 43 stations for the Patuxent sampling and 174 stations for the Chesapeake Bay sampling for ctenophores and fish; about 15 stations for zooplankton sampling. Field densities of ctenophores and fish were measured in the Chesapeake Bay in July and August 1995–1998 (data from the TIES project, summarized in Purcell et al. 2001, Tables 6 and 7) and the Patuxent River in June, July, and August 1992, 1993, 1999, and 2001 (Keister et al. 2000; Breitburg et al. 2003; Kolesar et al. 2010). Copepod nauplii, copepodite, and adult copepod field densities were from June, July, and August of 2000 and July 2001 for the Chesapeake Bay and from June, July, and August of 1999 and July 2001 for the Patuxent River in (Purcell et al. 2014 and unpublished data). For the field data, yolk sac larvae and feeding larval fish densities were combined and are reported under (fish larvae), and copepodite and adult copepod densities were combined (copepods). Ctenophore and fish data are presented for the whole water column; copepod data are presented for the surface layer only. Copepods are presented as thousands m^{-3}

	Model results		Field data	
	High DO	Low DO	Patuxent	Chesapeake Bay
Ctenophores	3.8 ± 0.2 (0.01–8.7)	3.7 ± 0.2 (0.02–7.9)	0.03–6.83	2.8–12.7
Fish eggs	2.5 ± 0.5 (0–20)	2.7 ± 0.5 (0–20)	0–41.3	1.2–28.8
Yolk sac larvae	1.4 ± 0.2 (0–10)	0.5 ± 0.1 (0–4)		
Fish larvae	1.5 ± 0.1 (0–6)	0.6 ± 0.03 (0–2.3)		
Yolk sac plus larvae			1.6–12.9	0.3–3.3
Nauplii	226.9 ± 7.4 (92.1–399.6)	190.8 ± 6.1 (66.5–393.6)	36.4–38.3	20.4–27.7
Copepodites	10.6 ± 0.4 (3.7–10.0)	8.7 ± 3.3 (2.6–19.7)		
Adult copepods	6.3 ± 0.3 (1.6–13.3)	4.9 ± 0.3 (1.0–13.1)		
Copepodites plus adult copepods			0.7–29.5	0.8–7.5

lengths and egg production rates to values reported in the literature. We then compared averaged densities of fish eggs, yolk sac plus larvae, ctenophores, and copepods (nauplii and copepodites plus adults), computed over a single replicate simulation for high and low DO conditions to field-measured densities for the Patuxent River estuary and Chesapeake Bay.

11.3.2 Predation, Competition, and DO Effects Within the IGP Food Web

To explore the effects of predation, competition, and hypoxia, three versions of the model were simulated under the high DO (normoxic) scenario and the low DO (hypoxic) scenarios for a total of six modeled conditions (Fig. 11.4). The first

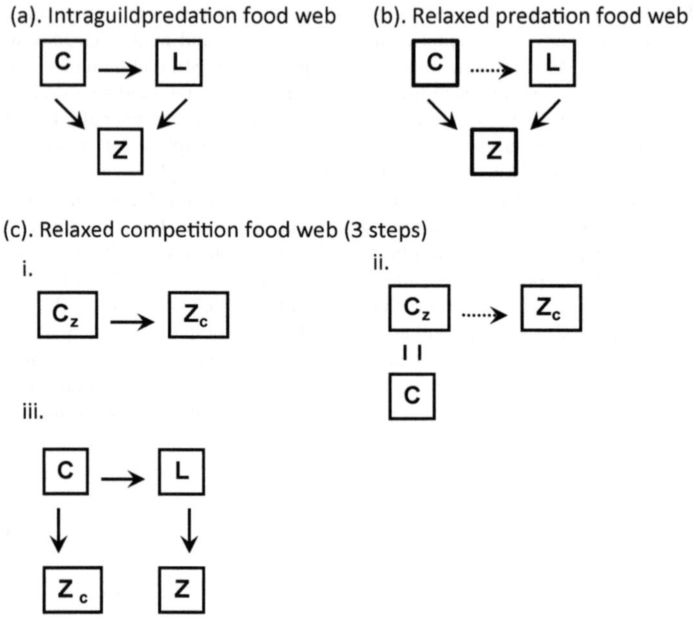

Fig. 11.4 Modeled simulations included three food webs, generated through five total iterations. **a** The baseline intra-guild predation (IGP) food web included ctenophores (C) as both predators on larval fish (L) and competitors for copepod prey (Z) (predation is designated by *solid arrows*), **b** the relaxed predation food web (RP) included ctenophore predation on larval fish, but larval fish were not removed from simulations (represented by a *dashed arrow*), and ctenophores and larval fish were competitors for copepod prey, and finally, **c** the relaxed competition food web (RC) had separate prey pools generated for ctenophores and larval fish through a series of steps. In the full RC food web, the zooplankton prey pool for ctenophores (Z_c) was generated from running simulations with a ctenophore-zooplankton-only model (i), a fitted density-independent model was run to calibrate ctenophores to baseline conditions (C) (ii), and the full RC model included two separate prey pools for ctenophore and larvae predators (Z_c and Z, respectively), with fitted ctenophores also preying on fish larvae (iii)

version of the model was the baseline version used in calibration and corroboration with the full predation and competitive interactions operating (Fig. 11.4a). The second version of the model relaxed the ctenophore predation effects on fish larvae (Fig. 11.4b). Ctenophore consumption depended on their encounters with fish larvae, but eaten larval fish were not removed from the simulation; therefore, ctenophores gained the appropriate prey resources, but fish larvae were not affected by ctenophore predation.

The third version of the model ultimately maintained the ctenophore predation effects on fish larvae, but relaxed the competition between ctenophores and fish larvae for copepods. Through a series of intermediate food webs, we created two sets of copepod densities (by stage, layer, and time step), one for ctenophores and one for fish larvae, to allow ctenophores to eat larvae and zooplankton, while larvae experienced a separate zooplankton prey pool (Fig. 11.4c). First, the baseline model was run with only ctenophores and copepods, and the predicted copepod numbers by life stage, layer, and time step were recorded (Fig. 11.4c.i.). We then used these copepod densities as input back into the ctenophore-copepod-only model but without ctenophore effects on copepods, and recalibrated the ctenophore egg and larval mortality rates (0.67–0.9 $(12\ \mathrm{h})^{-1}$; 0.74–0.9 $(12\ \mathrm{h})^{-1}$), so their dynamics closely resembled the baseline IGP version with the full interactions (Fig. 11.4c.ii.). Finally, using the recalibrated mortality rates, we ran the full model with ctenophores only preying upon their own pool of copepods (the output from ctenophore-copepod model) and fish larvae consuming their own pool of copepods (Fig. 11.4c.iii.). In this relaxed competition version, ctenophore dynamics resembled the dynamics in the IGP baseline with their consumption affecting fish larvae mortality, but their consumption of copepods not affecting the availability of copepods to fish larvae.

We first examined the baseline food web under both high and low DO conditions for general model behavior beyond the calibration and corroboration checks and for the effects of low DO on model dynamics. Second, we compared the predicted larval fish survival and growth among the three food webs for the high DO condition to determine the importance of ctenophore predation versus ctenophore competition on larval fish survival and growth. The third comparison was among the three food webs for high DO versus low DO to determine whether low DO altered the importance of predation versus competition obtained under high DO in the second comparison.

Model output variables averaged over the five replicate simulations include: (1) number of fish larvae surviving and their average duration from first feeding (i.e., introduced as model individuals) to 15 mm, and (2) percent survival of fish from egg to hatch, hatch to first feeding, first feeding to 15 mm, and egg to 15 mm. For simplicity and because of consistency among replicates, we used a single replicate simulation for each of the six conditions and examined for every 12-h time step over the 100 d of simulation: larval lengths and ctenophore weights over time for selected model individuals (every 50th model individual), and time series plots of larval, ctenophore, and adult-stage copepod densities by water layer. We also report water column integrated densities (i.e., sum of all individuals divided by volume of water column); these were output daily, and then the 100 values summarized as the minimum, average (±SE), and maximum values. Diets of ctenophores and larval fish (broken down by small (>5 mm), medium (5–10 mm), and large sized (>15 mm)) were summarized as the averaged proportion by biomass of nauplii, copepodites, and adult copepods over a single replicate simulation. Finally, as an aid for interpreting model results, we computed the average vertical overlap between ctenophores and fish larvae, ctenophores and copepods, and fish larvae and copepods for the duration of a single replicate simulation.

11.4 Results and Discussion

11.4.1 Model Calibration and Corroboration

We examined how DO could interact with food web structure to affect larval fish survival and growth. While both DO and food web structure altered outcomes for modeled larval fish, the effect of DO was not substantially altered under the three different food webs tested. Our IGP food webs included: baseline in which all predation and competitive interactions were operating, a version with relaxed competition between ctenophores and fish for zooplankton, and a version with relaxed predation of ctenophores on fish larvae. The ctenophore-fish larvae-copepod food web in this model typifies Chesapeake Bay and other temperate estuaries, and differs from more frequently examined IGP food webs in that the IGP predator (the ctenophore, *Mnemiopsis leidyi*) is also a superior competitor to its prey (fish larvae).

Using the calibrated parameter values, average larval growth rates in both the high and low DO baseline IGP food web model were similar to bay anchovy growth rates reported from field studies. Average growth rates of larval fish surviving to 15 mm in the baseline IGP food web model were 0.46 mm d^{-1} at high DO and 0.61 mm d^{-1} at low DO. Rilling and Houde (1999) reported field growth rates of larval bay anchovy ranging from 0.53 to 0.78 mm d^{-1}, and bay anchovy larvae from North Carolina were estimated to grow at about 4% d^{-1} or equal to about 0.48 mm d^{-1} (Fives et al. 1986).

Modeled ctenophore lengths and ctenophore egg production remained within the bounds observed in field samples and laboratory studies. Simulated ctenophore lengths under high and low DO ranged over the summer from 20 to 92 mm, the maximum being slightly smaller than the largest ctenophore length (100 mm) observed in the Chesapeake Bay system. Ctenophore egg production averaged about 1000 eggs $ctenophore^{-1}$ d^{-1} (range of 0–11,360), which was similar to the range of 0–14,000 eggs $ctenophore^{-1}$ d^{-1} reported in Purcell et al. (2001).

Simulated summertime densities of ctenophores and fish eggs and larvae were similar to values reported for Chesapeake Bay and the Patuxent River (Table 11.3). Simulated ctenophore densities averaged about 3.8 individuals m^{-3} over the summer (with a maximum daily value of 8.7), compared to summertime means of 0.03 to 6.83 in the Patuxent and 2.8 to 12.7 in the mainstem Chesapeake Bay. Averaged fish egg densities in the baseline simulations were about 2.5 eggs m^{-3}, with a daily maximum of 20, while field data showed a range of summertime means of 0 to 41.3 in the Patuxent and 1.2 to 28.8 in the Chesapeake Bay. Simulated fish larval densities were also within the range of the field data; model averaged values were about 1.5 individuals m^{-3} for both yolk sac and feeding larvae in high DO and about 0.6 individuals m^{-3} in low DO, versus 1.6 to 12.9 for yolk sac and feeding larvae combined in the Patuxent and 0.3 to 3.3 in the Chesapeake Bay.

Simulated copepodite and adult copepod densities were similar to those reported for the Patuxent River estuary, while simulated copepod nauplii densities were higher than reported values. Simulated values of copepodite and adult copepod densities averaged 106,000 individuals m^{-3} and 63,000 in high DO (summed value of about 160,000) and 87,000 and 49,000 in low DO (summed value of about 140,000), which were reasonable given the relatively wide range of summed observed values of 700–295,000 in the Patuxent and 800–75,000 in the Chesapeake Bay.

Simulated nauplii densities were almost 10 times higher than reported densities (Table 11.3). Simulated nauplii densities were 226,900 individuals m^{-3} under high DO and 190,800 individuals m^{-3} under low DO, while reported mean densities were less than 36,000. This is likely a combination of simulated densities being too high and field samples being underestimates due to difficulties in accurately sampling nauplii based on their small size and high variability. In other studies, nauplii densities of 100,000 individuals m^{-3} in the surface layer were reported for both the Patuxent River estuary (Heinle 1966) and the Chesapeake Bay (Purcell et al. 1994b), and Purcell et al. (1994b) even reported occurrences of copepodite and adult copepod densities approaching 100,000 individuals m^{-3} at some stations in August (implying nauplii densities were even higher). However, while these high densities occurred, they were extreme values rather than averaged values. Both simulated and observed copepods (Purcell et al. 1994b) were at their lowest densities during the summer period.

11.4.2 Baseline Model Behavior Under High DO

Ctenophore densities peaked during mid-summer coincident with the time that larval fish densities and zooplankton densities showed depressed values. Ctenophores peaked between days 180 and 200 at a water column integrated average density of about 9.0 individuals m^{-3}, and with most individuals in the bottom layer and secondarily in the pycnocline layer, more so during the day than at night (Fig. 11.5a). Larval fish densities reached their peak early in the summer (column integrated density of 6.0 individuals m^{-3}), as larval fish numbers accumulated from frequent spawning (Fig. 11.6a). Larval fish densities then declined during the middle of the summer and rebounded with a second, lower peak near the end of the summer. In the absence of ctenophore predation, a peak in density of larval fish would be expected in the middle of the summer, as a result of the build-up of repeated spawning events and removal of individuals as they reached 15 mm. Larval fish were spread between the bottom and pycnocline layers, with few larvae occurring in the surface layer at any time (Fig. 11.6a). Adult copepod densities showed a minimum during the mid-summer (2,000 individuals m^{-3} relative to an equilibrium value of 10,000), with high densities in the surface layer during the day and in the pycnocline during the nighttime (Fig. 11.7a). Nauplii and copepodites (not shown) had very similar temporal patterns as adult copepods.

Fish survival was low during the first feeding to 15-mm stage relative to egg and yolk sac larval stages due to ctenophore predation and extended exposure to

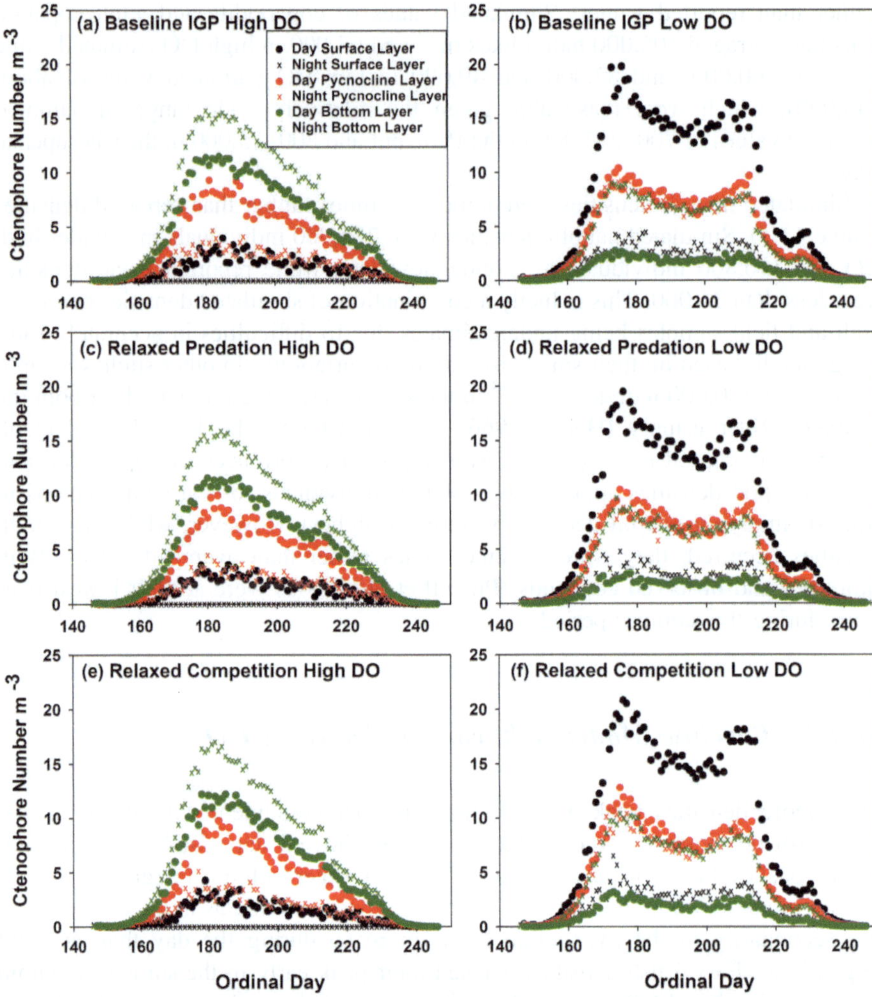

Fig. 11.5 Ctenophore number m^{-3} by layer plotted against ordinal day during both day (●) and night (x) for a representative simulation for each of six different food webs: **a** baseline IGP high DO, **b** baseline IGP low DO, **c** relaxed predation high DO, **d** relaxed predation low DO, **e** relaxed competition high DO, **f** relaxed competition low DO. *Black dots* denote the surface layer, *red dots* the pycnocline, and *green dots* the bottom layer

predation because of the long stage duration. Averaged survival (over the five replicate simulations) from egg production to hatching was 40%, hatch to first feeding was 16.8%, and first feeding to 15 mm was 2.1%, resulting in cumulative cohort survival of 0.14% and an average of 14.1 survivors to 15 mm (Table 11.4). First feeding to 15 mm survival due to constant mortality only (i.e., not from ctenophores and there was no DO-related mortality) was 20% (26 days at 3% per

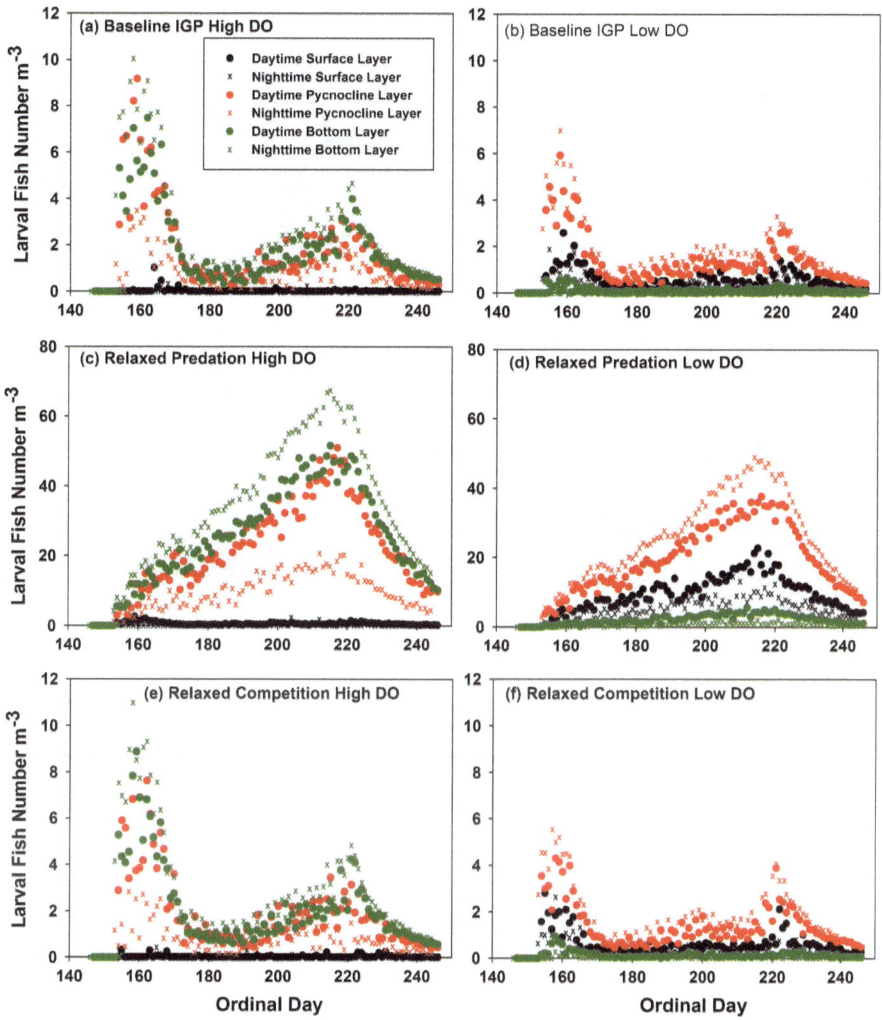

Fig. 11.6 Fish larvae number m^{-3} by layer plotted against ordinal day during both day (●) and night (x) for a representative simulation for each of six different food webs: **a** baseline IGP high DO, **b** baseline IGP low DO, **c** relaxed predation high DO, **d** relaxed predation low DO, **e** relaxed competition high DO, **f** relaxed competition low DO. *Black line* denotes the surface layer, *red line* the pycnocline, and *green line* the bottom layer. Note different scales

12 h, $0.97^{26 \cdot 2}$); thus, ctenophore predation reduced larval survival by an order of magnitude from 20 to 2.1%.

Ctenophore predation was very important to the survival of early life stages of fish in our modeled food webs. Movement in our model resulted in high overlap between ctenophores and larval fish, especially in food webs at high DO. The high consumption rates of ctenophores, coupled with their potential for rapid increase in

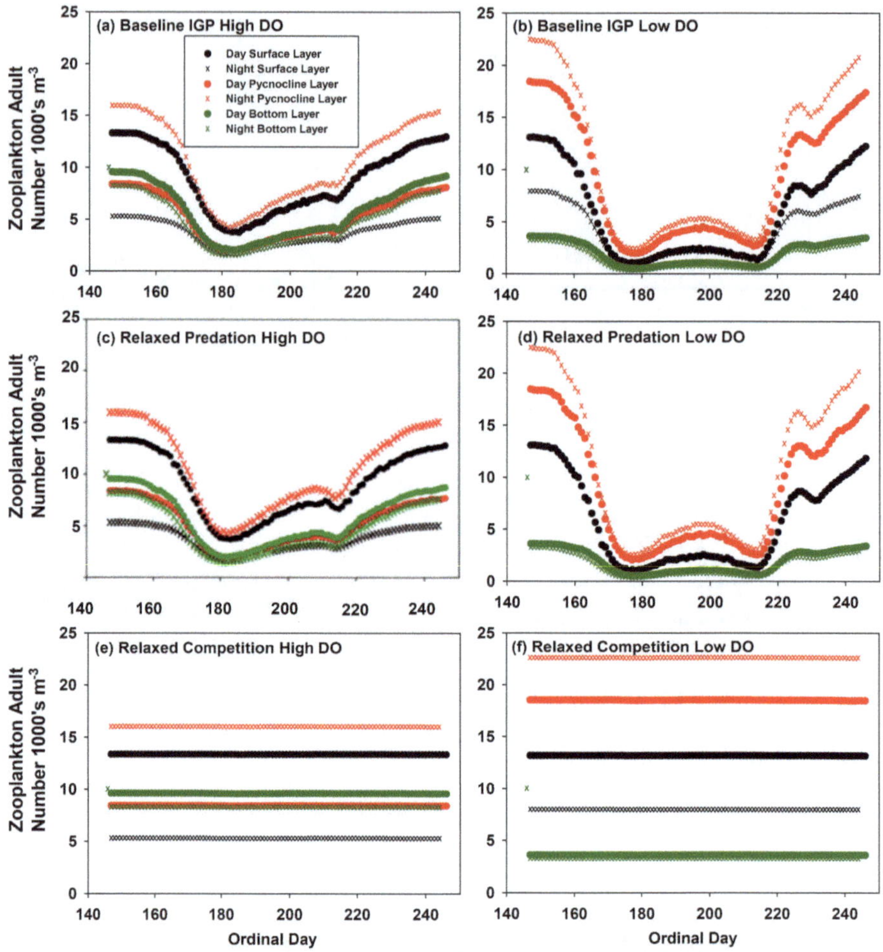

Fig. 11.7 Copepod number m^{-3} by layer plotted against ordinal day for the adult life stage during both day (●) and night (x) for a representative simulation for each of six different food webs: **a** baseline IGP high DO, **b** baseline IGP low DO, **c** relaxed predation high DO, **d** relaxed predation low DO, **e** relaxed competition high DO, **f** relaxed competition low DO. *Black dots* denote the surface layer, *red dots* the pycnocline, and *green dots* the bottom layer. Mean densities of copepodites and nauplii are reported for the baseline IGP food web (Table 11.6) and distribution patterns are similar to those of adult copepods in all food web scenarios. Single dots on the first day of simulations are an artifact of the initial density

biomass, makes them voracious planktonic predators (Monteleone and Duguay 1988; Purcell and Decker 2005). Predation is thought to be the largest source of mortality for the early life stages of fish (Bailey and Houde 1989). Slower growing larval fish are vulnerable to size-specific predation longer than are faster growing

Table 11.4 Results from five replicate runs of model simulations in the baseline IGP, relaxed predation, and relaxed competition food webs at both high and low DO. Reported values are the mean, minimum, and maximum total number of fish larvae reaching 15 mm and mean, minimum, and maximum total number of days for fish larvae to reach 15 mm during the 100 d simulation. No fish larvae less than 15 mm remain at the end of the simulations. Survivals are reported as percents. Minimum and maximum values based on the five replicate simulations are shown in parentheses for number of survivors to 15 mm and for larval duration

Variable	Baseline		Relaxed predation		Relaxed competition	
	High DO	Low DO	High DO	Low DO	High DO	Low DO
Number of survivors to 15 mm	14.1 (13.8–14.5)	9.9 (9.7–10.2)	352.0 (349.9–354.0)	243.5 (238.8–246.9)	41.5 (38.1–44.8)	19.4 (14.6–24.4)
Larval duration (days)	26.0 (25.5–26.4)	19.7 (19.3–20.1)	30.2 (30.1–30.4)	23.9 (23.7–24.3)	19.0 (18.9–19.0)	15.6 (15.4–15.9)
Egg to hatch survival	40.0	13.0	50.0	23.0	40.0	13.0
Hatch to first feeding survival	16.8	17.2	50.0	48.0	16.3	18.8
First feeding to 15 mm survival	2.1	4.4	14.2	22.2	6.4	7.7
Egg to 15 mm survival	0.14	0.098	3.55	2.45	0.417	0.189

larval fish (Bailey and Houde 1989). These model results support the importance of ctenophore predation to the survival of early life stages of fish suggested by experimental and field studies (Cowan and Houde 1993; Purcell et al. 1994a, b).

Larval fish growth rate (inversely related to duration) in the baseline model under high DO was affected indirectly via competition with ctenophores for copepod prey. Average larval growth rate to 15 mm was 0.46 mm d^{-1}, which corresponded to an average duration of 26 days from first feeding to 15 mm (Table 11.4). In general, larval fish lengths during the middle of the summer did not increase as rapidly as larval fish lengths during the early and late portions of the simulation (Fig. 11.8a). Slowed growth was due to competition and coincided with low copepod densities (Fig. 11.7a) and high ctenophore densities (Fig. 11.6a).

Larval fish diets in the baseline simulation were composed mostly of copepodites, with smaller proportions of copepod nauplii and adults (Table 11.5). The model restricted diets of small larvae (<5 mm) to nauplii; medium-sized larvae (5–10 mm) included copepodites, and large-sized larvae (>10 mm) further added adult copepods to their diet.

Growth of smaller ctenophores (<400 mg dw) was slowed during the middle summer by lowered copepod densities, while once ctenophores reached 400 mg dw, their growth rate was rapid throughout the summer (Fig. 11.9a). Ctenophores consumed mostly nauplii (Table 11.6). Weights of smaller ctenophores increased

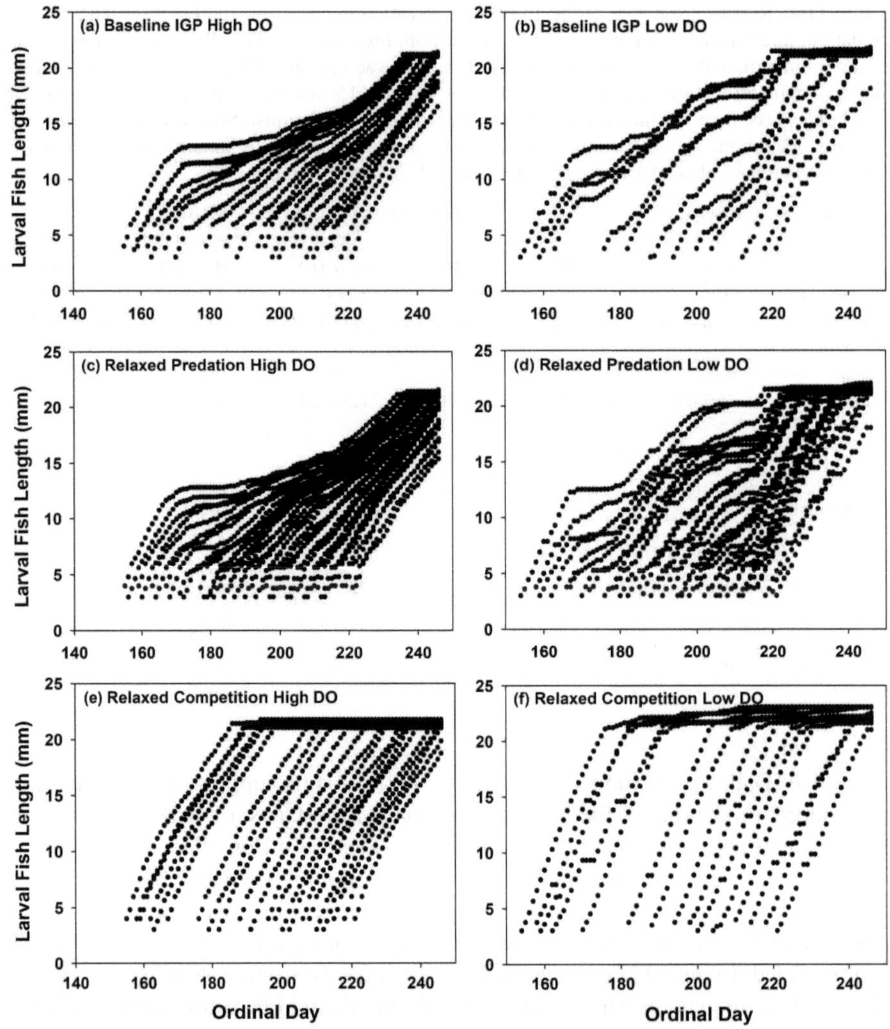

Fig. 11.8 Larval fish length (mm) plotted against ordinal day for six different simulations: **a** baseline IGP high DO, **b** baseline IGP low DO, **c** relaxed predation high DO, **d** relaxed predation low DO, **e** relaxed competition high DO, **f** relaxed competition low DO. Each *line* represents a cohort of individual feeding fish larvae throughout the simulation with each cohort entering the model at a different time step. The trajectory of size through time provides a representation of larval fish growth rates

rapidly early and late in the summer when nauplii densities were relatively high and showed slowed growth during the middle of the summer (days 50–150, Fig. 11.9a) when nauplii densities were low (Fig. 11.7a).

Table 11.5 Values for the mean proportion of biomass (mg dw) of each copepod life stage in modeled larval fish diets for three size classes of larval fish. Larval fish size classes were small (larval fish length <5 mm), medium (10 mm ≥ larval fish length ≥ 5 mm), and large (larval fish length >10 mm) for all three food webs: baseline IGP, relaxed predation, and relaxed competition at both high and low DO

	Small larvae		Medium larvae		Large larvae	
	High DO	Low DO	High DO	Low DO	High DO	Low DO
Baseline						
Copepod nauplii	1.00	1.00	0.13	0.36	0.26	0.38
Copepodites	0.00	0.00	0.87	0.64	0.60	0.35
Adult copepods	0.00	0.00	0.00	0.00	0.14	0.27
Relaxed predation						
Copepod nauplii	1.00	1.00	0.15	0.35	0.26	0.44
Copepodites	0.00	0.00	0.85	0.65	0.59	0.34
Adult copepods	0.00	0.00	0.00	0.00	0.15	0.22
Relaxed competition						
Copepod nauplii	1.00	1.00	0.26	0.88	0.32	0.49
Copepodites	0.00	0.00	0.74	0.12	0.25	0.15
Adult copepods	0.00	0.00	0.00	0.00	0.43	0.36

11.4.3 Effect of Low DO in the Baseline Food Web

Low DO caused a 30% reduction in overall survival of fish (9.9 vs. 14.1 survivors, Table 11.4), with the effect of DO on decreased egg survival partially offset by increased larval survival. Survival of fish eggs to hatch was lower in the low DO simulations than in high DO due to direct DO mortality on eggs. Thirteen percent of spawned fish eggs hatched to reach the yolk sac larvae stage at low DO as compared to 40% survival at high DO. Percent survival from hatch to first feeding was similar under both high and low DO (16.0 and 17.2%). In contrast, survival of larval fish from first feeding to 15 mm was higher under low DO (4.4% vs. 2.1%).

Higher larval fish survival under low DO was due to the effects of low DO reducing the overlap between fish larvae and ctenophores and the resulting reduction in encounter rates and predation mortality (two leftmost set of bars in Fig. 11.10). Potential encounter rates of later survivors were also reduced due to the higher mortality of fish eggs under hypoxia. The temporal patterns of densities between the low and high DO simulations were similar for ctenophores (Fig. 11.5a vs. b), larval fish (Fig. 11.6a vs. b), and adult copepod densities (Fig. 11.7a vs. b); however, the spatial overlap among the three vertical layers was altered. Peak densities of ctenophores shifted from the bottom and pycnocline layers under high DO (green in Fig. 11.5a) to the pycnocline and especially the surface layer during

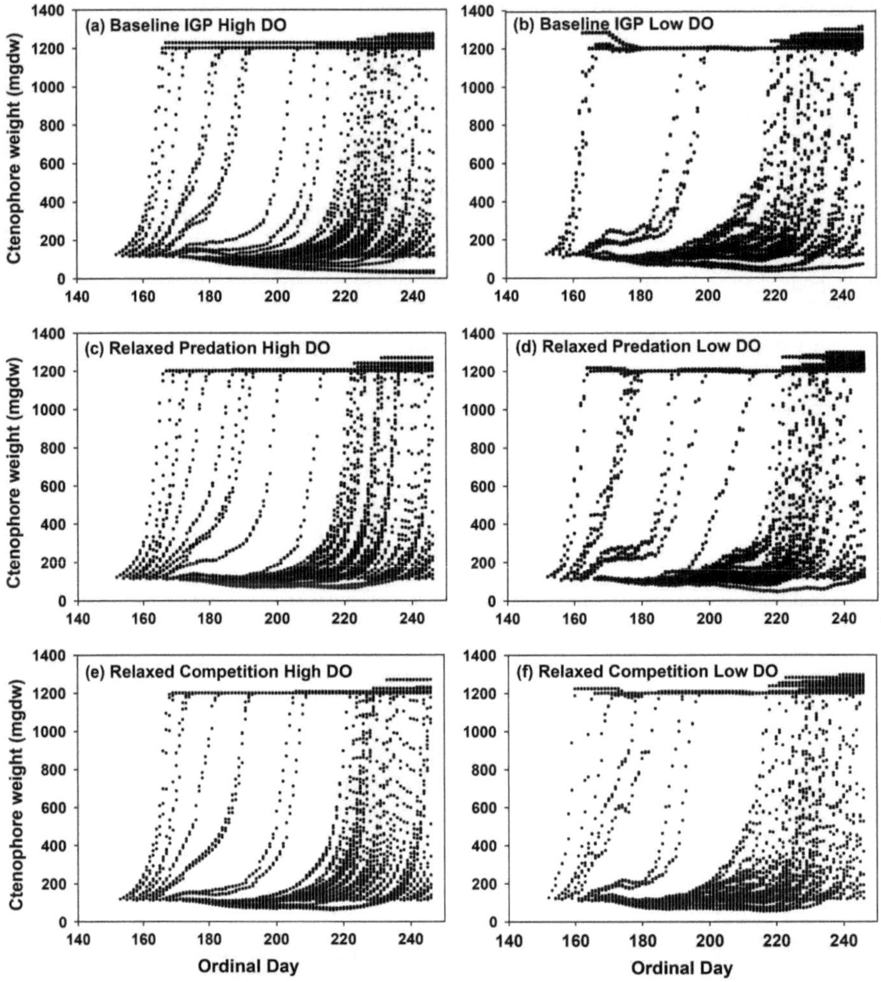

Fig. 11.9 Ctenophore weight (mg dw) plotted against ordinal day for six different simulations: **a** baseline IGP high DO, **b** baseline IGP low DO, **c** relaxed predation high DO, **d** relaxed predation low DO, **e** relaxed competition high DO, **f** relaxed competition low DO. Each line represents a cohort of individual ctenophores throughout the simulation with each cohort entering the model at a different time step. The trajectory of weight through time provides a representation of ctenophore growth rates

the day (black and red in Fig. 11.5b). Larval densities, which were highest in the bottom and pycnocline layers at high DO shifted mostly to the pycnocline (but not surface) layers (red in Fig. 11.6b). This resulted in the overlap between ctenophores and fish larvae being lowered from about 0.9 under high DO to about 0.5 under low DO (two leftmost bars in Fig. 11.10), resulting in reduced encounter rates and less predation by ctenophores on larval fish.

Table 11.6 Mean proportion of biomass (mg dw) due to each copepod life stage in ctenophore diets for the baseline, relaxed predation, and relaxed competition under both high and low DO scenarios. Copepods comprise the majority of ctenophore diets during the 100 d model simulation

	High DO	Low DO
Baseline		
Copepod nauplii	0.78	0.81
Copepodites	0.08	0.07
Adult copepods	0.14	0.12
Relaxed predation		
Copepod nauplii	0.78	0.81
Copepodites	0.07	0.07
Adult copepods	0.14	0.12
Relaxed competition		
Copepod nauplii	0.78	0.81
Copepodites	0.08	0.07
Adult copepods	0.14	0.12

It was important to include a spatial component in the model because changes in vertical distribution in response to low DO triggered indirect effects. Spatial dynamics increases model complexity (Polis and Strong 1996), but including a spatial component can capture important features of food web interactions such as habitat refuges that increase food web persistence and reduce the likelihood of local extinctions (Keitt 1997; Fulton et al. 2004; Amarasekare 2007). Although larval fish survival decreased with low DO in all food webs, the response was not solely due to higher DO-related mortality. At low DO, larval survival from first feeding to 15 mm increased, and larval growth rate increased. Thus, even within a limited part of the life cycle (eggs and larvae), the response of fish to low DO can be complicated and include indirect effects mediated by changes in spatial overlaps with other members of the food web.

Larval fish growth rates were faster in the low DO simulation than in the high DO simulation, despite the negative direct effect imposed (Eq. 11.11), resulting in shorter larval stage duration (Table 11.4). Growth rates of fish larvae from first feeding to 15 mm was 0.61 mm d^{-1} in the low DO simulation, which corresponded to an average duration of 19.7 days, more than 6 days faster than the 26.0 days in the high DO baseline food web. Lowered larval fish densities due to increased egg mortality relaxed some of the predation pressure on copepods, and low DO caused slightly more overlap between larvae and copepods (middle set of bars in Fig. 11.10), resulting in faster growth for the remaining larvae. Reduced overlap between ctenophore predators and copepod prey at low DO also increased zooplankton prey densities available to larval fish.

Low DO caused a shift in larval fish diets away from copepodites (Table 11.5), but had no effect on ctenophore diets (Table 11.6). Small larval fish (length <5 mm) ate only copepod nauplii, so diets shifts due to low DO were not possible. Medium-sized larvae (5–10 mm) shifted to more nauplii, and large-size larvae shifted to more nauplii and more adult copepods in lower DO regions. Low DO did

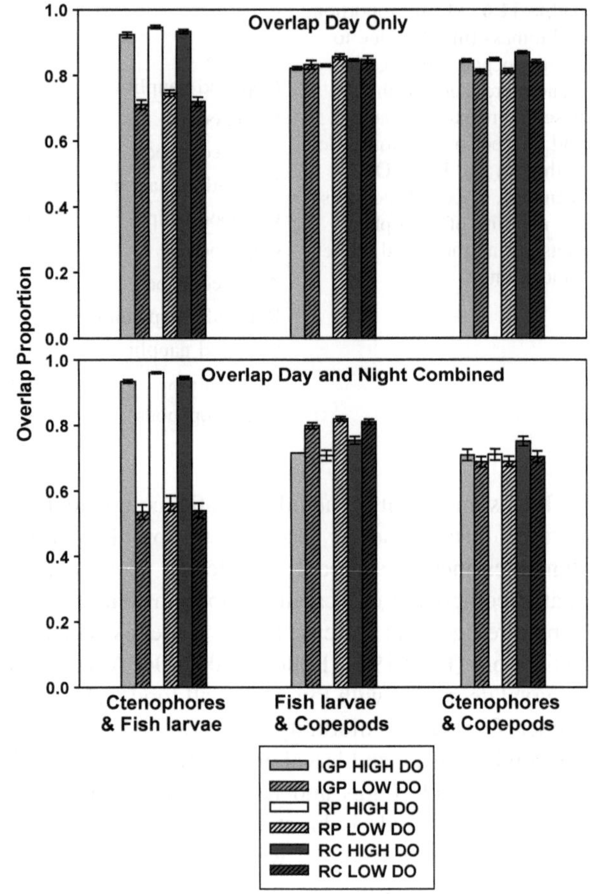

Fig. 11.10 Vertical habitat overlap for three predator–prey pairs in the six different simulations: baseline intra-guild predation high DO (IGP HI DO, *gray bars*), baseline IGP low DO (IGP LO DO, *gray striped bars*), relaxed predation high DO (RP HI DO, *white bars*), relaxed predation low DO (RP LO DO, *white striped bars*), relaxed competition high DO (RC HI DO, *dark gray bars*), relaxed competition low DO (RC LO DO, *dark gray striped bars*). The three predator–prey pairs are: ctenophores and fish larvae, fish larvae and copepods, and ctenophores and copepods. Overlap for all three copepod life stages (copepod nauplii, copepodites, and adult copepods) was combined since their vertical habitat distribution was the same. Daytime values are shown for larval fish predation and day and night combined are shown for ctenophores predators. Data are mean overlap ±95% confidence interval

not affect the relative proportions of the three copepod life stages in ctenophore diets (Table 11.6).

Uncoupling the influence of a heterogeneous habitat from predation can be difficult (Anholt and Werner 1995). Spatial distributions of predators, competitors, and prey in the environment may be important for food web persistence and species coexistence (Rosenheim et al. 2004), with habitat complexity leading to food web complexity (Angel and Ojeda 2001). For example, the presence of a motile, omnivorous predator such as a ctenophore may stabilize complex food webs by increasing energy flow through weak links that promote species coexistence (McCann et al. 2005; Morris 2005). Under this premise, we attribute enhanced growth of larval fish in our modeled low DO food webs to greater spatial overlap of the larvae with copepod prey. But overall survival of larval fish cannot be explained by the same mechanism.

11.4.4 Importance of Predation Versus Competition to Fish Larvae Under High DO

Ctenophore predation had a much larger effect on larval fish survival than competition, and competition had a larger effect on larval growth; however, both predation and competition affected larval survival and growth. Under high DO conditions, survival of fish eggs to hatch increased from 40% in the baseline and relaxed competition food webs to 50% in the relaxed predation food web, hatch to first feeding survival increased from 16.8 to 50%, and first feeding to 15 mm survival increased from less than 6.4 to 14.2% (Table 11.4). Cumulative survival from egg to 15 mm was 24 times higher than baseline survival under relaxed predation (3.55/0.14), but only three times higher than baseline under relaxed competition (0.417/0.14).

Larval fish growth rates under high DO conditions were fastest in the relaxed competition food web and, due to less predation resulting in higher larval densities and increased density-dependent effects, slowest in the relaxed predation food web (Table 11.4). Larval duration was 7 days shorter than baseline when competition was relaxed and 4 days longer when predation was relaxed. Corresponding average growth rates of survivors were 0.63 mm d^{-1} under relaxed competition, 0.46 mm d^{-1} under baseline, and 0.40 mm d^{-1} under relaxed predation. The slopes of lines for larval fish length versus day, an indication of growth rates, were very high throughout the relaxed competition food web (Fig. 11.8e), corresponding to high copepod densities throughout the summer (Fig. 11.7e). Slopes under baseline and relaxed predation flattened during mid-summer (Fig. 11.8a, c), which was when ctenophore densities were high (Fig. 11.6a, c) and copepod densities were low (Fig. 11.7a, c). Relaxed competition resulted in a shift of large-sized larvae to eat more adult copepods (Table 11.5). But the faster growth under relaxed competition resulted in a much smaller increase in survival than the relaxed predation scenario (Table 11.4).

Temporal patterns of larval densities showed a peak in mid-summer under relaxed predation (Fig. 11.6c), rather than a depression in mid-summer under baseline and relaxed competition (Fig. 11.6a, e). Water column integrated peak densities were 40 individuals m^{-3} under relaxed predation compared to less than 6 individuals m^{-3} in both baseline and relaxed competition. Relaxed predation and relaxed competition did not alter the vertical distribution of the larvae, with most larvae in the bottom and pycnocline layers as in baseline (red and green in Fig. 11.6 a, c, e). Thus, the drop in mid-summer larval densities under baseline was due to predation effects, rather than competition for zooplankton leading to slowed growth of larvae, longer duration, and higher cumulative mortality.

Ctenophores were generally unaffected by the relaxed competition and predation food webs. The magnitude and temporal pattern of ctenophore densities (Fig. 11.5a, c, e), growth rates (Fig. 11.9a, c, e), and diets (Table 11.6) were similar under high DO in all three food webs.

11.4.5 Interaction of Low DO with Different Predation and Competition Conditions

We used a factorial design in model simulations to isolate the interaction effect between DO level and food web type and found little evidence of an interaction effect. The effects of high versus low DO on larval survival and growth was consistent across the three food webs (Table 11.4), suggesting that there was not a strong interaction between DO conditions and food web type. The only exception was that DO had a larger effect on larval survival in the relaxed competition food web compared to the other food webs. Survival from egg to 15 mm under relaxed predation was 3.55% under high DO and 2.45% under low DO (a 31% decrease), which was very similar to the decrease under baseline (0.14% vs. 0.098%, or a 30% decrease). However, under relaxed competition, survival under high DO was 0.417% vs. 0.189%, or a 54% decrease, compared to a 30% decrease in survival under baseline. The difference was that first feeding to 15 mm survival increased much less (and therefore offset less of the increased egg mortality) under relaxed competition (6.4 to 7.7%), than the more than doubling under baseline (2.1 to 4.4%) and the 50% increase (14.2% to 22.2%) under relaxed predation (Table 11.4). The temporal pattern of larval densities, while different in the relaxed predation food web, all showed similar general reductions in densities from high to low DO and a shift from the bottom layer to the pycnocline and surface layers (Fig. 11.6a vs. b, c vs. d, e vs. f).

Going from high to low DO also had a consistent effect on larval growth and diets across the three food webs. Larval durations were 18–24% shorter at low DO compared to high DO for all three food webs (Table 11.4). Larval lengths over time showed consistently steeper slopes (faster growth and shorter duration) during the mid-summer under low DO (Fig. 11.8a vs. b, c vs. d, e vs. f). As with going from high to low DO under baseline, diets of the medium-sized larvae shifted away from copepodites toward nauplii and diets of large-sized larvae shifted to greater consumption of nauplii and adult copepods (Table 11.5).

Low DO affected ctenophore densities, growth, and diets similarly in all three food webs. Water column ctenophore densities peaked at around 9 individuals m^{-3} under high DO in the three food webs and at about 8 individuals m^{-3} in the low DO food webs and showed similar shifts by layer (Fig. 11.5). Ctenophore growth was also similarly slowed under low DO for all three food webs (Fig. 11.9). Ctenophore diets were very similar across all food webs and for low and high DO conditions (Table 11.6).

DO also had only small effects on the relative importance of competition versus predation to larval survival and growth (Table 11.4). The increase in larval survival from baseline to relaxed predation was the same for high DO (0.14 to 3.55%, 25x higher) as for low DO (0.098 to 2.45%, 25x higher). Similarly, the increase in larval survival was similar for baseline to relaxed competition for high DO (0.417/0.14, or 3x) versus low DO (0.189/0.098, or 2x). Thus, the effect of DO (high vs. low) on larval survival and growth did not depend on the food web and DO did not greatly change the relative importance of competition versus predation within the food web.

The relaxed predation food web results showed that larval densities would peak in mid-summer but that the ctenophore predation in the baseline IGP food web and in the relaxed competition food web caused a mid-summer dip in larval densities. Note that model simulations were conducted without sea nettle predators, which can depress ctenophore densities in some field populations. Relaxing predation resulted in a 25x increase in larval survival, but relaxing competition resulted in the smaller but still important 3x increase in survival. Relaxing competition resulted in shorter larval stage durations, and relaxed predation (because of high larval densities) resulted in longer larval stage durations.

In our analysis, the DO effects were consistent across food webs and DO did not greatly affect the importance of competition versus predation. These results suggest that the effects of DO we found are robust and likely apply to a broad set of field situations. The two extreme food webs (competition completely relaxed; predation completely relaxed) bound the conditions observed in many estuaries and the differences in the food webs possible among spatial subregions and during different time periods within a system.

11.5 Conclusion

Intraguild predation food webs are thought to persist due to the superiority of the IGP prey in exploiting shared resources, or because IGP prey has a resource subsidy unavailable to the IG predator (Polis 1984). But in our modeled food web, and in the Chesapeake Bay system, ctenophores were both a predator on fish larvae and a superior competitor for copepod prey. Our result, of lowest survival of early life stages of fish in the IGP food web, provides evidence that this particular IGP food web would likely not facilitate persistence of larval fish. However, the modeled system represents a subset of the complete food web structure of many temperate estuaries. Factors such as the age structure and seasonality of the food web, as well as the effects of DO on vertical habitat overlap, limit ctenophore predation on fish egg and larval stages to a brief period during the summer months. Temporal and spatial patchiness of ctenophores due to predation by *Chrysaora quinquecirrha* medusa, not considered in these simulations, can also contribute to larval fish survival in the field.

Using a modeling approach to address questions about food web structure and the effects of low DO on trophic interactions had advantages as well as limitations. The individual-based, spatially explicit food web model enabled us to simulate the effects of ctenophore predation and competition with fish larvae; the ability to simulate competition was especially valuable because competition is difficult to isolate in either the field or laboratory. In constructing the food web model, we made certain simplifying assumptions (e.g., constant temperature, DO levels fixed through time, simple behavioral rules for feeding and movement) to keep a moderate level of simplicity in a complicated model. Next steps for the food web model could include adding more trophic levels for both prey (e.g., phytoplankton and microzooplankton) and predators (*Chrysaora quinquecirrha* medusae), and considering traits of larval fish other than bay anchovy to test whether our predictions are species specific or generally robust, and therefore applicable to other estuarine food webs. Myriad variations can include additional within-summer variation in environmental variables and in the phenologies of the zooplankton, ctenophores, and larval fish. Low DO is often associated with eutrophication, and additional simulations might include eutrophication effects on the food web. Our results demonstrate that our ability to assess hypoxia effects on fish is improved by models that allow for both indirect effects via the food web and alteration of spatial distributions.

Acknowledgements We thank W. Boynton, T. Miller, and J. Purcell for comments on a previous version of this manuscript and G. Waldbusser for assistance with figure preparation. Comments from two anonymous reviewers greatly improved this submission, and we are grateful for their efforts. Research was partially supported (KAR) by the National Oceanographic and Atmospheric Administration, Center for Sponsored Coastal Ocean Research (CSCOR) CHRP Grant numbers NA10NOS4780157 awarded to Louisiana State University and NA10NOS4780138 awarded to the Smithsonian Environmental Research Center (DLB). This is publication number 220 of the NOAA's CSCOR NGOMEX and CHRP programs.

Appendix A. Stage-Based Matrix Projection Models for Fish Eggs and Yolk Sac Larvae, and Ctenophore Eggs and Larvae

Six stage-based matrix projection models were used to update fish eggs and yolk sac larvae, and ctenophore eggs and larvae. There was a model for fish and for ctenophores for each of the three layers. The models were 2 × 2 and operated on a 12-h time step. For each model, we computed the diagonal and subdiagonal elements from stage survival and duration every 12 h. We first computed from survival over 12 h for the ith stage from specified daily instantaneous survival rates as $S_i = e^{-M_i/2}$. For fish eggs only, DO was used to compute *SurEggDO* (Eq. 11.13), and S_i for eggs was then adjusted as $S_i \cdot SurEggDO$. We then computed φ_i, survival for each time period, from S_i and duration (D_i, number of 12 h):

$$\varphi_i = \frac{S_i^D - S_i^{D-1}}{S_i^D - 1}$$

The diagonal and off-diagonal elements were then:

$$D_{i,i} = S_i \cdot (1 - \varphi_i)$$

$$D_{i,i+1} = S_i \cdot \varphi_i$$

Fecundity (usually the top row of the matrices) was dealt with by simply adding newly entering eggs to those already present in each layer every 12 h. Number of eggs added was computed based on day of year, and dynamically each 12 h for ctenophores based on growth and summed over individual ctenophores.

At the beginning of each 12-h time step, the matrices were specified and the numbers of individuals in each stage were updated. Newly entering eggs for fish and ctenophores were then added to their egg abundances. Then during the next 12 h, consumption of fish eggs and yolk sac larvae by ctenophores was subtracted from the total number of individuals in each layer. The decreased numbers of individuals in each life stage in each layer were then used to start the next time step.

The mortality rates, durations, and fecundity rates are shown in Table 11.7. Typical matrices for each of the taxa were:

Fish:

$$\begin{bmatrix} 0.75 & 5 \\ 0.25 & 0.75 \end{bmatrix}$$

Ctenophores:

$$\begin{bmatrix} 0.402 & 20 \\ 0.27 & 0.402 \end{bmatrix}$$

Table 11.7 Mortality, stage duration, and fecundity rates for stage-based matrix projection models for fish eggs and yolk sac larvae, and ctenophore eggs and larvae

Taxa	Stages	Duration (12 h)	Mortality (d^{-1})	Fecundity (entering) or exiting
Fish	Eggs	2.0	0.001	Added by layer as 100 eggs m^{-3}; every 3 days for days 150–189 and 213–220; daily for days 190–212
	Yolk sac larvae	4.0	0.001	Become individual feeding larvae
Ctenophores	Eggs	2.0	0.8	Added by layer by summing over eggs produced by individual ctenophores
	Larvae	20.0	0.6	Become lobates; then after 5–7 days became 25-mm individuals

References

Adamack AT, Rose KA, Breitburg DL, Nice AJ, Lung WS (2012) Simulating the effects of hypoxia on bay anchovy egg and larval mortality in the Patuxent River using coupled watershed, water quality, and individual-based predation models. Mar Ecol Prog Ser 445:141–161

Angel A, Ojeda FP (2001) Structure and trophic organization of subtidal fish assemblages on the northern Chilean coast: the effect of habitat complexity. Mar Ecol Prog Ser 217:81–91

Amaraskare P (2007) Trade-offs, temporal variation, and species coexistence in communities with intraguild predation. Ecology 88:2720–2728

Anholt BR, Werner EE (1995) Interaction between food availability and predation mortality mediated by adaptive behavior. Ecology 76:2230–2234

Arim M, Marquet PA (2004) Intraguild predation: a widespread interaction related to species biology. Ecol Lett 7:557–564

Bailey KM, Houde ED (1989) Predation on eggs and larvae of marine fishes and the recruitment problem. Adv Mar Biol 25:1–67

Breitburg DL, Adamack A, Rose KA, Kolesar SE, Decker MB, Purcell JE, Keister JE, Cowan JH Jr (2003) The pattern and influence of low dissolved oxygen in the Patuxent River, a seasonally hypoxic estuary. Estuaries 26:280–297

Breitburg DL, Craig JK, Fulford RS, Rose KA, Boynton WR, Brady DC, Ciotti BJ, Diaz RJ, Friedland KD, Hagy JD III, Hart DR, Hines AH, Houde ED, Kolesar SE, Nixon SW, Rice JA, Secor DH, Targett TE (2009) Nutrient enrichment and fisheries exploitation: interactive effects on estuarine living resources and their management. Hydrobiologia 629:31–47

Breitburg DL, Loher T, Pacey CA, Gerstein A (1997) Varying effects of low dissolved oxygen on trophic interactions in an estuarine food web. Ecol Monogr 67:489–507

Brownlee DC, Jacobs F (1987) Mesozooplankton and microzooplankton in the Chesapeake Bay. In: Majumdar SK, Hall LW, Austin HM (eds) Contaminant problems and management of living chesapeake bay resources. The Pennsylvania Academy of Science, Philadelphia, Pennsylvania, pp 217–269

Caddy JF (1993) Toward a comparative evaluation of human impacts on fishery ecosystems of enclosed and semi-enclosed areas. Rev Fish Sci 1:57–95

Cloern JE (2001) Our evolving conceptual model of the coastal eutrophication problem. Mar Ecol Prog Ser 210:223–253

Cowan JH Jr, Houde ED (1993) Relative predation potentials of scyphomedusae, ctenophores and planktivorous fish on ichthyoplankton in Chesapeake Bay. Mar Ecol Prog Ser 95:55–65

Cowan JH Jr, Rose KA, Houde ED, Wang SB, Young J (1999) Modeling effects of increased larval mortality on bay anchovy population dynamics in the mesohaline Chesapeake Bay: evidence for compensatory reserve. Mar Ecol Prog Ser 185:133–146

Diaz RJ, Rosenberg R (2008) Spreading dead zones and consequences for marine ecosystems. Science 32:926–929

Decker MB, Breitburg DL, Purcell JE (2004) effects of low dissolved oxygen on zooplankton predation by the ctenophore *Mnemiopsis leidyi*. Mar Ecol Prog Ser 280:163–172

Diehl S (1995) Direct and indirect effects of omnivory in a littoral lake community. Ecology 76:1727–1740

Diehl S, Feißel M (2000) Effects of enrichment on three-level food chains with omnivory. Am Nat 155:200–218

Dorsey SE, Houde ED, Gamble JC (1996) Cohort abundances and daily variability in mortality of eggs and yolk-sac larvae of bay anchovy, *Anchoa mitchilli*, in Chesapeake Bay. Fish Bull 94:257–267

Ehler LE (1996) Structure and impact of natural enemy guilds in biological control of insect pests. In: Polis GA, Winemiller KO (eds) Food webs: integration of patterns and dynamics. Chapman and Hall, New York, NY, pp 337–342

Ekau W, Auel H, Pörtner H-O, Gilbert D (2010) Impacts of hypoxia on the structure and processes in pelagic communities (zooplankton, macro-invertebrates and fish). Biogeosciences 7:1669–1699

Fagan WF (1997) Omnivory as a stabilizing feature of natural communities. Am Nat 150:554–567

Fives JM, Warlen SM, Hoss DE (1986) Aging and growth of larval bay anchovy, *Anchoa mitchilli*, from the Newport River Estuary, North Carolina. Estuaries 9:362–367

Fulton EA, Smith ADM, Johnson CR (2004) Effects of spatial resolution on the performance and interpretation of marine ecosystem models. Ecol Model 176:27–42

Gerritsen J, Strickler JR (1977) Encounter probabilities and community structure in zooplankton: a mathematical model. J Fish Res Board Can 34:73–82

Gilbert D, Rabalais NN, Díaz RJ, Zhang J (2010) Evidence for greater oxygen decline rates in the coastal ocean than in the open ocean. Biogeosciences 7:2283–2296

Grove M, Breitburg DL (2005) Growth and reproduction of gelatinous zooplankton exposed to low dissolved oxygen. Mar Ecol Prog Ser 301:185–198

Gurevitch J, Morrison JA, Hedges LV (2000) The interaction between competition and predation: a meta-analysis of field experiments. Am Nat 155:435–453

Harris RP, Wiebe P, Lenz J, Skjoldal HR, Huntley, M (2000) ICES zooplankton methodology manual. Academic Press London, UK, 705 pps

Heinle DR (1966) Production of a calanoid copepod, *Acartia tonsa*, in the Patuxent River estuary. Chesap Sci 7:59–74

Heithaus MR (2001) Habitat selection by predators and prey in communities with asymmetrical intraguild predation. Oikos 92:542–554

Holt RD, Polis GA (1997) A theoretical framework for intraguild predation. Am Nat 149:745–764

Hunter JR, Leong R (1981) The spawning energetics of female Northern Anchovy, *Engraulis mordax*. Fish Bull 79:215–230

Janssen A, Sabelis MW, Magalhães S, Montserrat M, van der Hammen T (2007) Habitat structure affects intraguild predation. Ecology 88:2713–2719

Keister JE, Houde ED, Breitburg DL (2000) Effects of bottom-layer hypoxia on abundances and depth distribution of organisms in Patuxent River, Chesapeake Bay. Mar Ecol Prog Ser 205:43–59

Keitt TH (1997) Stability and complexity on a lattice: coexistence of species in an individual-based food web model. Ecol Model 102:243–258

Kemp WM, Boynton WR, Adolf JE, Boesch DF, Boicourt WC, Brush G, Cornwell JC et al (2005) Eutrophication of Chesapeake Bay: historical trends and ecological interactions. Mar Ecol Prog Ser 303:1–29

Kidwell DM, Lewitus AJ, Jewett EB, Brandt S, Mason DM (2009) Ecological impacts of hypoxia on living resources. J Exp Mar Biol Ecol 381:S1–S3

Kimmel DG, Roman MR (2004) Long-term trends in mesozooplankton abundance in Chesapeake Bay, USA: influence of freshwater input. Mar Ecol Prog Ser 267:71–83

Kolesar, SE (2006) The effects of low dissolved oxygen on predation interactions between *Mnemiopsis leidyi* ctenophores and larval fish in the Chesapeake Bay ecosystem. PhD dissertation, University of Maryland, College Park, MD

Kolesar SE, Breitburg DL, Purcell JE, Decker MB (2010) Effects of hypoxia on *Mnemiopsis leidyi*, ichthyoplankton and copepods: clearance rates and vertical habitat overlap. Mar Ecol Prog Ser 411:173–188

Kremer P (1976) Population dynamics and ecological energetics of a pulsed zooplankton predator, the ctenophore *Mnemiopsis leidyi*. In: Wiley ML (ed) Estuarine Processes. Academic Press, NY, 1:197–215

Kremer P (1979) Predation by the ctenophore *Mnemiopsis leidyi* in Narragansett Bay, Rhode Island. Estuaries 2:97–105

Kremer P, Reeve MR (1989) Growth dynamics of a ctenophore (*Mnemiopsis*) in relation to variable food supply. II. Carbon budgets and growth model. J Plankton Res 11:535–552

Kreps TA, Purcell JE, Heidelberg KB (1997) Escape of the ctenophore *Mnemiopsis leidyi* from the scyphomedusa predator *Chrysaora quinquecirrha*. Mar Biol 128:441–446

Laurence GC (1976) Caloric values of some North Atlantic calanoid copepods. Fish Bull 74:218–220

Levin LA, Ekau W, Gooday AJ, Jorissen F, Middelburg JJ, Naqvi SWA, Neira C, Rabalais NN, Zhang J (2010) Effects of natural and human-induced hypoxia on coastal benthos. Biogeosciences 6:2063–2098

McCann K, Hastings A, Huxel GR (1998) Weak trophic interactions and the balance of nature. Nature 395:794–798

McCann KS, Rasmussen JB, Umbanhowar J (2005) The dynamics of spatially coupled food webs. Ecol Lett 8:513–523

Monteleone DM, Duguay LE (1988) Laboratory studies of predation by the ctenophore *Mnemiopsis leidyi* on the early stages in the life history of the bay anchovy, *Anchoa mitchilli*. J Plankton Res 10:359–372

Morris DW (2005) Paradoxical avoidance of enriched habitats: have we failed to appreciate omnivores? Ecology 86:2568–2577

Navarrette SA, Menge BA, Daley BA (2000) Species interactions in intertidal food webs: prey or predation regulation of intermediate predators? Ecology 81:2264–2277

Polis GA (1984) Age structure component of niche width and intraspecific resource partitioning: can age groups function as ecological species? Am Nat 123:541–564

Polis GA (1998) Stability is woven by complex webs. Nature 395:744–745

Polis GA, Holt RD (1992) Intraguild predation: the dynamics of complex trophic interactions. Trends Ecol Evol 7:151–154

Polis GA, Myers CA, Holt RD (1989) The ecology and evolution of intraguild predation: potential competitors that eat each other. Annu Rev Ecol Syst 20:297–330

Polis GA, Strong DR (1996) Food web complexity and community dynamics. Am Nat 147:813–835

Purcell JE, Arai MN (2001) Interactions of pelagic cnidarians and ctenophores with fish: a review. Hydrobiolgia 451:27–44

Purcell JE, Breitburg DL, Decker MB, Graham WM, Youngbluth MJ, Raskoff KA (2001) Pelagic cnidarians and ctenophores in low dissolved oxygen environments: a review. In: Rabalais NN, Turner RE (eds) Coastal hypoxia: consequences for living resources and ecosystems, vol 58. American Geophysical Union, Coastal and Estuarine Studies, pp 77–100

Purcell JE, Decker MB (2005) Effects of climate on predation by ctenophores and scyphomedusae on copepods in Chesapeake Bay during 1987–2000. Limnol Oceanogr 50:376–387

Purcell JE, Decker MB, Breitburg DL, Broughton KJ (2014) Fine-scale vertical distributions of *Mnemiopsis leidyi* ctenophores: predation on copepods relative to stratification and hypoxia. Mar Ecol Prog Ser 500:103–120

Purcell JE, Nemazie DA, Dorsey SE, Houde ED, Gamble JC (1994a) Predation mortality of bay anchovy (*Anchoa mitchilli*) eggs and larvae due to scyphomedusae and ctenophores in Chesapeake Bay. Mar Ecol Prog Ser 114:47–58

Purcell JE, White JR, Roman MR (1994b) Predation by gelatinous zooplankton and resource limitation as potential controls of *Acartia tonsa* copepod populations in Chesapeake Bay. Limnol Oceanogr 23:740–751

Rabalais NN, Díaz RJ, Levin LA, Turner RE, Gilbert D, Zhang J (2010) Dynamics and distribution of natural and human-caused hypoxia. Biogeosciences 7:585–619

Reeve MR, Syms MA, Kremer P (1989) Growth dynamics of a ctenophore (*Mnemiopsis*) in relation to variable food supply I. Carbon biomass, feeding, egg-production, growth and assimilation efficiency. J Plankton Res 11:535–552

Rilling GC, Houde ED (1999) Regional and temporal variability in growth and mortality of bay anchovy, *Anchoa mitchilli*, larvae in Chesapeake Bay. Fish Bull 97:555–569

Rose KA, Adamack AT, Murphy CA, Sable SE, Kolesar SE, Craig JK, Breitburg DL, Thomas P, Brouwer MH, Cerco CF, Diamond S (2009) Does hypoxia have population-level effects on coastal fish? Musings from the virtual world. J Exp Mar Biol Ecol 381:S188–S203

Rose KA, Cowan JH Jr, Clark ME, Houde ED, Wang SB (1999) An individual-based model of bay anchovy population dynamics in the mesohaline region of Chesapeake Bay. Mar Ecol Prog Ser 185:113–132

Rosenheim JA, Glik TE, Goeriz RE, Rämert B (2004) Linking a predator's foraging behavior with its effects on herbivore population suppression. Ecology 85:3362–3372

Rosenheim JA (2007) Intraguild predation: new theoretical and empirical perspectives. Ecology 88:2679–2680

Scheffer M, Baveco JM, DeAngelis DL, Rose KA, van Nes EH (1995) Super-individuals: a simple solution for modelling large populations on an individual basis. Ecol Model 80:161–170

Snyder WE, Wise DH (2001) Contrasting trophic cascades generated by a community of generalist predators. Ecology 82:1571–1583

Tester PA, Turner JT (1988) Comparative carbon-specific ingestion rates of phytoplankton by *Acartia tonsa*, *Centropages velificatus* and *Eucalanous pileatus* grazing on natural phytoplankton assemblages in the plume of the Mississippi River (northern Gulf of Mexico continental shelf). Hydrobiologia 167(168):211–217

Thompson RM, Hemberg M, Starzomski BM, Shurin JB (2007) Trophic levels and trophic tangles: the prevalence of omnivory in real food webs. Ecology 88:612–617

Tucker JW Jr (1989) Energy utilization in bay anchovy, *Anchoa mitchilli*, and black sea bass, *Centropristis striata striata*, eggs and larvae. Fish Bull 78:279–293

Vandermeer J (2006) Omnivory and the stability of food webs. J Theor Biol 238:497–504

Waggett R, Costello JH (1999) Capture mechanisms used by the lobate ctenophore, *Mnemiopsis leidyi*, preying on the copepod *Acartia tonsa*. J Plankton Res 21:2037–2052

Wang SB, Houde ED (1994) Energy storage and dynamics in bay anchovy *Anchoa mitchilli*. Mar Biol 121:219–227

Wissinger S (1992) Niche overlap and the potential for competition and intraguild predation between size-structured populations. Ecology 73:1431–1444

Wissinger S, McGrady J (1993) Intraguild predation and competition between larval dragonflies: direct and indirect effects on shared prey. Ecology 74:207–218

Zhang J, Gilbert D, Gooday AJ, Levin L, Naqvi SWA, Middelburg JJ, Scranton M, Ekau W, Peńa A, Dewitte B, Oguz T, Monteiro PMS, Urban E, Rabalais NN, Ittekkot V, Kemp WM, Ulloa O, Elmgren R, Escobar-Briones E, Van der Plas AK (2010) Natural and human-induced hypoxia and consequences for coastal areas: synthesis and future development. Biogeosciences 7:1443–1467

Chapter 12
Simulating the Effects of Nutrient Loading Rates and Hypoxia on Bay Anchovy in Chesapeake Bay Using Coupled Hydrodynamic, Water Quality, and Individual-Based Fish Models

Aaron T. Adamack, Kenneth A. Rose and Carl F. Cerco

Abstract Water quality in the Chesapeake Bay has decreased since the 1950s due to an increase in nutrient loadings that have increased the extent and duration of hypoxic conditions. Restoration via large-scale reductions in nutrient loadings is now underway. How reducing nutrient loadings will affect water quality is well predicted; however, the effects of reduced nutrients (reduced food availability) and associated reduced hypoxia on fish are generally unknown as most water quality models do not include trophic levels higher than zooplankton. We dynamically coupled a spatially explicit, individual-based population dynamics model of juvenile and adult anchovy to the three-dimensional Chesapeake Bay eutrophication model. Growth rates of individual anchovy were calculated using a bioenergetics equation. Anchovy consumption rates were forced by zooplankton densities from the water quality model, and anchovy consumption of zooplankton was added as an additional mortality term on zooplankton in the eutrophication model. Anchovy mortality was size dependent and their movement depended on water temperature, dissolved oxygen, and zooplankton concentrations. Multi-year simulations with fixed annual recruitment were performed under decreased, baseline, and increased nutrient loadings scenarios. The results of our analyses show that anchovy

A.T. Adamack (✉) · K.A. Rose
Department of Oceanography and Coastal Sciences, Louisiana State University, Energy, Coast, and Environment Building, Baton Rouge, LA, USA
e-mail: aaron.adamack@gmail.com

K.A. Rose
e-mail: karose@lsu.edu

C.F. Cerco
U.S. Army Engineer Research and Development Center, Vicksburg, MS, USA
e-mail: carl.f.cerco@usace.army.mil

Present Address:
A.T. Adamack
Institute for Applied Ecology, University of Canberra, Bruce, ACT, Australia

D. Justic et al. (eds.), *Modeling Coastal Hypoxia*,
DOI 10.1007/978-3-319-54571-4_12

responses to changed nutrient loadings are dominated by changes in productivity, including simultaneous changes in growth and mortality rates, and spatial distribution, and depend on life stage. As such, we recommend using full life cycle, spatially explicit population models that are dynamically coupled to water quality models as a tool for predicting the effects of changes in nutrient loadings on fish population dynamics.

Keywords Nutrient loading • Hypoxia • Bay anchovy • Numerical modeling • Population dynamics • Individual-based model • Chesapeake Bay

12.1 Introduction

Understanding the link between water quality and fish population dynamics is especially important with the widespread efforts, often at considerable cost, to reduce eutrophication in coastal waters (e.g., Cloern 2001; Rabalais et al. 2002; Breitburg et al. 2003; Conley et al. 2009). Changes in how an ecosystem is managed, such as altering the rates of nutrient loadings, can have large and complex impacts on the ecosystem (Rabalais et al. 2002) including changes in the timing and spacing of ecosystem production dynamics. Historically, water quality models (freshwater) and nitrogen–phytoplankton–zooplankton (marine) models simulated the lower trophic level food web in order to predict chlorophyll-a and nutrient cycling (Rose et al. 2017). Zooplankton were included to get a realistic mortality term for the phytoplankton (i.e., closure term), and predicting realistic temporal and spatial variation in the zooplankton biomasses was a second-order consideration (Runge et al. 2004, Lett et al. 2009). Distinct from these models were the extensive efforts to simulate upper trophic level population dynamics (e.g., fish population) and food web dynamics (e.g., Ecopath with Ecosim, Christensen and Walters 2004). Many fish growth and population models focus on fish dynamics and do not include water quality and trophic levels lower than zooplankton (e.g., Luo and Brandt 1993; Rose et al. 1999; Lett et al. 2009).

Assessment of the effects of nutrient loadings on fish population dynamics is complicated by not only increasing nutrients stimulating the lower trophic level food web (potentially more food for fish), but also triggering hypoxia (dissolved oxygen (DO) < 2 mg L^{-1}) that has potentially negative effects on fish growth, reproduction, survival, and distribution (Rose et al. 2009). Earlier modeling studies (Brandt and Mason 2003; Breitburg et al. 2003; Adamack et al. 2012) that linked output from a water quality model for the Patuxent River to models of fish populations showed that the responses of populations to the changes in nutrient loading rates can be complex and different across life stages. Quantifying the effects of low DO at the population level further complicates the analysis because most effects beyond individuals are likely indirect as a result of shifts in other species (prey and predators of the fish of interest) or arise from spatial displacement of the species to areas of lower habitat quality.

The need to link fish bioenergetics and population models with water quality is particularly important in the Chesapeake Bay (Bay) ecosystem. Historically, episodic hypoxia and anoxia occurred in deeper portions of the water column of the Chesapeake Bay (Cooper and Brush 1991; Cronin and Vann 2003; Hagy et al. 2004). Since precolonial times, total nitrogen and total phosphorus loadings to the Bay are estimated to have increased 6.2-fold and 17.1-fold, respectively (Boynton et al. 1995), with much of the increase having occurred over the last century (Hagy et al. 2004; Kemp et al. 2005; Fisher et al. 2006); bottom-water hypoxia is now a persistent, annual occurrence. Nutrient loadings have increased because of a threefold increase in human population size within the Bay watershed over the past 100 years, changing land-use patterns (initially forested, followed by large-scale clearing for agriculture, today agricultural lands are decreasing as land becomes urbanized or reverts to forests), and an increase in the use of agricultural fertilizers with the use of nitrogen-based fertilizer in Maryland doubling between 1960 and 2000 (Kemp et al. 2005). The Bay watershed is now undergoing a costly, multi-decadal restoration effort (Chesapeake Bay Program 2013) and had a total maximum daily load (TMDL) implemented for nutrients and sediments by the US Environmental Protection Agency in 2010 (US EPA 2010). The objective of the restoration program is to "Correct the nutrient- and sediment-related problems in the Chesapeake Bay and its tidal tributaries sufficiently to remove the Bay and the tidal portions of its tributaries from the list of impaired waters under the Clean Water Act." The cost of this plan has been estimated as 13–15 billion dollars to Maryland alone (Gray 2013). Quantification of the benefits of achieving these nutrient reduction goals to fish and the food web at the population and higher levels was recognized early on as a challenge (Kemp et al. 2005) that continues on to today.

Bay anchovy (*Anchoa mitchilli*) in Chesapeake Bay is a well-studied species (e.g., Houde et al. 1989; Houde and Zastrow 1991; Jung and Houde 2004a) that is a good candidate species for linking water quality to fish growth and population dynamics. Bay anchovy are one of the dominant fish species in the Bay in terms of both abundance and biomass (Baird and Ulanowicz 1989; Houde et al. 1989; Jung and Houde 2003), and bay anchovy are a major trophic link between zooplankton and piscivores (Baird and Ulanowicz 1989; Hartman and Brandt 1995). Bay anchovy consume 15–18% of zooplankton production during the summer and fall (70–90% of all zooplankton consumed by planktivorous fish), and in turn, they are the source of 60–90% of the energy intake of the piscivorous fish that fed upon bay anchovy during the summer, fall, and spring seasons.

Here, we dynamically couple a version of the three-dimensional Chesapeake Bay water quality model (Cerco and Cole 1993) with a spatially explicit individual-based population dynamics model for bay anchovy. The individual-based anchovy model simulates the growth, mortality, and spatial distribution of anchovy on the same three-dimensional spatial grid as the water quality model. The coupled models are used to predict the effects of increasing and decreasing the nutrient loadings to Chesapeake Bay on bay anchovy growth, biomass, and spatial distribution. We performed simulations for low and high levels of bay anchovy recruitment and for wet, normal, and dry water years. Additional simulations were performed to examine how increased mortality during the egg and larval stages from exposure to hypoxia and how increased mortality rates on juveniles and adults due to increased habitat overlap of anchovy and their predators would affect the results.

12.2 Methods

12.2.1 Chesapeake Bay Water Quality Model

The Chesapeake Bay Environmental Modeling Package (CBEMP) was developed more than 20 years ago to assist in the management of eutrophication in the Chesapeake Bay ecosystem. The CBEMP has undergone continuous revisions and additions since then to respond to evolving management needs in the Chesapeake. Most recently, the CBEMP provided technical support for the development of the Chesapeake Bay TMDL (Batiuk et al. 2013). The CBEMP is comprised of three models: a three-dimensional hydrodynamics (CH3D-WES) model (Kim 2013), an eutrophication (CE-QUAL-ICM) model (Cerco et al. 2010), and a sediment diagenesis model (DiToro 2001).

The most recent version of the CBEMP operates on a three-dimensional grid of 50,000 elements and simulates the period from 1985 to 2005 (Cerco and Noel 2013). These levels of spatial detail and temporal extent are in contrast to the original 4,073 cell grid and 1984–1986 simulation period used here. The original grid and application period are retained, however, for educational purposes and as a "test bed" for model development. The model code for this smaller grid has been updated to keep track with current developments made to the most recent finer resolution grid version. In this analysis, we used the original grid version for the 1984–1986 time period as the simpler model grid was more amenable to carrying out the model development process (e.g., modifying the water quality model code and testing model behavior after the addition of fish), the necessary input model runs were readily available, and it sufficiently captured the major dynamics needed for linkage to fish with quicker run times. The initial application of the version of the CBEMP used here for the years 1984–1986 was described by Cerco and Cole (1993).

For the original grid (4,073 cell grid which we used here), the Bay's surface was divided into a horizontal grid of 729 cells with horizontal side lengths of ~5 by 10 km (Fig. 12.1). Vertically, each column of cells is 2–15 cell layers (each cell is about 1.5 m thick) deep. The hydrodynamics model is run separately from the eutrophication and diagenesis models and employs curvilinear coordinates in the surface plane and a Z-grid in the vertical direction to produce three-dimensional predictions of velocity, diffusion, surface elevation, salinity, and temperature for each cell on an intra-tidal (about 5 min) timescale. Outputs from the hydrodynamics model that were used in the water quality model include cell volume, flows between cells in the axial, transverse, and vertical directions, and vertical turbulent diffusion. A Lagrangian processor was used to filter out intra-tidal details from the

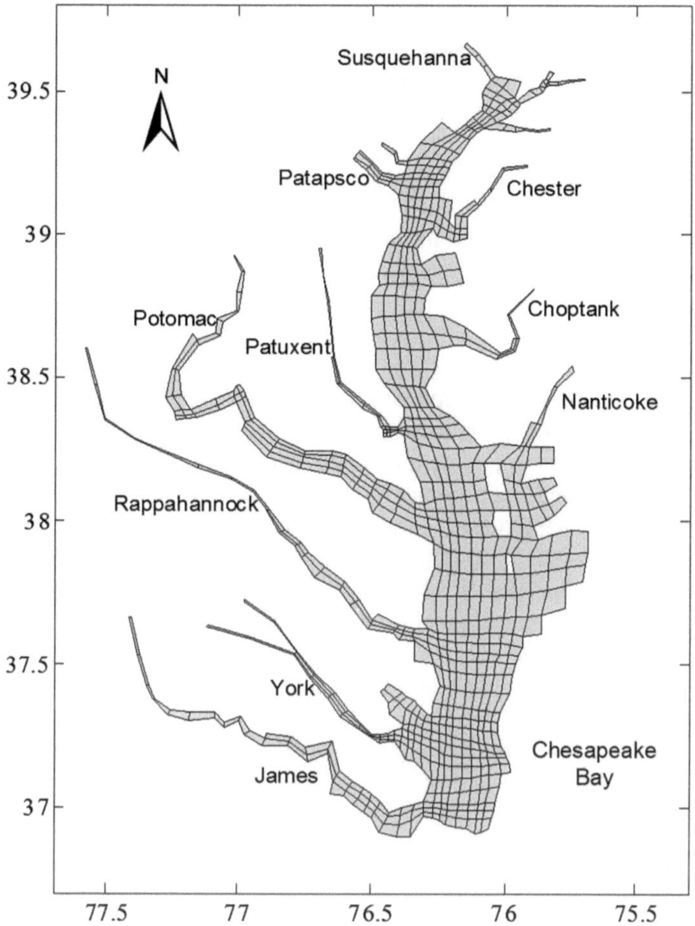

Fig. 12.1 Chesapeake Bay water quality model grid and the location of some key tributaries

hydrodynamic output, while maintaining the intertidal scale (about 12 h) transport (Dortch et al. 1992). Inputs to the hydrodynamics model include wind speed, air temperature, tributary freshwater inflows, surface heat exchange, tides, and the time-varying vertical distributions of temperature and salinity at the open boundary. Calibration and validation of the original model are described by Johnson et al. (1993).

The eutrophication and diagenesis models are run simultaneously. The eutrophication model simulates the nutrient, phytoplankton, and zooplankton dynamics in Chesapeake Bay on the same three-dimensional model grid as the hydrodynamics model (Kim 2013). Additionally, the eutrophication model provides boundary conditions in the water column including dissolved oxygen, temperature, and nutrient concentration, for the sediment diagenesis model (DiToro 2001). The diagenesis model computes sediment-water fluxes of nitrate, ammonium, phosphate, dissolved oxygen, and silica based on these boundary conditions and on computed particle deposition. The concentrations of 23 constituents are tracked in the eutrophication and diagenesis models, including temperature, salinity, DO, several forms of dissolved and particulate carbon, nitrogen, and phosphorus, summer and winter phytoplankton groups, and micro- and mesozooplankton biomasses. The primary forcing functions for the eutrophication model were transport information from the hydrodynamic model, fall-line nutrient loads from the Susquehanna River and other major tributaries, non-point-source loads from below the fall line, point-source loads from municipal and industrial sources, atmospheric loads, open-mouth boundary conditions, solar radiation, and meteorological conditions (Cerco et al. 2010).

Model constituents were computed on a mass/unit volume basis in each cell on the three-dimensional computational grid. Biological constituents, including phytoplankton and zooplankton, were quantified as carbonaceous biomass. Constituent dynamics were updated approximately every 15 min for each model cell using a three-dimensional mass-conservation equation that was solved using the finite-difference method.

In addition to improvements and enhancements to the hydrodynamics and eutrophication models since the original version used here, another significant development has been the addition of living resources to the model. These living resources include bivalve filter feeders (Cerco and Noel 2007, 2010) and vertebrate species such as Atlantic menhaden (Dalyander and Cerco 2010). Indeed, the anchovy analysis described here inspired the subsequent development of the menhaden addition to the recent version of the hydrodynamics–eutrophication model.

12.2.2 Bay Anchovy Model

The bay anchovy model is a spatially explicit individual-based population model (IBM) that tracks the growth, mortality, and movement of individual anchovy in the same three-dimensional grid as the hydrodynamics and eutrophication models. The anchovy model was inserted directly into the eutrophication model code so that it operates on the same time steps as the eutrophication model and is able to directly interact with the eutrophication model. Anchovy consumption rates are dependent on the micro- and mesozooplankton densities and temperature generated by the eutrophication model for each cell and time step, bioenergetics (respiration and egestion) are dependent on temperature, horizontal movement of anchovies is related to zooplankton densities and temperature, vertical movement depends on temperature and DO, and growth and mortality are affected by low DO. Zooplankton biomass consumed by the anchovy is returned to the eutrophication model in the form of particulate and dissolved nutrients. Parameter values and a brief description of parameters for anchovy bioenergetics and movement are shown in Table 12.1.

Table 12.1 Parameter values used in the model equations for bay anchovy growth and consumption from Luo and Brandt (1993) and for the bay anchovy kinesis movement model. Values of the eight K feeding efficiency parameters were determined during calibration of the model. Values of the vulnerability parameters (v) were all set to one and so did not affect the proportion of micro- versus mesozooplankton consumed

Symbol	Description	Value
Consumption		
a_c	Intercept for maximum consumption (g prey (g wet weight)$^{-1}$ d^{-1}	0.41
b_c	Exponent for maximum consumption	−0.33
Q	Slope for temperature dependence on consumption	2.22
T_o	Optimum temperature for consumption (°C)	27
T_m	Maximum temperature for consumption (°C)	33
K_{11}	Half saturation for microzooplankton for anchovy < 43 mm (g wet weight m^{-3})	4.8
K_{12}	Half saturation for mesozooplankton for anchovy < 43 mm (g wet weight m^{-3})	0.04
K_{21}	Half saturation for microzooplankton for anchovy 43–60 mm (g wet weight m^{-3})	10.0
K_{22}	Half saturation for mesozooplankton for anchovy 43–60 mm (g wet weight m^{-3})	2.0
K_{31}	Half saturation for microzooplankton for anchovy 60–90 mm (g wet weight m^{-3})	20.0
K_{32}	Half saturation for mesozooplankton for anchovy 60–90 mm (g wet weight m^{-3})	5.0
K_{41}	Half saturation for microzooplankton for anchovy > 90 mm (g wet weight m^{-3})	80.0
K_{42}	Half saturation for mesozooplankton for anchovy > 90 mm (g wet weight m^{-3})	20.0
v_1	Vulnerability of microzooplankton to anchovy	1.0
v_2	Vulnerability of mesozooplankton to anchovy	1.0
Respiration		
a_r	Intercept for maximum standard respiration (g O_2 (g wet weight)$^{-1}$ d^{-1})	0.0115
b_r	Exponent for maximum standard respiration	−0.346
Q_r	Slope for temperature dependence of standard respiration	2.25
T_{or}	Temperature for standard respiration (°C)	30
T_{mr}	Maximum temperature for standard respiration (°C)	36
S	Specific dynamic action coefficient	0.10
A_c	Activity parameter	2.0
Egestion and excretion		
A	Intercept for temperature dependence of egestion	0.77
B	Exponent for temperature dependence of egestion	−0.40
a_u	Proportion of assimilated food excreted	0.15
Caloric density		
Cal_z	Caloric density of micro- and mesozooplankton (calorie (g prey)$^{-1}$)	610

(continued)

Table 12.1 (continued)

Symbol	Description	Value
Cal_F	Caloric density of bay anchovy (calorie (g wet weight)$^{-1}$)	1000
Mortality		
q	Multiplier of length-dependent mortality rate	1.17
Movement		
H_1	Height parameter for the inertia component of movement	0.75
H_2	Height parameter for the random component of movement	0.9
Φ	Maximum swimming velocity (BL s^{-1})	1.0
Q_{MO}	optimum temperature for horizontal (°C)	27
σ	standard deviation of temperature for horizontal (°C)	2
Q_{MO}	prey availability (fraction of C_{MAX}) for horizontal	0.8
σ	standard deviation of prey availability for horizontal	0.05
Q_{MO}	optimum temperature for vertical (°C)	27
σ	standard deviation of temperature for vertical (°C)	4
Q_{MO}	optimum dissolved oxygen (mg L^{-1}) for vertical	5.0
σ	standard deviation of dissolved oxygen (mg L^{-1}) for vertical	1.5

In order to conveniently track the movement of individual anchovy, the grid coordinates were converted from the latitude and longitude coordinates used in the eutrophication model to Universal Transverse Mercator (UTM) coordinates, which are given in terms of meters (m). The conversion was done using the equations of Snyder (1987) and assumed that the shape of the Earth conformed to the dimensions assumed by the 1980 Geodetic Reference System/World Geodetic System 1984 ellipsoid. To confirm that model cell sizes were consistent for the two grids, the horizontal side lengths of the anchovy model grid cells were compared to the eutrophication model grid cells. Differences in side lengths for the two grids are less than 0.1 m per side.

12.2.2.1 Annual Recruitment of Juveniles

A fixed number (low, median, or high) of 23 mm long, 30-day-old bay anchovy were added to the model each year as weekly cohorts for the duration of the simulation. There are about 10^9 to 10^{14} bay anchovy individuals in Chesapeake Bay (Jung and Houde 2004a), and modeling each individual separately is not possible computationally. To solve this, we used a super-individual approach (Scheffer et al. 1995). Each super-individual being simulated was given an initial worth, the number of identical population individuals that each super-individual represented.

The initial worth of each super-individual within a recruitment scenario was set to a constant value that was determined as the total number of 23 mm recruits being added in a simulation year divided by the number of super-individuals (about 10^5 super-individuals were used). The initial worth of each super-individual was therefore 3.6×10^6 individuals for the low recruitment scenario, 1.2×10^7 individuals for the median recruitment scenario, and 1.584×10^7 individuals for the high recruitment scenario.

Use of super-individuals affected how anchovy mortality, consumption, and almost all model outputs were computed. In a true individual-based approach, the mortality rate would be converted to a probability of death and compared to a random number to determine whether the model individual was either removed or left in the population. Super-individuals remain in the simulation; mortality was simulated by decrementing the super-individual's worth by the total mortality rate (M_T, d^{-1}):

$$Worth_{i,t+1} = Worth_{i,t} e^{-M_T(\Delta t)} \tag{12.1}$$

where $\Delta t = \sim 0.0104$ days (15/1440 min). When a super-individual reached old age (1095 days) or its worth dropped below 0.001, it was removed from the simulation. The total amount of prey consumed by a super-individual during a time step was determined by multiplying the amount of prey consumed by a single model individual by the super-individual's worth. Model predictions of anchovy densities, lengths, weights, and other outputs were weighted by the worth (in the statistical sense) of each super-individual. For example, a mean length on a given day was the weighted average of the lengths of the super-individuals, with the weights for averaging being the worths of the super-individuals when they were output.

The number of individuals added each week as part of annual recruitment had a triangular distribution with the smallest cohorts being added in early June and mid-October and the largest cohort being added in mid-August (Luo and Musick 1991; Zastrow et al. 1991; Jung and Houde 2004a). Newly generated super-individuals were randomly placed in model cells in the spatial grid that had DO concentrations greater than 3.0 mg L^{-1} and zooplankton concentration greater than 0.005 g C m^{-3}. The probability of a newly generated super-individual being placed in a particular cell was set to the volume of each cell as a proportion of the total volume of all 4,073 cells. If the minimum DO and zooplankton concentrations were not met, a new initial cell was randomly selected for the anchovy.

12.2.2.2 Growth and Bioenergetics

Growth of each individual was based upon the Luo and Brandt (1993) bay anchovy bioenergetics model and was evaluated each eutrophication model time step (every 15 min):

$$\frac{dW}{dt} = [CON - (R + SDA + F + E + REP)] \cdot \frac{Cal_z}{Cal_f} \cdot W \cdot f(DO_{cell}) \tag{12.2}$$

where W is the wet weight (g wet weight) of an individual, CON is the amount of prey consumed, R is the respiration, SDA is the specific dynamic action, F is the egestion, E is the excretion, REP is the reproduction, Cal_z is the caloric density of prey (cal (g prey)$^{-1}$), Cal_f is the caloric density of anchovy (cal (g wet weight)$^{-1}$), and $f(DO_{cell})$ is the low DO effect on growth. Consumption and the loss terms were all in units of g prey (g wet weight)$^{-1}$ d^{-1}. The ratio of the caloric densities converts g prey into g anchovy for the terms within the brackets to obtain the units of g anchovy per g anchovy per day.

The values of temperature and zooplankton densities that affected anchovy growth and consumption were those predicted for the cell currently occupied by an individual by the eutrophication model. Consumption was the sum of the amount of the microzooplankton and mesozooplankton consumed by an anchovy during each time step and depended on a maximum possible consumption rate and a type 2 functional response:

$$CON = \sum_{j=1}^{2} C_j \tag{12.3}$$

$$C_j = \frac{CON_{\max}(PD_j v_j / K_{cj})}{1 + \sum_{k=1}^{2} (PD_k v_k / K_{ck})} \tag{12.4}$$

where C_j is the consumption of prey type j by the individual anchovy, PD_j are the densities of microzooplankton (j = 1) and mesozooplankton (j = 2) in the cell of the individual, v_j is the vulnerability (all assumed 1.0), and K_{cj} are the half-saturation parameters based on size interval c of the anchovy. Zooplankton densities were converted from g C m^{-3} in the eutrophication model to g wet weight m^{-3} for use in the anchovy model by multiplying by 12.5 (dry weight was 20% of wet weight and carbon weight was 40% of dry weight; Mauchline 1998). Maximum consumption rate was an allometric function of anchovy weight and temperature:

$$CON_{\max} = a_c W^{b_c} \cdot f(T) \tag{12.5}$$

where a_c and b_c determine the weight effect; $f(T)$ is bell-shaped function that is one at the optimum temperature, T_o; zero at the upper temperature, T_m; and Q is a parameter that approximates a Q_{10} relationship (a measure of the change in biological rates as a consequence of increasing temperature by 10 °C) within the CON_{max} function for temperatures below T_O.

Respiration was modeled as a power function of weight and then adjusted for temperature using the same temperature effect function as with CON_{max} but with different parameter values for the optimum and upper temperatures and for the Q_{10} relationship:

$$R = a_r W^{b_r} f(T) A_c \tag{12.6}$$

where a_r and b_r determine the weight effect; T_{or}, T_{mr}, and Q_r are specified for $f(T)$ for respiration; and A_c is the activity multiplier. The values of T_{or} and T_{mr} were set so that respiration rate increased over the range of simulated temperatures.

Egestion was represented as a fraction of consumption, while excretion and specific dynamic were represented as fractions of assimilated energy (consumption minus egestion).

$$F = CON \cdot A \cdot T^B \tag{12.7}$$

$$E = (CON - F) \cdot a_u \tag{12.8}$$

$$SDA = (CON - F) \cdot S \tag{12.9}$$

where A and B determine the temperature effect on egestion, a_u is the fraction of assimilated energy lost to excretion, and S is the fraction of assimilated energy lost to specific dynamic action.

Energy related to reproduction (REP in Eq. 12.2) was computed as the grams/day expended during days 115–246 of each year by anchovy that were at least 43 mm long and that had a positive net energy intake for the time step. The energy used for reproduction was set to half of the net energy intake, and reproduction costs were applied to all individuals (males and females).

A logistic sigmoidal function, developed originally by Luo et al. (2001) for Atlantic menhaden, was used to simulate the physiological effects of exposure to low dissolved oxygen on anchovy growth rate.

$$f(DO_{cell}) = 1/(1 + e^{(-2.1972DO_{cell} + 6.5916)}) \tag{12.10}$$

The function $f(DO_i)$ used the DO in the cell to determine the multiplier, and then the multiplier was applied to the predicted change in weight of the anchovy for that time step (Eq. 12.2). Growth begins to be reduced at a DO of about 6 mg L^{-1} is 50% of normal growth at 3 mg L^{-1} and approaches zero at DO concentrations less than 1 mg L^{-1}.

Anchovy weights were converted into lengths using a weight–length relationship modified from Jung and Houde 2004a:

$$L_{new} = \left(\frac{W}{e^{-11.799}}\right)^{\frac{1}{3}} \tag{12.11}$$

where L_{new} is the new length of the anchovy and $e^{-11.799}$ is a constant derived from a length–weight function fitted to summer (July) anchovy lengths and weights. An anchovy's length only changes if its new length is longer than its length during the previous time step.

12.2.2.3 Mortality

Four sources of mortality were included in the bay anchovy model: general mortality (M_G), starvation mortality (M_{starv}), mortality due to exposure to low DO (M_{DO}), and mortality due to old age. Once each 15-min time step, the general, DO-related, and starvation mortality rates were summed to get total mortality (M_T), and used to update the worth of the super-individual (Eq. 12.1). Old age mortality was threshold based on age (i.e., not a rate) and resulted in the elimination of the super-individual with all of its worth.

General mortality represented all causes of anchovy mortality except mortality due to low DO, starvation, and old age. The rate was a simple constant during winter (October through March) and decreased with anchovy length during the summer (April through September).

$$M_G = q \cdot L^{-1} \tag{12.12}$$

where q is the multiplier for length-dependent mortality. Jung and Houde (2004a) estimated wintertime and length-dependent summer mortality rates based upon 6 years (1995–2000) of field data. As low DO, starvation, and old age mortality are added to the general rate in the model, we used the lowest summer (q = 1.17) and winter mortality rates (0.005 d^{-1}) estimated by Jung and Houde (2004a) for general mortality.

Mortality rate due to exposure to hypoxia was a function of the DO concentration in the cell.

$$M_{DO} = 0.093487 + 70.11894 \cdot (\ln[\mathrm{DO_{cell}}])^2 \tag{12.13}$$

This function was fit to experimental data on Atlantic menhaden (*Brevoortia tyrannus*) reported by Burton et al. (1980). As observed mortality rates become extremely high at low DO concentrations, and as anchovy only moved vertically once every 4 eutrophication model time steps (each eutrophication model time step simulates ~15 min of real time; 4 eutrophication model time steps ≈ 1 h of real time), the maximum mortality rate due to exposure to very low DO was set to a very high value (33.78 d^{-1}). This high mortality rate resulted in almost complete mortality for the super-individual after 4 time steps, and thus, if such long exposure occurred, the individual would be removed from the simulation due to near zero worth.

Starvation mortality (0.1 d^{-1}) was applied to anchovy model individuals whose weight was 70% or less of the expected weight given their length for each time step that an anchovy was below its expected weight. Mortality due to old age (100% mortality; super-individual removed from simulation) was applied to all model individuals upon reaching an age of 1,095 days (3 years old) based on Newberger et al. (1989), who found that bay anchovy generally did not live past 3 years of age.

12.2.2.4 Movement

A kinesis approach (Humston et al. 2000; Humston 2001; Watkins and Rose 2013) was used to simulate the horizontal and vertical movement of each anchovy super-individual. Anchovy positions were tracked in continuous x, y, and z space. The x and y coordinates corresponded to an individual's UTM coordinates, while the z coordinate was the distance from the water surface. Horizontal and vertical movements were evaluated separately at fixed time intervals (x and y every 12 h; z hourly) to improve model run time, to account for differences in the distances involved and the time required for an anchovy to cross model cells in the horizontal and vertical planes, and to allow for the establishment of an inertial gradient (described below). Horizontally, model cells are several kilometers across and could require more than a day for an adult anchovy (e.g., ~5 cm TL) to cross the cell if they took the most direct path across the cell. Vertically, model cells are only 1.5 meters and an adult anchovy could swim from the bottom of a cell to the top of the cell in ~30 s. The movement model added together an inertia component ($f(V_{t-1})$) and a random component ($g(\varepsilon)$) to produce a net velocity for each of the x, y, and z dimensions each time step.

$$V_t = f(V_{t-1}) + g(\varepsilon) \quad (12.14)$$

where $f(V_{t-1})$ is a function that was based upon the anchovy's velocity during the previous movement step, and g(ε) is a randomly generated velocity. The net velocities were then multiplied by the time since the last movement step to determine changes in distance for x, y, and z, which then were added to the current location to get a new x, y, and z location and a cell location.

The relative contribution of inertial versus random components was dependent on how close a movement cue was to its optimum level. When conditions in a cell were close to the optimum, the inertial component dominated movement. When conditions in a cell were far from the optimum, the random component dominated movement. Both functions were described using Gaussian functions:

$$f(V_{t-1}) = V_{t-1} \times H_1\left[e^{(-0.5)[(Q_M - Q_{MO})/\sigma]^2}\right] \quad (12.15)$$

$$g(\varepsilon) = \varepsilon \times H_2\left[e^{(-0.5)[(Q_M - Q_{MO})/\sigma]^2}\right] \quad (12.16)$$

where H_1 and H_2 control the height of the function and were restricted to the range 0–1.0, σ controlled the width of the Gaussian function, ε is a random number deviate drawn from a normal distribution with a mean and standard deviation based on an individual's swimming speed, Q_M is the actual value of the movement cue, and Q_{MO} is the optimum value of the movement cue. For the two components of the movement model to work properly, model individuals needed to shift between cells somewhat frequently (in terms of movement steps) in order to generate cue gradients that will drive the inertia component of the movement model. As vertical distances across cells were short, vertical movement could be evaluated frequently (hourly) while the relatively long horizontal distances across cells required longer movement time steps (12 h).

Horizontal (x and y) movement depended on the cues from water temperature and zooplankton density, while vertical (z) movement depended on water temperature and DO. Water temperature and prey availability were used as cues because they were the key factors affecting anchovy growth, while DO was used as a cue because it affects the vertical distribution of anchovy and other fish species (e.g., Constantini et al. 2008 and Zhang et al. 2009) and can cause direct mortality. The optimal water temperature (Q_{MO} for temperature) was set to 27 °C, the optimal temperature for consumption by anchovy (Luo and Brandt 1993). Densities of micro- and mesozooplankton were combined into a single measure of overall prey availability using the functional response portion of the equation for consumption, and the optimum value for prey availability (Q_{MO} for food) was set to 80% of CON_{MAX} (Eq. 12.5). To avoid mortality due to low DO, while at the same time preventing anchovy from aggregating at the surface of the water column, we set the optimum DO concentration (Q_{MO} for DO) to 5.0 mg L^{-1}.

Deviates from optimal were computed separately for each cue, and the smaller of the deviates for water temperature and prey availability was used to determine the net velocities for x and y movement, while the product of the deviates for water temperature and DO was used for vertical movement. Vertical movement was restricted to a maximum change of 3 m per vertical movement step in order to simplify the tracking of anchovy movement, as anchovy can potentially swim many times the thickness of the water column between the 1 h movement time steps.

Once the distances that an anchovy moved along x, y and z had been determined, a two-part procedure was used to update the anchovy's position. In the first part of the procedure, the anchovy's new x and y values were determined by adding the distances moved along the x and y coordinates to the anchovy's previous position. A point-in-polygon subroutine (Burkadt 2014) was used to determine whether the anchovy's new position was outside of the cell that it had begun the time step in. If the anchovy finished in the same cell that it started in, the anchovy's horizontal movement was complete for the current movement step. If the anchovy's new position was outside the initial cell, then the anchovy's position was checked to determine whether or not it was on the model grid for the current depth layer. If the individual's new position was not on the grid, then the individual was reflected back onto the grid. Finally, the individual's new cell was identified and updated.

The second part of the movement updating procedure was to update the individual's vertical position. Updating vertical position was simpler as z values were constrained to the interval between the surface and bottom of the water column, and all cells in the vertical dimension were in had the same shape and thickness. If the z value was above the surface or below the bottom, then the z value was reset to place the anchovy just below the surface or just above the bottom. Once the anchovy's final position was determined, its velocities along the x- and y-axes for the current movement step were recalculated using the anchovy's starting and actually realized final positions. The recalculated velocities were used to set V_{t-1} for the next movement step.

12.2.3 Modifications to the Eutrophication Model to Accommodate Bay Anchovy

The eutrophication model had been calibrated previously without anchovy predation on zooplankton. To accommodate the addition of anchovy to the eutrophication model, we reduced the mortality rates of summer algal biomass from 1.00 d^{-1} to 0.15 d^{-1}, and predation mortality on micro- and mesozooplankton by 70% based on model calibration results.

12.3 Simulations

Three sets of model simulations were performed: calibration and baseline to examine the effects of water year and recruitment level, effects of increased and decreased nutrient loadings, and the effects of forced increased mortality rates to offset the benefits of increased nutrient loadings (e.g., increases in zooplankton prey). The results of single model runs were reported, as runs that used different random number seeds generated very similar model predictions of anchovy growth, densities, and spatial distributions. We used eutrophication model input files and hydrodynamics output for the years 1984, 1985, and 1986; these were arranged to obtain a single 10-year sequence. The 3 years can be assigned water year types (e.g., normal, wet, and dry) by using the US Geological Survey (2014) classification of annual-mean stream-flow entering Chesapeake Bay: 1984 was a wet year, 1985 was a dry year, and 1986 was a normal year. Calibration runs used a repeating series of the input files for the 1986 conditions (normal years). For all other simulations, we linked the input files for the 1984 (wet), 1985 (dry), and 1986 (normal) water years in order to match the sequence of wet, normal, and dry water year types that were observed between 1984 and 1993 (Fig. 12.2). This was done to use a real sequence of water year types in model simulations. Settings for all of the model runs are summarized in Table 12.2.

12.3.1 Bay Anchovy Recruitment

Annual recruitment levels for the low and high recruitment scenarios were based on estimates of the total number of eggs produced each year by anchovy, and age-dependent growth and size-dependent mortality rates from Jung and Houde (2004a). The number of eggs produced was combined with growth and mortality rates to estimate the total number of 23 mm long bay anchovy (the initial size of anchovy added to the model; eggs and larvae were not directly included in our simulations) produced each year. Low (3.6×10^{11} individuals) and high (1.58×10^{12} individuals) recruitment levels were set to the years with the lowest

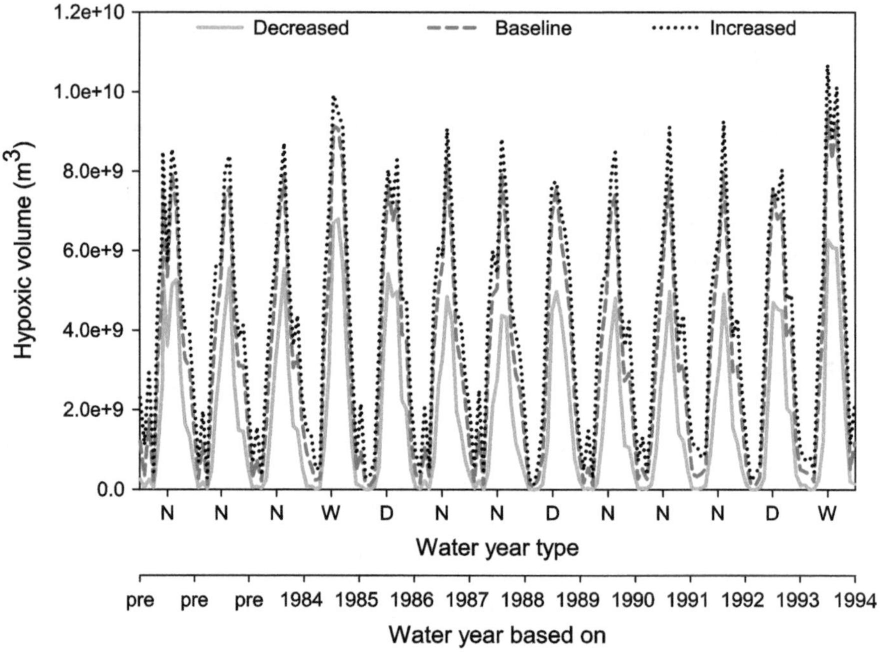

Fig. 12.2 Volume of hypoxic water for decreased, baseline, and increased nutrient loadings during the 3 prerun years and the 10 years after that used the same water year type as observed in 1984–1993. Spin-up years are indicated by the label "pre." Water year type is indicated by the initials: D = dry year, W = wet year, and N = normal year. Water year types were obtained from the USGS (http://www.md.water.usgs.gov/monthly/bay.html). Note that the calender year starts on the ticks for "Water year based on"

and highest numbers of 23 mm long individuals produced across the six annual estimates. Our low recruitment condition was probably truly low as data from the Maryland Department of Natural Resources Bay-wide index (2017) indicated that age-0 bay anchovy catch per unit effort has declined since 1959. Our high recruitment condition, while high for the 1995–2000 period, may be closer to average recruitment relative to the long-term record for anchovy in Chesapeake Bay.

12.3.2 Nutrient Loading Scenarios

Three nutrient loadings were used: baseline, decreased, and increased. The baseline nutrient loadings scenario used the nutrient loads estimated for the years 1984, 1985, and 1986. The decreased scenario reduced nutrient loads by 50% from baseline levels in each year. A 50% reduction in nutrient loadings was roughly equivalent to that required under the old Chesapeake 2000 Agreement, which required reductions of 48 and 53% (based on 1985 levels) for total nitrogen and

Table 12.2 Summary of the model runs analyzed in this paper. All simulations were structured 23 years long and started with 10 years of spin-up for nutrient loadings (no anchovy), then 3 years of spin-up for anchovy (normal water years), and then 10 years of water years that matched the water years of 1984–1993. The first 10 years for spin-up used all normal water years and the same nutrient loadings as was used for the 1984–1993 period. The 3 years of spin-up for anchovy also used normal water years and the low or high recruitment level used in 1984–1993

Run	Purpose	Nutrient loadings	Recruitment level	q in general mortality
1	Re-calibration and effects of water year and recruitment level	Baseline	Low	1.17
2		Baseline	High	1.17
3	Effects of increased and decreased nutrient loadings	Decreased	Low	1.17
4		Increased	Low	1.17
5		Decreased	High	1.17
6		Increased	High	1.17
7	Effects of decreased recruitment (increased egg or larval mortality) due to increased nutrient loadings	Increased	86% of Low	1.17
8		Increased	49% of Low	1.17
9		Increased	12% of Low	1.17
10		Increased	76% of High	1.17
11		Increased	52% of High	1.17
12		Increased	23% of High	1.17
13	Effects of increased general mortality rate of juveniles and adults due to increased nutrient loadings	Increased	Low	1.3
14		Increased	Low	1.43
15		Increased	High	1.3
16		Increased	High	1.43
17		Increased	High	2.3
18		Increased	High	2.725

phosphorus inputs (Kemp et al. 2005; Chesapeake Bay Program 2013). In the increased loadings scenario, nutrient loads are 50% higher than the baseline loads. The increased loadings scenario provides a contrasting scenario to the decreased loadings scenario. It may reflect future conditions if no action is taken, but no such land use and loadings projections have been made. Increases and decreases in nutrient loadings were applied through the use of a simple multiplier of the nutrient loading rates (e.g., nutrient loadings × 0.5 for the decreased nutrient loadings scenario and × 1.5 for the increased nutrient loadings scenario). In the standard 10-year simulations, the volume of hypoxic water was typically 20–40% less under decreased nutrient loadings and about 10–15% larger under increased nutrient loadings (Fig. 12.2). The model represents peak hypoxia in summer, coincident with peak primary production of organic matter and temperature-induced respiration. There can be some modeled residual hypoxia in isolated deep holes during winter. The residual hypoxia is an artifact of the relatively coarse grid and is absent in later, more highly resolved grids (e.g., Cerco and Noel 2013). The isolated persistent hypoxic volumes do not impact the anchovy, as the location and small spatial extent of any residual winter hypoxia did not affect anchovy habitat.

12.3.3 Initial Conditions from Spin-Up

All simulations were structured as 23-year simulations. Each simulation started with 10 years of spin-up for nutrient loadings (no anchovy), then 3 years of spin-up for anchovy (normal water years), and then 10 years of water years that matched the water years of 1984–1993. The first 10 years for spin-up used all normal water years and the same nutrient loadings scenarios as was used for the 1984–1993 period (e.g., 50% increase, baseline, or 50% decrease). The 3 years of spin-up for anchovy also used normal water years and the same low or high recruitment level as used in the 1984–1993 portion of the simulation. For example, run 13 (Table 12.2) had 10 years of no anchovy under normal water years and increased nutrient loadings, then 3 years with anchovy under normal water years, increased nutrient loadings and low recruitment, and then 1984–1993 water years under increased nutrient loadings, low recruitment, and q increased from 1.17 to 1.3.

12.3.4 Increased Mortality Rates

Changes in nutrient loadings can have several potential effects on bay anchovy that were not directly simulated in our bay anchovy model. Chesney and Houde (1989) used laboratory studies to show that anchovy egg hatchability declined significantly at DO concentrations less than 3 mg L^{-1}, while Adamack et al. (2012) used a simulation model and showed that increased spatial extent and duration of low DO conditions due to increases in nutrient loadings could result in significant increases in egg mortality rates. Costantini et al. (2008) and Ludsin et al. (2009) found that increases in the extent of hypoxia in Chesapeake Bay could increase the mortality rates of forage fish by increasing the degree of vertical spatial overlap between the forage fish and their predators resulting in increased encounter rates. The above studies were focused on the negative effects of increased hypoxia from increased nutrient loadings on anchovy. However, the reverse situation of decreased nutrient loadings reducing hypoxia resulting in less vertical spatial overlap between forage fish and their predators reducing their encounter rates and thereby reducing mortality is also possible and of potential importance.

To investigate those mortality-related effects not explicitly covered in the anchovy model, we ran two sets of simulations for the low and the high recruitment scenarios with increased nutrient loadings. The first set of simulations examined the potential effects of increased egg mortality due to increases in the vertical extent and intensity of hypoxia under the increased nutrient loadings scenario. The egg stage is not explicitly simulated in our model because we add recruits each year. We represented possible changes in egg (and larval) mortality rates by adjusting the recruitment levels. For the low and high recruitment scenarios, we tested three levels of recruitment reductions from the original recruitment levels with the goal of identifying how much larval recruitment had to be reduced in order to result in no net gain in anchovy production relative to the baseline nutrient loading scenario. The minimum recruitment level tested for the low recruitment scenario was arbitrarily set to 12% of the baseline low recruitment level, while for the high recruitment scenario the minimum was set to the low recruitment level (23% of high recruitment). Two additional recruitment levels that were spaced evenly between the minimum recruitment level and the original recruitment level were tested for each recruitment scenario (low and high).

The second set of simulations examined the effects of increases in bay anchovy juvenile and adult mortality rate due to potential increases in the degree of vertical spatial overlap between anchovy and their predators. While we simulated anchovy vertical movement, predators were represented as a mortality rate. As an initial approach to simulating predator responses that would cause increased overlap, we increased q, the length-specific mortality coefficient (Eq. 12.12), from our original value of 1.17 to 1.3 and to 1.43. These values were the median and maximum values of q measured by Jung and Houde (2004a). The value of q was further increased to 1.725 and to 2.3, which were equivalent to instantaneous mortality rates that were about 50% and about 100% higher than the instantaneous mortality rates when q was set the baseline value of 1.17. These higher values of q were to ensure that mortality would be sufficiently high to offset the increased anchovy production under increased nutrient loadings.

12.3.5 Model Outputs

Model outputs are presented for the three sets of simulations: re-calibration and the baseline simulations, the effects of water years and nutrient loading on anchovy, and how changes in recruitment and mortality due to increased nutrient loads affect anchovy biomass. All model outputs and how they were computed from super-individuals in the simulations are documented in Table 12.3.

Results summarizing the baseline and re-calibration results include time series plots of daily total biomass, total abundance, and mean length, and comparisons of daily nutrient, chlorophyll-a, and micro- and mesozooplankton concentrations before and after the addition of anchovies to the model. We then summarize the baseline results using YOY individuals only in October from the last years (normal, dry, and wet) of the simulations: box plots of lengths, biomass, and abundance. Finally, we compared box plots of the latitudes of individuals between each water year type (last 3 years) and reported field data.

The effects of water year type and nutrient loadings also use mean length, biomass, and abundance for YOY in October. Two additional outputs are two-dimensional spatial maps of all individuals in July by water year type (last 3 years) and high or low recruitment, and the percentage of mortality due to the possible causes over all years in the simulations. The effects of additional mortality were assessed by comparing YOY biomass in October under increased loadings with various values of reduced recruitment and increased values of q to October

Table 12.3 Calculations performed to go from super-individuals to the model outputs reported in tables and figures in this paper

Model output	Calculations	Runs and Years	Figures
Daily biomass, abundance, and mean length of all individuals	Output all individuals on first time step of each day; compute summed weight x worth (biomass) and worths (abundance); compute mean length (weighted by worth)	Runs 1 and 2 All 13 years for each run	Figure 12.3
October YOY median length, growth rate, abundance, and biomass	Output all YOY individuals on first time step after midnight on October 15; compute the median length (weighted by worth), mean growth rate (weighted by worth) from 23 mm in length to current length of super-individual, summed worths (abundance), and summed weight x worths (biomass)	Runs 1 and 2 Last 3 years (dry, normal, and wet) of each run	Figure 12.4 (also Table 12.4)
Mean latitude in mid-October	Output all individuals on first time step after midnight on October 15; assign individuals to 10 mm size classes; compute box plots of latitudes (weighted by worth)	Run 2 Last 3 years (dry, normal, and wet) of the run	Figure 12.4
Box plot of lengths and biomass in mid-October, and percent survival (initial recruitment to October)	Output all YOY individuals on first time step after midnight on October 15; compute box plots of lengths (weighted by worth) and biomass (weights x worth); and compute survival as summed worths of all individuals on October 15 divided by initial total worth of 23 mm recruits	Runs 1–6 Last 3 years (dry, normal, and wet) of each run	Figure 12.5
Depth-integrated densities of age-1 and older anchovy	Output all age-1 and older individuals on first time step after midnight on July 15; compute summed worths of all individuals over depth by surface grid cell	Runs 3–6 Last 3 years (dry, normal, and wet) of each run	Figure 12.6
Mean latitude by month of age-1 and older individuals	Output the latitude and worth of each age-1 and older individual on midnight of the midpoint day in each of April, May, June, July, and August; compute the mean latitude (weighted by worth)	Runs 3–6 Last 3 years (dry, normal, and wet) of each run	Figure 12.7

(continued)

Table 12.3 (continued)

Model output	Calculations	Runs and Years	Figures
Proportion of total mortality due to each cause	Cumulative sum of all worths due to mortality, and by each mortality source (hypoxia, starvation, old age, and natural); compute the proportions as sum of deaths by source divided by total deaths	Runs 1–6 All 13 years for each run	Figure 12.8
YOY biomass in mid-October	Output all YOY individuals on first time step after midnight on October 15; compute biomass as summed weight × worths	Runs 7–18 Last 3 years (dry, normal, and wet) of each run	Figure 12.9

biomass predicted under baseline conditions. The patterns in the model results that rely on snapshot outputs in October were consistent across water years, regardless of day selected for outputting, but the magnitude of the individual model results varied across output days.

12.4 Results

12.4.1 Baseline Simulations

In the baseline simulations (Runs 1 and 2), total (all ages) anchovy biomass (Fig. 12.3a) and abundance (Fig. 12.3b) show repeating annual cycles that reflect the effects of recruitment level more than water year type. Under high recruitment, biomass initially increased rapidly each year as recruits were added to the population and then dropped very rapidly due to the rising biomass of anchovy rapidly reducing their prey causing slower growth and higher mortality rates relative to the low recruitment scenarios. While biomass and abundance were much higher under high recruitment compared to low recruitment, mean length was lower under high recruitment confirming a density-dependent effect of abundance on growth (Fig. 12.3c). Abundance under high recruitment (Fig. 12.3b) initially dropped in tandem with a drop in biomass each year (Fig. 12.3a), suggesting that slowed growth led to increased mortality. Under low recruitment, both biomass and abundance showed smoother increases as juveniles were added and decreases after that each year, and the lower predation pressure on their zooplankton prey resulted in faster growth and longer mean lengths (gray lines in Fig. 12.3).

Within each of the baseline simulations, the annual cycles of biomass, abundance, and mean length did not show large differences among the dry, normal, and wet water years (Fig. 12.3). Use of total population, which included all ages, acted

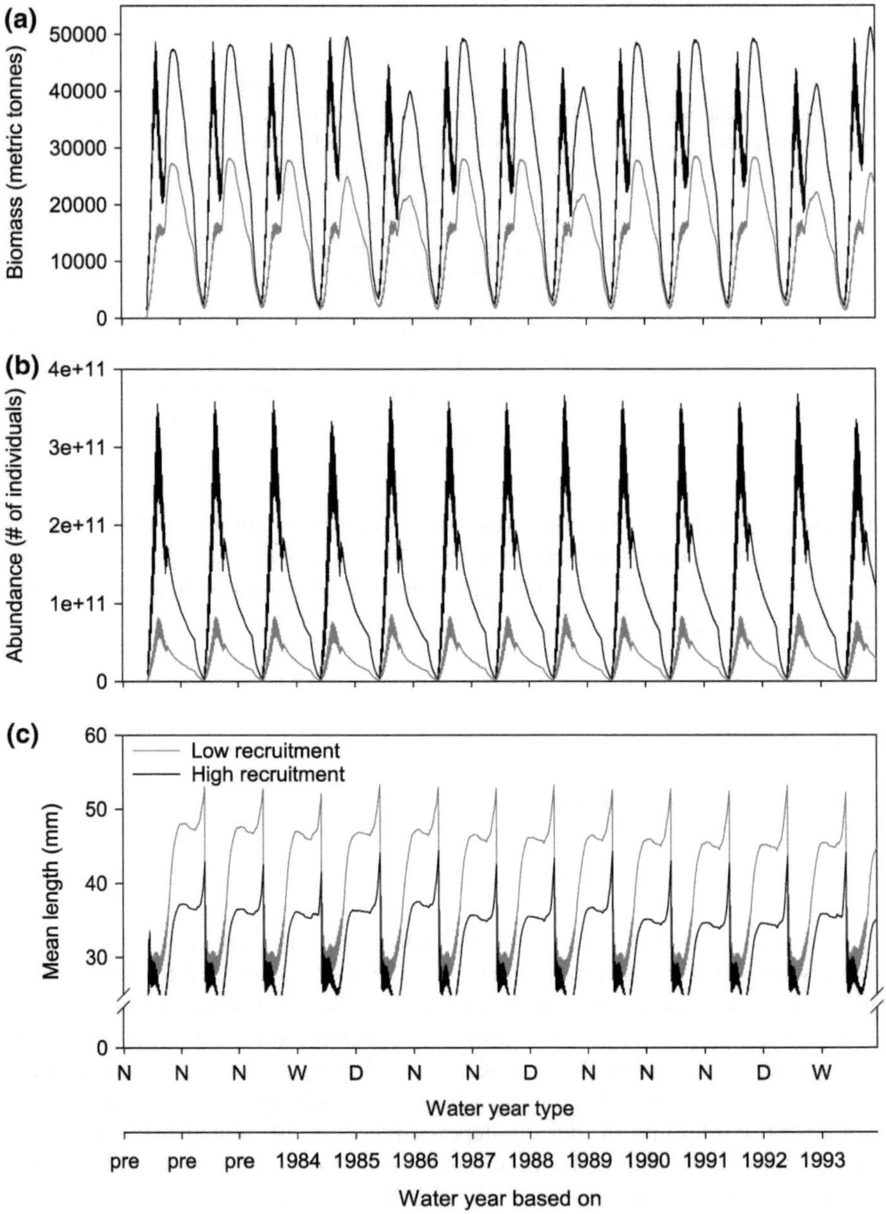

Fig. 12.3 Total daily population biomass (**a**), abundance (**b**), and mean length (**c**) for the baseline simulations under low and high recruitment

to smooth differences from year-to-year. When only YOY were examined, water year type had somewhat larger effects, but they were still small compared to the effects of recruitment level (discussed below in Fig. 12.6).

12.4.2 Assessment of Re-calibration Using Baseline Simulations

Following the re-calibration of the coupled models after adding bay anchovy, chlorophyll-a concentrations and mesozooplankton densities in the two simulations (Runs 1 and 2) were 1-fold to 3-fold higher than predicted values without anchovy. Predicted chlorophyll-a (based on the daily values) ranged from 0.85 to 18.65 g C m^3 with anchovy versus 0.67 to 14.10 g C m^3 without anchovy, and mesozooplankton ranged from 0.0037 to 0.12 g C m^3 with anchovy versus 0.0037 to 0.046 g C m^3 without anchovy. In contrast, microzooplankton densities ranged from a third of to being similar in magnitude to the predicted densities obtained without anchovy. The temporal and spatial trends in micro- and mesozooplankton densities were generally consistent with the densities from model runs without anchovy. Nitrate and phosphate concentrations were almost identical between simulations without and with anchovy. Adamack (2007) further demonstrated the similarity of micro- and mesozooplankton densities (with and without anchovy) to the historical field data (roughly monthly) collected at station CB5.2 from the long-term monitoring program (see Magnien 1987).

Model predictions of anchovy lengths and growth rates for YOY individuals that survived to the end of October for the baseline simulations (Runs 1 and 2) are somewhat comparable with the October field observations of Jung and Houde (2004a, their Table 2). The median length of anchovy under low recruitment was about 34 mm for all 3 water year types (last 3 years of the simulation), which is lower than the range of mean lengths (46.7–51.1 mm) reported by Jung and Houde (2004a). The shorter simulated lengths were driven by a combination of the date that model results were output and the pattern of recruitment that we used. Delaying output by 14 days resulted in median lengths being ~5 mm longer, while ignoring recruits added in September and October produced additional increases in the simulated lengths. Simulated median lengths in the last 3 years of the high recruitment baseline simulation were about 27 mm, which is approximately half of the shortest mean lengths observed (46.7–51.1 mm, Table 12.4). Predicted YOY anchovy growth rates for both the low and high recruitment simulations (about 0.38 mm d^{-1}) are similar to Jung and Houde's values (0.36 and 0.40 mm d^{-1}), and also similar to the juvenile anchovy growth rates reported by others (0.20–0.33 mm d^{-1}) for mid-Chesapeake Bay (Morin and Houde 1989; Newberger et al. 1989).

Predicted abundances of YOY anchovy in late October for the two baseline runs are within the range of abundances observed by Jung and Houde (2004a) (Table 12.4). Anchovy abundances for the low recruitment scenario (44.4 to 48.4 × 10^9

Table 12.4 Predicted median length in October, mean growth rate of survivors, abundance, and biomass of YOY anchovy in dry, normal, and wet years of the baseline simulation under low and high anchovy recruitment compared to observed values from Jung and Houde (2004a). The years of field data are the lowest, median, and highest values of 23 mm anchovy recruits reported by Jung and Houde in order to provide a range of values to compare to the model results

Year type	Median length (mm)	Growth rate (mm d^{-1})	Abundance ($\times 10^9$)	Biomass (metric tonnes)
Low recruitment				
Dry	32.4	0.34	45.0	18,393
Normal	33.8	0.38	48.4	22,502
Wet	33.8	0.39	44.4	19,535
High recruitment				
Dry	26.1	0.15	169.1	28,346
Normal	27.3	0.14	187.6	39,094
Wet	28.3	0.14	177.9	39,684
Jung and Houde (2004a)				
Low (1996)	49.1	0.36	29.2	27,000
Median (1997 and 1999)	51.1	0.36	98.2	~100,000
High (1998)	46.7	0.40	273.8	193,000

individuals) are similar to the observed abundance in 1995 (44.5 $\times$ 10^9 individuals), which was the year with the second lowest observed YOY recruitment. Predicted abundances for the high recruitment scenario (169.1–187.6 $\times$ 10^9 individuals) are between the median and highest levels of recruitment observed by Jung and Houde (98.2–273.8 $\times$ 10^9 individuals).

October YOY biomass tended to be low relative to the October standing stock biomass of anchovy reported by Jung and Houde (Table 12.4). Anchovy biomasses predicted for low recruitment (18,393–22,502 metric tonnes) are similar to the fall peak in biomasses observed by Jung and Houde during years with low estimated anchovy biomass (~25,000–50,000 metric tonnes; their Fig. 4a). However, predicted biomasses for the high recruitment scenario (28,346–39,684 metric tonnes, Table 12.4) are less than half of the reported field-based estimates (~100,000–150,000 metric tonnes) during years with high anchovy biomass.

Model predictions of the mean latitude of age-1 and older anchovy (not YOY) by length category are generally comparable to field data from the CHESFIMS project (Miller et al. 2008). When each of the simulated water years were matched with field data for years of similar water type, the predicted mean latitude of anchovy overlaps with the observed latitudes for the intermediate length classes (Fig. 12.4)

We examined the vertical distributions of anchovy for two cross sections of Chesapeake Bay (not shown) and found that anchovy were generally concentrated in waters above the hypoxic layer, consistent with Zhang et al. (2014) who showed many pelagic fish species moving vertically in response to hypoxic conditions.

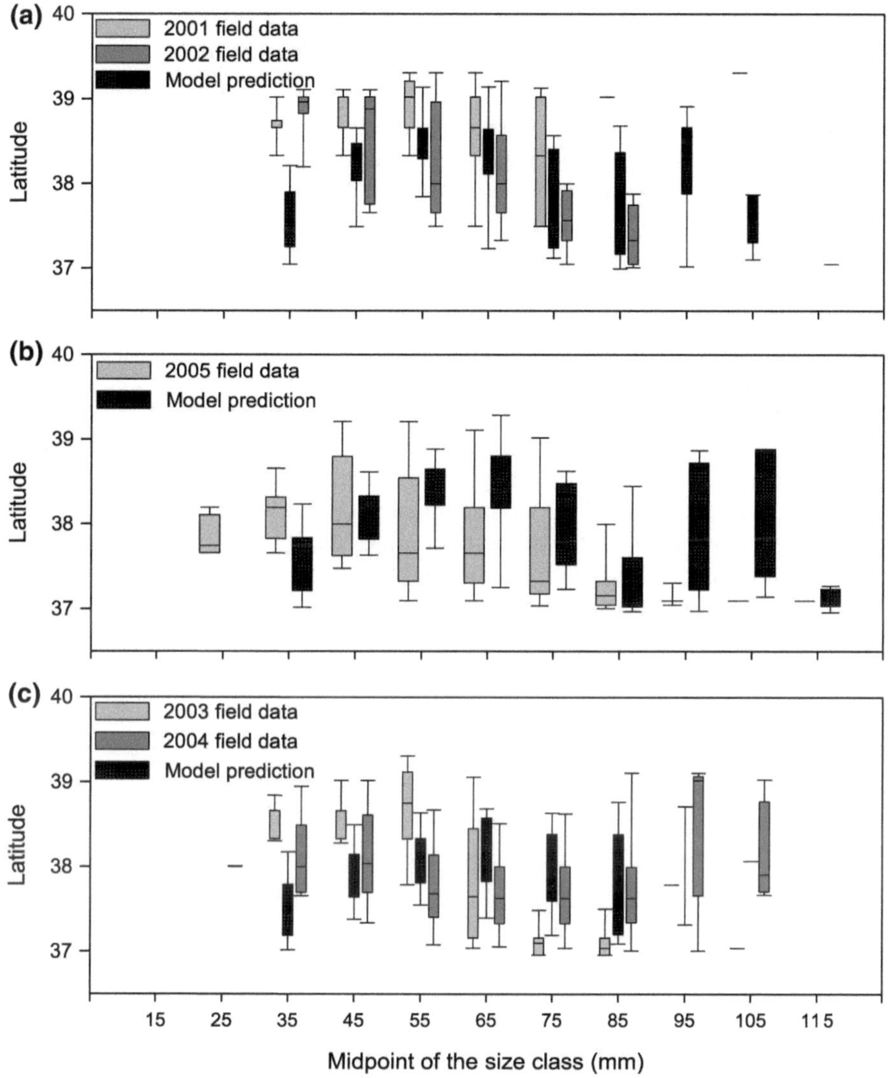

Fig. 12.4 Predicted latitudes (°N) of age 1 and older anchovy in October for the last 3 years water year types: (dry (**a**), normal (**b**), wet (**c**)) of the baseline simulation under high recruitment compared to field observations of the distributions of anchovy in the fall. Years of the field data were matched to the simulated years based on being dry, normal, or wet water years. Anchovy field distributions are from the CHESFIMS project

12.4.3 Effects of Water Year and Nutrient Loadings on Salinity, Temperature, and Zooplankton

Salinity and water temperature were only affected by water years, while mesozooplankton was affected by both water years and nutrient loadings scenarios. For salinity, going from dry years to wet years resulted in lower overall salinities and increased the southward extent of low salinity waters (e.g., salinity < 15; Fig. 12.5 top row). Water temperatures were somewhat similar for the dry year and the normal year, but were somewhat cooler in wet years (Fig. 12.5 middle row). Spatially, water temperatures tended to be warmest along the edges of the Bay and in tributaries and coolest along the main channel, particularly in the northern half of the Bay. Mesozooplankton density increased going from dry years to wet years and from decreased nutrient loadings scenarios to increased nutrient loadings scenarios in response to increases in nutrient loadings rates to the Bay (Fig. 12.5 bottom row). Spatially, the main channel of the Bay had the highest mesozooplankton densities, but this is in part a result of the mesozooplankton densities being depth-integrated and the main channel being the deepest portion of the Bay. Microzooplankton (not shown) had similar patterns to mesozooplankton.

The effects of water years and nutrient loadings scenarios on salinity, temperature, zooplankton, and other water quality variables are covered in detail in Cerco and Cole (1993), Cerco et al. (2010), and Cerco and Noel (2013).

12.4.4 Effects of Water Year and Recruitment Level in Baseline Simulations

Recruitment level affected YOY anchovy lengths more than water years (Fig. 12.6). Anchovy in the low recruitment scenario (Run 1) generally reached longer lengths than anchovy in the high recruitment scenario (Run 2) (Fig. 12.6a vs. b). However, there was little difference in the length distributions of anchovies across water years (middle bars for each water year in Fig. 12.6a, b). Similar results were seen for October biomasses (Fig. 12.6c, d) and abundances (Table 12.4) of YOY anchovy, though there was more variability across water years for biomass and abundance than for lengths. Across water years, YOY anchovy biomasses were similar for all three water years under low recruitment (18,393–22,502 metric tonnes), but higher for the normal and wet years than for dry year (about 39,000 vs. 28,346 metric tonnes) under high recruitment (Table 12.4; Fig. 12.6c, d). The same pattern was predicted for YOY abundance—similar for all years under low recruitment (44.4–48.4 $\times 10^9$ individuals) and higher for the normal and wet years than for the dry year (187.6 and 177.9 vs. 169.1 $\times 10^9$ individuals) under high recruitment.

Water year had a larger effect on anchovy spatial patterns than recruitment level. As would be expected, anchovy densities (all ages) were generally higher under high recruitment than low recruitment (note scale of color bar in Fig. 12.7). Across

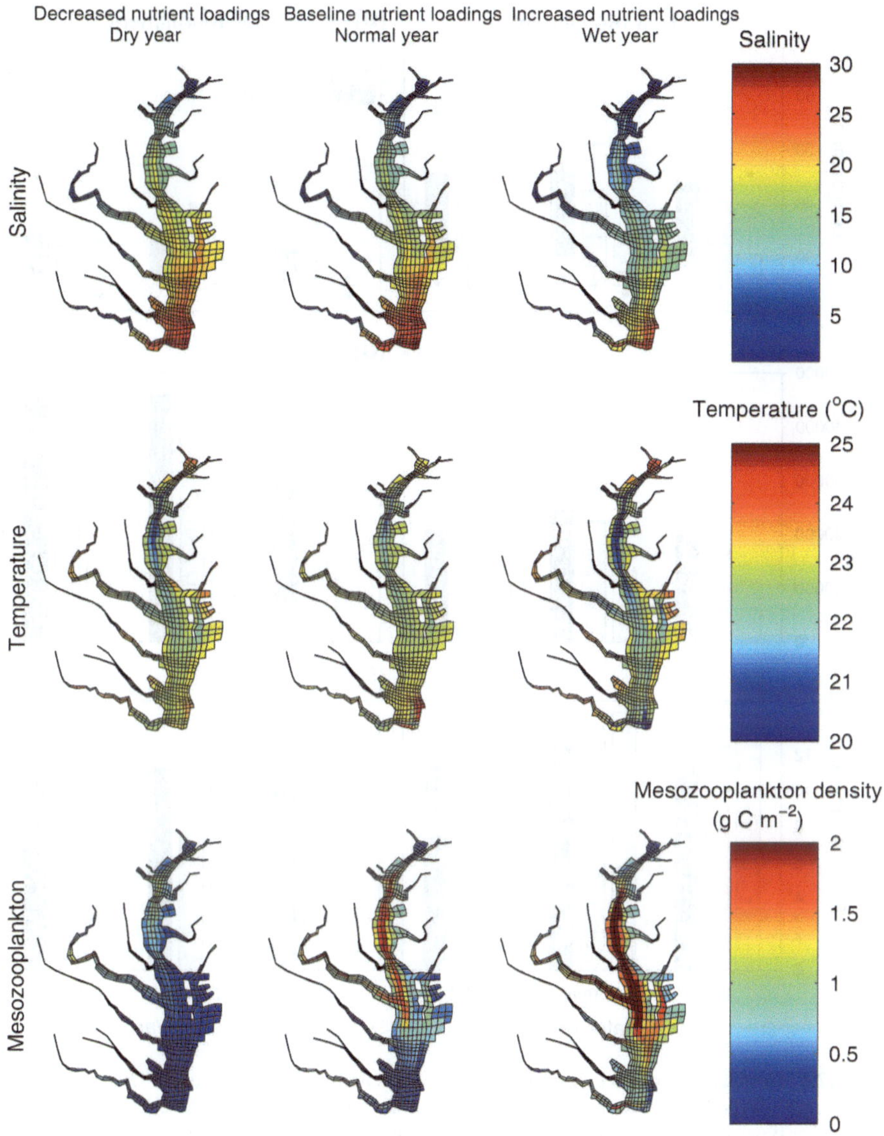

Fig. 12.5 July spatial patterns of depth-integrated salinity (*top row*), temperature (*middle row*), and mesozooplankton densities (*bottom row*) for the dry year of the decreased nutrient loadings scenario, the normal year of the baseline nutrient loadings scenario, and the wet year of the increased nutrient loadings scenario. The three scenarios were selected to show the full range of the nutrient loadings scenarios (e.g., the extreme minimum, "typical," and maximum nutrient loadings scenarios)

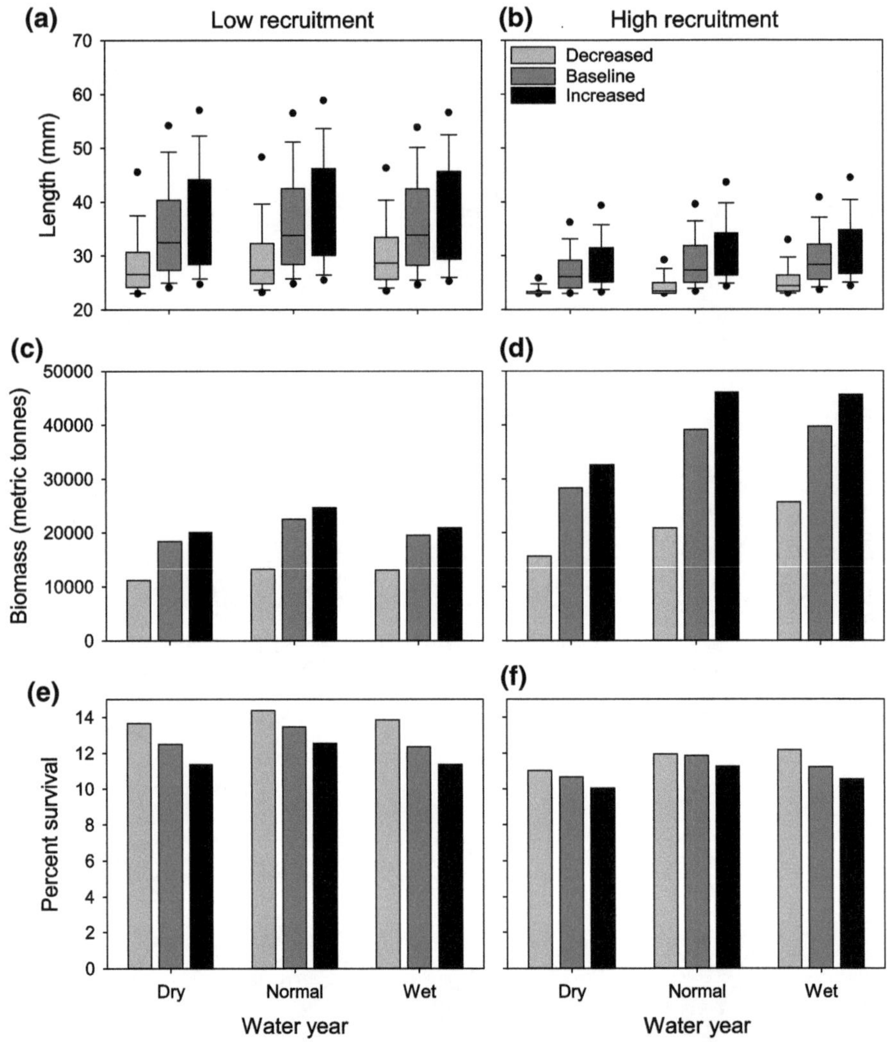

Fig. 12.6 Comparison of YOY bay anchovy October length distributions (**a**, **b**), October biomass (**c**, **d**), and survival from recruitment to October (**e**, **f**) across recruitment levels (*low* and *high*), water year types (dry, normal, and wet), and nutrient loadings (decreased, baseline, and increased). The last 3 years in each simulation were used as dry, normal, and wet year types. For anchovy length *box plots*, *circles* show the 5th and 95th percentiles of the distributions, *whiskers* show the 10th and 90th percentiles, the *box* shows the 25th and 75th percentiles, and the *solid line* in the *box* shows the median

water years, the mean latitude of anchovies in July was farther south in the wet year compared to the normal and dry years, likely in response to changes in the extent of hypoxia between wet and dry years. In the wet year, bottom layer hypoxia extended further south and was more intense than in normal and dry years (not shown). The

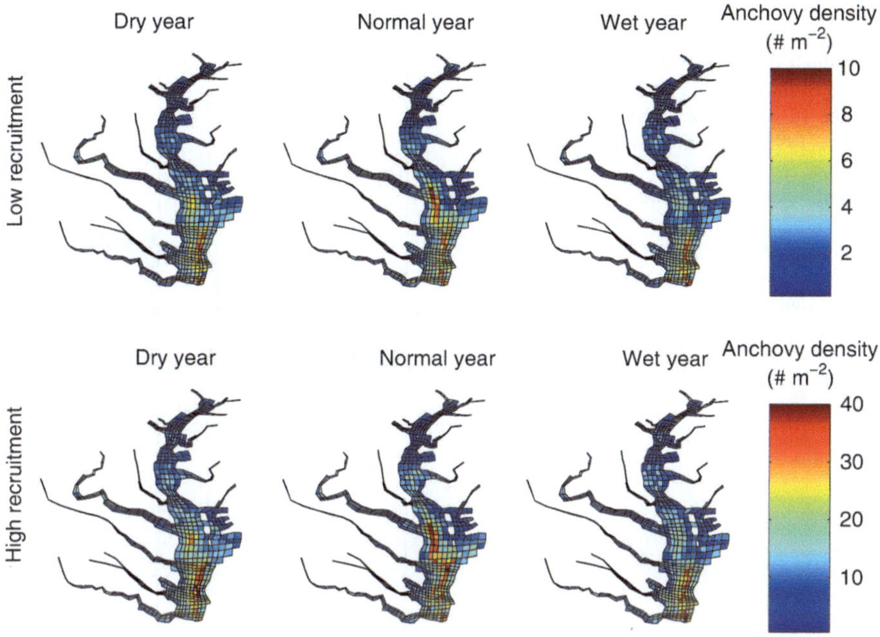

Fig. 12.7 July depth-integrated (# m^{-2}) anchovy densities for the last 3 years (dry, normal, and wet) of the baseline simulation under low recruitment (*top row*) and under high recruitment (*bottom row*). Note the difference in the scale on the *color bars*

effect of water year type on latitude in July was roughly consistent under low and high recruitment levels. Under low recruitment, mean latitude of age-1 and older individuals in July was 37.27°N for the wet year versus 37.35°N and 37.33°N for the dry and normal years. Under high recruitment, mean latitude in July was also lowest (37.33°N) for the wet year compared to the normal and dry years (37.43°N and 37.41°N, respectively).

12.4.5 Effects of Nutrient Loadings

Changing nutrient loadings (Runs 3–6) affected dissolved oxygen concentrations and zooplankton densities in the eutrophication model. Increasing nutrient loads by 50% had a relatively small effect on hypoxic volume (Fig. 12.2). Reducing nutrient loadings had a much larger effect on hypoxic volumes, with a 50% reduction in nutrient loadings resulting in a 20–40% reduction in hypoxic volume (Fig. 12.2). Higher nutrient loadings resulted in higher zooplankton densities throughout the Bay (and vice versa).

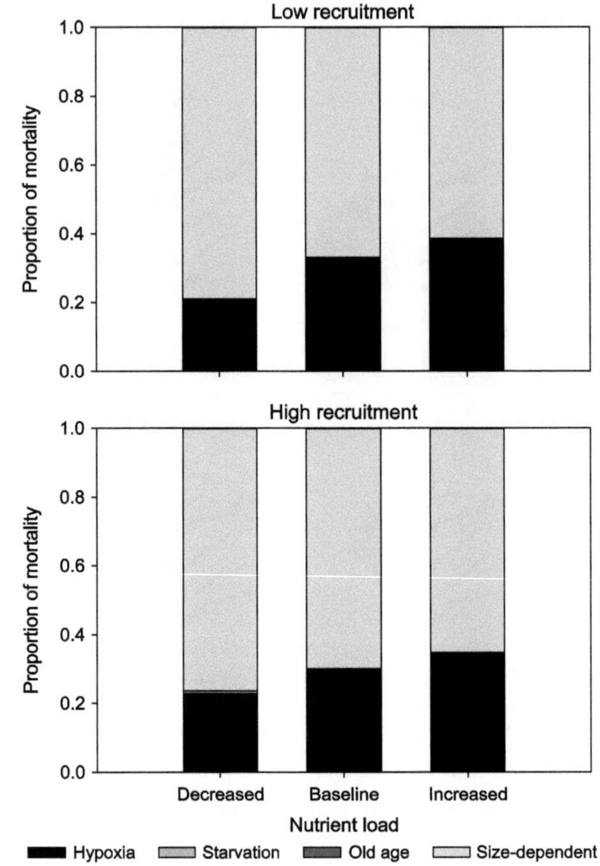

Fig. 12.8 Causes of anchovy mortality for all anchovy individuals dying over the course of the 3 prerun years and the 10 water years from 1984 to 1993 for decreased, baseline, and increased nutrient loads and low and high recruitment. Causes of mortality include exposure to low DO (hypoxia), starvation, old age, and general (size-dependent) mortality

The effects of increasing or decreasing nutrient loads on bay anchovy were consistent across the two recruitment levels and the three water year types. Increased nutrient loadings resulted in increases in YOY anchovy lengths and biomasses, while decreased nutrient loadings resulted in decreases in lengths and biomasses (Fig. 12.6a–d). However, YOY anchovy survival rates were inversely related to nutrient loadings, with increased nutrient loads resulting in lower survival rates for both recruitment levels (Fig. 12.6e, f), primarily due to increases in mortality due to increasing rates of exposure to hypoxic conditions (Fig. 12.8) as the volume of hypoxic water increased (Fig. 12.2). Changes in anchovy survival rates (Fig. 12.6e, f) were small relative to the changes in anchovy lengths (Fig. 12.6a, b), which resulted in increased nutrient loadings producing higher biomasses of YOY bay anchovy (Fig. 12.6c, d) despite the higher YOY mortality rates. We note that going from baseline nutrient loadings to increased nutrient loadings had a smaller effect on anchovy lengths, biomass, and survival than going from baseline to decreased nutrient loadings.

12.4.6 Effects of Additional Hypoxia-Related Mortality

For the low recruitment level (Runs 7–9), reducing recruitment to 86% of the baseline low recruitment level was sufficient to reduce October biomass of YOY anchovy in the increased nutrient loadings scenario to the same level as the biomass obtained under baseline nutrient loadings (Fig. 12.9a). This same reduction in recruitment resulted in the offsetting of the nutrient-fueled increased high biomass for all three water year types. Under high recruitment (Runs 10–12), recruitment had to be reduced to 52% of the baseline high recruitment level to result in October YOY anchovy biomasses that were approximately equal to biomasses under baseline nutrient loadings (Fig. 12.9b).

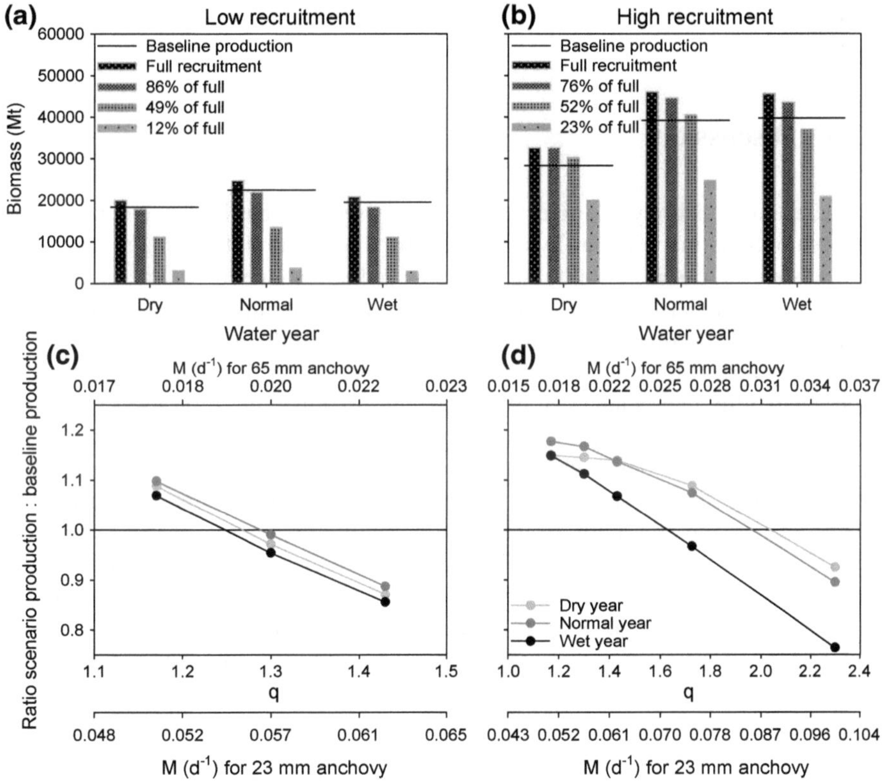

Fig. 12.9 Effects of reduced recruitment (**a**, **b**) and increased juvenile and adult mortality rate (**c**, **d**) under increased nutrient loadings on bay anchovy YOY biomass in October by water year type and for the low and high recruitment levels. The October YOY biomass values for the last three years (dry, normal, and wet) of each simulation were used. Panels **a** and **b** show the October biomass values obtained from simulation; the *horizontal lines* show the biomass for baseline nutrient loadings. Panels **c** and **d** show the ratio of October biomasses (changed q biomass/baseline) for increasing values of q under increased nutrient loadings. In Panels **c** and **d**, the *horizontal line* at 1.0 indicates when the biomasses were the same under increased q and baseline simulations

For the low recruitment level (Runs 13 and 14), increasing q from 1.17 to 1.30 resulted in October YOY anchovy biomass under increased nutrient loadings dropping to the same level as under baseline nutrient loadings (Fig. 12.9c). Increasing q from 1.17 to 1.30 translated into the mortality rate of a 23 mm long individual increasing from 0.05 d^{-1} to 0.057 d^{-1} and for a 65 mm long adult to increase from 0.018 d^{-1} to 0.020 d^{-1}. Much larger increases in q were needed to reduce October YOY biomasses back down to baseline levels under high recruitment (Runs 15–18). During wet years, an increase of q from 1.17 to ~1.6 was needed (black line in Fig. 12.9d), while for normal and dry years q had to increase even more to about 2.0. Increasing q from 1.17 to 1.6 resulted in mortality rate of a 23 mm individual increasing from 0.05 to 0.070 d^{-1} and for a 65 mm individual increasing from 0.018 to 0.025 d^{-1}. A q value of 2.0 resulted in about a 75% increase in mortality rate of juveniles and adults (from 0.05 to 0.087 d^{-1} and from 0.018 to 0.031 d^{-1}).

12.5 Discussion

The effects on anchovy of increased and decreased nutrient loadings depended on the assumptions made about hypoxia causing direct mortality on eggs and larvae and causing increased mortality on juveniles and adults due to assumed changes in anchovy and predator vertical spatial overlap. If we assume that mortality would only be caused by direct exposure of juveniles and adults to hypoxia, then a 50% change in nutrient loadings would have the expected "fertilizer" effect of changing food availability and changes in growth would dominate the response of anchovy. Decreased nutrient loadings would result in smaller YOY anchovy lengths and lower biomass in October, while increased nutrient loadings would result in larger YOY anchovy lengths and slightly higher biomass. These results include low DO effects on juvenile and adult growth and mortality arising from direct exposure to low DO and also reflect avoidance behavior by juveniles and adults of low DO. The simulations ignored DO effects on egg and larval mortality and any mortality changes related to habitat overlap between anchovy and their predators.

However, if we assume that egg or larval mortality (i.e., reduced recruitment to 23 mm) or juvenile and adult mortality are related to the nutrient loadings via hypoxia, then the predicted positive effects on anchovy of increased nutrient loadings can be offset and even reversed by simultaneous changes in hypoxia extent and intensity. Under low recruitment, a 14% reduction in the number of recruits was sufficient to eliminate the benefits of increased nutrient loadings on anchovy biomass, while even larger reductions in recruitment could reduce anchovy biomass to levels that were less than half of those under baseline nutrient loads (Fig. 12.9a). Within the prerecruit phase, an effect on eggs and yolk-sac larvae is more likely than on larvae. Adamack et al. (2012) found that egg mortality was 2-fold to 7-fold higher when nutrient loads were increased, while the response of early larvae was complicated because of shifting degrees of vertical overlap with their invertebrate predators. Alternatively, juvenile and adult mortality rate would only need to increase from 0.051 d^{-1} to 0.056 d^{-1} for 23 mm anchovy and from 0.018 d^{-1} to 0.020 d^{-1} for 65 mm anchovy due to more spatial overlap with predators to also offset the benefits of increased nutrient loadings (Fig. 12.9c). While the same general result was obtained under high recruitment, larger changes in mortality related to increased nutrient loadings effects were needed.

We carried out some additional model simulations under increased nutrient loadings with both sources of hypoxia-related mortality (direct egg/larval mortality; predation on juveniles and adults due to changes in the degree of vertical spatial overlap) changed simultaneously. While the reductions in recruitment and increased mortality needed to offset increased biomass were reasonable when examined alone for low recruitment, much larger changes were needed under high recruitment. We did not do an exhaustive search of all combinations of reduced recruitment and increased q values that would offset nutrient-fueled higher production under high recruitment; rather, we tried a few combinations. For example, a combination of a 16% increase in juvenile and adult mortality rates combined with a 14% reduction in recruitment in wet years and a 50% reduction in dry years was sufficient to offset the benefits of increased nutrient loadings. Another viable combination was a 22% increase in juvenile and adult mortality rates combined with a 14% reduction in recruitment in wet years and a 25% reduction in dry years.

We also performed some additional simulations to explore the changes in mortality needed to offset the reduced growth predicted under decreased nutrient loadings. Predicted October YOY biomass under low recruitment with decreased nutrient loadings was about 60% of the biomass under low recruitment with baseline nutrient loadings across the 3 year types (Fig. 12.6c). One combination that offsets the reduced biomass under decreased nutrient loadings was assuming that reduced hypoxia spatial extent with decreased nutrient loadings would result in lowered egg and larval mortality and therefore high recruitment (rather than low recruitment) and less overlap of anchovy with their predators and therefore q would decrease from 1.17 to 0.585. Predicted October YOY biomass under decreased nutrient loadings but with these improved conditions was higher than the biomass predicted under baseline loadings for the three water year types. There are other combinations of egg and larval mortality (recruitment) and juvenile and adult mortality (q values) that would also offset the reduced biomass predicted with decreased nutrient loadings.

Thus, the apparent benefit of increased nutrient loadings (and by analogy of decreased loadings) depends on both the increased zooplankton predicted (more food) and possible increased mortality due to the direct and indirect effects hypoxia. Under low recruitment, relatively small reductions in recruitment or increased juvenile and adult mortality, and even smaller effects if both are operating simultaneously, were sufficient to offset or even reverse the effects of changes in nutrient loadings. Larger reductions in recruitment or increased mortality were needed under high recruitment conditions. Relatively large, but not impossible, reductions in mortality due to less hypoxia offset the reduced growth under decreased nutrient loadings. Determining the likelihood and realism of the needed changes in egg to larva and juvenile and adult mortality rates due to variation in hypoxia will require a combination of field data collection coordinated with modeling in the future.

We included high and low recruitment and different water year types to bound model predictions of responses to changes in nutrient loadings and the associated changes in zooplankton and hypoxia. We used different simulated years to represent dry, normal, and wet year types; however, we used only single years to represent each of these types. In reality, there is also variability among years within dry, normal, and wet year types. Recruitment level had the expected effect on biomass and abundances, with the contrast between low and high recruitment showing the effects of density-dependent growth and survival of YOY. Mean length of YOY in October was almost 50% larger under low recruitment compared to high recruitment. The slowed growth results in consistent, but relatively small, changes in YOY survival from individuals being smaller with a mortality rate that decreased with length (Rose et al. 2001). Wang et al. (1997) and Cowan et al. (1999) also found density-dependent growth and survival using an individual-based anchovy population model configured for a single, well-mixed box for the mesohaline region of Chesapeake Bay. However, Jung and Houde (2004b) hypothesized that anchovy should have higher survival rates during wet years than dry years due to their higher concentrations in a smaller area satiating predators. This potential response of anchovy mortality rates to water years cannot be seen in our simulations, as predators are only included as a general size-dependent mortality term. Our predicted changes in survival were for individuals from recruitment at 23 mm to October; whether the effect would be amplified if the dynamics of the larval stage or predators were simulated is not clear.

Water year type, for a given level of recruitment, had smaller effects on biomass, abundance, and lengths, but did cause a southward movement in wet years (Figs. 12.4 and 12.7). The change in latitudinal distributions is consistent with Jung and Houde (2004b). They suggested that during wet years, anchovy would be located further south as their northward movement would be blocked by low DO in the deeper waters of the mid-Bay region.

The idea that eventually the negative effects of hypoxia associated with increased nutrient loadings result in reduced fish production has been widely discussed. Cross-system comparisons have shown that estuarine systems with higher nutrient loads generally have higher biological production (Caddy 1993, Nixon and Buckley 2002, Breitburg et al. 2009a, b). Caddy (1993) suggested that eutrophication in a coastal ecosystem will initially increase the production of pelagic and demersal fish species, but predicted that further increases in eutrophication would eventually cause the production of demersal fish to decline while pelagic fish production would continue to increase or plateau. Diaz and Rosenberg (2008) offer a similar argument of increasing nutrient loadings eventually leading to a decline in fish populations via hypoxia decreasing benthic energy production that results in decreased consumer production. Often in these reviews (e.g., Breitburg et al. 2009b), the potential for the negative effects of hypoxia from increased nutrient loadings causing a decline in fish populations and fisheries, as suggested by Caddy, is discussed; however, the empirical evidence for the “tipping point” is often limited with a few case studies offered as possible examples. Our analysis offers a case study using a common pelagic species in Chesapeake Bay and illustrates how coupled biophysical modeling can be used to quantify the negative direct and indirect effects needed to offset the increased production from increasing nutrient loadings. Our results can also be used to quantify the positive direct and indirect effects needed to offset the reduction in production under decreased nutrient loadings scenarios.

Our approach used here was a bottom-up analysis of the population dynamics of a pelagic species. Despite hypoxia being a lower water column phenomenon, there are many examples of analyses of hypoxia effects on pelagic fish species (Vanderploeg et al. 2009; Dylander and Cerco 2010; Pothoven et al. 2012), including studies that specifically dealt with bay anchovy (Ludsin et al. 2009; Adamack et al. 2012; Zhang et al. 2014). Our analysis also focused on how nutrient loadings affected phytoplankton and zooplankton as part of the water quality modeling (i.e., bottom-up), but any other responses of the food web were assumed unaffected by nutrient loadings and associated hypoxia. We did account for prerecruit (egg to larval) mortality and juvenile and adult mortality, but as a sensitivity analysis with forced changes in recruitment and mortality rates and in a non-spatial manner. Increased juvenile and adult mortality would likely have a strong spatial aspect as it would emerge from altered movement of anchovy and their predators (Breitburg et al. 2009b; Ludsin et al. 2009). Other possible responses outside of the population approach are direct and indirect effects on zooplankton (Ludsin et al. 2009; Roman et al. 2012; Elliott et al. 2013); the interplay among zooplankton, ctenophores (*Mnemiopsis leidyi*), and larval fish (Kolesar et al. 2010, 2017); and spatially explicit losses of benthic production (Diaz and Rosenberg 2008). These top-down and food web effects, which were mostly ignored or dealt with by forcing changes in mortality in our model simulations, can modify the predicted anchovy responses to changes in nutrient loadings we obtained here using a population-based approach driven by a hydrodynamics–eutrophication model.

We have successfully demonstrated that we can dynamically couple the three-dimensional Chesapeake Bay water quality model with a population dynamic model of bay anchovy. Most analyses of the relationships among nutrient loadings, hypoxia, and fish have focused on specific life stages (e.g., Adamack et al. 2012), individuals at local scales (e.g., Rose et al. 2009), or changes in habitat (e.g., Constantini et al. 2008). Our approach attempted to operate at the population level for the entire Chesapeake Bay, albeit not full life cycle because we forced recruitment of early juveniles into the model every year. We believe that there are three high-priority expansions needed to the present model. These include continuing to develop the model for bay anchovy in Chesapeake Bay by adding the egg and larval stages as individuals in the simulations, exploring different options for the avoidance behavior, and adding individual predators so they can respond to changes in anchovy distributions and to hypoxia. Making these changes to the model would help us to better understand the effects of changes in nutrient loadings, and concomitant changes in hypoxia, on bay anchovy population dynamics.

Acknowledgements The authors would like to thank T.J. Miller for providing data on the spatial distribution of bay anchovy in Chesapeake Bay and M. Noel for providing post-processing scripts for the analysis of CE-QUAL-ICM model output files. Additionally, we would like to thank D. Justic, J.H. Cowan Jr., J. Geaghan, and R. Malone for comments on an earlier version of this manuscript. The Chesapeake Bay eutrophication model was developed with the support of the US Army Engineer District Baltimore and the US Environmental Protection Agency Chesapeake Bay Program. ATA was supported by a graduate assistantship from the Department of Oceanography and Coastal Sciences at Louisiana State University and was partially supported by the Cooperative Institute for Limnology and Ecosystems Research at the University of Michigan and by the University of Canberra Postdoctoral Fellowship Scheme while completing this manuscript. This research was partially supported (KAR) by the National Oceanographic and Atmospheric Administration, Center for Sponsored Coastal Ocean Research (CSCOR) NGOMEX09 Grant number NA09NOS4780179 awarded to the University of Texas, and CHRP Grant number NA10NOS4780157 awarded to Louisiana State University. This is publication number 219 of the NOAA's CSCOR NGOMEX and CHRP programs.

References

Adamack AT (2007) Predicting water quality effects on bay anchovy (*Anchoa mitchilli*) growth and production in Chesapeake Bay: linking water quality and individual-based fish models. Dissertation, Louisiana State University

Adamack AT, Rose KA, Breitburg DL et al (2012) Simulating the effect of hypoxia on bay anchovy egg and larval mortality using coupled watershed, water quality, and individual-based predation models. Mar Ecol Prog Ser 445:141–160

Baird D, Ulanowicz RE (1989) The seasonal dynamics of the Chesapeake Bay ecosystem. Ecol Monogr 59:329–364

Batiuk R, Linker L, Cerco C (2013) Featured collection introduction: Chesapeake Bay total maximum daily load development and application. J Am Water Resour Assoc 49:981–985

Boynton WR, Garber JH, Summers R, Kemp WM (1995) Inputs, transformations, and transport of nitrogen and phosphorus in Chesapeake Bay and selected tributaries. Estuaries Coasts 18:285–314

Brandt SB, Mason DM (2003) Effect of nutrient loading on Atlantic menhaden (*Brevoortia tyrannus*) growth rate potential in the Patuxent River. Estuaries 26:298–309

Breitburg DL, Adamack A, Rose KA et al (2003) The pattern and influence of low dissolved oxygen in the Patuxent River, a seasonally hypoxic estuary. Estuaries 26:280–297

Breitburg DL, Craig JK, Fulford RS et al (2009a) Nutrient enrichment and fisheries exploitation: interactive effects on estuarine living resources and their management. Hydrobiologia 629:31–47

Breitburg DL, Hondorp DW, Davias LA et al (2009b) Hypoxia, nitrogen and fisheries: integrating effects across local and global landscapes. Annu Rev Mar Sci 1:329–349

Burkadt J (2014) Geometry—geometric calculations. http://people.scs.fsu.edu/~burkardt/cpp_src/geometry/geometry.html. Accessed 8 Aug 2014

Burton DT, Richardson LB, Moore CJ (1980) Effect of oxygen reduction rate and constant low dissolved oxygen concentrations on two estuarine fish. T Am Fish Soc 109:552–557

Caddy JF (1993) Toward a comparative evaluation of human impacts on fishery ecosystems of enclosed and semi-enclosed seas. Rev Fish Sci 1:57–95

Cerco C, Cole T (1993) Three-dimensional eutrophication model of Chesapeake Bay. J Environ Eng-ASCE 119:1006–1025

Cerco C, Kim S-C, Noel M (2010) The 2010 Chesapeake Bay eutrophication model. A report to the US Environmental Protection Agency Chesapeake Bay Program and to the US Army Engineer Baltimore District. http://www.chesapeakebay.net/content/publications/cbp_55318.pdf. Accessed 23 July 2014

Cerco C, Noel M (2007) Can oyster restoration reverse cultural eutrophication in Chesapeake Bay? Estuar Coast 30:331–343
Cerco C, Noel M (2010) Monitoring, modeling, and management impacts of bivalve filter feeders in the oligohaline and tidal fresh regions of the Chesapeake Bay system. Ecol Model 221:1054–1064
Cerco CF, Noel MR (2013) Twenty-one-year simulation of Chesapeake Bay water quality using the CE-QUAL-ICM eutrophication model. J Am Water Resour Assoc 49:1119–1133
Chesapeake Bay Program (2013) Chesapeake Bay Program http://www.chesapeakebay.net. Accessed 15 Dec 2013
Chesney EJ, Houde ED (1989) Laboratory studies on the effect of hypoxic waters on the survival of eggs and yolk-sac larvae of the bay anchovy, *Anchoa mitchilli*. In: Houde ED, Chesney EJ, Newberger TA et al (eds) Population biology of bay anchovy in mid-Chesapeake Bay. Solomons, Maryland, pp 98–107
Christensen V, Walters CJ (2004) Ecopath with Ecosim: methods, capabilities and limitations. Ecol Model 172:109–139
Cloern JE (2001) Our evolving conceptual model for the coastal eutrophication problem. Mar Ecol Prog Ser 210:223–253
Conley DJ, Paerl NW, Howarth RW et al (2009) Controlling eutrophication: nitrogen and phosphorus. Science 323:1014–1015
Cooper SR, Brush GS (1991) Long-term history of Chesapeake Bay anoxia. Science 254:992–996
Costantini M, Ludsin SA, Mason DM et al (2008) Effect of hypoxia on habitat quality of striped bass (Morone saxatilis) in Chesapeake Bay. Can J Fish Aquat Sci 65:989–1002
Cowan JH Jr, Rose KA, Houde ED et al (1999) Modeling effects of increased larval mortality on bay anchovy population dynamics in the mesohaline Chesapeake Bay: evidence for compensatory reserve. Mar Ecol Prog Ser 185:133–146
Cronin TM, Vann CD (2003) The sedimentary record of climatic and anthropogenic influence on the Patuxent estuary and Chesapeake Bay ecosystems. Estuaries 26:169–209
Dalyander P, Cerco C (2010) Integration of a fish bioenergetics model into a spatially explicit water quality model: application to menhaden in Chesapeake Bay. Ecol Model 221:1922–1933
Diaz RJ, Rosenberg R (2008) Spreading dead zones and consequences for marine ecosystems. Science 321:926–929
Di Toro D (2001) Sediment flux modeling. Wiley, New York
Dortch MS, Chapman RS, Abt SR (1992) Application of three dimensional Lagrangian residual transport. J Hydraul Eng-ASCE 118:831–848
Elliott DT, Pierson JJ, Roman MR (2013) Copepods and hypoxia in Chesapeake Bay: abundance, vertical position and non-predatory mortality. J Plankton Res 35:1027–1034
Fisher TR, Hagy JD III, Boynton WR et al (2006) Cultural eutrophication in the Choptank and Patuxent estuaries of Chesapeake Bay. Limnol Oceanogr 51:435–447
Gray A (2013) Chesapeake Bay Fiscal 2014 budget overview. Department of Legislative Services, Office of Policy Analysis, Annapolis MD
Hagy JD, Boynton WR, Keefe CW et al (2004) Hypoxia in Chesapeake Bay, 1950-2001: long-term change in relation to nutrient loading and river flow. Estuaries 27:634–658
Hartman KJ, Brandt SB (1995) Trophic resource partitioning, diets and growth of sympatric estuarine predators. T Am Fish Soc 124:520–537
Houde ED, Chesney EJ, Newberger TA et al (1989) Population biology of bay anchovy in mid-Chesapeake Bay. Solomons, Maryland
Houde ED, Zastrow CE (1991) Bay anchovy. In: Funderburk SL, Mihursky JA, Jordan SJ et al (eds) Habitat requirements for Chesapeake Bay living resources, 2nd edn. Living Resources Subcommittee, Chesapeake Bay Program. Annapolis, Maryland, pp 8-1–8-14
Humston R, Ault JS, Lutcavage M et al (2000) Schooling and migration of large pelagic fishes relative to environmental cues. Fish Oceanogr 9:136–146
Humston R (2001) Development of movement models to assess the spatial dynamics of marine fish populations. Dissertation, University of Miami

Johnson BH, Kim KW, Heath RE et al (1993) Validation of three-dimensional hydrodynamic model of Chesapeake Bay. J Hydraul Eng-ASCE 119:2–20

Jung S, Houde ED (2003) Spatial and temporal variabilities of pelagic fish community structure and distribution in Chesapeake Bay, USA. Estuar Coast Shelf S 58:335–351

Jung S, Houde ED (2004a) Production of bay anchovy *Anchoa mitchilli* in Chesapeake Bay: application of size-based theory. Mar Ecol Prog Ser 281:217–232

Jung S, Houde ED (2004b) Recruitment and spawning-stock biomass distribution of bay anchovy (*Anchoa mitchilli*) in Chesapeake Bay. Fish B NOAA 102:63–77

Kemp WM, Boynton WR, Adolf JE et al (2005) Eutrophication of Chesapeake Bay: Historical trends and ecological interactions. Mar Ecol Prog Ser 303:1–29

Kim S-C (2013) Evaluation of a three-dimensional hydrodynamic model applied to Chesapeake Bay through long-term simulation of transport processes. J Am Water Resour Assoc 49:1078–1090

Kolesar SE, Breitburg DL, Purcell JE et al (2010) Effects of hypoxia on Mnemiopsis leidyi, ichthyoplankton and copepods: clearance rates and vertical habitat overlap. Mar Ecol Prog Ser 411:173–188

Kolesar SE, Rose KA, Breitburg DL (2017). Hypoxia effects within an intraguild predation food web of *Mnemiopsis leidyi* ctenophores, larval fish, and copepods. In: Justice D (ed) Modeling coastal hypoxia. Springer, Cham

Lett C, Rose KA, Megrey BA (2009) Biophysical models fish. In: Checkley D, Alheit J, Oozeki Y, Roy C (eds) Climate change and small pelagic fish. Cambridge University Press p 88–111

Ludsin SA, Zhang X, Brandt SB et al (2009) Hypoxia-avoidance by planktivorous fish in Chesapeake Bay: implications for food web interactions and fish recruitment. J Exp Mar Biol Ecol 381:S121–S131

Luo J, Brandt SB (1993) Bay anchovy *Anchoa mitchilli* production and consumption in mid-Chesapeake Bay based on a bioenergetics model and acoustic measures of fish abundance. Mar Ecol-Prog Ser 98:223–236

Luo J, Hartman KJ, Brandt SB et al (2001) A spatially-explicit approach for estimating carrying capacity: an application for the Atlantic menhaden (*Brevoortia tyrannus*) in Chesapeake Bay. Estuaries 24:545–556

Luo J, Musick JA (1991) Reproductive biology of the bay anchovy in Chesapeake Bay. T Am Fish Soc 120:701–710

Magnien RE (1987) Monitoring for management actions: Maryland Office of Environmental Programs Chesapeake Bay Water Quality Monitoring Program, First biennial report. Maryland Office of Environmental Programs, 61 pp

Maryland Department of Natural Resources, Annapolis (2017) Juvenile striped bass survey. http://dnr2.maryland.gov/fisheries/Pages/striped-bass/juvenile-index.aspx. Accessed 3 March 2017

Mauchline J (1998) The biology of calanoid copepods. Adv Mar Biol 33

Miller TJ, Nye JA, Loewensteiner DL (2008) Development and implementation of the Chesapeake Bay Fishery-Independent Multispecies Survey (CHESFIMS)—final report. Technical report series No TS-545-08-020 of the University of Maryland Center for Environmental Science, Solomons, Maryland

Morin LG, Houde ED (1989) Hatch-date frequencies and young-of-the-year growth rates of bay anchovy in mid-Chesapeake Bay. In: Houde ED, Chesney EJ, Newberger TA et al (eds) Population biology of bay anchovy in mid-Chesapeake Bay, Solomons, Maryland, pp 98–107

Newberger TA, Houde ED, Chesney EJ (1989) Relative abundance, age, growth and mortality of bay anchovy (*Anchoa mitchilli*) in the mid-Chesapeake Bay. In: Houde ED, Chesney EJ, Newberger TA et al (eds) Population biology of bay anchovy in mid-Chesapeake Bay. Solomons, Maryland, pp 17–77

Nixon SE, Buckley BA (2002) "A strikingly rich zone"—nutrient enrichment and secondary production in coastal marine ecosystems. Estuaries 25:782–796

Pothoven SA, Vanderploeg HA, Höök TO et al (2012) Hypoxia modifies planktivore–zooplankton interactions in Lake Erie. Can J Fish Aquat Sci 69:2018–2028

Rabalais NN, Turner RE, Dortch Q et al (2002) Nutrient-enhanced productivity in the northern Gulf of Mexico: past, present and future. Hydrobiologia 475:39–63

Roman MR, Pierson JJ, Kimmel DG et al (2012) Impacts of hypoxia on zooplankton spatial distributions in the northern Gulf of Mexico. Estuar Coast 35:1261–1269

Rose KA, Cowan JH Jr, Clark ME et al (1999) An individual-based model of bay anchovy population dynamics in the mesohaline region of Chesapeake Bay. Mar Ecol Prog Ser 185:113–132

Rose KA, Cowan JH Jr, Winemiller KO et al (2001) Compensatory density dependence in fish populations: importance, controversy, understanding and prognosis. Fish Fish 2:293–327

Rose KA, Adamack AT, Murphy CA et al (2009) Does hypoxia have population-level effects on coastal fish? Musings from the virtual world. J Exp Mar Biol Ecol 381:S188–S203

Rose KA, Justic D, Fennel K, Hetland R (2017) Numerical modeling of hypoxia and its effects: synthesis and going forward. In: Justice D (ed) Modeling coastal hypoxia. Springer, Cham

Runge JA, Franks PJS, Gentleman WC et al (2004) Diagnosis and prediction of variability in secondary production and fish recruitment processes: developments in physical-biological modeling. In: Robinson AR, Brink KH (eds) The global coastal ocean: multi-scale interdisciplinary processes, vol 13. The Sea, p 413–473

Scheffer M, Baveco JM, DeAngelis DL et al (1995) Super-individuals a simple solution for modeling large populations on an individual basis. Ecol Model 80:161–170

Snyder JP (1987) Map projections—a working manual. Geological Survey Professional Paper, 1395. U. S. Government Printing Office, Washington, D. C

U.S. Environmental Protection Agency (2010) Chesapeake Bay total maximum daily load for nitrogen, phosphorus and sediment. U.S. Environmental Protection Agency Chesapeake Bay Program Office. http://www.epa.gov/reg3wapd/tmdl/, http://www.epa.gov/reg3wapd/tmdl/ChesapeakeBay/tmdlexec.html. Accessed 16 Dec 2013

U.S Geological Survey (2014) Estimating stream flow entering Chesapeake Bay. http://md.water.usgs.gov/waterdata/chesinflow/. Accessed 11 Aug 2014

Vanderploeg HA, Ludsin SA, Ruberg SA et al (2009) Hypoxia affects spatial distributions and overlap of pelagic fish, zooplankton, and phytoplankton in Lake Erie. J Exp Mar Biol Ecol 381: S92–S107

Wang S-B, Cowan JH Jr, Rose KA et al (1997) Individual-based modeling of recruitment variability and biomass production of bay anchovy in mid-Chesapeake Bay. J Fish Biol 51 (Suppl. A):101–120

Watkins KS, Rose KA (2013) Evaluating the performance of individual-based animal movement models in novel environments. Ecol Model 250:214–234

Zastrow CE, Houde ED, Morin LG (1991) Spawning fecundity, hatch-date frequency and young-of-the-year growth of bay anchovy *Anchoa mitchilli* in mid-Chesapeake Bay. Mar Ecol-Prog Ser 73:161–171

Zhang H, Ludsin SA, Mason DM et al (2009) Hypoxia-driven changes in the behavior and spatial distribution of pelagic fish and mesozooplankton in the northern Gulf of Mexico. J Exp Mar Biol Ecol 381:S80–S91

Zhang H, Mason DM, Stow CA et al (2014) Effects of hypoxia on habitat quality of pelagic planktivorous fishes in the northern Gulf of Mexico. Mar Ecol Prog Ser 505:209–226

Chapter 13
Simulation of the Population-Level Responses of Fish to Hypoxia: Should We Expect Sampling to Detect Responses?

Kenneth A. Rose, Sean Creekmore and Shaye Sable

Abstract Our ability to use monitoring data to quantify the effects of hypoxia at the population level for fish remains elusive. We performed a simulation analysis similar to a power analysis to determine the probability that sampling would detect a known hypoxia effect on croaker recruitment for the northern Gulf of Mexico. We used 100 years of simulated annual recruitments of croaker from a population dynamics model under normoxic and hypoxia conditions to establish a credible magnitude of population response to historical hypoxia conditions. We also analyzed long-term monitoring data to determine realistic interannual variation in recruitment and used the fitted lognormal distributions to add variability to each year's recruitment value from the population model. Segments of the two time series were randomly selected as 5-, 10-, and 25-year sequences of continuous years, variability added to the recruitment values, and then a t-test used to determine whether the known hypoxia effect would have been detected. Under field-estimated sampling variability, and 25 years of sampling with a generous cutoff p-value for detection of 0.1, there was still only a 50% chance of properly detecting the roughly 20% reduction in average croaker recruitment. Shorter time samples and use of the 0.05 cutoff resulted in lower probabilities of detection. When we artificially reduced the variability generated from the lognormal distributions, the probabilities of detection increased as expected and under the best conditions approached 85%. Despite the low probability of detecting the hypoxia effect with sampling, a 20% reduction in long-term average recruitment would be considered by most to be an ecologically significant impact.

K.A. Rose (✉) · S. Creekmore
Department of Oceanography and Coastal Sciences,
Louisiana State University, Baton Rouge, LA 70803, USA
e-mail: karose@lsu.edu

S. Creekmore
e-mail: sbcreekmore@gmail.com

S. Sable
Dynamic Solutions, LLC, 450 Laurel StreetSuite 1650, Baton Rouge,
LA 70801, USA
e-mail: ssable@dsllc.com

D. Justic et al. (eds.), *Modeling Coastal Hypoxia*,
DOI 10.1007/978-3-319-54571-4_13

Keywords Atlantic croaker • Hypoxia • Recruitment • Population dynamics • Numerical modeling • Gulf of Mexico

13.1 Introduction

Our ability to use monitoring data to quantify the effects of hypoxia at the population level for fish remains elusive (Breitburg et al. 2009). The idea of using long-term monitoring data to derive population abundance or indices and to then relate the annual abundances to environmental drivers and interannual differences in the severity of hypoxia is attractive. The results would be deeply rooted in observational data and commonly accepted statistical analyses, and while the processing of the data and the statistical methods used can get complicated, the results are intuitive to many people. Observing population abundances over enough different conditions should allow for the fluctuations due to climate and other environmental conditions to be quantified, thereby allowing for clear attribution of variation in population abundance to measures of hypoxia.

While the statistical analysis and monitoring approach is typical for separating the effects of multiple factors on fish population dynamics (e.g., Kimmerer 2006), it is especially challenging with the factor of hypoxia (e.g., Chesney and Baltz 2001). Responses of individuals to hypoxia have been clearly documented in laboratory experiments (e.g., see Nielan and Rose 2014; Vaquer-Sunyer and Duarte 2008) and in the field (e.g., Thomas et al. 2007; Thomas and Rahman 2012); direct exposure to low DO can cause slowed growth, higher mortality, and reduced reproduction of individuals. However, a major response of mobile organisms is avoidance resulting in altered spatial distributions of individuals (Breitburg 2002; Craig et al. 2005; Brady and Targett 2013). How these altered spatial distributions then affect growth, mortality, and reproduction is harder to document because the effects of hypoxia become indirect. It is not the low exposure of DO causing the growth, mortality, or reproduction effects of individuals. Rather, low DO causing individuals to move, and the hypoxia effects are then how growth, mortality, and reproduction in the new location compares to the growth, mortality, and reproduction that would have occurred if the individuals had not moved. Getting to the population level then involves combining how these direct and indirect effects, mixed in with variation in population abundance caused by other factors, accumulate over individuals all of the way to the population level. Long-term monitoring data is a powerful way to document status and trends in fish and other aquatic species (e.g., Chesney et al. 2000; de Mutsert et al. 2008; Messer et al. 1991; Parr et al. 2003) and is fundamental to stock assessment and fisheries management (Hilborn and Walters 1992). Partitioning out the variation to different factors is more challenging (Rose 2000).

Power analysis is a common statistical method for determining what magnitude of changes can be reliably detected with monitoring-based data (Urquhart and Kincaid 1999). Power is inversely related to probability of making a Type II error (i.e., concluding no change or trend when one actually occurred; Fairweather 1991).

The higher the power, the more likely the effect is detected when the effect is present. There are a variety of forms of power analysis, but all share the same general result of quantifying the likelihood of detecting changes of some known magnitude. They often also include how additional stations or samples would improve the probability of detecting the known changes. Monitoring of fish is difficult because their populations show high variability, thereby making power analysis especially important (Peterman 1990). Power analysis in our context of long-term monitoring of fish population response to a stressor typically involves generating time series of values from an underlying model with a known change or trend, and then simulating sampling using Monte Carlo techniques from that time series to derive an observed time series of values. A statistical model (e.g., regression with time as variable) is then fit to the observed (realized) time series, and the fit of the statistical model (e.g., slope coefficient for trend) is recorded. This is repeated many times, either randomly generating new time series of the underlying model or a new realization of sampling, or both. One can then interpret how many times the known change or trend was successfully detected by the statistical method (Gerrodette 1987; Fairweather 1991; Wagner et al. 2013). Some analyses also include spatial variation in addition to the temporal variability (Larsen et al. 2001). Hobday and Evans (2013) recently used power analysis to determine the likelihood of detecting climate signals in long-term fish monitoring data and Dolan et al. (2016) used a similar approach as we will use here to determine the power of detecting a change in fish species richness and total fish density related the operations of a power plant. While the underlying models used to generate the simulated data (e.g., Dakos et al. 2012) and the sampling designs imposed (e.g., Urquhart and Kincaid 1999) can become quite complicated, the general approach remains the same.

In this chapter, we use the simulation results from a population model of croaker to assess the likelihood of monitoring being able to detect the hypoxia effect. The model was designed to simulate croaker population dynamics for the northern Gulf of Mexico, and specifically to generate and compare simulated annual recruitment under normoxic and hypoxic conditions. The population model results provide us with a realistic estimate of the magnitude of the population effect (i.e., change in recruitment) caused by hypoxia. We then use long-term monitoring data on croaker to derive estimates of interannual variability in recruitment. We apply this variation to the simulated recruitment values under normoxia and hypoxia to generate time series of croaker recruitment that reflect realistic interannual variation under normoxia and hypoxia. We then simulate sampling by taking 5-, 10-, and 25-year segments of the two time series, average the recruitment in each and determine how often a t-test would show differences in the average recruitment when we know the hypoxia effect is present. We conclude with a discussion about the robustness of our analysis, how the lack of detection of hypoxia should not be misinterpreted as hypoxia having no effects, and the power of using a modeling and monitoring data combined approach.

13.2 Methods

13.2.1 Croaker Population Model

The population model simulated the hourly growth, mortality, and movement, and daily reproduction, of thousands of individual croaker from release as eggs until age-8 for a period of 140 years on a 2-dimensional (300 × 800 1 km^2 cells) spatial grid. The model is fully described in Rose et al. (in review-a). The model grid is centered on coastal Louisiana, which encompasses the hypoxic zone and the major habitat of croaker (Fig. 13.1). Temperature affects growth and movement, and chlorophyll-a concentration was used as an index of food availability. The derivation of temperature and chlorophyll-a concentration for use in the population model started with mean monthly values from reported climatological values. These were then interpolated to the IBM grid. Temperature was further interpolated to daily values, which were used for all hours within each day, while chlorophyll-a concentration was maintained at cell-specific monthly values and used for all hours and days within a month.

Predicted hourly bottom-layer DO for the inner Louisiana-upper Texas shelf was taken from a coupled hydrodynamics-water quality model (Justić and Wang 2014) and used to specify hourly DO concentrations in each spatial cell for the entire year. The hydrodynamics model was an implementation of FVCOM (Finite Volume Community Ocean Model); the WASP (Water Quality Analysis Simulation Program) model was used to simulate water quality. The domain for the FVCOM model extended from Mobile Bay, AL to an area in the GOM south of the mouth of Galveston Bay, TX, and covered the area off the Louisiana–Texas Shelf affected by hypoxia. Validation of the DO predictions from the FVCOM-WASP model was carried out using 2002 as the reference year (see Justić and Wang 2014), and then two additional years (2001 and 2010) were simulated to obtain 3 years for use as

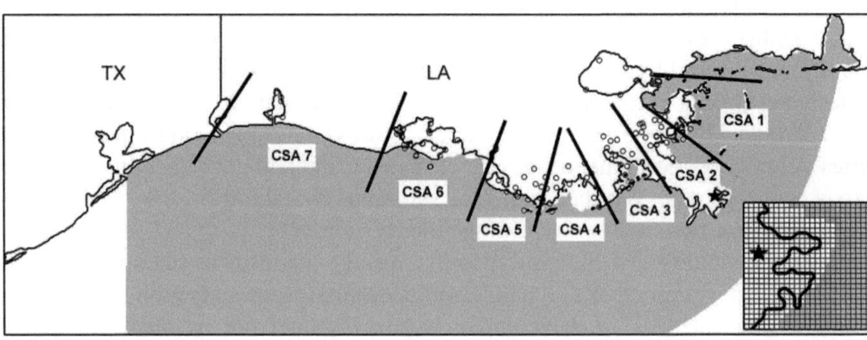

Fig. 13.1 Map showing the model grid for the croaker population model. The insert shows the 1 km^2 cells for a smaller region within the domain. The solid lines denote the Coastal Study Areas (CSAs) or regions, and the open circles the monitoring stations within each region, that were aggregated to obtain estimates of annual croaker recruitment

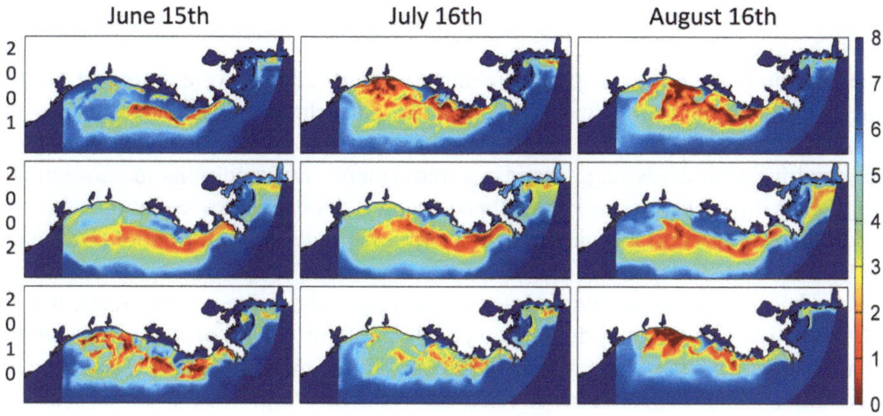

Fig. 13.2 Mid-month snapshots of DO used in the croaker population model based on FVCOM-WASP output for 2 severe years (2001 and 2002) and an intermediate year (2010). The darkest blue areas are not included in the model domain

input to the croaker population model. The three years represent intermediate (2010) and severe (2001 and 2002) hypoxia conditions. Mid-month snapshots of DO, as used within the croaker model, are shown in Fig. 13.2.

The climatological chlorophyll-a concentrations were assumed to be associated with years of intermediate hypoxia and were further modified to reflect mild and severe hypoxia years in simulations. We used multipliers applied to daily values of chlorophyll to simulate how nutrient loadings from the Mississippi River could lead to both more severe hypoxia and increased food for croaker. The multipliers were derived from simulation results of the ROMS water quality model for the Gulf of Mexico (Fennel et al. 2013). Multipliers were applied to coastal regions within the croaker model grid during spring and summer months, and multiplier values ranged from 0.85 (mild) to 1.2 (severe). The evidence for the correlations among hypoxia, primary production, and river flow is quite strong (e.g., Turner et al. 2012; Rabalais et al. 2007), while the next logical step from primary production to croaker food is more tenuous (Rose et al. in review-b).

Each individual croaker was tracked hourly as it progressed through the life stages of egg, yolk-sac larva, ocean larva, estuary larva, early juvenile, late juvenile, and adult. Individuals were started on each day there was spawning. Possible spawning was constrained between mid-September and a maximum degrees-day value (roughly March). Each individual was tracked as it experienced grid-wide temperatures that determined on what hour it switched from egg to yolk-sac larvae, yolk-sac larva to ocean larva, and ocean larva to estuary larva. When an individual was ready to go from estuary larva to early juvenile, it was placed on the 2-dimensional grid with a starting length of 32 mm and followed as an individual until age-8 (Diamond et al. 2013). Development rates of egg and larval stages (off-grid individuals: eggs, and yolk-sac, ocean, and estuary larvae) were temperature-dependent, while mortality rates were stage-dependent. Early juveniles were restricted to cells shallower than 5 m, while late juveniles were restricted to cells shallower than 20 m.

Individuals on the grid (early juveniles through adults) were followed hourly through growth, mortality, and movement, and daily through reproduction. Growth, mortality, reproduction, and movement depended on the local conditions in their cell, and in some cases (e.g., avoidance movement), on conditions in surrounding cells. Growth each hour was calculated using a bioenergetics equation with the chlorophyll-a concentration in the cell determining realized consumption. Mortality of each individual depended on a constant, stage-specific rate, old age (removed at age-8), and starvation (individual got too skinny). The stage-specific rates, generally reflective of predation, did not vary with length or weight of individuals within a stage. Maturity was determined based on length on September 1, and annual fecundity (weight-dependent) was released on the first hour of each day as series of batches over their month of spawning during the following late winter and spring. Hourly movement was simulated using a kinesis approach (Humston et al. 2004; Watkins and Rose 2013) that used temperature in the cell as the cue. When an individual experienced DO less than 2 mg/L, movement switched to avoidance behavior and the individuals moved to the cell in the surrounding neighborhood with DO greater than 2 and temperature closest to its optimal temperature for growth. The size of the neighborhood increased with the individual's length; computed swim speeds that included avoidance movement remained realistic (averaged about 3.5 body lengths/s). Hourly growth and survival, and annual fecundity, of individuals on the grid were reduced based on accumulated exposure to the hourly DO values using a set of exposure-effects submodels (Neilan and Rose 2014). The parameters of the exposure-effects submodels were estimated from constant exposure laboratory experiments, and when available, validated against the same experiments under fluctuating exposures.

13.2.2 Simulations of Abundance and Recruitment

Two 140-year simulations were used in this analysis. Both simulations used normoxic conditions and climatological chlorophyll-a concentrations for the first 40 years as a spin-up period and a random sequence of mild (normoxic), intermediate (2010), and severe (2001 or 2002) years using the FVCOM-WASP output as year types for the remaining 100 model years. The first simulation used normoxic conditions with chlorophyll concentrations adjusted by the type of hypoxia each year for the final 100 model years to simulate the different food conditions associated with the severity of hypoxia, but without the corresponding effects of low DO. The second simulation used chlorophyll and DO concentrations based on the severity of hypoxia to impose both food and hypoxia effects. The probability of the severity of hypoxia was estimated by classifying each model year based on the reported areal late July extent of hypoxia as either mild (<10,000 km^2), intermediate (10,000–20,000 km^2), or severe (>20,000 km^2). Both simulations used a

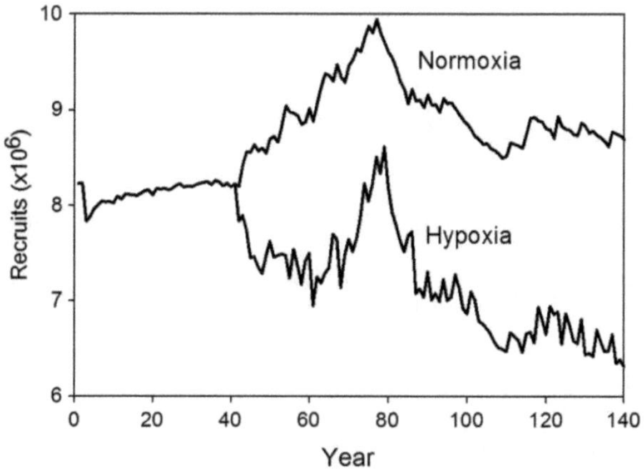

Fig. 13.3 Simulated annual recruitment (number of young-of-the-year individuals surviving to September 1) by the croaker population model using a realistic sequence of mild, intermediate, and severe hypoxia years. The normoxia simulation included the interannual variation in chlorophyll concentrations related to hypoxia but had no low DO effects, while the hypoxia simulation included both interannual variation in chlorophyll-a concentrations and hypoxia effects

sequence of mild, intermediate, and severe hypoxia year types in approximately the same proportions (19 mild, 55 intermediate, and 26 severe out of 100 years) as the hypoxia area estimates reported in Obenour et al. (2013) for years 1986–2011 (0.19 mild, 0.52 intermediate, and 0.29 severe). The second simulation represents a realistic simulation of croaker population responses to historical hypoxia conditions. We report the annual recruitment each year (number of young-of-the-year (YOY) individuals on September 1) in both simulations (Fig. 13.3). The two simulations used in this chapter are described in detail in Rose et al. (in review-b).

13.2.3 Recruitment Variability from Monitoring Data

We used the long-term monitoring data on croaker reported as part of the Louisiana Department of Wildlife and Fisheries (LDWF) coastal sampling program (LDWF 2014) as the basis for representing recruitment variability of croaker. The monitoring includes multiple gear types (e.g., otter trawls, gill nets, and bag seines) at dozens of stations that span decades; the stations are generally outside of the shelf-based hypoxia zone. We used the gill net data for July, August, and September to form an annual index of croaker recruitment. Gill nets captured 190–200 mm croaker, which is the approximate length of new age-1 croaker around our defined date of recruitment of September 1. Other studies have used the same approach taken here of using CPUE for months corresponding to recruitment sizes to represent an index of annual recruitment (e.g., Taylor et al. 2009).

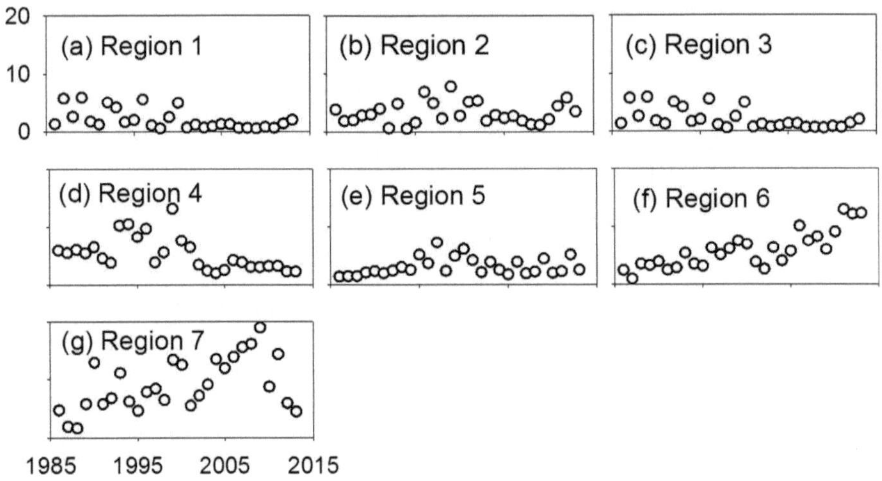

Fig. 13.4 Annual recruitment index (CPUE) of croaker based on July, August, and September gill net sampling for each of the seven CSA regions. CPUE is in units of number of individuals per gill net set

We computed monthly catch-per-unit-effort (CPUE) values using the gill net data for stations within each of the seven LWDF coastal study areas (CSAs) or regions for 1986–2013 (see Fig. 13.1). The gill net was 228 m, with 5 panels each 45.7 m long and 2.4 m in depth. The gill nets were deployed and the boat circled the gill net at least twice to "force" fish to strike the net, after which the net was retrieved and catch enumerated. We summed catch (number of individuals per gill net set) over stations for each month, and divided by the number of stations that contributed to the catch to obtain a single CPUE value per month. Except for 3 years, all months had 10 or more stations in each of their 3 months. We then averaged the 3 monthly values to obtain a single estimate of annual recruitment by CSA regions (Fig. 13.4). For analyses, we then divided each annual recruitment value by the mean over all years to obtain each year's estimated recruitment as a multiplier of the long-term mean. We refer to our CPUE-based annual index as recruitment for convenience, but it is not truly recruitment to the population. We analyzed the index by spatial (CSA) region, even though the regions constitute a single stock, the index has not been correlated to adult abundances, and because it includes multiple sources of variability. However, the variability in the index is based on field data and is what would be expected to be encountered with field sampling targeted as developing an index of approximately age-1 individuals.

Lognormal probability distributions were fit to the annual recruitment multiplier (Fig. 13.5), and these distributions were used to generate deviates of recruitment and applied to the simulated annual recruitment values from the two model simulations. Coefficient-of-variation (CV) values (standard deviation of annual recruitment/mean annual recruitment) based on 5000 deviates from the lognormal distributions varied from 0.41 (CSA region 1) to 0.88 (CSA region 3) across the 7

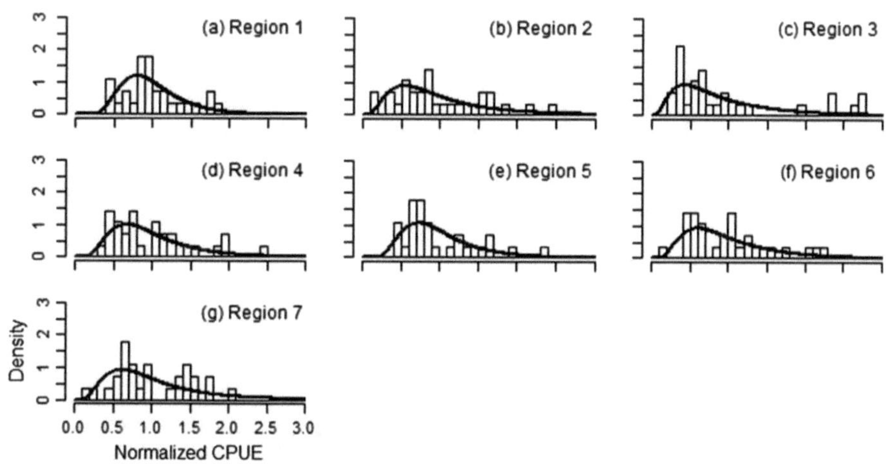

Fig. 13.5 Fitted lognormal distributions to the normalized CPUE values (unitless) representing annual croaker recruitment for each of the 7 CSA regions

CSA regions. The simulated CVs were similar to the observed CVs of the normalized CPUE data for region 1 (0.441 vs. 0.443), region 2 (0.782 vs. 0.701), and region 5 (0.493 vs. 0.465). Distribution-generated CV was lower for region 3 (0.879 vs. 1.73), mostly because the CPUE data showed an almost bimodal distribution (Fig. 13.5c). Generated CVs were generally about twice or less the observed for the remaining regions (0.564 vs. 0.348 for region 4; 0.659 vs. 0.38 for region 6; 0.653 vs. 0.293 for region 7). The general fits of the lognormal distributions to the CPUE data were reasonable, and the fit across the CSA regions spanned the range of the generated CVs being similar, smaller, and larger than the CVs from the CPUE data. We used the same distributions for years in the normoxia and hypoxia simulations; we assumed that hypoxia would not affect the interannual variation in recruitment because of the general low overlap between young-of-the-year and the main shelf-based hypoxic zone. We also partially assessed how different fits would affect the results by reducing the SD coefficient of the fitted lognormal probability distributions (see below).

13.2.4 Simulating Sampling

Sampling was simulated by taking the same segments (same years) of the two simulated recruitment time series (Fig. 13.6). We used 5-, 10-, and 25-year segments. We first generated multipliers from the fitted lognormal distribution for a region and then applied them to each year in a time segment (each year's value got its own multiplier). We then computed the mean and variance of recruitment for that segment based on the annual values in the time segment. Each segment

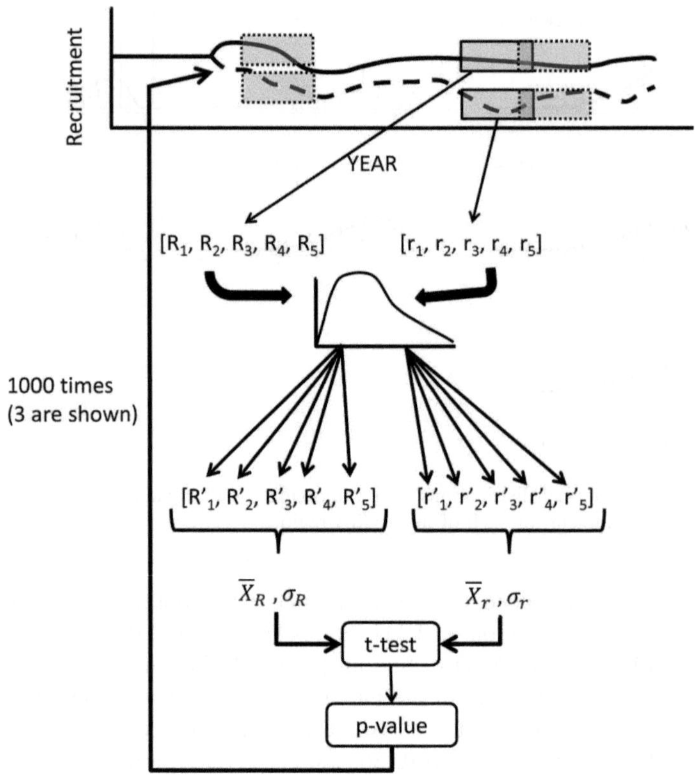

Fig. 13.6 Steps in the simulation analysis showing how the two recruitment time series were sampled (R for normoxia and r for hypoxia effect included) using a 5-year segment of each, variability added to each year's value, and then the two samples compared using a t-test. This was repeated 1000 times with randomly selected 5-year segments. The same method was used for the 10- and 25-year segments

represents a possible sample from the long-term time series of recruitments. With estimates of mean and variance of recruitment for the same time segment and region for the normoxic and hypoxia recruitment time series, we performed a t-test, the average recruitment values between normoxia and hypoxia and recorded the p-value. We did this for 1000 realizations of 5-year segments, 10-year segments, and 25-year segments. Three possible segments of a fixed time length are shown in Fig. 13.6. We report the proportion of the 1000 p-values that were less than 0.05 and 0.1.

We repeated the analysis using the lognormal distributions applied to the annual recruitment estimates but with the variability of the fitted distributions reduced. We simply multiplied the scale (standard deviation, SD) parameter of the lognormal distributions by 0.7 and then by 0.4 (Fig. 13.7). The CVs from the reduced variability lognormal distributions were lower than the CVs from the original fitted distributions by the expected ratios (average ratios of 0.67 and 0.37). The purpose

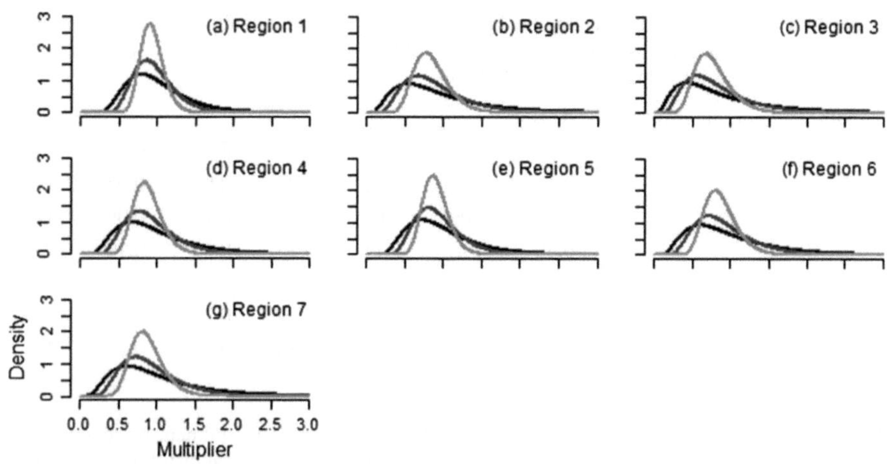

Fig. 13.7 Comparison of the fitted and reduced variability lognormal distributions that generated interannual variability in recruitment for each of the seven CSA regions. The distributions fitted to the normalized CPUE data (*black*) are wider and centered on lower values than the reduced variability distributions achieved by multiplying the SD parameter by 0.7 (*dark gray*) and 0.4 (*light gray*)

was to see how reduced variability in the monitoring data would affect the detectability of the known hypoxia effects.

13.3 Results

13.3.1 Realistic Recruitment Variability

The probability of detecting a difference between averaged recruitments (hypoxia effect) was generally low and roughly doubled for the cutoff of 0.05 upon going from the 5 to 10 to 25-year windows and increased by 50% each step for the cutoff of 0.1 (Table 13.1). Probabilities of correct detection with the 0.05 cutoff ranged from 0.05 (regions 2, 6, and 7) to 0.11 (region 1) for the 5-year window, from 0.08 (regions 3 and 7) to 0.18 (region 1) for the 10 year window, and from 0.11 (region 3) to 0.39 (region 1) for the 25-year window. A very similar pattern was observed for the 0.1 cutoff, except that the increase in probabilities from 5 to 10 to 25 years was smaller. For example, with a 0.1 cutoff, probabilities for region 1 increased from 0.21 to 0.29 to 0.49, or about 50% each time. CSA region 1 always had the highest probabilities of detection; the fitted lognormal distribution was centered near one with a slightly higher peak than the distributions at the other regions (Fig. 13.5a). At best, a 25-year window with a p-value cutoff of 0.1 generated probabilities of detection of 0.21–0.49 (i.e., less than 50% chance of correctly detecting the hypoxia effect).

Table 13.1 Probabilities of detecting the known hypoxia effect on recruitment by CSA region for 5-, 10-, and 25-year samples using p-value cutoffs of 0.05 and 0.1. Variability in recruitment was generated from the fitted lognormal distributions

Cutoff	Years in samples	Region						
		1	2	3	4	5	6	7
0.05	5	0.11	0.05	0.06	0.07	0.07	0.05	0.05
	10	0.18	0.09	0.08	0.17	0.14	0.09	0.08
	25	0.39	0.16	0.11	0.25	0.30	0.19	0.21
0.1	5	0.21	0.11	0.11	0.13	0.15	0.10	0.10
	10	0.29	0.17	0.14	0.19	0.25	0.16	0.18
	25	0.49	0.27	0.21	0.38	0.42	0.29	0.31

Table 13.2 Probabilities of detecting the known hypoxia effect on recruitment by region for 5-, 10-, and 25-year samples for variability generated from the fitted lognormal distributions (realistic) and the lognormal distributions with their variability reduced by multiplying the SD by 0.7 and by 0.4. Results are shown for the cutoff value of 0.1

Region	5 years			10 years			25 years		
	1xSD	0.7xSD	0.4xSD	1xSD	0.7xSD	0.4xSD	1xSD	0.7xSD	0.4xSD
1	0.21	0.30	0.56	0.29	0.46	0.82	0.49	0.71	0.86
2	0.11	0.16	0.29	0.17	0.25	0.50	0.27	0.42	0.74
3	0.11	0.14	0.26	0.14	0.21	0.43	0.21	0.35	0.67
4	0.13	0.20	0.38	0.19	0.33	0.66	0.38	0.57	0.82
5	0.15	0.22	0.47	0.25	0.40	0.73	0.42	0.64	0.85
6	0.10	0.15	0.32	0.16	0.26	0.57	0.29	0.48	0.79
7	0.10	0.15	0.31	0.18	0.29	0.57	0.31	0.48	0.77

13.3.2 Reduced Recruitment Variability

Reducing variability of recruitment by multiplying the SD parameter of the lognormal distributions by 0.7 and by 0.4 (Table 13.2) caused the expected pattern of generally increasing the probabilities of detection. We show the results for the cutoff of 0.1 because they were very similar to the results based on the 0.05 cutoff. Using 5-year samples, probabilities of detection under the lognormal distributions with a 0.7 multiplier were about 1.5x those using the fitted (realistic) distributions and about 3x higher with the 0.4 multiplier. For example, the probability of detection for CSA region 2 was 0.11 with the realistic variation and increased to 0.16 (1.45x) for the 0.7 multiplier and to 0.29 (2.64x) for the 0.4 multiplier. The 10- and 25-year samples also showed similar increases the probabilities of detection (about 1.5x for the 0.7 multiplier and 3x for the 0.4 multiplier). Under the best conditions (0.4xSD, 25 years, 0.1 cutoff), probabilities of detection exceeded 0.67 for all regions and were greater than 0.8 for some regions. CSA region 3 consistently had the lowest probabilities of detection corresponding to having a lognormal distribution with a peak toward the left (multipliers less than one) and a relatively thick, long tail toward the right (more frequent multipliers greater than one).

13.4 Discussion

Detection of population-level hypoxia effects on croaker in the northern Gulf of Mexico, and by extension to coastal fish species with a similar life history as croaker, using monitoring data alone is difficult. Under realistic variability (untreated field data), and 25 years of sampling with a generous cutoff of 0.1, there is still only a 50% chance of properly detecting the roughly 20% reduction in average croaker recruitment. While a change of 20% for a year or two is small relative to interannual variation, a 20% reduction in long-term average recruitment would be considered by most to be an ecologically significant impact. In the model simulations, this corresponded to a 20% reduction in long-term adult (age-1 and older) population abundance. Thus, the lack of detection of a hypoxia effect in the long-term fish monitoring data should not be misinterpreted as definitive evidence that hypoxia has little effect at the population level.

Our analysis seems robust and used some optimistic (i.e., more likely to detect effects) conditions. We obtained similar results keeping the three monthly recruitment values separate in the analyses (lognormal fitted to 3 estimates of recruitment per year), and by using a gamma distribution in place of the lognormal distribution. Other factors besides hypoxia, such as temperature, did not vary from year to year, which, in nature, would add variability to the two recruitment time series. The variability in recruitment generated by the fitted lognormal distributions was realistic but could be even considered low. The simulated CVs based on 5000 recruitment multipliers from each of 7 regionally based distributions ranged from 0.41 to 0.88. Others report CVs of annual recruitment (or SD of log-transformed recruitment) of about 0.4–1.0 (Mertz and Myers 1996; Leggett and Frank 1997; Hennemuth et al. 1980; Rose et al. 2001); some of the reported CVs include statistical removal of stock effects but still remain within this general range.

We showed that reducing variability leads to the expected higher probabilities of detection; probability values were near or above 0.67 for 25 years and 0.1 cutoff. How to achieve this is unclear. CVs from the reduced variability lognormal distributions were lower than the CVs from the original fitted distributions by the expected ratios (average ratios of 0.67 and 0.37). The resulting CVs of annual recruitment were 0.28–0.51 for the 0.7 multiplier and from 0.16 to 0.31 for the 0.4 multiplier. Such interannual variability in recruitment in coastal fish species would be considered low, especially the values for the 0.4 multiplier. For example, of the 51 CVs reported by Mertz and Myers (1996), only five were less than 0.35. Thus, some type of variance reduction approach would be needed to achieve such CVs.

There are several ways to improve the power of monitoring for detecting hypoxia effects. One approach would be to filter the recruitment variability and remove the effects of other factors, leaving the effects of hypoxia in the filtered recruitment values but now imbedded in lower general variability. The filtering could be done using statistical methods (e.g., LOWESS—Rago 2005) or life cycle considerations. For example, recruitment values are often filtered by removing the effects of spawning stock (e.g., use the residuals from the fitted spawner-recruit relationship) to look for how variation in recruitment can be attributed to environmental and other factors (e.g., Mertz and Myers 1996). Another approach would be to use a comparative method based on a space-for-time substitution, such that local recruitment or other measures of population-level responses are compared across regions that experience different degrees of hypoxia but similar other conditions. One could also attempt to gain power by using a fish community approach (Fausch et al. 1990), and by using multiple indices that together provide more explanatory power than single indices (Perez-Dominguez et al. 2012); the LDWF data is typical in that multiple species are captured and enumerated in all samples. Finally, collecting biomarkers that show the exposure of individuals to the factor of interest (e.g., hypoxia-inducible factors—Thomas and Rahman 2009) would enable further refinement of the cause-and-effect aspect in the analysis of the monitoring data.

We did not evaluate the LDWF sampling design or data in this analysis; we used their extensive long-term data to derive realistic variation in recruitment. We then simulated very simple sampling—extracting segments of recruitment from the normoxic and hypoxic time series and comparing mean recruitment between the two samples. More extensive analyses are possible that would examine how the placement of stations within regions and the frequency of their sampling (weekly, monthly) would affect the power of the design (e.g., Pearson and Rose 2001). In our situation, we would simulate sampling of croaker within the spatial grid of the population model, which provides the location of each individual croaker on an hourly basis. One can then evaluate trade-offs of repeated sampling at fewer stations with less sampling but at more stations, and also examine how the placement of stations on the grid, affects the power of detecting population effects of hypoxia. We opted to use data to derive recruitment variability because the data-based approach is more realistic than getting the croaker population model itself to generate realistic variability in recruitment.

Modeling offers another approach to quantifying individual and population-level responses to hypoxia (see Rose et al. 2009). Population and food web modeling allows for the explicit representation of growth, mortality, reproduction, and movement, and specification of how DO concentrations affects these processes. Simulations can be done under normoxia and with low DO values as inputs to the model, and model predictions compared between the two simulations to isolate the population (or food web, depending on the model) response to low DO. A variety of models have been used to simulate hypoxia effects on the population (e.g., Rose et al. in review-a, b) and food web levels (de Mutsert et al., this book). The use of modeling to quantify the population responses to hypoxia addresses some of the issues that hinder the analysis of long-term monitoring data. However, modeling introduces other issues, such as high demands on data, potentially high uncertainty in the specifics of model formulation and therefore in predictions, and difficulties in validation.

Some combination of long-term data, modeling, and also process studies, is the best approach for quantifying hypoxia effects at the population level. Here, we used the output of a population model to generate realistic effects of hypoxia and long-term data to characterize realistic variability around those model predictions to assess the likelihood of detecting changes in recruitment due to hypoxia. The added value in increased credibility of using an integrative approach far outweighs the use of each of these approaches alone, even with use of a single approach generating much quicker results. The croaker model itself that was used here is another example of a team effort that relied heavily on all three approaches. In addition to the modeling (Rose et al. in review-a, b), laboratory experiments on growth effects and new experiments done specifically done to determine reproductive effects and biomarkers of exposure were performed in conjunction with the modeling (Thomas and Rahman 2009; Rahman and Thomas 2011). There was also extensive use of existing field monitoring data and new field collected (e.g., Thomas and Rahman 2010, 2012) to assess the fraction of the population to which the laboratory effects would apply (i.e., exposed). While such an integrated modeling and data collection approach requires sufficient time for the feedback between the modeling needs and the new data to be collected, the higher certainty of the resulting predictions of population responses to hypoxia make the effort worth it.

Acknowledgements Funding for this project was provided by the National Oceanic and Atmospheric Administration, Center for Sponsored Coastal Ocean Research (CSCOR), NGOMEX06 grant number NA06NOS4780131 and NGOMEX09 grant number NA09NOS4780179 awarded to the University of Texas. This is publication number 222 of the NOAA's CSCOR NGOMEX program.

References

Brady DC, Targett TE (2013) Movement of juvenile weakfish *Cynoscion regalis* and spot *Leiostomus xanthurus* in relation to diel-cycling hypoxia in an estuarine tidal tributary. Mar Ecol Prog Ser 491:199–219

Breitburg D (2002) Effects of hypoxia, and the balance between hypoxia and enrichment, on coastal fishes and fisheries. Estuaries 25(4):767–781

Breitburg DL, Craig JK, Fulford RS, Rose KA, Boynton WR, Brady DC, Ciotti BJ, Diaz RJ, Friedland KD, Hagy JD, Hart DR, Hines AH, Houde ED, Kolesar SE, Nixon SW, Rice JA, Secor DH, Targett TE (2009) Nutrient enrichment and fisheries exploitation: interactive effects on estuarine living resources and their management. Hydrobiologia 629:31–47

Chesney EJ, Baltz DM (2001) The effects of hypoxia on the northern Gulf of Mexico coastal ecosystem: a fisheries perspective. In: Rabalais NN, Turner RE (eds) Coastal hypoxia: consequences for living resources and ecosystems. American Geophysical Union, Washington, DC, pp 321–354

Chesney EJ, Baltz DM, Thomas RG (2000) Louisiana estuarine and coastal fisheries and habitats: perspective from a fish's eye view. Ecol Appl 10(2):350–366

Craig JK, Crowder LB, Hemwood TA (2005) Spatial distribution of brown shrimp (*Farfantepenaeus aztecus*) on the northwestern Gulf of Mexico shelf: effects of abundance and hypoxia. Can J Fish Aquat Sci 62:1295–1308

Dakos V, Carpenter SR, Brock WA, Ellison AM, Guttal V, Ives AR, Kefi S, Livina V, Seekell DA, van Nes EH, Scheffer M (2012) Methods for detecting early warnings of critical transitions in time series illustrated using simulated ecological data. PLoS ONE 7(7):e41010

de Mutsert K, Cowan JH, Essington TE, Hilborn R (2008) Reanalyses of Gulf of Mexico fisheries data: landings can be misleading in assessments of fisheries and fisheries ecosystems. Proc Natl Acad Sci USA 105(7):2740–2744

de Mutsert K, Steenbeek J, Cowan JH, Christensen V (this book) The use of ecosystem-based modeling in determining hypoxia effects on fish and fisheries. In: Justic D, Rose KA, Hetland R, Fennel, K (eds) Modeling coastal hypoxia: numerical simulations of patterns, controls and effects of dissolved oxygen dynamics. Springer, New York

Diamond SL, Murphy CA, Rose KA (2013) Simulating the effects of global climate change on Atlantic croaker population dynamics in the mid-Atlantic region. Ecol Model 264:98–114

Dolan TE, Lynch PD, Karazsia JL, Serafy JE (2016) Statistical power to detect change in a mangrove shoreline fish community adjacent to a nuclear power plant. Environ Monit Assess 188(3):1–16

Fairweather PG (1991) Statistical power and design requirements for environmental monitoring. Mar Freshwater Res 42(5):555–567

Fausch KD, Lyons J, Karr JR, Angermeier PL (1990) Fish communities as indicators of environmental degradation. Amer Fish Soc Symp 8:123–144

Fennel K, Hu J, Laurent A, Marta-Almeida M, Hetland R (2013) Sensitivity of hypoxia predictions for the northern Gulf of Mexico to sediment oxygen consumption and model nesting. J Geophy Res Oceans 118(2):990–1002

Gerrodette T (1987) A power analysis for detecting trends. Ecology 68(5):1364–1372

Hennemuth RC, Palmer JE, Brown BE (1980) A statistical description of recruitment in eighteen selected fish stocks. J Northwest Atl Fish Sci 1:101–111

Hilborn R, Walters CJ (1992) Quantitative fisheries stock assessment: choice, dynamics, and uncertainty. Kluwer Academic, Norwell, MA

Hobday AJ, Evans K (2013) Detecting climate impacts with oceanic fish and fisheries data. Clim Change 119(1):49–62

Humston R, Olson DB, Ault JS (2004) Behavioral assumptions in models of fish movement and their influence on population dynamics. Trans Am Fish Soc 133:1304–1328

Justić D, Wang L (2014) Assessing temporal and spatial variability of hypoxia over the inner Louisiana–upper Texas shelf: application of an unstructured-grid three-dimensional coupled hydrodynamic-water quality model. Cont Shelf Res 72:163–179

Kimmerer WJ (2006) Response of anchovies dampens effects of the invasive bivalve *Corbula amurensis* on the San Francisco estuary foodweb. Mar Ecol Prog Ser 324:207–218

Larsen DP, Kincaid TM, Jacobs SE, Urquhart NS (2001) Designs for evaluating local and regional scale trends. Bioscience 51(12):1069–1078

Leggett WC, Frank KT (1997) A comparative analysis of recruitment variability in North Atlantic flatfishes—testing the species range hypothesis. J Sea Res 37(3):281–299

LDWF (Louisiana Department of Wildlife and Fisheries) (2014) Experimental gill net samples coast-wide 1986-2013 [Data set]. Accessed Feb 2014 from LDWF FTP Download

Messer JJ, Linthurst RA, Overton WS (1991) An EPA program for monitoring ecological status and trends. Environ Monit Assess 17(1):67–78

Mertz G, Myers RA (1996) Influence of fecundity on recruitment variability of marine fish. Can J Fish Aquat Sci 53(7):1618–1625

Obenour DR, Scavia D, Rabalais NN, Turner RE, Michalak AM (2013) Retrospective analysis of midsummer hypoxic area and volume in the northern Gulf of Mexico, 1985–2011. Environ Sci Technol 47(17):9808–9815

Neilan RM, Rose K (2014) Simulating the effects of fluctuating dissolved oxygen on growth, reproduction, and survival of fish and shrimp. J Theor Biol 343:54–68

Parr TW, Sier AR, Battarbee RW, Mackay A, Burgess J (2003) Detecting environmental change: science and society—perspectives on long-term research and monitoring in the 21st century. Sci Total Environ 310(1):1–8

Pearson SM, Rose KA (2001) The effects of sampling design on estimating the magnitude and distribution of contaminated sediments in a large reservoir. Environmetrics 12(1):81–102
Pérez-Domínguez R, Maci S, Courrat A, Lepage M, Borja A, Uriarte A, Neto JM, Cabral H, Raykov VS, Franco A, Alvarez MC (2012) Current developments on fish-based indices to assess ecological-quality status of estuaries and lagoons. Ecol Indic 23:34–45
Peterman RM (1990) Statistical power analysis can improve fisheries research and management. Can J Fish Aquat Sci 47(1):2–15
Rabalais NN, Turner RE, Sen Gupta BK, Boesch DF, Chapman P, Murrell MC (2007) Hypoxia in the northern Gulf of Mexico: does the Science support the plan to reduce, mitigate, and control hypoxia? Estuaries Coasts 30:753–772
Rago PJ (2005) Fishery independent sampling: survey techniques and data analyses. In: Musick JA, Bonfil R (eds) Management techniques for elasmobranch fisheries. FAO Fisheries Technical Paper 474, FAO, Rome, pp 201–215
Rahman MS, Thomas P (2011) Characterization of three IGFBPs in Atlantic croaker and their regulation during hypoxic stress: potential mechanisms of their upregulation by hypoxia. Am J Physiol—Endoc M 301:E637-E648
Rose KA (2000) Why are quantitative relationships between environmental quality and fish populations so elusive? Ecol Appl 10(2):367–385
Rose KA, Adamack AT, Murphy CA, Sable Kolesar SE, Craig JK, Breitburg DL, Thomas P, Brouwer MH, Cerco CF, Diamond S (2009) Does hypoxia have population-level effects on coastal fish? Musings from the virtual world. J Exp Mar Biol Ecol 381:S188–S203
Rose KA, Cowan JH, Winemiller KO, Myers RA, Hilborn R (2001) Compensatory density dependence in fish populations: importance, controversy, understanding and prognosis. Fish Fish 2:293–327
Rose KA, Creekmore S, Thomas P, Craig JK, Miller-Neilan R, Rahman MS (in review-a). Modeling the population effects of hypoxia on Atlantic croaker (*Micropogonias undulatus*) in the northwestern Gulf of Mexico: Part 1—model description and idealized hypoxia. Estuaries Coasts
Rose KA, Creekmore S, Wang L, Justic'D, Thomas P, Craig JK, Miller-Neilan R, Rahman MS (in review-b) Modeling the population effects of hypoxia on Atlantic croaker (*Micropogonias undulatus*) in the northwestern Gulf of Mexico: Past 2—Realistic hypoxia and sensitivity analysis. Estuaries Coasts
Taylor JC, Mitchell WA, Buckel JA, Walsh HJ, Shertzer KW, Martin GB, Hare JA (2009) Relationships between larval and juvenile abundance of winter-spawned fishes in North Carolina. USA Mar Coast Fish 1(1):12–21
Thomas P, Rahman MS, Khan IA, Kummer JA (2007) Widespread endocrine disruption and reproductive impairment in an estuarine fish population exposed to seasonal hypoxia. Proc Roy Soc B-Biol Sci 274:2693–2702
Thomas P, Rahman MS (2009) Chronic hypoxia impairs gamete maturation in Atlantic croaker induced by progestins through nongenomic mechanisms resulting in reduced reproductive success. Environ Sci Technol 43(11):4175–4180
Thomas P, Rahman MS (2010) Region-wide impairment of Atlantic croaker testicular development and sperm production in the northern Gulf of Mexico hypoxic dead zone. Mar Environ Res 69:S59–S62
Thomas P, Rahman MS (2012) Extensive reproductive disruption, ovarian masculinization and aromatase suppression in Atlantic croaker in the northern Gulf of Mexico hypoxic zone. Proc Roy Soc B-Biol Sci 279:28–38
Turner RE, Rabalais NN, Justić D (2012) Predicting summer hypoxia in the northern Gulf of Mexico: Redux. Marine Poll Bull 64:319–324
Vaquer-Sunyer R, Duarte CM (2008) Thresholds of hypoxia for marine biodiversity. Proc Natl Acad Sci USA 105(40):15452–15457

Wagner T, Irwin BJ, Bence JR, Hayes DB (2013) Detecting temporal trends in freshwater fisheries surveys: statistical power and the important linkages between management questions and monitoring objectives. Fisheries 38(7):309–319

Watkins KS, Rose KA (2013) Evaluating the performance of individual-based animal movement models in novel environments. Ecol Model 250:214–234

Urquhart NS, Kincaid TM (1999) Designs for detecting trend from repeated surveys of ecological resources. J Agric Biol Envir S 4(4):404–414

Chapter 14
Using Ecosystem Modeling to Determine Hypoxia Effects on Fish and Fisheries

Kim de Mutsert, Jeroen Steenbeek, James H. Cowan and Villy Christensen

Abstract The effects of coastal hypoxia on fish biomass and fisheries landings in the northern Gulf of Mexico have been difficult to quantify. A main complicating factor is the fact that nutrient loading from freshwater discharge is not only the main contributor to the formation of the hypoxic zone, but also a driver of secondary productivity through bottom-up processes. Other complicating factors include food web interactions, movement of nekton to more suitable habitat, and temporal and spatial variability in hypoxic area. Through case studies using Ecopath with Ecosim and Ecospace, we show that ecosystem modeling can provide a tool to evaluate population-level effects on nekton biomass as well as changes in fisheries landings due to hypoxia. Fitting model simulations to time series (observations) in Ecosim reveals that including hypoxia improves the fit of the model to observations. These findings led to the development of a spatially and temporally dynamic Ecospace model, coupled to a physical-biological model with high skill in replicating dissolved oxygen and Chl *a* levels. The results of simulations with this coupled modeling approach suggest that, for most species, the positive effects of increased phytoplankton as a result of nutrient enrichment from the Mississippi River outweigh the negative effect of bottom hypoxia. Decoupling enrichment from hypoxia

K. de Mutsert (✉)
Department of Environmental Science and Policy, George Mason University, 4400 University Drive, Fairfax, VA 22030, USA
e-mail: kdemutse@gmu.edu

J. Steenbeek
Ecopath International Initiative, C/Sant Eduard 22-A-2-1, Sant Cugat del Valles SP, 08172 Barcelona, Spain
e-mail: jeroen@ecopathinternational.org

J.H. Cowan
Department of Oceanography, Louisiana State University, 1002-Y Energy, Coast and Environment Building, Baton Rouge, LA 70803, USA
e-mail: jhcowan@lsu.edu

V. Christensen
Institute for the Oceans and Fisheries, University of British Columbia, V6T 1Z4a Vancouver, B.C, Canada
e-mail: v.christensen@fisheries.ub.ca

D. Justic et al. (eds.), *Modeling Coastal Hypoxia*,
DOI 10.1007/978-3-319-54571-4_14

also showed that hypoxia does reduce biomass and landing as compared to enrichment alone, and that there are winners and losers: Some species such as red snapper decrease in biomass even with enrichment. Future directions include simulating nutrient reduction scenarios to inform management.

Keywords Hypoxia · Fisheries · Ecospace · Ecosystem modeling · Ecopath with Ecosim · Food webs · Gulf of Mexico

14.1 Introduction

While the existence of the hypoxic zone off the coast of Louisiana has been well established (Rabalais et al. 2001 and earlier chapters), the effect it has on fish and fisheries has been more elusive (Rose 2000; O'Connor and Whitall 2007; Rose et al. 2009). There are several ecosystem modeling approaches that can be used to consider environmental effects on living marine resources (e.g., Ecopath with Ecosim (EwE, Christensen and Walters 2004); Trosim (Fulford et al. 2010), CASM (Bartell et al. 1999), and ATLANTIS (Kaplan et al. 2012)). In this chapter, we use the EwE modeling approach and its spatial module, Ecospace, to determine potential effects of hypoxia on fish and fisheries.

New developments in EwE described in Christensen et al. (2014a) and De Mutsert et al. (2015a) allow for the inclusion of avoidance of hypoxia by marine nekton and reduced feeding rates of organisms in response to low oxygen levels; these effects can cascade through the food web through trophic interactions. Inclusion of ports and fishing fleets allows for the simulation of the added effect of catches on the abundance of living marine resources, while simultaneously providing an estimate of landings and revenue changes during different simulation scenarios. The virtual representation allows for the decoupling of the bottom-up effect of high levels of primary production on higher trophic levels, and the negative effects of hypoxic events on these consumers related to these same high levels of primary productivity that occur in the northern Gulf of Mexico (Rabalais et al. 2002). Output of a coupled physical-biological model (Fennel et al. 2011, 2013; Laurent et al. 2012; Laurent and Fennel 2014) provides the scenarios of primary productivity on the coastal shelf with and without summer hypoxia.

In this chapter, we describe these modifications to EwE to simulate hypoxia effects in the northern Gulf of Mexico, and illustrate the modeling approach with three case studies that use simulations to evaluate effects of hypoxia on fish and fisheries. The results remind us to not underestimate the extent to which the Mississippi River outflow fuels the productivity of the northern Gulf of Mexico ecosystem, and give insight into the difficulty in finding negative empirical correlations between landings and hypoxia. Through these case studies, the rapid evolution of the capabilities of Ecopath with Ecosim and Ecospace to simulate effects of environmental drivers on fish and fisheries is highlighted.

14.1.1 The Louisiana Coastal Ecosystem

Perhaps the most characteristic feature of the Louisiana continental shelf region is the presence of two major sources of freshwater, the Mississippi and Atchafalaya rivers, that strongly influence the physics (Wiseman et al. 2004), biology (Hanson 1982; Wiseman et al. 1986), and chemistry (Ho and Barrett 1977). The freshwater runoff, which averages 1.83×10^4 m^3 s^{-1}, contributes to a low salinity, highly turbid, near shore water mass within a westward flowing coastal current, which is constrained within a steep, horizontal, salinity front on the mid-shelf about 20–50 km offshore (Crout 1983; Cochrane and Kelly 1986).

The nutrient-rich waters entering the Gulf of Mexico in this manner lead to high primary production on the coastal shelf, which in turn stimulates the secondary production of this marine ecosystem (Nixon and Buckley 2002; Livingston 2002). The Louisiana coastal area has indeed been referred to as the *Fertile Fisheries Crescent* (Gunter 1963); fisheries landings in Louisiana are the highest of the Gulf states and contribute significantly to the total commercial and recreational catch in the USA (Chesney et al. 2000).

High levels of primary production on the Louisiana shelf and the resulting bacterial respiration during the decay of these large amounts of organic matter, in combination with summer stratification of coastal waters, are the cause of an extensive area with hypoxic bottom waters each summer (Rabalais and Turner 2001). The areal extent of the affected region is positively related to Mississippi River discharge and has had an average size of 13,650 km^2 over the past 30 years (1985–2014, http://www.gulfhypoxia.net/Overview/). This seasonal hypoxic zone affects the living marine resources of the northern Gulf of Mexico.

14.1.2 Ecological Effects of Hypoxia on the Gulf of Mexico

It has been suggested that the formation of the hypoxic zone could lead to altered food web dynamics on the Louisiana shelf (Chesney et al. 2000; Rabalais and Turner 2001). Effects may be both direct via increased mortality through prolonged exposure to low dissolved oxygen (DO) concentrations (Breitburg et al. 1999; Turner 2001; Breitburg et al. 2003) or indirect via alteration of benthic (Turner 2001) and water column (Breitburg et al. 1999; Chesney et al. 2000; Turner 2001), habitat availability, and food web structure. An example of altered food web structure would be increased abundances of gelatinous zooplankton predators that consume zooplankton and larval fish (Graham 2001; Grove and Breitburg 2005).

Several studies have described the effects of hypoxia on feeding, growth, behavior, and mortality of fishes from a variety of taxonomic groups. In particular, sublethal effects of hypoxia have been shown to result in decreased feeding (Chabot and Dutil 1999; Tallqvist et al. 1999; Pichavant et al. 2001) and growth rate (Bejda et al. 1992; Secor and Gunderson 1998; Chabot and Dutil 1999; Taylor and Miller

2001), changes in activity level (Crocker and Chech 1997; Schurmann and Steffensen 1992), and spatial distribution (Pihl et al. 1991; Breitburg et al. 1999, 2003; Keister et al. 2000; Wannamaker and Rice 2000). Studies also have demonstrated direct effects of severe or chronic hypoxia on mortality (Schurmann and Steffensen 1992; Tallqvist et al. 1999; Miller et al. 2002); specific DO levels that can elicit sublethal effects have been shown to be species-specific (see reviews by Davis 1975; USEPA 2001; Miller et al. 2002).

14.1.3 Potential Implications for Gulf of Mexico Fisheries

The hypoxic zone could have economic consequences, if hypoxia reduces production of commercially and recreationally valuable fish and shellfish (Diaz and Rosenberg 1995; Breitburg 2002; O'Connor and Whitall 2007). Aggregation near hypoxic edges has been shown for gulf shrimp and finfish, which may enhance their susceptibility to commercial shrimp trawls (Craig 2012). This "edge effect" can result in a positive correlation between hypoxia and fisheries landings, and create a false sense of increased abundances when fisheries dependent data are used to determine stock size. Positive correlations could also occur as a result of the bottom-up effect that nutrient enrichment has on higher trophic levels (Nixon and Buckley 2002; Breitburg et al. 2009).

14.1.4 Ecosystem Modeling

The concept of using ecosystem models in fisheries science and ecology is to include effects of environmental parameters, trophic interactions of multiple species, and fishing on the biomass of all species included in such models. The ecosystem models presented in this chapter have been developed with the open-source Ecopath with Ecosim modeling suite, freely available from http://www.ecopath.org. The advantage of creating a virtual representation of the northern Gulf of Mexico ecosystem is that the factors affecting living marine resources can be evaluated separately and together.

14.2 Constructing an Ecopath Model

An Ecopath model is a virtual representation of an ecosystem, with a focus on higher trophic levels. It was originally developed by Polovina (1984), and subsequently modified to replace the assumption of steady state by an assumption of mass balance over a period of time, usually a year (Christensen and Pauly 1992). Other modifications include the addition (and subsequent modification) of Ecosim, a time-dynamic module, and Ecospace, a spatially explicit module (Walters et al. 1997, 1999). These dynamic modules of EwE will be discussed in Sects. 14.3 and 14.4.

There are two master equations at the base of Ecopath. The first equation describes the production term and can be expressed as follows:

$$B_i \cdot (P/B)_i \cdot EE_i - \sum_{i=1}^{n} B_j \cdot (Q/B)_j \cdot DC_{ji} - Y_i - E_i - BA_i = 0 \quad (14.1)$$

where B_i and B_j are the biomasses of the prey (i) and predators (j), respectively; $(P/B)_i$ the production/biomass ratio; EE_i the ecotrophic efficiency, which is the proportion of the production that is utilized in the system; $(Q/B)_j$ the consumption/biomass ratio; DC_{ji} the fraction of prey (i) in the diet of predator (j); Y_i the total fishery catch rate of (i); E_i the net migration rate (emigration–immigration); and BA_i the biomass accumulation rate for (i).

The second master equation ensures energy balance within each group as follows:

$$\text{Consumption} = \text{production} + \text{respiration} + \text{unassimilated food} \quad (14.2)$$

To construct a model, the parameters B, P/B, Q/B, DC_{ij}, Y, E, and BA need to be specified for each group (Y only if there is a fishery), while the model solves for EE. While this is the recommended approach, the model can also solve for missing B, P/B, and Q/B parameters since Ecopath can handle solving for different and multiple unknowns while balancing the model. Missing parameters are estimated by linking the production of each group to the consumption of all groups, based on the mass-balance requirement of Eq. 14.1.

Each group in an Ecopath model can be a single species, an aggregation of multiple species of similar role in the ecosystem, (e.g., a functional group), or a single life history stage of a species with a complex trophic ontogeny (i.e., a multi-stanza group). With the multi-stanza approach, different diets, predation levels, tolerance ranges (e.g., to DO), and fisheries landings can be assigned to different life stages of a species, which increase the realism of the model. The stanzas of a species are linked with a von Bertalanffy growth model. A model like this was constructed for the Gulf of Mexico (Fig. 14.1), serving as a base model for simulations described in this chapter.

14.3 Temporal Dynamic Modeling with Ecosim

Ecosim provides temporal dynamic simulation capabilities building on the Ecopath base model. Biomass dynamics in Ecosim are expressed through a series of coupled differential equations derived from the Ecopath master equation (Eq. 14.1). The resulting differential equation for biomass is as follows:

$$dB_i/dt = gi \sum_j Q_{ji} - \sum_j Q_{ij} + I_i - (MO_i + F_i + e_i)B_i \quad (14.3)$$

where dB_i/dt represents the growth rate during the time interval of group (i) in terms of its biomass (B_i), g_i is the net growth efficiency (production to consumption ratio), $\sum Q_{ji}$ the total consumption by group (i), $\sum Q_{ij}$ the predation by all predators on the same group (i), I_i the immigration rate, MO_i the non-predation natural mortality rate, F_i the fishing mortality rate, and e_i the emigration rate.

The set of differential equations is solved in Ecosim using an Adams-Bashforth integration routine. The consumption rates are calculated based on the foraging arena theory, where the biomass of each group is divided into vulnerable and invulnerable components (Walters et al. 1997; Ahrens 2012). The transfer rate between these two components (ν_{ij}) determines the relative contribution of top-down (biomass of predator impacts how much prey is consumed) and bottom-up control (being caught is a function of the prey's productivity).

A useful feature in Ecosim is the ability to fit simulations to time series (Christensen and Walters 2011). The main purpose is to calibrate the model to observation, but can also be used to determine whether the model explains more variability with the addition of environmental drivers. This feature is demonstrated in the first case study described in Sect. 14.3.3.

14.3.1 Including Hypoxia Effects

The flow of biomass in Ecosim from prey biomass (i) to predator biomass (j) follows the function:

$$Flow(biomass/time) = a_{ij} \cdot V_{ij} \cdot P_j \tag{14.4}$$

where a_{ij} is the rate of effective search, V_{ij} is vulnerable prey biomass, and P_j is effective predator biomass. P_j is the group biomass for simple groups, but calculated differently when a group has ontogenetic diet splits using the multi-stanza function (see Sect. 14.2). For multi-stanza groups, P_j is the sum of biomass over ages in that group to the 2/3 power, an index of per-predator search rate.

The effective search rate a_{ij} is reduced in the model during unfavorable DO conditions using species-specific response curves. Response curves of individual fish species to environmental parameters are determined by plotting catch rate versus the environmental parameter for each group in the model, using large datasets (Fig. 14.2). This method has previously been used to include effects of salinity (De Mutsert et al. 2012). For the DO response curves in the NGOMEX Ecosim scenarios, catch rates from SEAMAP groundfish data were plotted against bottom DO values as measured during catch. The resulting parameters (the minimum DO value that is not causing limitations and the standard deviation from that point) were used to create sigmoidal curves that serve as multipliers on effective

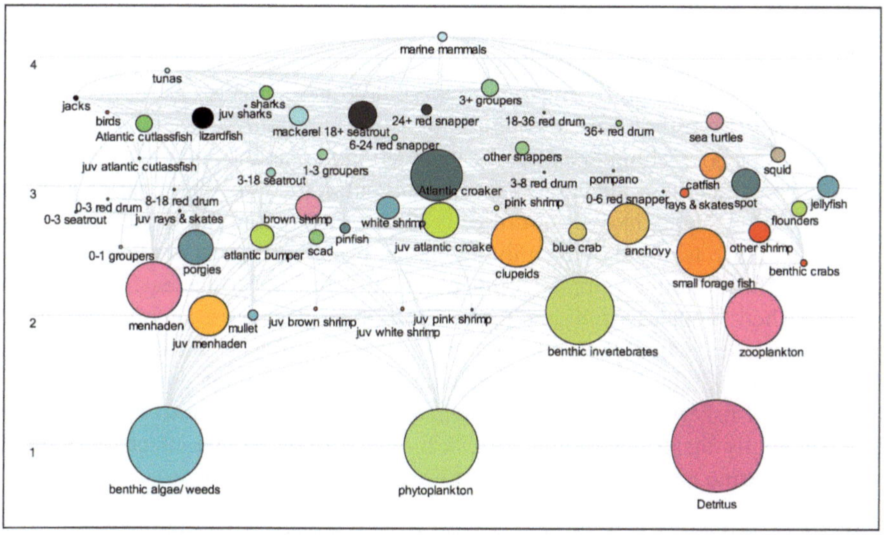

Fig. 14.1 NGOMEX Ecopath model. *Y-axis* indicates trophic level and the size of the nodes indicates the relative size of the biomass pool. Trophic interactions are indicated with the *gray lines* and represent the flow of energy in the model

Fig. 14.2 Oxygen response curves of five example species in the model. Reproduced from De Mutsert et al. (2015) under a creative commons license

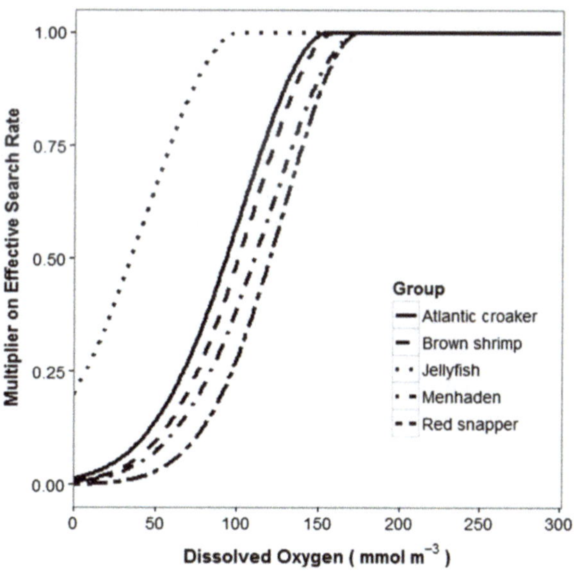

search rate (Fig. 14.2). The response (y value between 0 and 1) that corresponds with the environmental DO value (x value) is then used to multiply a_{ij} in Eq. 14.4, with DO time series data serving as the forcing function.

14.3.2 Including Fisheries

Fisheries can be affected by hypoxia as well, and these effects cannot be extrapolated directly from the average biomass of living marine resources in the system (see Sect. 14.1.3). Fishing fleets can be included in the model, and the catch of each of these fleets can be simulated and form an output of the Ecosim scenarios just like group biomass. Hypoxia does not directly affect the fleets in the model, but, more realistically, effects of hypoxia on living marine resources have an indirect effect on fisheries. These indirect effects could potentially include a decrease in fisheries catches when biomass of living marine resources decreases, or an increase in fisheries catches when living marine resources congregate at the edge of the hypoxic zone (Craig 2012). The inclusion of fleets not only ensures the direct simulation of fisheries (and effects of hypoxia on fisheries) but also includes catch as a potentially major determinant of fish biomass; to determine effects of hypoxia on living marine resources, the effects of fishing cannot be ignored.

14.3.3 Case Study 1: Hypoxia in Ecosim

An Ecopath model was developed for the northern Gulf of Mexico (NGOMEX; Fig. 14.1 and Table 14.1), based on the biomass (t km^{-2} wet weight) of species representative of the NGOMEX using SEAMAP data (http://sero.nmfs.noaa.gov/operations_management_information_services/state_federal_liaison_branch/seamap/index.html). The model and expansions on it as described in two more case studies are referred to as the NGOMEX model. Species were divided into multiple life history stages where needed to consider ontogenetic shifts in the life history of a species. Mass balance was achieved after multiple iterations involving adjustments of input parameters, resulting in a plausible virtual representation of the northern Gulf of Mexico ecosystem.

In Ecosim, hypoxia was added, and a fishery for menhaden, shrimp, recreational, snapper/grouper, crab pots, squid, and longlines was included. Landings data used are published online by NOAA: http://www.st.nmfs.noaa.gov/st1/commercial/landings/annual_landings.html.

Two scenarios were simulated with the NGOMEX version of Ecosim in this case study: one without hypoxia as a forcing function and one with a time series of spatially averaged summer bottom hypoxia (mg/l) measured by the Louisiana Department of Wildlife and Fisheries within the area of the hypoxic zone from 1998 to 2002 as a forcing function. We ran a simulation from 1982 to 2008, which were the years for which we had SEAMAP data to which the simulations were fitted. Annual averaged biomass data derived from SEAMAP and landings documented by NOAA were used as observations to fit the predictions to. For this case study, the dissolved oxygen pattern measured from 1998 to 2002 by the Louisiana Department of Wildlife and Fisheries (LDWF) was repeated in the years of the

Table 14.1 Groups in the NGOMEX Ecopath model and their start biomass

Group name	Biomass in habitat area (t km^{-2})
Marine mammals	0.069
Tunas	0.024
Jacks	0.018
Birds	0.011
Juv Atlantic cutlassfish	0.003
Atlantic cutlassfish	0.228
Lizardfish	0.384
Juv sharks	1.2E-04
Sharks	0.148
Mackerel	0.300
0–3 sea trout	2.5E-04
3–18 sea trout	0.072
18 + sea trout	0.647
0–6 red snapper	0.001
6–24 red snapper	0.032
24 + red snapper	0.090
0–1 groupers	0.008
1–3 groupers	0.090
3 + groupers	0.226
Other snappers	0.141
0–3 red drum	4.4E-06
3–8 red drum	1.2E-04
8–18 red drum	0.001
18–36 red drum	0.003
36 + red drum	0.029
Juv rays and skates	0.001
Rays and skates	0.082
Flounders	0.202
Pompano	0.002
Atlantic bumper	0.434
Scad	0.182
Juv Atlantic croaker	1.303
Atlantic croaker	4.344
Catfish	0.582
Spot	0.690
Squid	0.168
Pinfish	0.094
Porgies	1.223
Anchovy	2.032
Juv menhaden	1.891
Menhaden	6.240

(continued)

Table 14.1 (continued)

Group name	Biomass in habitat area (t km^{-2})
Clupeids	4.448
Mullet	0.100
Sea turtles	0.030
Small forage fish	3.715
Jellyfish	0.360
Blue crab	0.244
Juv brown shrimp	0.007
Brown shrimp	0.558
Juv white shrimp	0.004
White shrimp	0.300
Juv pink shrimp	2.6E-04
Pink shrimp	0.020
Other shrimp	0.369
Benthic crabs	0.045
Benthic invertebrates	12.08
Zooplankton	7.642
Benthic algae/weeds	29.8
Phytoplankton	25
Detritus	100

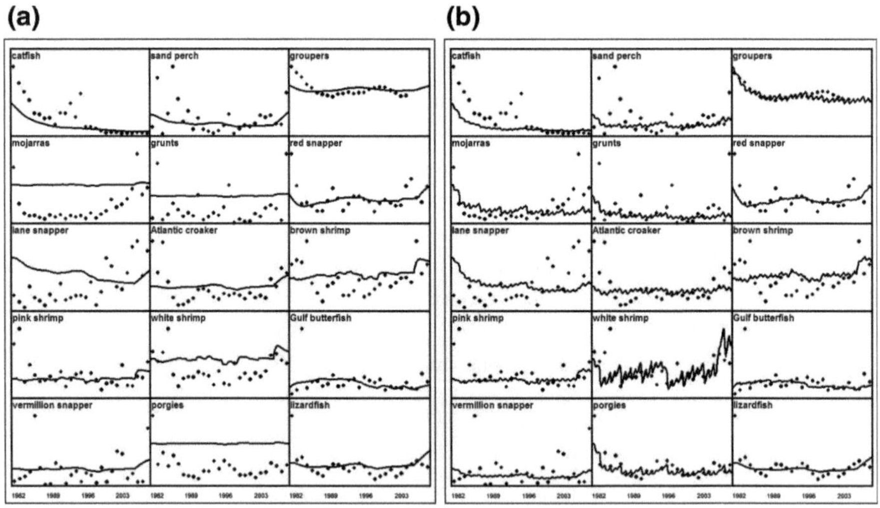

Fig. 14.3 Ecosim scenario without hypoxia forcing function (**a**) and with summer hypoxia as a forcing function (**b**)

Ecosim run from 1982 to 1997 and 2003 to 2008. Nutrient loading to the Gulf of Mexico, represented by NO_x loads measured by USGS at Tarbert Landing, was kept the same in both model runs. In the first case study presented here, scenario runs with the NGOMEX model demonstrate the improvement of fit during calibration in Ecosim with the inclusion of hypoxia as an environmental factor affecting fish biomass (Fig. 14.3). The total sum of squares (SS) of the fit of the simulation to observed biomass data decreased from 780 to 670 when the hypoxia forcing function was activated. By adding hypoxia, the model explained more of the variation in the data than the model did with trophic interactions and fishing alone (Fig. 14.3).

While the Ecosim scenarios provide biomass and fisheries output, the lack of spatial dynamics in the model puts restriction on the interpretation of the outcomes. Without a spatial component, the organisms in the model do not have the ability to move away from hypoxia. In other words, hypoxia is either "on" or "off" in the model, and the only relief organisms receive from hypoxia is in time, since the DO is only reduced to hypoxic levels in the summer months. Effects of biomass aggregation at the edge of the hypoxic zone and subsequent effects on fisheries will thereby also not be simulated. A solution to this problem is provided by Ecospace, which provides spatially explicit simulations detailed in the next section. When using Ecospace, it is still useful to first use Ecosim, since model calibration by fitting simulations to time series as described in the first case study is performed in Ecosim (which notably entails incorporating density-dependent effects), and model robustness is tested in Ecosim by running Monte Carlo simulations (Christensen et al. 2004). In the above presented case study, Ecosim provides an indication that hypoxia could indeed affect biomass of living marine resources (Fig. 14.3), and that further exploration of the NGOMEX model in Ecospace is warranted.

14.4 Ecospace

14.4.1 Use of GIS in Ecospace

Up to version 6.3 of the EwE software suite, a major shortcoming of the Ecospace model has been its lack of facilities to integrate external spatial forcing data into its computations, and to deliver its outputs in geospatial data formats for model interoperability. Although the Ecopath and Ecosim models have been successfully linked to other models, the spatial model Ecospace had seen no use in this regard due to the complexity involved, including lack of capabilities to easily exchange data. Continued popularity of the EwE approach, increasing demand for the ability to use the Ecospace model in conjunction with spatial analytical tools, specialist models, and planning tools such as Marxan (Loos, 2011) resulted in increasing demand to integrate varying environmental conditions into the Ecospace model. EwE source code was migrated in 2006 to the .NET programming environment to which gave rise to the idea of a flexible spatial-temporal data framework to solve the data connectivity shortcomings of Ecospace. This spatial-temporal data framework is an abstraction layer onto the Ecospace model that facilitates the

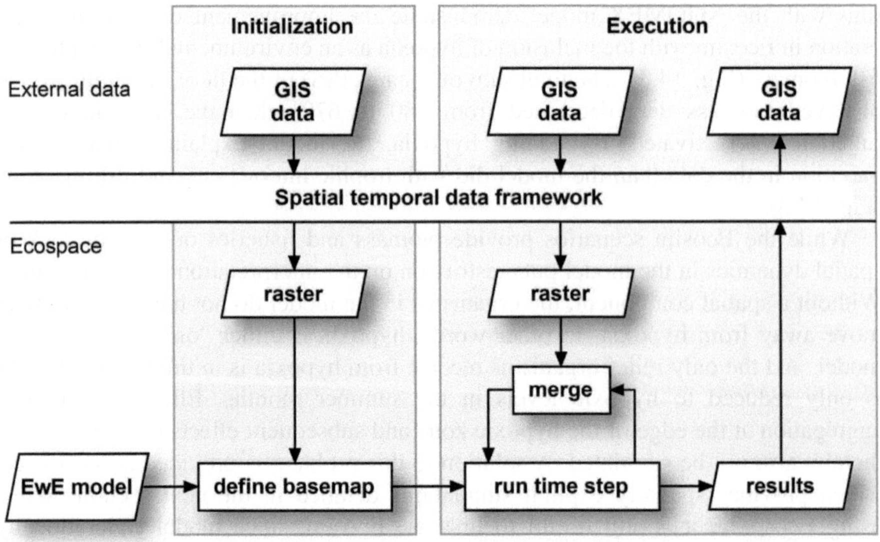

Fig. 14.4 Conceptual overview of the spatial-temporal data framework, which provides external GIS data to Ecospace model initialization and at runtime, and provides Ecospace results in spatial data formats when the model executes. Reproduced from Steenbeek et al. (2013) with permission

process of locating, loading, and adapting geospatial data to the Ecospace internal data formats, and facilitates the process of converting Ecospace result maps to geospatial data formats (Fig. 14.4).

To date, the spatial-temporal framework of Ecospace has been applied in only a few case studies. One such study demonstrated that the ability of Ecospace to hindcast observed trends in species occurrences significantly improved when driven by monthly and annual SeaWiFS-derived primary production data (Steenbeek et al. 2013). Other case studies use the spatial-temporal data framework to drive the habitat foraging capacity model with altering patterns in oxygen, temperature, salinity, turbidity, substrate, or depth for exploring the impacts of environmental change on marine food webs (De Mutsert et al. 2015a, b, 2016).

14.4.2 The Habitat Capacity Model

The Ecospace model can consider cell-specific habitat descriptions or environmental parameters (level of, e.g., salinity and DO). In versions of the Ecospace model before EwE release 6.3, large-scale habitat structure with attendant impact on biomass distributions and trophic interactions was represented only by a binary habitat use pattern for each group, with each spatial cell being either suitable or not for each group. Biomass dynamics in unsuitable cells was modified by predicting higher rates of emigration, lower feeding rates, and/or higher vulnerability to predation, and there was a complex gradient calculation to modify dispersal rates so as to direct biomass toward suitable cells (Walters et al. 1999).

In the new habitat capacity model used here for hypoxia, the relative habitat capacity by group and by cell is estimated from a vector of habitat attributes—which can be made to consider any environmental parameter, e.g., DO, salinity, or temperature (Christensen et al. 2014a). The habitat capacities can be updated for each time step (in a computationally efficient manner) and can typically be obtained from output from physical or biogeochemical models.

The habitat capacities, C, are linked to trophic interactions in the foraging arena model so that it impacts the size of the foraging arena, i.e.,

$$V = vB/(2v + aP/C) \tag{14.5}$$

The predation activity is thereby concentrated in a smaller area when C decreases which in turn impacts the vulnerable prey densities (V) more rapidly if predator density (P) increases because of a decrease in the size of the foraging arena. This in effect makes spatial patterns of biomass proportional to their habitat capacities. Furthermore, the habitat capacity model modifies spatial mixing rates so as to obtain movements toward preferred cells and avoid dispersing excessive biomass into unsuitable spatial cells.

Using the habitat capacity model, multiple layers can be added to Ecospace models. Users can define the response of organisms to each habitat attribute with response curves (shape and parameters of each response curve are user-specified), or by indicating relative habitat use. An example of the latter would be to define water cells as estuarine and offshore, and then exclude certain organisms from entering the estuaries (habitat use of the estuary would be 0, and use of offshore water cells would be 1). Habitat use can be any value between 0 and 1, and the levels of habitat could be two or more.

14.4.3 Spatial Considerations for Fishing Fleets

In Ecospace, the fishing mortality rates (F) are distributed using a simple "gravity model" where the proportion of the total effort (as defined in the Ecopath base model) allocated to each cell is assumed proportional to the sum over groups of the product of the biomass, and the profitability of fishing the target groups (Christensen and Walters 2004). The profitability of fishing includes the price per pound and the cell-specific cost of fishing (determined by price of gas and distance from port). The NGOMEX Ecospace model used in the second and third case studies incorporates the cost of fishing by including two ports with the highest fish landings in Louisiana: Empire-Venice and Intracoastal City (www.oceaneconomics.org; Fig. 14.5). Distance from port in addition to target species biomass in a particular cell and species-specific price per pound determines movement of fleets.

14.4.4 Case Study Two: Hypoxia in Ecospace

To pursue spatial modeling with the NGOMEX Ecospace model, a base map was loaded in based on a GIS bathymetry map of coastal Louisiana. The model area of the NGOMEX Ecospace model has a 5 × 5 km grid and covers the Louisiana coastal shelf, totaling 8978 model cells (Fig. 14.5). The bathymetry of the area included in the base map has a resolution of 1 m. Bathymetry (with response curves), salinity-based habitat description (the three "levels" of habitat are fresh, brackish, and marine), and DO (with response curves) were included as habitat attributes.

The effect of hypoxia was tested in the Ecospace simulations with a stylized (and static) hypoxic zone (Fig. 14.6). Output of this case study demonstrated a decrease in biomass within the hypoxic area of organisms directly affected by low bottom oxygen (as defined through response curves) while others were unaffected or even showed an increase as an indirect effect of bottom hypoxia through trophic interactions (Fig. 14.7). Trophic interactions can result in neutral or positive effects of

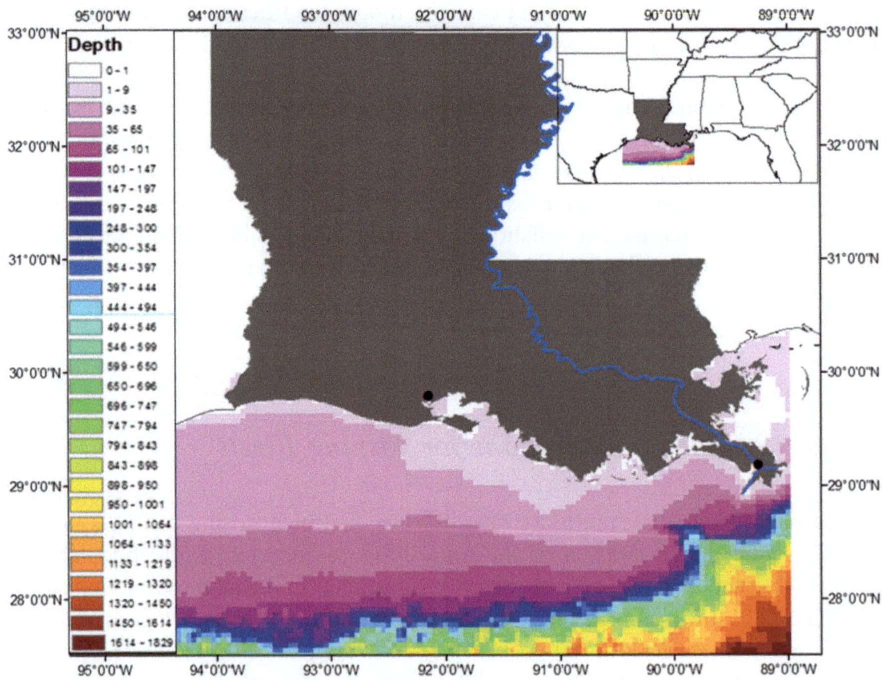

Fig. 14.5 Model area of the NGOMEX (northern Gulf of Mexico) ecosystem model. Louisiana (USA) is indicated in *gray*, and the Mississippi River in *blue*. The coloration in the northern Gulf of Mexico indicates the bathymetry. The two ports with the most landings in Louisiana, Intracoastal City and empire-venice, are indicated with a *black dot* (source of bathymetry data: the fish and wildlife research institute). Reproduced from De Mutsert et al. (2015) under a creative commons license

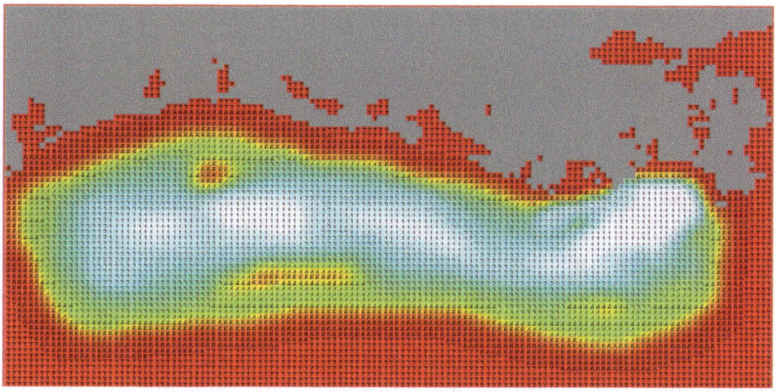

Fig. 14.6 Hypoxic zone used in the second case study. DO (mg/l) is indicated in each cell. The color scheme indicates increasing DO concentrations with warmer colors from *white* (hypoxic) to *red* (normoxic)

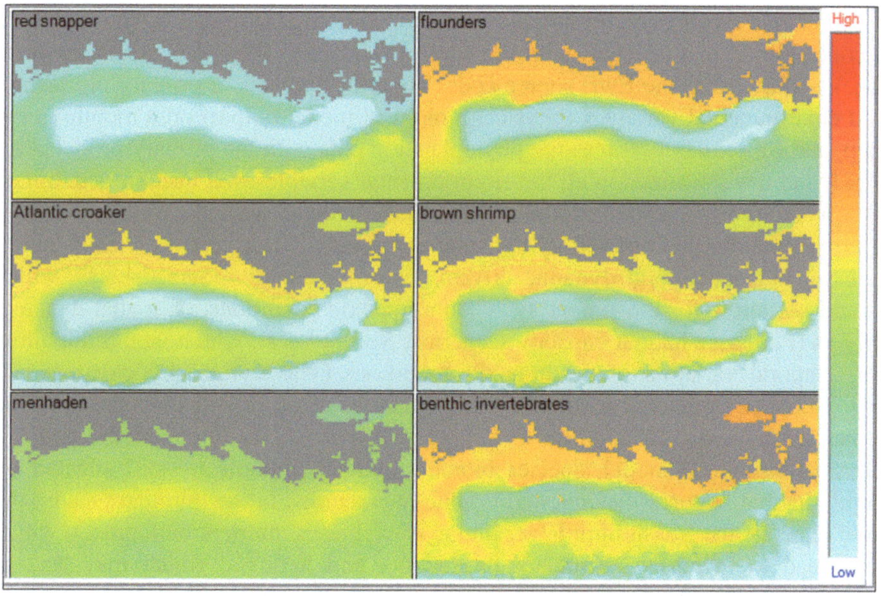

Fig. 14.7 Biomass (relative) of six groups in the NGOMEX Ecospace model

hypoxia on biomass of some species through release of competition for food, or release of predation pressure, as has been shown in Altieri (2008). The location of areas where biomass aggregation occurred outside the hypoxic zone was influenced by species preferences for depth and habitat in addition to hypoxia.

14.4.5 Including Spatial and Temporal Dynamic Forcing Functions

Including spatial and temporal dynamic forcing functions, which is essential to simulate the effect of summer hypoxia on living marine resources, is very data intensive. A value of bottom DO and a measure of primary production are needed for each grid cell at each time step (monthly) for the duration of the model run (typically for multiple decades). Output from hydrodynamic or physical-chemical models would provide data at this resolution, and simultaneously provide the ability to run exploratory and management-based scenarios. Generally, averaging is needed, since most hydrodynamic models provide output on shorter time steps than months, plus spatial re-averaging may be needed if the grid cells of the two models are of different size and/or shape. Whether the re-averaging of physical-chemical information makes sense ecologically needs to be determined on a case-by-case basis, and additional plug-ins can be designed to evaluate effects on shorter time steps than months if effects are known to manifest on shorter time scales (not performed in the case studies presented here, but see De Mutsert et al. 2016).

To couple the physical-biological model to the Ecospace model, a "plug-in" (i.e., a code snippet that interacts with the core model, data layers, and scientific interface) was added to the EwE source code. The plug-in reads in primary production and environmental variables (physical-biological model output) on a monthly time step, and subsequently forces the phytoplankton data with the primary productivity input, and the consumers with the environmental variables in a fashion determined by species-specific response curves (Fig. 14.2). This allowed for oxygen and Chl *a* to be spatially and temporally dynamic environmental drivers. Previous simulations included temporal dynamic drivers (Ecosim scenarios; Sect. 14.3.1), which would not allow for movement away from hypoxic conditions, or spatial dynamic drivers (Ecospace habitat capacity scenarios; Sect. 14.4.4), which would not allow for temporal variation of the hypoxic conditions (The stylized hypoxic zone of Fig. 14.6 was present year-round). Both situations result in an exaggeration of the effects of hypoxia. In addition to bottom DO and Chl *a* output of the coupled physical-biological model, species biomass and distribution are still affected by depth, habitat features, fishing, and trophic interactions. In the third case study, the dual effect of the Mississippi River plume is evaluated; nutrient enrichment, which may lead to increased secondary production, and hypoxia, which may lead to decreased secondary production are simulated together and separately.

14.4.6 Case Study 3: Hypoxia as a Spatial-Temporal Driver in Ecospace

The third case study is from De Mutsert et al. (2015a). To test effects of spatial and temporal dynamic forcing functions on the biomass of all groups in the model,

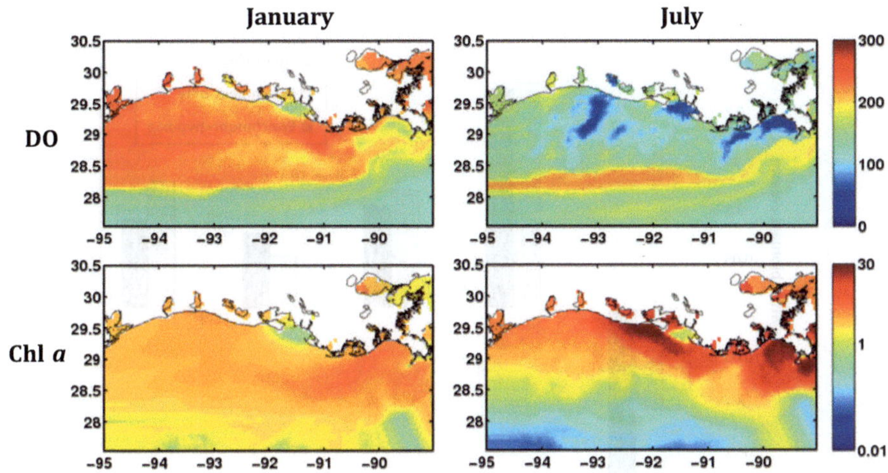

Fig. 14.8 Example output of dissolved oxygen in mmol m^{-3} (*top*) and Chl a in mg m^{-3} (*bottom*) from Fennel et al. (2011) in a month without hypoxia (*January*) and a month with hypoxia (*July*). Monthly "maps" of this output are used as spatial-temporal forcing functions in the NGOMEX ecosystem model. Output was extrapolated in the estuaries as shown in the figure. Reproduced from De Mutsert et al. (2015) under a creative commons license

bottom DO and Chl *a* output of a coupled physical-biological model developed in the same geographic area (Fennel et al. 2011, 2013; Laurent et al. 2012, Laurent and Fennel 2014) were used to force the NGOMEX Ecospace model. The simulated DO and Chl *a* output from 1990 to 2007 were first matched to the grid and time step of the NGOMEX Ecospace model (Fig. 14.8). The model output had to be extrapolated into the estuaries to provide Chl *a* and DO values for each of the cells in the NGOMEX model, and was extended to 2010 by repeating the last 3 years of model output. The groups in the model responded to DO as determined by the group-specific response curves (Fig. 14.2). The Chl *a* data were normalized and drove the biomass of phytoplankton group only with a linear function.

Three scenarios were tested with the NGOMEX Ecospace model with spatial-temporal dynamic drivers. In scenario 1 ("no forcing"), the coupled physical-biological model (Fennel et al. 2011) is not linked to Ecospace, which results in no phytoplankton forcing and no effects of bottom DO on species in the model. This, in a way, represents a scenario where the Mississippi River is "turned off." In scenario 2 ("enrichment only"), the coupled physical-biological model is linked to Ecospace, but species in the model are not affected by bottom DO (i.e., not negatively affected by DO levels below normoxic conditions). In this scenario, the phytoplankton biomass is driven by the coupled physical-biological model Chl *a* output, but the bottom DO output has no effect on the groups in the model. In scenario 3 ("enrichment + hypoxia"), phytoplankton biomass is driven by the coupled physical–biological model, and species are affected by DO levels in a way determined by their species-specific response curve (Fig. 14.2). Some results of

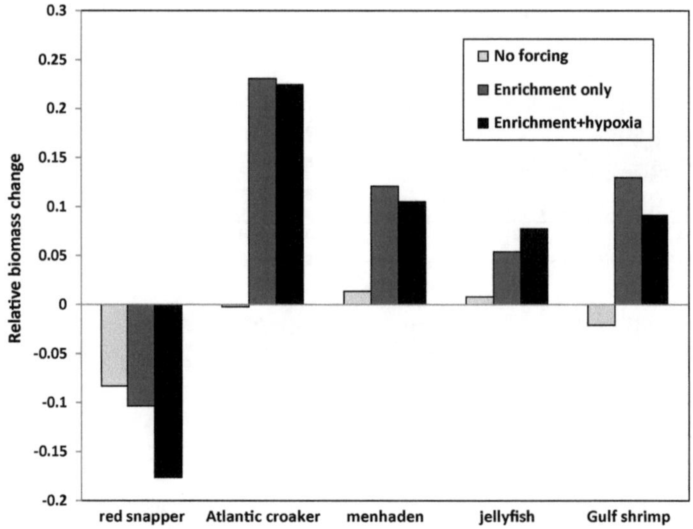

Fig. 14.9 Relative change in biomass (=(end biomass/start biomass)−1) of five example species in scenario "no forcing," "enrichment only," and "enrichment + hypoxia." Gulf shrimp are *brown*, *white*, and *pink* shrimp

these simulation outputs are shown in Figs. 14.9, 14.10 and 14.11. The simulations ran from 1950 to 2010.

The results of these simulations suggest that, for most species, the positive effects of increased phytoplankton as a result of nutrient enrichment from the Mississippi River outweigh the negative effect of bottom hypoxia as a result of the same nutrient enrichment from the Mississippi River (see De Mutsert et al. (2015a) for more details). This is a result of the trophic dynamics in Ecospace (bottom-up effect of increased productivity). Similar results have been found in previous studies (Breitburg et al. 2009). A study using a similar approach in the Baltic Sea found biomass reductions when simulating nutrient reduction scenarios (Niiranen et al. 2008), although they did not consider changes to the level or area of hypoxia in response to nutrient reductions. Other mechanisms such as release from predation pressure and/or competition play a role as well (Altieri 2008), which are mechanisms well represented with the trophic dynamics in Ecospace. The bottom-up effects of nutrient enrichment plus indirect effects of trophic dynamics trickle through into the landings of the simulated fleets (Fig. 14.11).

Exceptions occur as is shown in the case of red snapper (Fig. 14.9), which was one of the least hypoxia-tolerant species in the model (Fig. 14.2), and did suffer reduced biomass with hypoxia. The red snapper simulations show an interesting case where biomass is more affected than landings, providing an example of the edge effect; fleets obtain landings at the edge of the hypoxic zone and exacerbate biomass reductions (note that there is no policy or quotas part of these simulations).

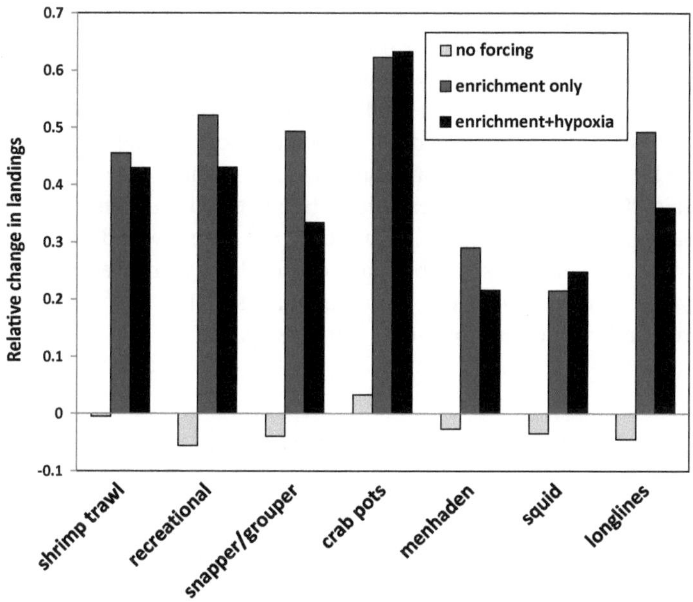

Fig. 14.10 Relative change in landings (=(end landings/start landings)−1) of fisheries in the model in scenario "no forcing," "enrichment only," and "enrichment + hypoxia"

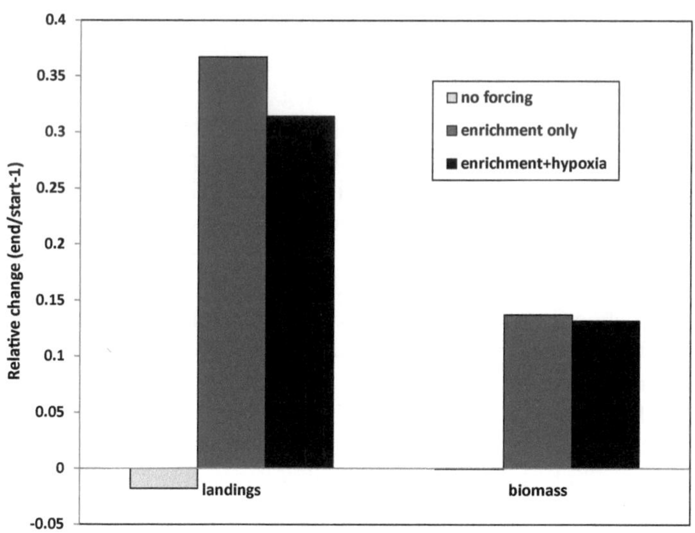

Fig. 14.11 Relative change in total landings and total biomass (=(end/start)−1) of all nekton species/groups in the model in scenario "no forcing," "enrichment only," and "enrichment + hypoxia"

The results of model simulations as presented here can have important management implications. However, before these results are used to inform management, additional nutrient reduction scenarios need to be simulated with the model. There may be an optimum nutrient loading from the river that is lower than current levels, which needs to be identified via simulation because there is no direct linear relationship between phytoplankton biomass increase and oxygen decrease (Walker and Rabalais 2006). Simulating realistic nutrient reduction scenarios is important, because our current "no forcing" scenario is not a real-world scenario, and the comparatively small decrease in biomass caused by hypoxia as compared to the increase of enrichment (versus no forcing) may still be ecologically significant. Ongoing research is looking into these very issues.

14.4.7 Conclusions and Future Directions

In summary, the new Ecospace capabilities in combination with model output from a coupled physical-biological model can provide more realistic simulations of long-term effects of summer hypoxia on the northern Gulf ecosystem and its living marine resources. As shown in Sect. 14.4.5, connections between EwE and physical-biological models that deliver oxygen, Chl *a*, and other environmental drivers into the habitat foraging capacity model can be created with relative ease. True coupling of models can be accomplished on a case-by-case basis, as has recently been done in de Mutsert et al. (2015b). Other advances in Ecospace are the improved representation of the ecosystem within the model area, and the inclusion of ports, which will lead to better simulations of the fishing fleets (and socioeconomic parameters, not considered here, see, e.g., Christensen et al. 2014b) that are included in the spatial model. These tools could be very useful to inform management and can be used to simulate effects of Mississippi River nutrient reduction scenarios on living marine resources in the northern Gulf of Mexico. The spatial temporal data framework functions as a flexible data exchange engine, linking the input driver layers of the Ecospace model to external spatial-temporal data from an open-ended number of data sources and data formats. This functionality not only offers the foundation for a new way of model interoperability to EwE itself, but also aims to contribute to the discussion of standardized model interoperability for food web models in general.

Through the data access tier described in Sect. 14.4.1, bidirectional data exchange linkages to other running models can be facilitated. Data access components can forge links to other models, delivering content valid for select driver maps of the Ecospace model, and in return can deliver estimates of biomass, catch, and other typical Ecospace outputs back to a linked model. Issues such as centralized time stepping and model data conversions will have to be solved, in particular for components that are represented on both sides of a bidirectional model interoperability link.

The results of the simulations provided with the case studies presented in this chapter suggest that nutrient reductions aimed at reducing the size of the hypoxic zone may decrease secondary production, and thereby fisheries biomass and landings. Still, they also suggest that hypoxia has an effect on fisheries species, because model fits improved when dissolved oxygen was included as an environmental factor, and that effects of hypoxia are species-specific, resulting in "winners and losers." This is another example of how trophic interactions can provide potential counterintuitive results of restorative actions (Walters et al. 2008), which can only be found when an ecosystem food web model is used to simulate potential outcomes. The results of the case studies presented here should be considered exploratory, and simulations of actual proposed nutrient reduction scenarios are needed to inform management; these efforts are underway.

Acknowledgements KdM would like to acknowledge Carl Walters and Joe Buszowski for their advice and technical assistance during model development, Katja Fennel for providing output from her coupled physical-biological model, and Arnaud Laurent for preparing this output for use in the Ecospace model. We also like to thank two anonymous reviewers and the editors who improved the chapter with their thoughtful comments. This work was supported by National Oceanic and Atmospheric Administration Award NA09NOS4780233.

References

Ahrens RN, Walters CJ, Christensen V (2012) Foraging arena theory. Fish and fisheries 13: 41–59. Altieri, A. H. 2008. Dead zones enhance key fisheries species by providing predation refuge. Ecology 89:2808–2818. doi:10.1890/07-0994.1

Bartell SM, Lefebvre G, Kaminski G, Carreau M, Campbell KR (1999) An ecosystem model for assessing ecological risks in Quebec rivers, lakes, and reservoirs. Ecol Model 124:43–67

Bejda AJ, Phelan BA, Studholme AL (1992) The effect of dissolved oxygen on the growth of young-of-the-year winter flounder, *Pseudopleuronectes americanus*. Environ Biol Fishes 34:321

Breitburg D (2002) Effects of hypoxia, and the balance between hypoxia and enrichment, on coastal fishes and fisheries. Estuaries 25:767–781

Breitburg D, Craig J, Fulford R, Rose K, Boynton W, Brady D, Ciotti BJ, Diaz R, Friedland K, Hagy J III (2009) Nutrient enrichment and fisheries exploitation: interactive effects on estuarine living resources and their management. Hydrobiologia 629:31–47

Breitburg DL, Adamack A, Rose KA, Kolesar SE, Decker B, Purcell JE, Keister JE, Cowan JH (2003) The pattern and influence of low dissolved oxygen in the Patuxent River, a seasonally hypoxic estuary. Estuaries 26:280–297

Breitburg DL, Rose KA, Cowan J (1999) Linking water quality to larval survival: predation mortality of fish larvae in an oxygen-stratified water column. Mar Ecol Prog Ser 178:39–54

Chabot D, Dutil JÄ (1999) Reduced growth of Atlantic cod in non-lethal hypoxic conditions. J Fish Biol 55:472–491

Chesney EJ, Baltz DM, Thomas RG (2000) Louisiana estuarine and coastal fisheries and habitats: perspectives from a fish's eye view. Ecol Appl 10:350–366

Christensen V, Coll M, Steenbeek J, Buszowski J, Chagaris D, Walters CJ (2014a). Representing variable habitat quality in a spatial food web model. Ecosystems 17:1397–1412

Christensen V, de la Puente S, Sueiro JC, Steenbeek J, Majluf P (2014b) Valuing seafood: the Peruvian fisheries sector. Mar Policy 44:302–311

Christensen V, Pauly D (1992) ECOPATH II—a software for balancing steady-state ecosystem models and calculating network characteristics. Ecol Model 61:169–185

Christensen V, Walters CJ (2004) Ecopath with Ecosim: methods, capabilities and limitations. Ecol Model 172:109–139

Christensen V, Walters CJ (2011) Progress in the use of ecosystem modeling for fisheries management. In: Christensen V, Maclean JL (eds) Ecosystem approaches to fisheries: a global perspective. Cambridge University Press, Cambridge, pp 189–205. Cochrane J, Kelly F (1986) Low-frequency circulation on the Texas-Louisiana continental shelf. J Geophys Res Oceans (1978–2012) 91: 10645–10659

Craig JK (2012) Aggregation on the edge: effects of hypoxia avoidance on the spatial distribution of brown shrimp and demersal fishes in the Northern Gulf of Mexico. Mar Ecol Prog Ser 445:75–95

Crocker CE, Cech JJ Jr (1997) Effects of environmental hypoxia on oxygen consumption rate and swimming activity in juvenile white sturgeon, *Acipenser transmontanus*, in relation to temperature and life intervals. Environ Biol Fishes 50:383–389

Crout RL (1983) Wind-driven, near-bottom currents over the west Louisiana inner shelf. Louisiana State University Baton Rouge

Davis JC (1975) Minimal dissolved oxygen requirements of aquatic life with emphasis on Canadian species: a review. J Fish Board Can 32:2295–2332

De Mutsert K, Cowan JH Jr, Walters CJ (2012) Using Ecopath with Ecosim to explore nekton community response to freshwater diversion into a Louisiana estuary. Mar Coast Fish 4:104–116

De Mutsert K, Steenbeek J, Lewis KA, Buszowski J, Cowan JH Jr, Christensen V (2015a) Exploring effects of hypoxia on fish and fisheriges in the northern Gulf of Mexico using a dynamic spatially explicit ecosystem model. Ecol Model. doi:10.1016/j.ecolmodel.2015.10.013

De Mutsert K, Lewis KA, Buszowski J, Steenbeek J, Milroy S (2015b) 2017 coastal master plan: ecosystem outcomes, community model description (4.5). Final report. (99 pp). Baton Rouge, Louisiana: Coastal Protection and Restoration Authority

De Mutsert K, Lewis KA, Buszowski J, Steenbeek J, Milroy S (2016) Delta management fish and shellfish ecosystem model: ecopath with ecosim plus ecospace (ewe) model description. Final report. (79 pp). Baton Rouge, Louisiana: Coastal Protection and Restoration Authority

Diaz RJ, Rosenberg R (1995) Marine benthic hypoxia: a review of its ecological effects and the behavioural responses of benthic macrofauna. Oceanogr Mar Biol Annu Rev 33:203–245

Fennel K, Hetland R, Feng Y, DiMarco S (2011) A coupled physical-biological model of the northern Gulf of Mexico shelf: model description, validation and analysis of phytoplankton variability. Biogeosci Discuss 8

Fennel K, Hu J, Laurent A, Marta-Almeida M, Hetland R (2013) Sensitivity of hypoxia predictions for the Northern Gulf of Mexico to sediment oxygen consumption and model nesting. J Geophys Res Oceans 118:990–1002. doi:10.1002/jgrc.20077

Fulford RS, Breitburg DL, Luckenbach M, Newell RI (2010) Evaluating ecosystem response to oyster restoration and nutrient load reduction with a multispecies bioenergetics model. Ecol Appl 20:915–934

Graham W (2001) Numerical increases and distributional shifts of *Chrysaora quinquecirrha* (Desor) and *Aurelia aurita* (Linne) (Cnidaria: Scyphozoa) in the northern Gulf of Mexico. In: Jellyfish Blooms: Ecological and Societal Importance. Springer, pp 97–111

Grove M, Breitburg DL (2005) Growth and reproduction of gelatinous zooplankton exposed to low dissolved oxygen. Mar Ecol Prog Ser 301:185–198

Gunter G (1963) The fertile fisheries crescent. J Mississippi Acad Sci 9:286–290

Hanson RB (1982) Influence of the Mississippi River on the spatial distribution of micrheterotrophic activity in the Gulf of Mexico. Contrib Mar Sci 23:181–198

Ho CL, Barrett BB (1977) Distribution of nutrients in Louisiana's coastal waters influenced by the Mississippi River. Estuar Coast Mar Sci 5:173–195

Kaplan IC, Horne PJ, Levin PS (2012) Screening California current fishery management scenarios using the Atlantis end-to-end ecosystem model. Prog Oceanogr 102:5–18
Keister JE, Houde ED, Breitburg DL (2000) Effects of bottom-layer hypoxia on abundances and depth distributions of organisms in Patuxent river Chesapeake bay. Mar Ecol Progress ser 205:43–59
Laurent A, Fennel K (2014) Simulated reduction of hypoxia in the northern Gulf of Mexico due to phosphorus limitation, Elementa 2:000022, doi:10.12952/journal.elementa.000022
Laurent A, Fennel K, Hu J, Hetland R (2012) Simulating the effects of phosphorus limitation in the Mississippi and Atchafalaya river plumes. Biogeosci 9:4707–4723. doi:10.5194/bg-9-4707-2012
Livingston SRJ (2002) Eutrophication processes in coastal systems: origin and succesion of plankton blooms and effetcs on secondary production in Gulf Coast Estuaries. CRC Press, Boca Raton, Florida
Loos SA (2011) Marxan analyses and prioritization of the central interior ecoregional assessment. J Ecosyst Manag 12
Miller D, Poucher S, Coiro L (2002) Determination of lethal dissolved oxygen levels for selected marine and estuarine fishes, crustaceans, and a bivalve. Mar Biol 140:287–296
Niiranen S, Stipa T, Pääkkönen JP, Norkko AK, Kaitala S (2008) Modelled impact of changing nutrient conditions on the Baltic Sea food web. In: Conference Proceedings of the 2008 ICES ASC
Nixon SW, Buckley BA (2002) "A strikingly rich zone"—nutrient enrichment and secondary production in coastal marine ecosystems. Estuaries 25:782–796
O'Connor T, Whitall D (2007) Linking hypoxia to shrimp catch in the northern Gulf of Mexico. Mar Pollut Bull 54:460–463
Pichavant K, Person-Le-Ruyet J, Bayon NL, Severe A, Roux AL, Boeuf G (2001) Comparative effects of long-term hypoxia on growth, feeding and oxygen consumption in juvenile turbot and European sea bass. J Fish Biol 59:875–883
Pihl L, Baden SP, Diaz RJ (1991) Effects of periodic hypoxia on distribution of demersal fish and crustaceans. Mar Biol 112:349–361
Polovina JJ (1984) Model of a coral reef ecosystem. Coral Reefs 3:1–11
Rabalais NN, Turner RE (2001) Coastal hypoxia: consequences for living resources and ecosystems. Washington, DC. 2001: American Geophysical Union
Rabalais NN, Turner RE, Dortch Q, Justic D, Bierman VJ Jr, Wiseman WJ Jr (2002) Nutrient-enhanced productivity in the northern Gulf of Mexico: past, present and future. Hydrobiologia 475(476):39–63
Rabalais NN, Turner RE, Wiseman WJ (2001) Hypoxia in the Gulf of Mexico. J Environ Qual 30:320–329
Rose KA (2000) Why are quantitative relationships between environmental quality and fish populations so elusive? Ecol Appl 10:367–385
Rose KA, Adamack AT, Murphy CA, Sable SE, Kolesar SE, Craig JK, Breitburg DL, Thomas P, Brouwer MH, Cerco CF, Diamond S (2009) Does hypoxia have population-level effects on coastal fish? musings from the virtual world. J Exp Mar Biol Ecol 381(Supplement):S188–S203
Schurman H, Steffensen JF (1992) Lethal oxygen levels at different temperatures and the preferred temperature during hypoxia of the Atlantic cod *Gadus morhua* L. J Fish Biol 41:927–934
Secor DH, Gunderson TE (1998) Effects of hypoxia and temperature on survival, growth, and respiration of juvenile Atlantic sturgeon, *Acipenser oxyrinchis*. Fish Bull, 603–613
Steenbeek J, Coll M, Gurney L, Melin F, Hoepffner N, Buszowski J, Christensen V (2013) Bridging the gap between ecosystem modeling tools and geographic information systems: driving a food web model with external spatial-temporal data. Ecol Model 263:139–151
Tallqvist M, Sandberg-Kilpi E, Bonsdorff E (1999) Juvenile flounder, *Platichthys flesus* (L.), under hypoxia: effects on tolerance, ventilation rate and predation efficiency. J Exp Mar Biol Ecol 242:75–93

Taylor JC, Miller JM (2001) Physiological performance of juvenile southern flounder, *Paralichthys lethostigma* (Jordan and Gilbert, 1884), in chronic and episodic hypoxia. J Exp Mar Biol Ecol 258:195–214

Turner RE (2001) Some effects of eutrophication on pelagic and demersal food webs. In: Rabalais NN, Turner RE (eds) Coastal hypoxia: consequences for living resources and ecosystems. Washington, DC: American Geophysical Union

USEPA (2001) Ambient aquatic life water quality criteria for dissolved oxygen (saltwater); cape cod to cape hatteras. Technical report EPA-882-R-00-012. USEPA, Washington, DC

Walker ND, Rabalais NN (2006) Relationships among satellite chlorophyll a, river inputs, and hypoxia on the Louisiana continental shelf, Gulf of Mexico. Estuaries Coasts 29:1081–1093

Walters C, Christensen V, Pauly D (1997) Structuring dynamic models of exploited ecosystems from trophic mass-balance assessments. Rev Fish Biol Fish 7:139–172

Walters C, Martell SJD, Christensen V, Mahmoudi B (2008) An Ecosim model for exploring Gulf of Mexico ecosystem management options: implications of including multistanza life history models for policy predictions. Bull Mar Sci 83:251–271

Walters C, Pauly D, Christensen V (1999) Ecospace: prediction of mesoscale spatial patterns in trophic relationships of exploited ecosystems, with emphasis on the impacts of marine protected areas. Ecosystems 2:539–554

Wannamaker CM, Rice JA (2000) Effects of hypoxia on movements and behavior of selected estuarine organisms from the southeastern United States. J Exp Mar Biol Ecol 249:145–163

Wiseman Jr WJ, Turner R, Justic D, Rabalais N (2004) Hypoxia and the physics of the Louisiana coastal current. In: Dying and Dead Seas Climatic Versus Anthropic Causes. Springer, pp 359–372

Wiseman WJ, Turner RE, Kelly FJ, Rouse LJ, Shaw RF (1986) Analysis of biological and chemical associations along a turbid coastal front during winter 1982. Contrib Mar Sci 29:141–151

Chapter 15
Numerical Modeling of Hypoxia and Its Effects: Synthesis and Going Forward

Kenneth A. Rose, Dubravko Justic, Katja Fennel and Robert D. Hetland

Abstract Numerical models can provide the needed information for understanding hypoxia and ensuring effective management, and this book provides a snapshot of representative modeling analyses of hypoxia and its effects. In this chapter, we used the modeling and analyses across the other 14 chapters to illustrate 8 themes that relate to the general strengths, uncertainties, and future areas of focus in order for modeling of hypoxia and its effects to continue to advance. These themes are role of physics; complexity of the dissolved oxygen (DO) models; oxygen minimum zones (OMZs) and shallow coastal systems; observations; vertical dimension; short-term forecasting; possible futures; and ecological effects of hypoxia. Modeling the dynamics and causes of hypoxia has greatly progressed in recent decades, and modern models routinely simulate seasonal dynamics over 0.1–1 km scales. Despite these advances, prevailing model limitations include uncertain specification of boundary conditions and forcing functions, challenges in representing the sediment-water exchange and multiple nutrient limitation, and the limited availability of observations for multiple contrasting years for model calibration and validation. A major challenge remains to effectively link the water quality processes to upper trophic levels. A variety of approaches are illustrated in this book and show that quantifying this linkage is still in the formative stages. There will be

K.A. Rose (✉) · D. Justic
Department of Oceanography and Coastal Sciences, Louisiana State University, Energy, Coast, and Environment Building, Baton Rouge, LA, USA
e-mail: karose@lsu.edu

D. Justic
e-mail: djusti1@lsu.edu

K. Fennel
Department of Oceanography, Dalhousie University, Halifax, Canada
e-mail: Katja.Fennel@dal.ca

R.D. Hetland
Department of Oceanography, Texas A&M University, College Station, TX, USA
e-mail: hetland@tamu.edu

D. Justic et al. (eds.), *Modeling Coastal Hypoxia*,
DOI 10.1007/978-3-319-54571-4_15

increasing demands for predicting the ecological responses to hypoxia in order to quantify the ecological benefits and costs of management actions and to express the simulated effects of coastal management and climate change in terms of direct relevance to managers and the public.

Keywords Hypoxia • Nutrients • Simulation • Modeling • Ecological effects • Food webs • Climate change • Fisheries • Management

15.1 Introduction

Hypoxia is increasing in coastal and oceanic waters worldwide (Stramma et al. 2008; Zhang et al. 2010). Increasing hypoxia is manifested in more locations showing low dissolved oxygen (DO) conditions (e.g., Chan et al. 2008) and hypoxia becoming more severe in terms of its magnitude (e.g., areal extent) and duration (Diaz and Rosenberg 2008). Modeling is an important tool for understanding the timing and spatial dynamics of hypoxia, and how hypoxia affects key biological populations and the food web. Field observations provide the empirical basis for model development and for assessing the skill of physical and biological simulations, but using field data to directly quantify the spatial and temporal dynamics of hypoxia and to decipher the contribution of the multiple physical and biological factors that contribute to hypoxia development is difficult. Once hypoxic conditions are documented (either through field data or model predictions), using field data to quantify the effects on individual fish and other upper trophic level organisms is possible, but critical issues prevent the direct extrapolation of the individual effects to the population level (Rose et al. 2009). Restoration actions are often costly and therefore knowing the causes of hypoxia (e.g., nitrogen versus phosphorus; riverine versus shelf waters), and quantifying the ecological benefits of management actions designed to reduce hypoxia (e.g., increased fish populations) helps design effective management actions.

Numerical models can provide the needed information for understanding hypoxia and ensuring effective management via predictive simulations and diagnostic model-based experiments. This book provides a snapshot of representative modeling analyses of hypoxia and its effects. Collectively, these models allow us to assess where we are and discuss where, perhaps, we should be going. In this chapter, we discuss 8 topics or themes that span across multiple chapters. These themes are as follows: (1) role of physics; (2) complexity of the DO models; (3) oxygen minimum zones (OMZs) and shallow coastal systems; (4) observations; (5) vertical dimension; (6) short-term forecasting; (7) possible futures; and (8) ecological effects of hypoxia. We use the models presented in the various chapters, supplemented with other examples from the literature, to illustrate each of these themes (Table 15.1) and to offer our prognosis for future efforts.

Table 15.1 Major features of the models used in the chapters of this book. The themes that referred to each chapter are shown

Chapters	System	Physics	Water quality/ecology	Selected analyses	Theme
Chapter 1 Allahdadi and Li	Louisiana shelf	FVCOM-3-D		Effects of solar heating on stratification during June 2009	1, 4
Chapter 2 Bravo et al.	Green Bay	3-D model nested in POM		Multi-year averages of the effects of wind, circulation, and boundary exchanges on thermal regime and stratification	1, 4
Chapter 3 Hetland and Zhang	Louisiana—Texas shelf	ROMS-3-D		Fraction of shelf waters from each river for 2008	1
Chapter 4 Brush and Nixon	Greenwich Bay	7 boxes, each divided into 2 layers	C, N, and P in phytoplankton and macroalgae, DIN, DIP, DO BOD	Surface chlorophyll, DIN, DIP, and bottom DO compared to data for May 1996–May 1997	2, 3, 4
Chapter 5 Testa et al.	Chesapeake Bay	ROMS-3-D or 17 boxes	4 or 23 compartments	Box with 23, ROMS with 23, ROMS with 4, all for 1996–2005, compared for spatial and temporal dynamics of DO	2, 4
Chapter 6 Wiggert et al.	Chesapeake Bay	ROMS-3-D	Nitrogen forms, P, chlorophyll, zooplankton, DO, benthic	Detailed comparison (horizontal and vertical) for calibration and sensitivity analysis by station using goodness-of-fit measures to 1999	2, 4, 5, 6
Chapter 7 Laurent and Fennel	Mississippi River plume on shelf (plus review of other systems)	ROMS-3-D	Nitrogen forms, DIP, phytoplankton, zooplankton, detritus	Conceptual model on role of P limitation in affecting hypoxia dynamics, illustrated with Louisiana shelf simulations	2
Chapter 8	Gulf of Mexico	NCOM-3-D	N, P, Si forms, POM, DOM,	Response of stratification,	4, 7

(continued)

Table 15.1 (continued)

Chapters	System	Physics	Water quality/ecology	Selected analyses	Theme
Lehrter et al.			phytoplankton, zooplankton (micro, macro), DO	chlorophyll, and DO to climate change imposed on 2006	
Chapter 9 Koch et al.	Upwelling off coastal Oregon	ROMS-3-D	Nitrogen forms, phytoplankton, zooplankton, detritus, DO	Sensitivity analysis and DO budgets to determine role of physical factors governing hypoxia in 2002 and 2006	1, 3, 5
Chapter 10 LaBone et al.	Louisiana shelf	FVCOM	Version of WASP and particle tracking	Effects of different movement behaviors on the percent of individuals exposed to DO values (0–2, 2–3 mg/L, etc.) over time for 7 days during summer of 2002	4, 5, 8
Chapter 11 Kolesar et al.	Mesohaline portion of Patuxent River	3-layer box without water exchange	Zooplankton, fish larvae, ctenophore food web	Effects of hypoxia on ctenophore competition versus predation on fish larvae = l growth and survival during May to September	5, 8
Chapter 12 Adamack et al.	Chesapeake Bay	Individuals imbedded into 3-D model	Population dynamics (growth, mortality, movement) of bay anchovy	Hypoxia and food effects on bay anchovy abundance under changed nutrient loadings over 10 years	8
Chapter 13 Rose et al.	Louisiana shelf	DO and chlorophyll from 3-D models	Recruitment effect from 2-D individual-based population model	Probability of detecting hypoxia effect with 5, 10, and 25 years of simulated monitoring of croaker	4, 8
Chapter 14 de Mutsert et al.	Gulf of Mexico	DO and Chlorophyll from a 3-D water quality model	Predator and competitive interactions for a complete food web—2-D	Upper food web responses (% change in averaged biomasses) to increased nutrient loadings versus worsen hypoxia for 1990–2007	4, 8

15.2 Emerging Themes

15.2.1 Theme 1: Role of Physics

Hypoxia models that include hydrodynamics have progressed over the past decade, and regional models are now capable of simulating physics on the scale of a few hours and kilometers. Global and regional physical models have been trending toward consideration of finer and finer spatial resolution (Thomas et al. 2008; Kirtman et al. 2012; Holt et al. 2014). Regional and subregional models now can simulate velocities, temperature, and salinity with sufficient detail to capture mesoscale (10's of kilometers) and some submesoscale (kilometer) phenomena (e.g., Hetland 2017) and provide a sound basis for including variables related to biogeochemical processes and water quality (including DO). All of the hydrodynamic models used in this book resolve horizontal scales on the order of a few kilometers.

Further resolving the physics may improve model skill in some cases and help improve simulation of DO concentrations. However, mesoscale and submesoscale variability, when resolved in high-resolution models, can create stochastic variability (sensu Lorenz's butterfly effect) that leads to uncertainty in simulated distributions of physical properties (Marta-Almeida et al. 2013) and DO (Mattern et al. 2013). Submesoscale phenomena are increasingly being considered for their effects on phytoplankton and productivity (Levy et al. 2012; Mahadevan 2016). We expect that the higher resolution simulation of physical dynamics will allow for more detail in spatiotemporal variation of hypoxia, but will not result in dramatic improvement in model skill for DO at the scale of 0.1–1 km. The physics simulated in the present models appear sufficient to capture the major effects on hypoxia formation and dynamics at the kilometer scale.

Boundary conditions and forcings for the hydrodynamics (and water quality) are critical for accurately simulating the dynamics of hypoxia and understanding the causes of hypoxia formation (e.g., Monteiro et al. 2011; Fennel et al. 2013), and uncertainties about specification of boundary conditions and forcings remain an issue. The hydrodynamics of both OMZs and coastal systems can be sensitive to the fluxes at their boundaries with the adjacent open ocean (Blumberg and Kantha 1985; Koch et al.—Chap. 9), and coastal systems are often heavily influenced by riverine inputs at their landward boundary (e.g., Hetland and DiMarco 2012). Allahdadi and Li (Chap. 1) used a 1-month simulation of rising temperatures in a 3-D hydrodynamic model (FVCOM) to examine how solar heating would lead to stronger stratification on the Louisiana shelf; simulated DO declined during the same time period.

Bravo et al. (Chap. 2) used a model of Green Bay nested in a POMS model of Lake Michigan and 5-year averaged results for the summertime to compare the relative importance of heat flux, wind, and circulation (including at the Lake Michigan boundary) on stratification. They show that the wind shift in direction, typically in June, sets up a stable period during which conditions are conducive to formation of hypoxia until mixing in September.

Hetland and Zhang (Chap. 3) make innovative use of a dye (tracer) simulation study (summer 2008) to quantify the contribution from two rivers (Mississippi and Atchafalaya) to Louisiana shelf waters in the same location of historical hypoxia. The two rivers contribute water with different constituent characteristics (e.g., inorganic nitrogen concentration), and their modeling results show that the relative contribution of the two rivers varies temporally by month and spatially across the hypoxic region.

Koch et al. (Chap. 9) used a ROMS-NAPZD (nitrate, ammonium, phytoplankton, zooplankton, and detritus) model for the Oregon shelf and show the importance of accurate boundary conditions. They discuss how the commonly used approach of using larger-scale models to generate boundary conditions for their model was especially challenging because of the resolution needed by their local shelf model.

The ever-increasing computing power will allow for multiple simulations and use of model ensembles to better capture within-year and interannual variability as well as inherent uncertainty in the simulated physics and hypoxia. Increasing computing power will also help with conducting Monte Carlo style uncertainty analysis of the physics and associated water quality via input variation and use of alternative process formulations. To date, the limited formal quantification of uncertainty with hypoxia models can prevent the use of modeling results by the management community because interpretation of results across scenarios is hindered by not including realistic variability around model predictions. Numerically efficient methods for quantifying uncertainty in 3-D hypoxia models (e.g., Mattern et al. 2013) are needed.

15.2.2 Theme 2: Complexity of the DO Models

Determining the appropriate complexity of the water quality models to generate DO dynamics remains an ongoing issue in water quality and nutrient–phytoplankton–zooplankton (NPZ) model development (Denman 2003; Litchman et al. 2006; Friedrichs et al. 2007). While there are differences among the models reported here, there seems to be a consensus about the general form of the models needed to generate sufficiently realistic DO dynamics. In line with previous simple DO models (e.g., Hetland and DeMarco 2008; Scully 2013; Li et al. 2015; Yu et al. 2015; Fennel et al. 2016), Brush and Nixon (Chap. 4) offer a relatively simple DO model that they term "hybrid empirical-mechanistic." They use this term to reflect the fact that the model tracks relatively few state variables and uses relationships roughly fitted to data derived from across system studies to represent some processes. In this sense, they sacrifice mechanistic representation in favor of a simpler (and easier to apply) statistically based relationship.

Testa et al. (Chap. 5) directly address the complexity issue by comparing the performance of three models designed to simulate DO in Chesapeake Bay. The models were a 17-box (9 surface and 8 bottom) system with a detailed water quality model (23 compartments), a detailed 3-D hydrodynamic model (ROMS) coupled to a simple DO submodel, and ROMS with the 23-compartment model. All models simulated 1996–2005, and Testa et al. (Chap. 5) offer insights into the factors controlling hypoxia by each of the models. They also highlight how the box model version reproduces seasonal and regional patterns in DO but is limited in resolving lateral dynamics, how the ROMS with the simple DO model is useful for examining wind and tidal mixing and freshwater input effects but is limited in its ability to simulate interannual variation in hypoxia, and how the ROMS with the complex DO model dynamics is capable of addressing these shortcomings but requires significant computing and the most observations to develop sufficient model confidence. Their analysis emphasizes that the complexity of the modeling must be tailored to the specific questions being asked: regional patterns in biogeochemistry (box model with simple DO model), climate effects on DO (ROMS with simple DO model) and interannual variability, biological responses to nutrient loadings, or fine-scale physical effects (ROMS with complex DO model).

Finally, Wiggert et al. (Chap. 6) used sensitivity analyses with a 3-D ROMS-NPZD (NPZ plus detritus) model applied to 1999 data for the Chesapeake Bay to examine how alternative parameter values emphasize and understate key processes that affect model skill. For example, they increased the sinking velocity of large detritus that promoted the flux of organic matter to the benthos, and reduced the half-saturation coefficient that affected denitrification and thus amplified DO consumption within the water column. They used the agreement among the predictions of the alternative versions with data to demonstrate how improvements in the degree of correspondence with one variable (e.g., hypoxia persistence) by certain parameter changes involve some loss of agreement with other variables (e.g., poorer fit for dissolved inorganic nitrogen).

While the complexity of the model being dictated by the questions is a well-known best practice (Wainwright and Mulligan 2005; Rose et al. 2015a), it is sometimes not given enough consideration because there are existing codes and models available that make their use convenient. The reasoning behind the final model structure that is used in an application should be clearly explained and documented. This is needed for transparency (why models differ within and across systems) and ultimately will affect the credibility of the modeling results.

The models also highlight two long-standing critical uncertainties in modeling hypoxia: multiple nutrient limitation (Howarth 1988; Flynn 2003; Moore et al. 2013) and processes at the sediment-water interface (Fennel et al. 2013; Testa et al. 2013). Laurent and Fennel (Chap. 7) directly address the role of phosphorus limitation in areas of coastal hypoxia. They review several well-studied systems and derive a conceptual model that views phosphorus limitation effects on hypoxia as spatial and timing shifts by categorizing ecosystems as dominated by flow-through versus open-dispersive physical processes. By use of a 3-D ROMS coupled to a NPZD-type model of the Louisiana shelf, which is dominated by river inputs, they use simulations for multiple years (highlighted with 2004) based on nitrogen-only limitation and with both nitrogen limitation and phosphorus limitation to demonstrate how consideration of phosphorus limitation resulted in a reduced and westward shift of the hypoxic zone.

The sediment-water exchanges often play a large role in affecting nutrient and DO dynamics (Middleburg and Levin 2009; Lehrter et al. 2012), and Voss et al. (2013) documented the importance of the sediment-water exchanges for the nitrogen budget (and thus DO dynamics) in both shelf and coastal systems. Kemp et al. (1992) and Heip et al. (1995) have emphasized that sediments have strong impacts on the water column in shallow coastal ecosystems. Fennel et al. (2013) used a similar model setup as Laurent and Fennel (Chap. 7) and found that projections of summertime hypoxia for the Gulf of Mexico were very sensitive to the formulation used for sediment oxygen consumption. Most of the chapters that focus on water quality and DO at least mention the uncertainty associated with representing the sediment-water interface and exchange (e.g., Brush and Nixon—Chap. 4; Testa et al.—Chap. 5; Laurent and Fennel—Chap. 7).

Two emerging areas in plankton ecology that were not represented in the DO models but may be worth considering are more refined representation of the microbial loop (Treseder et al. 2012) and the idea of internal stoichiometry of phytoplankton (Glibert et al. 2013; Bonachela et al. 2015). The field data collection for microbial organisms and processes, including those that affect nutrient biogeochemistry, is rapidly advancing, and Reed et al. (2014) provide a strategy for incorporating these types of data into water quality and plankton models. For example, Ayata et al. (2013) incorporated different formulations for internal stoichiometry into a 1-D vertical NPZD model. How improving the representation of microbial processes and nutrient dynamics of phytoplankton will help the simulation of DO is not clear for present-day conditions in many systems, but may need to be considered under management scenarios of reduced nutrient loadings (coastal systems) and for climate change (coastal and OMZs).

15.2.3 Theme 3: Ocean and Shallow Ecosystems

Koch et al. (Chap. 9) discuss how the past emphasis on hypoxia in coastal systems has resulted in a focus on the role played by nutrient inputs via large rivers. The hypoxia of shelf-oriented oxygen minimum zones (OMZs) is gaining attention (Stramma et al. 2008; Altieri and Gedan 2015; Levin and Breitburg 2015). Koch et al. (Chap. 9) used a ROMS-NAPZD model (NPZD plus ammonium) for the Oregon shelf, where hypoxia is seasonal and linked to upwelling favorable winds. They simulated the 2002–2006 period focusing on vertical profiles and report the results of a sensitivity analysis that varied initial and boundary conditions of nitrate and used budgets to assess the relative importance of physical versus biological drivers on DO dynamics.

In addition to OMZs and major coastal hypoxic zones, there is increasing attention to modeling hypoxia in small shallow coastal systems that show rapid changes in DO, often influenced by local conditions (Shen et al. 2008; Soetaert and Middleburg 2009; Tyler et al. 2009). Brush and Nixon (Chap. 4) make a strong case for modeling eutrophication and hypoxia in shallow coastal systems. While the major coastal hypoxic zones such as the Chesapeake Bay and the Louisiana shelf garner a lot of attention, the shallow subestuaries often experience very high rates of nutrient loading that can be mediated by local management decisions. Given the limited site-specific data for these shallow systems, Brush and Nixon (Chap. 4) present a reduced complexity model developed to allow for easy application and portability among different systems. They then applied the model to Greenwich Bay (mean depth of 2.6 m), a subestuary of Narragansett Bay, and performed a one-year simulation that matched monitoring data. The bay was divided into 7 spatial elements and two vertical layers, and water and material exchanges were based on salinity distributions and freshwater inputs rather than on hydrodynamics. The model was successfully calibrated to surface chlorophyll-a biomass, dissolved inorganic nitrogen (DIN), and dissolved inorganic phosphorus (DIP) by spatial segment and to bottom layer DO by spatial segment. They also compared simulated and observed process rates, such as water column production and respiration.

These models represent a broadening of the types of systems being investigated and offer both new opportunities for model development and testing and for using models for management. OMZs bring in the dynamics of the middle and deeper water column on the continental shelf, and the shallow systems provide a test bed for diurnal DO cycles and the role of sediments in shallow systems. Greater cooperation between the seemingly distinct OMZs and coastal hypoxia research is needed (Levin and Breitburg 2015), and this also applies to the modeling. While the major estuarine and coastal systems (e.g., Chesapeake Bay and Louisiana shelf) involve large systems and major monetary investments for restoration, the OMZs are an indicator of ocean conditions (Stramma et al. 2008; Paulmier and Ruiz-Pino 2009). There is also a growing number of small shallow systems experiencing hypoxia that, when combined, can add up to significant cumulative environmental impacts. While DO models can be implemented relatively easily in the shallow systems, a limiting factor is the availability of data for model configuration (e.g., bathymetry), calibration, and validation.

15.2.4 Theme 4: Observations

Maintaining long-term monitoring efforts for water quality and DO remains a challenge. Several models used observations from station-based monitoring for calibration and validation. Allahdadi and Li (Chap. 1) make extensive use of a set of monitoring stations (WAVCIS) located on the Louisiana shelf to derive boundary conditions (e.g., wind data) and assess model skill (currents, SST). Bravo et al. (Chap. 2) illustrate how historical observations, strategically augmented with new measurements, can provide a sound basis for model evaluation. Testa et al. (Chap. 5) provide an illustration of the value to modeling of the long-term monitoring data available for the Chesapeake Bay; they used the data to generate Taylor diagrams and examined interannual variation to permit skill assessment of 3 models of differing complexity. Wiggert et al. (Chap. 6) similarly make use of the extensive monitoring data for the Chesapeake Bay by focusing on monthly measurements (salinity, chlorophyll, and nutrients) from a series of stations sampled for one year (1999) that represented "average" conditions. They used an index of agreement that included both vertical gradients and 27 stations in the summary index of the scaled differences between predicted and observed values. Lehrter et al. (Chap. 8) in their analyses of the Gulf of Mexico using a 3-D hydrodynamic NPZ model assess model skill based on 3 cruises (April, June, and September) at a grid of stations for 2006. They computed bias, RMSE, and model efficiency metrics for temperature, nitrate, chlorophyll, salinity (pycnocline), and DO on a station basis and also with the station observations aggregated into subregions (median for surface and bottom layers). They also compared model results to observations for two stations in detail and used the LUMCON-based mid-July estimate of hypoxia to assess total area and its spatial distribution. Some models made use of intensive monitoring studies that were of 1- to 2-year duration (e.g., Brush and Nixon, Chap. 4—1996–1997, Bravo et al., Chap. 2—1989, Lehrter et al., Chap. 8—2006). These clearly demonstrate how observations inform the development and validation of the models, which is required to establish model credibility for management use.

Whereas water quality and DO monitoring data are well suited for numerical model evaluation because the purpose of monitoring and modeling is similar, the monitoring of fish and shellfish is done for a variety of other reasons than detecting hypoxia effects. Long-term fish monitoring is typically done to make tactical management decisions (e.g., opening an area for harvest and consumption advisories) or for developing indices of abundance used for fisheries stock assessment and management. The Ecospace application to the Louisiana shelf (de Mutsert et al. Chap. 14) used fish biomass time series derived from long-term fishery-independent monitoring for model calibration. They showed improved model fit to the data when hypoxia effects were included. However, in general, such time series reflect the effects of many factors and so it is difficult to use the data to calibrate or validate a model specifically to confirm (cause-and-effect) simulated hypoxia effects (Chesney and Baltz 2001). Rose et al. (Chap. 13) discuss difficulties in using monitoring data for fish to detect hypoxia effects at the level of recruitment or higher. They used the simulated effect of hypoxia on croaker recruitment in the Gulf of Mexico to establish a realistic "effect" of a 20% reduction under historical hypoxia conditions. Using long-term field data on croaker, the interannual variability in recruitment was estimated and a Monte Carlo simulation used to determine the likelihood of detecting the hypoxia effect (knowing it is present) with sampling (5, 10, and 25 years) that included the variability estimated from the field data. The probability of detecting the known hypoxic effect with reasonable sampling and realistic interannual variability in recruitment is very low. Thus, long-term monitoring of water quality and DO must be maintained, and we need to develop creative ways to use the fish and upper trophic level data, designed and collected for other purposes, for model development and testing. The Gulf of Mexico Hypoxia Watch (https://www.ncddc.noaa.gov/hypoxia/) is an example where data from multiple water quality and fish surveys are merged.

In general, focused studies that are limited to a few years or less generate the observations to investigate the effects of hypoxia on fish. These studies are immensely useful for modeling. For example, Thomas et al. (2015) measured how croaker exposure in the summer affected their fecundity in the fall, and these results were critical to the development of a croaker population dynamic model (Rose et al. in review) that projected the population-level responses of croaker to hypoxia in the Gulf of Mexico. A several year intensive study allowed clear documentation of the response of zooplankton abundance and spatial distribution to hypoxia (Roman et al. 2012). Such studies could be even more helpful to modelers if they were repeated for multiple, contrasting years and their quantified effects were scalable beyond the study area to an area relevant to the population and food web levels.

A major gap is the lack of observations on the movement patterns of fish and other biota. As the modeling moves toward linking water quality to the upper trophic levels, the representation of movement of individuals in the spatially and temporally dynamic environment generated by the hydrodynamics-water quality models will become critical. A major pathway of hypoxia effects is determined by the effectiveness of the avoidance movement of individuals (determines direct effects via exposure), and where they get displaced to and what environmental and ecological conditions they experience (indirect effects) in the new locations. While the data on animal movement in general are constantly improving (McClintock et al. 2013; Bestley et al. 2016; Sippel et al. 2015), LaBone et al. (Chap. 10) reports on the difficulties in trying to use individual movement data presently available for testing model predictions of fish avoidance of hypoxia.

15.2.5 Theme 5: Vertical Dimension

Simulating hypoxia and its effects on biota requires accuracy in both the horizontal and vertical dimensions. This is needed to generate reliable estimates of hypoxia extent (area, volume, see Obenour et al. 2013) and also because many organisms show vertical movements (Cohen and Forward 2009; Gutowsky et al. 2013). Avoidance behavior can occur both by horizontal and by vertical movement, with the mix being dependent on the species involved. Wiggert et al. (Chap. 6) make extensive use of the vertical aspects of the monitoring data available for the Chesapeake Bay in their analysis by including the vertically resolved data in their goodness-of-fit measure. Koch et al. (Chap. 9) use a ROMS-NAPZD model for the Oregon shelf, where vertical resolution is critical because of the deep-water location of the hypoxia zone.

In a subsequent analysis, LaBone (2016) extended the 2-D simulation of avoidance of fish within FVCOM for the Gulf of Mexico reported here by LaBone et al. (Chap. 10) by allowing individuals to also move vertically. For the 3-D scenario that allowed for vertical avoidance and the simulated exposure of the individuals to lethal DO concentrations as they moved within the FVCOM grid over 10 days was greatly reduced, while the exposure to sublethal DO concentrations under 3-D remained similar to the 2-D (horizontal) scenario. LaBone (2016) also noted that the model assumes that there are no additional costs (e.g., swimming and predation risk) associated with moving vertically.

Kolesar et al. (Chap. 11) specifically focus on how movement in the vertical dimension creates changes in the overlap between zooplankton, fish larvae, and ctenophores, with larvae and ctenophores both eating zooplankton and ctenophores also eating larvae. They showed with summertime simulations in a 3-layer vertical box model that hypoxia-induced shifts in movement (plus direct effects on growth and mortality) did not alter the relative importance of ctenophore competition versus predation affecting fish larvae, but that lower DO decreased fish larval survival and increased the growth rates of the surviving larvae.

15.2.6 *Theme 6: Short-Term Forecasting*

Wiggert et al. (Chap. 6) discuss the use of their model for short-term forecasting. We do not typically consider now-term forecasting of hypoxia as a pressing issue for management, as it is unclear whether hydrodynamic-based models for hypoxia are accurate enough to be effectively used for short-term forecasts of relatively fine-scale spatial dynamics of biological dynamics that occur over days to weeks. However, Wiggert et al. (Chap. 6) link, at least illustratively, their modeling to harmful algal blooms (HABs) and jellyfish abundance for which near-term forecasts are very useful for beach closings and water quality warnings. They present results from their water quality model imbedded in ROMS, with the coupled models being part of the larger Chesapeake Bay Ecological Prediction System (CBEPS) that can generate now-casts and 3 day forecasts. They illustrate how their output can be post-processed using habitat suitability to identify likely areas of high concentrations of sea nettles, the pathogenic bacterium *Vibrio*, and harmful algal species. Most HAB models use a statistical approach (e.g., Kim et al. 2014), while there is some effort toward using 3-D hydrodynamic-water quality models similar to hypoxia models (Andersen et al. 2015). The occurrence of HABs and hypoxic conditions shares many of the same environmental drivers (e.g., circulation and high nutrients) (O’Neil et al. 2012; Watson et al. 2016), and thus, coupled hydrodynamics-water quality models can play a role, with some modification, for simulating the timing and magnitude of HABS and effectiveness of management actions (see Paerl et al. 2016). HAB species can either be added to the models or, as illustrated by Wiggert et al. (Chap. 6), the models can be used to generate the environmental conditions that are then post-processed for possible habitat “hot spots.”

15.2.7 *Theme 7: Possible Futures*

OMZs and coastal hypoxia are both highly susceptible to climate change (Grantham et al. 2004; Rabalais et al. 2009; Altieri and Gedan 2015), and global climate change is expected to increase hypoxia in both shelf (OMZs) and coastal systems (Voss et al. 2013). OMZs can be used to understand and track biogeochemical cycling in the ocean and as an indicator of ocean health (Paulmier and Ruiz-Pino 2009). Because the management actions for reducing nutrient loadings in coastal systems are costly and require time to implement, and also because biological responses of upper trophic levels may involve delays, projections of how management actions may affect hypoxia cannot be made without knowing both present-day and anticipated future conditions (Justic et al. 2007). Justification for the investment involves ensuring that the benefits will be sufficiently large now and also into the future. Thus, scenarios of possible future conditions are needed and these should reflect both anticipated changes locally (e.g., land use) and regionally (climate change).

Lehrter et al. (Chap. 8) simulated a future scenario with a 3-D hydrodynamics model coupled to a NPZ-type model for 2006 conditions that was designed to reflect global climate change. They imposed on the 2006 simulation a 3 °C warmer air temperature, a 10% increase in river discharge, and adjusted ocean transport at the east, south, and west boundaries based on a broader-scale regional model itself run under the climate change conditions. Climate change was predicted to cause stronger stratification that leads to more hypoxia during the initial onset and generally lower DO concentrations where hypoxia occurred. They note some of the many factors that could be affected by climate change that were not accounted for in their climate change scenario: changes in hydrology, coastal winds, phytoplankton community structure, and the nutrient concentrations in the rivers.

Altieri and Gedan (2015) present a conceptual model of how the climate drivers (cloud cover, winds) can affect physical (e.g., stratification) and biological (e.g., primary productivity) factors that, in turn, would affect hypoxia. Justic et al. (1996) demonstrated how climate change could be incorporated into projections of hypoxia using a 2-box model; very telling is that 20 years later it is not clear that if we repeated their analyses now we could significantly reduce the uncertainties they faced in implementing a climate change scenario. Some progress is illustrated with the use of downscaled regional and GCM models and an ensemble approach for the hypoxia models with the well-studied Baltic Sea (Meier et al. 2011). Uncertainties remain about how to downscale from global and regional models to 100 m scales for use in hydrological and ecological models (e.g., Flint and Flint 2012) and also about the direction of changes in some key variables (e.g., precipitation, see Trenberth 2011). Acidification will also need to be considered in future scenarios because acidification and hypoxia affect each other in shelf and coastal ecosystems (Melzner et al. 2013; Miller et al. 2016; Laurent et al. 2017). Ocean warming will "very likely" lead to further declines in DO; however, in an update to the recent IPCC's Fifth Assessment Report (AR5), it was noted that the uncertainty about present-day conditions caused the assessment to determine it "as likely as it is unlikely" that hypoxic and subhypoxic zones will increase (Howes et al. 2015). Thus, projecting hypoxia into the future remains a challenge.

15.2.8 Theme 8: Ecological Effects of Hypoxia

While there has been some progress in linking the models for simulating physics and hypoxia to the models that simulate the response of the biota, significant gaps still remain. One critical determinant of exposure of biota to low DO is their movement, including their avoidance behavior. LaBone et al. (Chap. 10) used the particle-tracking bookkeeping in FVCOM for the Gulf of Mexico (Justic and Wang 2014) to simulate how individual fish would avoid low DO conditions. They performed 7-day simulations and with a collection of individuals (hundreds) not directly scalable to the population level. They show that exposure to lethal and sublethal DO concentrations is not greatly affected by the movement algorithm used for non-avoidance (default) movement behavior. Expressing the effects of hypoxia as the percent of an arbitrary number of total individuals is helpful for comparing effects on a relative basis but does not lend itself to easy use for management. The Lagrangian approach operating within the same grid as the hydrodynamics-water quality model allows for direct simulation of movement as water quality conditions change on the scales represented in the hydrodynamics (e.g., Ibarra et al. 2014). This approach can be expanded to simulate growth, mortality, and reproduction of individuals for a summer, year, or multiple years and using techniques such as superindividuals that allow scaling to large number populations (e.g., Rose et al. 2015b). Getting to the population or system level (e.g., an estuary, Gulf of Mexico) is critical for management so that long-term effects can be simulated or inferred and the predictions are the same biological scale as the resources are managed.

Two dominant approaches for simulating the system-level responses to hypoxia are illustrated in this book. Adamack et al. (Chap. 12) illustrate the approach of simulating population-level responses by imbedding the growth, mortality, reproduction, and movement of individual bay anchovy (using a superindividual approach) directly into the 3-D physics-water quality model for the Chesapeake Bay. The individual or agent-based approach has become very popular for simulating fish and other organisms (Grimm and Railsback 2013; DeAngelis and Grimm 2014). Several other models in this book also used the individual-based approach (Kolesar et al. Chap. 11, Rose et al. Chap. 13).

Adamack et al. simulated 10 years that matched the historical pattern of dry, normal, and wet years by stringing together years from a pool of hydrodynamics-water quality simulations. They assumed a recruitment of young anchovy each year entered the model grid and showed that changes in nutrient loadings had major effects on the population dynamics whether recruitment was assumed to be high or low each year. To accommodate possible low DO effects on prerecruit stages and increased overlap of juveniles and adults with predators, they determined how much recruitment would need to be reduced by hypoxia (higher egg mortality) or how much more predation on juveniles and adults would be needed to offset the benefits of more food under increased nutrient loadings. The magnitude of the increases needed in prerecruit mortality or more intense predation was deemed within the variability observed in the system and therefore ecologically feasible.

De Mutsert et al. (Chap. 14) demonstrate the alternative approach (i.e., Eulerian) and how the response of the entire upper-level food web can be simulated using a modified version of the popular EwE software, specifically by developing an Ecospace model that allows for a 2-D spatial grid and the inputting of environmental variables that can vary in time and space. A major modification was the specification of a new habitat capacity effect whereby multiple environmental variables (including DO) can be used to reduce the foraging area of the predators.

A major advantage of a Lagrangian approach for examining hypoxia effects is that it is easier to simulate movement, and especially avoidance behavior, and has the capability to allow for the accumulation of exposures of individuals over time. This is more difficult to formulate for the Eulerian approach that underlies Ecospace (biomass is exchanged among spatial cells). How to include avoidance in Eulerian models is challenging, and the history of individuals is not tracked, and thus, exposure is necessarily based on biomass in that location for that time step only. However, the bookkeeping involved with using an individual-based approach for large number of populations (e.g., use of superindividuals) requires some effort and custom coding, especially when the individuals are simulated, as with Adamack et al. (Chap. 12), within the same 3-D grid as the physics model.

More importantly, the individual-based approach is limited to simulating a few species, while the Eulerian approach of Ecospace allows for many species, and thus, the entire food web can be represented. There is a rich history of simulating population dynamics of well-studied fish species that use readily available data on growth, mortality, and reproduction, but then, the food web effects need to be represented implicitly (i.e., forced by changes in parameter values). Rose et al. (2009) suggested that indirect effects could be critical for simulating hypoxia effects on key populations, and representing the entire food web can be critical in some situations to minimize the chances of missing possible complicated indirect effects mediated through food web competition and predator–prey interactions. The food web approach illustrated by Ecospace has a clear advantage over the individual-based, multi-species approach for assessing possible food web-mediated indirect effects. De Mutsert et al. (Chap. 14) show with 1950–2010 simulations (monthly time step) of the Louisiana shelf ecosystem with and without hypoxia and food effects (chlorophyll) included that the positive effects of increased food as a result of increased nutrient loadings outweigh the negative effects on biomass of more extensive hypoxia. One comparison showed that there were higher average biomasses of key fish groups with enrichment and hypoxia than with no enrichment and no hypoxia, and another comparison (ignoring a food effect) showed that biomasses with hypoxia were lower in the hypoxia area but were not consistently lower compared to no hypoxia when viewed system-wide. However, specification of the many possible predator–prey interactions in food web models remains a challenge (Ainsworth et al. 2010).

The simulation of DO will not, within foreseeable future, get to a fine enough scale for explicitly representing the effects on individual fish (e.g., meters and minutes for direct mortality) so implicit approaches for representing fine-scale variability in DO are needed. Measurement methods can provide detailed DO data for short periods of time and are not comprehensive for the grid but can be sufficient to characterize fine-scale variability (e.g., Zhang et al. 2014). How to switch from explicit representation of DO fields to the variability expected within each of the spatial cells is a challenge. There may be approaches that can be borrowed from downscaling methods of broadscale models to finer-scale models (e.g., Tabor and Williams 2010; Flint and Flint 2012).

With coastal hypoxia, an added complication is the need to also deal with changes in food that are concomitant with hypoxia for the upper trophic levels (Breitburg et al. 2009). Reduced nutrient loadings involve weighing the costs of less food against the benefits of reduced hypoxia. At present, there is not an obvious optimal way to simulate physics, water quality, prey dynamics in response to nutrient changes, and upper tropic level dynamics in a single integrated model. De Mutsert et al. (Chap. 14) and Adamack et al. (Chap. 12) offer progress toward that goal. Several ongoing efforts are striving to achieve such an end-to-end solution, and trying different approaches is a good strategy. In order to ensure the results of the efforts can be compared, and thus results maximally leveraged, a coordinated approach should be part of the modeling efforts to more effectively link hypoxia to population and the food web dynamics.

15.3 Concluding Remarks

Modeling the dynamics and causes of hypoxia has greatly progressed to the point that models can simulate seasonal dynamics over scales of a few kilometers, but numerical description of the effects of hypoxia on biota at the population and higher levels is still in the formative stage. The models presented in this book should be considered examples of the current state of hypoxia modeling, but are not a true representative sample of all hypoxia modeling and thus this chapter cannot be considered a comprehensive review. When viewed together, however, the chapters highlight some areas of modeling strengths and also some critical uncertainties that suggest areas for future research efforts. The existing models coupling physics and water quality can simulate seasonal hypoxia on kilometer scales quite well, and they now cover the major coastal systems, OMZs, and many shallow estuarine subsystems. Increased computing power in the future will allow for better quantification of the uncertainty with model predictions. The limitations include uncertain specification of boundary and forcing functions (e.g., wind), challenges in representing the sediment-water exchanges and multiple nutrient limitation, and the limited availability of data for multiple contrasting years for calibration and validation. To increasingly utilize these models for practical applications toward restoration goals, we need to ensure that the water quality modeling is capable of accurately predicting responses to changed conditions (nutrient loadings and climate change) and that the model has sufficient skill in the vertical as well as horizontal dimensions.

A major challenge remains to effectively link the water quality to the upper trophic levels. A variety of approaches were illustrated in this book. These included individual-based modeling focused on a specific process (avoidance affects exposure), simulation of population dynamics over multiple years imbedded in the 3-D water quality, individual-based modeling focused on a few species and life stages (fish larvae and ctenophores) in 3 vertical layers, and a Eulerian representation of the full food web in 2-D. Data for calibration and testing of the upper trophic level models are much more limited than for water quality and, except for some special studies, are collected for other reasons than quantifying hypoxia effects. There will be increasing demands that the population and system responses of biota to hypoxia be predicted in order to quantify the ecological benefits and costs of changes in nutrient loadings (management of coastal systems) and to express the effects of ocean management and climate change in terms of direct relevance to managers and the public.

Acknowledgements Funding for the preparation of this chapter (KAR) was partially provided by the National Oceanic and Atmospheric Administration, Center for Sponsored Coastal Ocean Research (CSCOR), NGOMEX16 grant number NA16NOS4780204 awarded to Louisiana State University. This is publication number 218 of the NOAA's CSCOR NGOMEX program.

References

Ainsworth CH, Kaplan IC, Levin PS, Mangel M (2010) A statistical approach for estimating fish diet compositions from multiple data sources: Gulf of California case study. Ecol Appl 20:2188–2202

Altieri AH, Gedan KB (2015) Climate change and dead zones. Glob Change Biol 21:1395–1406

Anderson CR, Moore SK, Tomlinson MC, Silke J, Cusack CK (2015) Living with harmful algal blooms in a changing world: strategies for modeling and mitigating their effects in coastal marine ecosystems. Coastal and marine hazards, risks, and disasters. Elsevier BV, Amsterdam, pp 495–561

Ayata SD, Lévy M, Aumont O, Sciandra A, Sainte-Marie J, Tagliabue A, Bernard O (2013) Phytoplankton growth formulation in marine ecosystem models: should we take into account photo-acclimation and variable stoichiometry in oligotrophic areas? J Mar Syst 125:29–40

Bestley S, Jonsen I, Harcourt RG, Hindell MA, Gales NJ (2016) Putting the behavior into animal movement modeling: improved activity budgets from use of ancillary tag information. Ecol Evol 6:8243–8255

Blumberg AF, Kantha LH (1985) Open boundary condition for circulation models. J Hydraul Eng 111:237–255

Bonachela JA, Klausmeier CA, Edwards KF, Litchman E, Levin SA (2015) The role of phytoplankton diversity in the emergent oceanic stoichiometry. J Plankton Res p fbv087

Breitburg DL, Hondorp DW, Davias LA, Diaz RJ (2009) Hypoxia, nitrogen, and fisheries: integrating effects across local and global landscapes. Ann Rev Mar Sci 1:329–349

Chan F, Barth JA, Lubchenco J, Kirincich A, Weeks H, Peterson WT, Menge BA (2008) Emergence of anoxia in the California Current large marine ecosystem. Science 319:920

Chesney EJ, Baltz DM (2001) The effects of hypoxia on the northern Gulf of Mexico coastal ecosystem: a fisheries perspective. In: Rabalais NN, Turner RE (eds) Coastal hypoxia: consequences for living resources and ecosystems. American Geophysical Union, Washington, DC, pp 321–354

Cohen JH, Forward RB (2009) Zooplankton diel vertical migration—a review of proximate control Oceanog. Mar Biol Annu Rev 47:77–109

DeAngelis DL, Grimm V (2014) Individual-based models in ecology after four decades. F1000 prime reports 2014, 6:39. doi:10.12703/P6-39
Denman KL (2003) Modelling planktonic ecosystems: parameterizing complexity. Prog Oceanogr 57:429–452
Diaz RJ, Rosenberg R (2008) Spreading dead zones and consequences for marine ecosystems. Science 32:926–929
Fennel K, Laurent A, Hetland R, Justić D, Ko DS, Lehrter J, Murrell M, Wang L, Yu L, Zhang W (2016) Effects of model physics on hypoxia simulations for the northern Gulf of Mexico: a model intercomparison. J Geophys Res Oceans 121. doi:10.1002/2015JC011577
Fennel K, Hu J, Laurent A, Marta-Almeida M, Hetland R (2013) Sensitivity of hypoxia predictions for the northern Gulf of Mexico to sediment oxygen consumption and model nesting. J Geophys Res 118:990–1002
Flint LE, Flint AL (2012) Downscaling future climate scenarios to fine scales for hydrologic and ecological modeling and analysis. Ecol Proces 1:2
Flynn KJ (2003) Modelling multi-nutrient interactions in phytoplankton; balancing simplicity and realism. Prog Oceanogr 56:249–279
Friedrichs MA, Dusenberry JA, Anderson LA, Armstrong RA, Chai F, Christian JR, Doney SC, Dunne J, Fujii M, Hood R, McGillicuddy DJ, Moore JK, Schartau M, Spitz YH, Wiggert JD (2007) Assessment of skill and portability in regional marine biogeochemical models: role of multiple planktonic groups. J Geophys Res Oceans 112(C8)
Glibert PM, Kana TM, Brown K (2013) From limitation to excess: the consequences of substrate excess and stoichiometry for phytoplankton physiology, trophodynamics and biogeochemistry, and the implications for modeling. J Mar Syst 125:14–28
Grantham BA, Chan F, Nielsen KJ, Fox DS, Barth JA, Huyer A, Lubchenco J, Menge BA (2004) Upwelling-driven nearshore hypoxia signals ecosystem and oceanographic changes in the northeast Pacific. Nature 429:749–754
Grimm V, Railsback SF (2013) Individual-based modeling and ecology. Princeton University Press
Gutowsky LF, Harrison PM, Martins EG, Leake A, Patterson DA, Power M, Cooke SJ (2013) Diel vertical migration hypotheses explain size-dependent behaviour in a freshwater piscivore. Anim Behav 86:365–373
Heip CHR, Goosen NK, Herman PMJ, Kromkamp J, Middelburg JJ, Soetaert K (1995) Production and consumption of biological particles in temperate tidal estuaries. In: Ansell AD, Gibson RN, Barnes M (eds) Oceanography and marine biology: an annual review, vol 33. University College London Press, pp 1–149
Hetland RD (2017) Suppression of baroclinic instabilities in buoyancy-driven flow over sloping bathymetry. J Phys Oceanogr 47:49–68
Hetland RD, DiMarco SF (2008) How does the character of oxygen demand control the structure of hypoxia on the Texas-Louisiana continental shelf? J Mar Syst 70:49–62
Hetland RD, DiMarco SF (2012) Skill assessment of a hydrodynamic model of circulation over the Texas-Louisiana continental shelf. Ocean Model 43:64–76
Holt J, Allen JI, Anderson TR, Brewin R, Butenschön M, Harle J, Huse G, Lehodey P, Lindemann C, Memery L, Salihoglu B (2014) Challenges in integrative approaches to modelling the marine ecosystems of the North Atlantic: physics to fish and coasts to ocean. Prog Oceanogr 129:285–313
Howarth RW (1988) Nutrient limitation of net primary production in marine ecosystems. Annu Rev Ecol Syst 19:89–110
Howes EL, Joos F, Eakin M, Gattuso JP (2015) An updated synthesis of the observed and projected impacts of climate change on the chemical, physical and biological processes in the oceans. Front Mar Sci 2:36. doi:10.3389/fmars.2015.00036
Ibarra D, Fennel K, Cullen J (2014) Coupling 3-D Eulerian bio-physics (ROMS) with individual-based shellfish ecophysiology (SHELL-E): a hybrid model for carrying capacity and environmental impacts of bivalve aquaculture. Ecol Model 273:63–78

Justic D, Wang L (2014) Assessing temporal and spatial variability of hypoxia over the inner Louisiana-upper Texas shelf: application of an unstructured-grid three-dimensional coupled hydrodynamic-water quality model. Cont Shelf Res 72:163–179

Justić D, Bierman VJ, Scavia D, Hetland RD (2007) Forecasting Gulf's hypoxia: the next 50 years? Estuaries Coasts 30:791–801

Justic D, Rabalais NN, Turner RE (1996) Effects of climate change on hypoxia in coastal waters: a doubled CO_2 scenario for the northern Gulf of Mexico. Limnol Oceanogr 41:992–1003

Kemp WM, Sampou PA, Garber J, Tuttle J, Boynton WR (1992) Seasonal depletion of oxygen from bottom waters of Chesapeake Bay: roles of benthic and planktonic respiration and physical exchange processes. Mar Ecol Prog Ser 85:137–152

Kim DK, Zhang W, Watson S, Arhonditsis GB (2014) A commentary on the modelling of the causal linkages among nutrient loading, harmful algal blooms, and hypoxia patterns in Lake Erie. J Great Lakes Res 40:117–129

Kirtman BP, Bitz C, Bryan F, Collins W, Dennis J, Hearn N, Kinter JL, Loft R, Rousset C, Siqueira L, Stan C (2012) Impact of ocean model resolution on CCSM climate simulations. Clim Dyn 39:1303–1328

LaBone E (2016) Modeling the effects of hypoxia on fish movement in the Gulf of Mexico hypoxic zone. PhD dissertation, Louisiana State University, Baton Rouge

Laurent A, Fennel Cai W-J, Huang W-J, Barbero L, Wanninkhof R (2017) Eutrophication-induced acidification of coastal waters in the northern Gulf of Mexico: insights into origin and processes from a coupled physical-biogeochemical model. Geophys Res Lett 44. doi:10.1002/2016GL071881

Lévy M, Ferrari R, Franks PJ, Martin AP, Rivière P (2012) Bringing physics to life at the submesoscale. Geophys Res Lett 39:L14602. doi:10.1029/2012GL052756

Lehrter JC, Beddick DL, Devereux R, Yates DF, Murrell MC (2012) Sediment-water fluxes of dissolved inorganic carbon, O2, nutrients, and N2 from the hypoxic region of the Louisiana continental shelf. Biogeochemistry 109:233–252

Levin LA, Breitburg DL (2015) Linking coasts and seas to address ocean deoxygenation. Nat Clim Change 5:401–403

Li Y, Li M, Kemp WM (2015) A budget analysis of bottom-water dissolved oxygen in Chesapeake Bay. Estuaries Coasts 38:2132–2148

Litchman E, Klausmeier CA, Miller JR, Schofield OM, Falkowski PG (2006) Multi-nutrient, multi-group model of present and future oceanic phytoplankton communities. Biogeosci Discuss 3:607–663

Mahadevan A (2016) The impact of submesoscale physics on primary productivity of plankton. Ann Rev Mar Sci 8:161–184

Marta-Almeida M, Hetland RD, Zhang X (2013) Evaluation of model nesting performance on the Texas-Louisiana continental shelf. J Geophys Res Oceans 118:2476–2491. doi:10.1002/jgrc.20163

Mattern JP, Fennel K, Dowd M (2013) Sensitivity and uncertainty analysis of model hypoxia estimates for the Texas-Louisiana shelf. J Geophys Res Oceans 118:1316–1332

McClintock BT, Russell DJ, Matthiopoulos J, King R (2013) Combining individual animal movement and ancillary biotelemetry data to investigate population-level activity budgets. Ecology 94:838–849

Meier HM, Andersson HC, Eilola K, Gustafsson BG, Kuznetsov I, Müller-Karulis B, Neumann T, Savchuk OP (2011) Hypoxia in future climates: a model ensemble study for the Baltic Sea. Geophys Res Lett 38:L24608. doi:10.1029/2011GL049929

Melzner F, Thomsen J, Koeve W, Oschlies A, Gutowska MA, Bange HW, Hansen HP, Körtzinger A (2013) Future ocean acidification will be amplified by hypoxia in coastal habitats. Mar Biol 160:1875–1888

Middelburg JJ, Levin LA (2009) Coastal hypoxia and sediment biogeochemistry. Biogeosciences 6:1273–1293

Miller SH, Breitburg DL, Burrell RB, Keppel AG (2016) Acidification increases sensitivity to hypoxia in important forage fishes. Mar Ecol Prog Ser 549:1–8

Monteiro PM, Dewitte B, Scranton MI, Paulmier A, Van der Plas AK (2011) The role of open ocean boundary forcing on seasonal to decadal-scale variability and long-term change of natural shelf hypoxia. Environ Res Lett 6. doi:10.1088/1748-9326/6/2/025002

Moore CM, Mills MM, Arrigo KR, Berman-Frank I, Bopp L, Boyd PW, Galbraith ED, Geider RJ, Guieu C, Jaccard SL, Jickells TD (2013) Processes and patterns of oceanic nutrient limitation. Nat Geosci 6:701–710

Obenour DR, Scavia D, Rabalais NN, Turner RE, Michalak AM (2013) Retrospective analysis of midsummer hypoxic area and volume in the northern Gulf of Mexico, 1985–2011. Environ Sci Technol 47:9808–9815

O'Neil JM, Davis TW, Burford MA, Gobler CJ (2012) The rise of harmful cyanobacteria blooms: the potential roles of eutrophication and climate change. Harmful Algae 14:313–334

Paulmier A, Ruiz-Pino D (2009) Oxygen minimum zones (OMZs) in the modern ocean. Prog Oceanogr 80:113–128

Paerl HW, Gardner WS, Havens KE, Joyner AR, McCarthy MJ, Newell SE, Qin B, Scott JT (2016) Mitigating cyanobacterial harmful algal blooms in aquatic ecosystems impacted by climate change and anthropogenic nutrients. Harmful Algae 54:213–222

Rabalais NN, Turner RE, Diaz RJ, Justic D (2009) Global change and eutrophication of coastal waters. ICES J Mar Sci 66:1528–1537

Reed DC, Algar CK, Huber JA, Dick GJ (2014) Gene-centric approach to integrating environmental genomics and biogeochemical models. Proc Natl Acad Sci 111:1879–1884

Roman MR, Pierson JJ, Kimmel DG, Boicourt WC, Zhang X (2012) Impacts of hypoxia on zooplankton spatial distributions in the northern Gulf of Mexico. Estuaries Coasts 35:1261–1269

Rose KA, Adamack AT, Murphy CA, Sable SE, Kolesar SE, Craig JK, Breitburg DL, Thomas P, Brouwer MH, Cerco CF, Diamond S (2009) Does hypoxia have population-level effects on coastal fish? Musings from the virtual world. J Exp Mar Biol Ecol 381:S188–S203

Rose KA, Sable S, DeAngelis DL, Yurek S, Trexler JC, Graf W, Reed DJ (2015a) Proposed best modeling practices for assessing the effects of ecosystem restoration on fish. Ecol Model 300:12–29

Rose KA, Fiechter J, Curchitser EN, Hedstrom K, Bernal M, Creekmore S, Haynie A, Ito SI, Lluch-Cota S, Megrey BA, Edwards CA (2015b) Demonstration of a fully-coupled end-to-end model for small pelagic fish using sardine and anchovy in the California Current. Prog Oceanogr 138:348–380

Rose KA, Creekmore S, Thomas P, Craig JK, Rahman MS, Neilan RM (in review) Modeling the population effects of hypoxia on Atlantic croaker (Micropogonias undulatus) in the northwestern Gulf of Mexico: part 1—model description and idealized hypoxia. Estuaries Coasts

Scully ME (2013) Physical controls on hypoxia in Chesapeake Bay: a numerical modeling study. J Geophys Res Oceans 118:1239–1256

Shen J, Wang T, Herman J, Mason P, Arnold GL (2008) Hypoxia in a coastal embayment of the Chesapeake Bay: a model diagnostic study of oxygen dynamics. Estuaries Coasts 31:652–663

Sippel T, Eveson JP, Galuardi B, Lam C, Hoyle S, Maunder M, Kleiber P, Carvalho F, Tsontos V, Teo SL, Aires-da-Silva A (2015) Using movement data from electronic tags in fisheries stock assessment: a review of models, technology and experimental design. Fish Res 163:152–160

Soetaert K, Middelburg JJ (2009) Modeling eutrophication and oligotrophication of shallow-water marine systems: the importance of sediments under stratified and well-mixed conditions. Hydrobiologia 629:239–254

Stramma L, Johnson GC, Sprintall J, Mohrholz V (2008) Expanding oxygen-minimum zones in the tropical oceans. Science 320:655–658

Tabor K, Williams JW (2010) Globally downscaled climate projections for assessing the conservation impacts of climate change. Ecol Appl 20:554–565

Testa JM, Brady DC, Di Toro DM, Boynton WR, Cornwell JC, Kemp WM (2013) Sediment flux modeling: Simulating nitrogen, phosphorus, and silica cycles. Estuar Coast Shelf Sci 131:245–263

Thomas LN, Tandon A, Mahadevan A (2008) Submesoscale processes and dynamics. In: Hecht MW, Hasumi H (eds) Ocean modeling in an eddying regime. American Geophysical Union, Washington, DC, pp 17–38. doi:10.1029/177GM04

Thomas P, Rahman MS, Picha ME, Tan W (2015) Impaired gamete production and viability in Atlantic croaker collected throughout the 20,000 km^2 hypoxic region in the northern Gulf of Mexico. Mar Pollut Bull 101:182–192

Trenberth KE (2011) Changes in precipitation with climate change. Clim Res 47:123–138

Treseder KK, Balser TC, Bradford MA, Brodie EL, Dubinsky EA, Eviner VT, Hofmockel KS, Lennon JT, Levine UY, MacGregor BJ, Pett-Ridge J (2012) Integrating microbial ecology into ecosystem models: challenges and priorities. Biogeochemistry 109:7–18

Tyler RM, Brady DC, Targett TE (2009) Temporal and spatial dynamics of diel-cycling hypoxia in estuarine tributaries. Estuaries Coasts 32:123–145

Voss M, Bange HW, Dippner JW, Middelburg JJ, Montoya JP, Ward B (2013) The marine nitrogen cycle: recent discoveries, uncertainties and the potential relevance of climate change. Philos Trans R Soc Lond B 368:20130121. doi:10.1098/rstb.2013.0121

Wainwright J, Mulligan M (2005) Modelling and model building. In: Wainwright J, Mulligan M (eds) Environmental modelling: finding simplicity in complexity. Wiley, West Sussex, pp 7–73

Watson SB, Miller C, Arhonditsis G, Boyer GL, Carmichael W, Charlton MN, Confesor R, Depew DC, Höök TO, Ludsin SA, Matisoff G (2016) The re-eutrophication of Lake Erie: harmful algal blooms and hypoxia. Harmful Algae 56:44–66

Yu L, Fennel K, Laurent A (2015) A modeling study of physical controls on hypoxia generation in the northern Gulf of Mexico. J Geophys Res Oceans 120:5019–5039

Zhang H, Mason DM, Stow CA, Adamack AT, Brandt SB, Zhang X, Kimmel DG, Roman MR, Boicourt WC, Ludsin SA (2014) Effects of hypoxia on habitat quality of pelagic planktivorous fishes in the northern Gulf of Mexico. Mar Ecol Prog Ser 505:209–226

Zhang J, Gilbert D, Gooday A, Levin L, Naqvi SWA, Middelburg JJ, Scranton M, Ekau W, Pena A, Dewitte B, Oguz T, Monteiro PMS, Urban E, Rabalais NN, Ittekkot V, Kemp WM, Ulloa O, Elmegen R, Escobar-Briones E, Van der Plas AK (2010) Natural and human-induced hypoxia and consequences for coastal areas: synthesis and future development. Biogeosciences 7:1443–1467

Index

D. Justic et al. (eds.), *Modeling Coastal Hypoxia*,
DOI 10.1007/978-3-319-54571-4

O

MIX
Papier aus verantwortungsvollen Quellen
Paper from responsible sources
FSC® C105338

If you have any concerns about our products,
you can contact us on
ProductSafety@springernature.com

In case Publisher is established outside the EU,
the EU authorized representative is:
Springer Nature Customer Service Center GmbH
Europaplatz 3, 69115 Heidelberg, Germany

Printed by Libri Plureos GmbH
in Hamburg, Germany